CORROSION BASICS

An Introduction

National Association of Corrosion Engineers

Published by

National Association of Corrosion Engineers
1440 South Creek Drive
Houston, Texas 77084

Library of Congress Catalog Card Number: 84-061042
ISBN 0-915567-02-4

CONTENTS

INTRODUCTION

This book was originally issued as the *Basic Corrosion Course* in 1970. The utility of that publication can be appreciated when it is realized that 13 printings were necessary to supply the great number of persons interested in the subject.

As envisioned by the Editor in the Introduction of the original book, some revision is necessary with the passage of time. This has been accomplished with a minimum of change from the initial format and material content of the original. The authors of the first edition provided such an excellent coverage of the subject matter that only some rearrangement of the text and updating has been necessary.

Those responsible for the topical matter of the first edition were as follows:

Editor .... A. deS. Brasunas

Scope of the Subject .... A. deS. Brasunas and N. E. Hamner
Basics .... F. L. LaQue and N. D. Greene
Metallurgy .... R. F. Hochman
Materials .... M. G. Fontana and J. H. Peacock
Localized Corrosion .... H. P. Godard
Stress Corrosion Cracking .... H. L. Logan
Cathodic Protection and Soils .... A. W. Peabody and M. E. Parker
Inhibitors .... N. Hackerman and E. S. Snavely
Atmospheric Corrosion .... K. G. Compton
Coatings .... N. E. Hamner
High Temperature Corrosion .... A. deS. Brasunas and J. J. Moran, Jr.
Water Corrosion .... W. E. Berry
Testing .... B. W. Lifka and F. L. McGeary
Failure Analysis .... E. D. Verink

A great amount of the material written by these authors has been retained. Changes have been made where it was believed a better continuity, less repetition, and more recent data would improve the development of the subject. Those responsible for the revisions are C. P. Dillon, J. S. Snodgrass, L. S. Van Delinder, and H. A. Webster.

This book provides a general coverage of the wide field of corrosion control. It is designed to be helpful to those being initiated into the work, and consequently attempts to present each corrosion process or control procedure in the most basic terms. A comprehensive discussion of the topics has not been attempted. Certainly the book is not represented as providing the latest or most erudite discussion of the subject. Yet for those knowledgeable in the field of corrosion control, we hope that each chapter presents some aspect of the work in terms that are stimulating and helpful. We solicit suggestions for improvement of the work.

To assist the reader, a bibliography is provided at the end of each chapter for those who may want to pursue a topic further. Many of the books which are listed in more than one chapter are obviously regarded as prime sources for corrosion information. The appendix to Chapter 1 contains a list of definitions of corrosion-related terms. A comprehensive and thoroughly cross referenced subject index is located in the back of the book and should provide a ready guide for all persons wishing to find specific items.

Education in corrosion control is our primary concern. We sincerely hope this new edition will contribute even more capably in attaining that goal. As with the preponderance of the first edition, a portion of the Introduction must be repeated: "We trust that ... this book will be of considerable benefit to the nation, to numerous industries, and to individuals all over the world who will take advantage of the opportunity afforded them by this educational effort."

L. S. Van Delinder
Editor and Chairman of Subcommittee ETC-1
NACE Education and Training Committee
May, 1984

Chapter 1

Scope and Language of Corrosion

THE SCOPE AND LANGUAGE OF CORROSION

Purpose

When a study of corrosion is first undertaken, it is often assumed that corrosion is a single simple reaction, and that once understood, it can be turned off like a spigot.

If cost and availability were not factors in material selection, only the very best materials would ever be chosen. We must, however, dismiss consideration of materials like gold or platinum and think in terms of affordable substances for practical use in homes, industries, automobiles, etc.

Practical materials like steel, aluminum, and copper alloys, plastics, ceramics, wood, refractory metals, stainless steels, and many other modern alloys, all have certain advantages as well as disadvantages. Any one of these alloys or classes of alloy may serve as the best choice for a particular application. Learning when to choose which material comes with experience and knowledge, which is a part of what is presented in this text.

All corrosion is not the same. Therefore, the reasons why such a varied selection of materials degrades in certain environments are also enumerated, along with the various mechanisms of the attack experienced. The basic properties of specific materials are also described. A better understanding of all of these factors combining to produce corrosion is essential when analyzing problems and recommending countermeasures.

As with any other technical discipline, corrosion work utilizes certain words, methods of presenting data, and shorthand symbols that are peculiar to the field. The text reveals this jargon in a manner that should, in the future, allow you to read or discuss corrosion topics with greater confidence and understanding.

The text is intended to convey the scope of the field of corrosion control. Many persons work in only one area of this total discipline. It is important to realize the extent of the effort being made today in analyzing and combatting corrosion. So much of the experience and so many of the workable solutions developed in one area of corrosion work can be used to improve the control procedures of another area. There is, of course, always the chance that tomorrow may bring a change in position or interest and thus work in that "other" area.

The best techniques available for detecting corrosion, determining the resistance of a material, or evaluating the efficacy of a control procedure serve as daily tools for attacking the problems faced by thousands of persons engaged in corrosion work. This text will hopefully foster a better appreciation of these procedures.

Certainly anyone concerned with corrosion work is primarily interested in control procedures that allow the safe, economic use of a material in a specific environment. Thus, basic aspects of the most current control techniques are described. After learning the elements of each procedure, persons can further investigate the possibility of using these approaches to solve their own specific problems. For instance, a person may find the corrosion control technique of cathodic protection, which in many instances can halt corrosion by the throw of a switch (much like the spigot referred to in the opening of the chapter), applicable in the control of their corrosion problems (Chapter 9).

Corrosion

Most people are familiar with corrosion in some form or another, particularly the rusting of an iron fence or a "tin" can, the degradation of steel pilings or boats and boat fixtures, or the rusting of an iron nail (Figure 1.1).

Almost everyone has seen the erection of a bridge or tall, modern buildings with steel frameworks. These steel frames usually have a green, orange, or reddish color, which is the color of a protective coating placed on steel members even before they leave the steel mill. These coatings are applied to protect the frames against rusting. Then, after installation and particularly on bridge construction, these surfaces are recoated with other materials that may lend a better appearance or added corrosion protection. However, developments in low-alloy steel technology permit the erection of iron structures without using paints or coatings under certain circumstances. The special alloys used in the steel rust initially, but then appear to stop rusting almost completely (Figure 1.2).

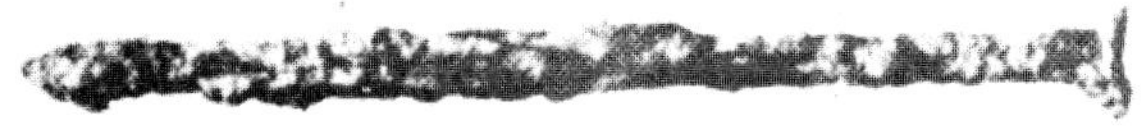

FIGURE 1.1 — A prevalent example of corrosion; the rusting of an ordinary iron nail. Although commonplace, numerous technical articles can be devoted to this natural, yet complex phenomenon.

FIGURE 1.2 — Downtown Chicago office building in 1965 with an unpainted low-alloy steel exterior. A pleasant dark brown rust has developed which effectively protects the underlying steel from significant additional corrosion. [SOURCE: U.S. Steel Corp.]

Piping is another major type of equipment subject to corrosion. This includes water pipes in the home, where corrosion attacks mostly from the inside, as well as the underground water, gas, and oil pipelines that crisscross our nation. Thus, it would appear safe to say that almost everyone is at least somewhat familiar with corrosion, which is defined as follows:

> *Corrosion is the deterioration of a substance (usually a metal) or its properties because of a reaction with its environment.*

Note that the words *chemical* or *electrochemical* are not used in this definition. This is because a few forms of corrosion that will be discussed do not involve these processes.

Furthermore, the above definition recognizes that materials other than metals may corrode. The deterioration of wood, ceramics, plastics, etc., must also be studied by the corrosion engineer and is included in the term *corrosion*.

Finally, attention is also directed to the idea that properties, as well as the material itself, can and do deteriorate. In some forms of corrosion, there is no weight change or visible deterioration, yet properties change and the material may fail unexpectedly because of certain changes within the material. Such changes may defy ordinary visual examination or weight change determinations. These are nevertheless important forms of corrosive action with which a corrosion engineer should be familiar.

This book is intended to acquaint individuals having little or no corrosion knowledge with the fundamentals of corrosion and commonly used practices in order to control corrosion in the home, out-of-doors, or in industry.

Many people become engaged in controlling or preventing corrosion by appointment rather than as the final step in a process of formal education with this as its original goal. This book is therefore designed to be helpful to that segment entering this field without the benefit of any extensive training in the basic sciences related to corrosion, but who may be called upon from time to time to take at least the first steps in anticipating, diagnosing, and otherwise dealing with corrosion problems, either on their own or in collaboration with others.

Lastly, the book provides a good review of corrosion basics which may serve as a refresher for those who have an education in the field.

In discussing corrosion, it is unfortunately necessary to use technical terms which are rarely used outside the field of corrosion. While the book defines some terms as they appear, it would also be useful to develop the habit of referring to the Glossary of Corrosion-Related Terms which appears at the end of this chapter as Appendix A. These definitions should not be regarded as the only correct ones; other definitions may serve as well or even better.

Since the modern corrosion engineer or technician must find suitable materials for a great variety of corrosive environments and temperature ranges, he must be familiar with the behavior of numerous alloys, as well as possible substitute materials like plastics,[2] wood, rubber, ceramics, leather, glass, or graphite. Thus, the meaning of the term *corrosion* includes the deterioration of nonmetallic materials as well as metals. However, since metals have and will undoubtedly continue to be the engineering materials of major concern, this text will concentrate on the corrosion behavior of metals.

In asking the general question, "In what environments does corrosion occur?," the only suitable answer is, "In just about any environment, depending on the material being used." To be more specific, a study of Figure 1.3, which attempts to list various environments in which corrosion has been known to occur, may prove more satisfying.

Another general question which often arises is, "How many forms of corrosion are there?" Again, the answer may be unexpected. Some authorities quote two or three, others propose eight, and still others more safely say, "Quite a few." Some of these varieties will be discussed later.

Economic Significance

Unfortunately, many people, outside of industry as well as in industry, often observe corrosion yet simply accept it as an inevitable problem. Actually, something can and should be done to prolong the life of many metals exposed to corrosive environments. This is only a common-sense approach, considering corrosion involves the gradual destruction of a material which costs money to replace. Thus, corrosion control becomes a major factor in the very important area of dollars and cents.

While corrosion processes form an interesting basis for scientific studies which are frequently undertaken as exercises in chemistry (and particularly electrochemistry), by far the greatest interest in, and concern for corrosion stems from its practical effects and how they may be avoided. Various estimates have been made of the annual economic loss resulting from corrosion. There is no general agreement as to just what should be included in calculating this loss. For example, should we include the coating on tin cans which would not be needed if the contents were not corrosive to steel? Although such examples may make it seem fruitless to argue over which figure should be used, there is ample evidence that annual losses attributable to corrosion in North America amount to many billions of dollars.[3]

As our products and manufacturing processes have become more complex and the penalties of failures from corrosion, including safety hazards and interruptions in plant operations, have become more costly and more specifically recognized, the attention that is being given the control and prevention of corrosion has increased.

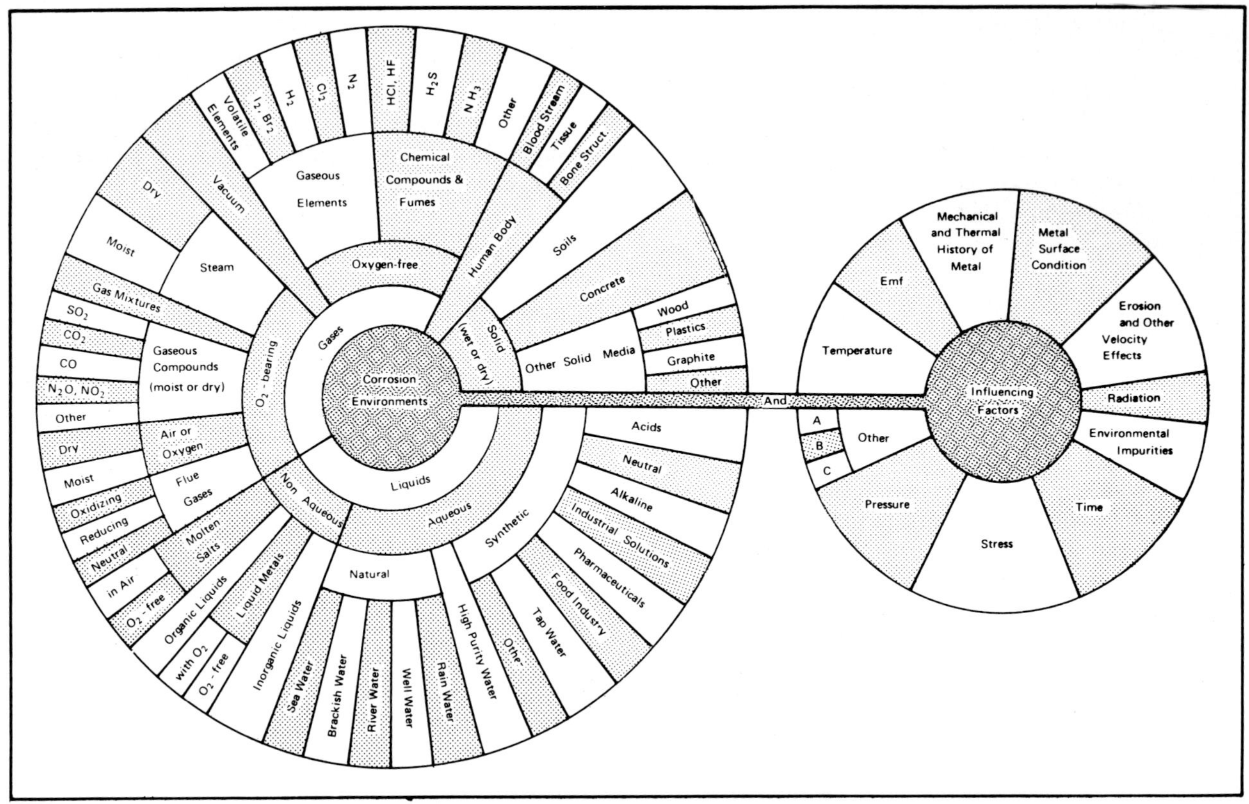

FIGURE 1.3 — A chart showing the various media in which corrosion is known to occur and the factors that can have an important influence. [SOURCE: Slight modification of a chart first published in: Symposium on Corrosion Fundamentals, Anton deS. Brasunas and E. E. Stansbury, eds., Univ. of Tennessee, Knoxville, TN, 1956.]

As further indication of its importance, studies on corrosion are being reported in many journals and books and it is receiving increasing attention from industry, government laboratories, and technical societies. The organization publishing this particular book, the National Association of Corrosion Engineers, was formed in 1943 to devote exclusive attention to this particular subject. This Society has grown to over 15,000 members, with its headquarters building and staff located in Houston, Texas.

Corrosion Around Us

All metals and alloys can and generally do corrode. In looking about, it is easy to find corrosion occurring all around us as well as to find some metals which do not appear to be corroding at all. The following are some examples of common objects and areas suceptible to corrosion.

1. The automobile is a veritable corrosion testing device. Materials must be selected for use for atmospheric, high-temperature, closed cycle water, wear, fatigue, and crevice exposures. Many different metals, plastics, paints, platings, and inhibitors are used to prolong the life of the parts. However, the body,[4] and certainly the muffler,[5] are generally seen in various stages of corrosion. In fact, when examining a car over five years old, it is not even necessary to look under the car; just looking at the car's body will probably reveal corrosion taking place in numerous locations. Even the once-bright bumper will show either corrosion pin holes or more serious attack.

2. In looking at the water pipes of an older home, particularly if they are made out of steel or even galvanized steel, some corrosion is usually evident, particularly at the joints where the galvanized coating might have been cut, or at a junction where a brass valve has caused heavy corrosion in the adjacent area.

3. Barnyard roofs made of steel or galvanized iron rust through in time, whereas the newer aluminum roofs (of the *proper* aluminum alloy) look bright and shiny and appear to be free of corrosion.

4. Kitchen utensils generally look bright and shiny, although their interiors are often stained below the water line and sometimes pitted. These also represent forms of corrosion.

5. The green coatings on copper or brass roofs of churches and municipal buildings develop as a result of corrosion. Copper-bearing metals that have reacted with certain components of the atmosphere, develop the green coating frequently called a *patina* or *aerugo,* which is actually considered a protective corrosion layer that retards continued reaction and is also pleasing to the eye.

6. Various industries, particularly petroleum

FIGURE 1.4 — River Queen succumbs to Ol' Man River on the St. Louis levee. Pitting corrosion developed a hole in the steel hull below the water line, thereby putting an end to this famous Mississippi riverboat. The well-known Eads bridge is in the background. [SOURCE: St. Louis Post-Dispatch, St. Louis, MO.]

and chemical companies, have numerous corrosion problems unique to their own operations. The average person is probably totally unaware of them. Such companies generally maintain a corrosion-conscious staff to keep the plants out of trouble.

7. In the case of an aluminum canoe, corrosion seems almost negligible, although a close look often reveals tiny pits of corrosive attack. This may not appear serious since it is limited to such a tiny spot, but even a small leak might be considered very serious. Those who have seen the old Mississippi River paddle boat, *River Queen,* in the film *Gone With the Wind* would be dismayed to learn that it developed a tiny hole in the bottom of its hull while tied up at the riverfront in St. Louis, Missouri. It slowly settled to the bottom during the pre-dawn hours of a December day in 1967, never to surface again. Figure 1.4 illustrates the predicament from which attempts to refloat the historic boat failed and resulted in a ''parts sale'' several weeks later.

8. Metals exposed to seawater may corrode heavily. Ship hulls (inside and out), piling, oil rig platforms, equipment for fresh water recovery, and other items exposed to this highly conductive medium must be protected in some manner. Luckily, most all types of corrosion control may be used in seawater.

9. Concrete degradation of bridges, highways, buildings, and drainage ducts is often caused by reaction with the environment rather than by mechanical forces. When corrosion control is not exercised during initial construction, tax payer's money is often required for the restoraton of the failed facilities.

10. Many houses in the nation have cold water lines, vent and drain ducts, moldings, and other construction components which are made of plastic materials. These have been successful, but how many people are aware of their limitations, how to use them, and the differences between the available plastic materials?

11. A person may drive along a roadway and see a metal box at the fence line along the road right-of-way marked something like, ''XYZ Gas

Co. Do Not Disturb.'' While this may not be much of a clue to most people, it probably indicates a test point installation which allows the corrosion control personnel, by electrical means, to measure the corrosion protection being provided for a steel pipe many feet below the surface through a process called *cathodic protection* (Chapter 9).

Other common examples of corrosion, either occurring or under obvious control, are observed by everyone each day. This book will hopefully aid in more intelligent observation of these phenomena.

Various Metals

Of the 105 elements known to man (Table 1.1), about 80 are metals. Each has different mechanical, chemical, and physical properties, and although all can corrode, they may each corrode in a given situation to a different degree and probably in a different manner. Furthermore, about half of these 80 metals have been alloyed to make more than 40,000 different alloys. Undoubtedly, many more will be made as the years go by.

Obviously, a person cannot memorize the composition of such a multitude of alloys. Reference texts such as *Engineering Alloys* or the *Alloy Digest* are available to provide such information. The corrosion characteristics of most of the alloys can then be inferred from the analysis and the material placed in a broader category of materials, such as those discussed in Chapter 4.

A Natural Process

The fact that corrosion does occur should not be cause for surprise. Almost all material *should* be expected to deteriorate with time when exposed to ''the elements.'' Corrosion is a perfectly natural process, as natural as water flowing downhill. If water flowed uphill or remained stationary on a hillside, there may be cause for surprise, yet our human ingenuity can accomplish this by putting water in a closed container (pipe) and closing the bottom end, or by merely freezing it.

Similarly, if iron were exposed to air and water, rust would be expected to develop within a matter of hours. In fact, it would be surprising if the exposed iron did *not* corrode or rust. Of course, if copper, brass, aluminum, or stainless steel were substituted for iron, a given degree of corrosion might take longer, but some corrosion would still be anticipated. Instead of forming rust (a form of iron oxide), some oxide of copper, aluminum, or chromium may form very slowly and coat the bare metal. This oxide coating, even if extremely thin, could form a partial barrier to continued attack and slow down the rate of corrosion almost to a standstill.

A surface layer formation, whether it is oxide, carbonate, sulfate, or any other compound, is a major factor in corrosion resistance, particularly if the layer effectively separates the underlying metal from its environment. Such a naturally formed coating must be diffusion- and moisture-resistant to be effective. Ordinary iron does not naturally form an effective barrier; its rust permits oxygen and moisture to penetrate and continue rusting. Thus, unless precautions are taken, failure will eventually occur.

Precautions to prevent iron and its alloys from corroding constitute a major effort in corrosion control. It is here that we often resort to painting, electroplating, etc. to form artificial protective layers over the iron surface and thus prolong its useful life. Other techniques which form the basis of modern corrosion control in addition to these are discussed in later chapters.

Some metals, like stainless steel, titanium, or aluminum, are frequently left unpainted. This is not because these metals are inert, but because the oxygen of the air helps develop a protective surface layer of chromium oxide, titanium oxide, and aluminum oxide, respectively, and although they are so thin as to be invisible to the eye, these layers can be detected and their presence verified. If such coatings did not form naturally over the entire surface of these normally corrosion-resistant metals, they would be susceptible to corrosive attack. Certain instances are known where such unexpected attack has occurred, and although these are natural processes, we would do well to learn Nature's way.

Some environments are more corrosive than others. While there are exceptions, the following statements are generally accepted as facts.

1. Moist air is more corrosive than dry air.
2. Hot air is more corrosive than cold air.
3. Hot water is more corrosive than cold water.
4. Polluted air is more corrosive than clean air.
5. Acids are more corrosive than bases (alkalies).
6. Salt water is more corrosive than fresh water.
7. Stainless steel will outlast ordinary steel.
8. No corrosion will occur in a vacuum, even at very high temperatures, etc.

While it may be a surprise to some, there are instances where every one of the above statements, including the last one, are incorrect. This would indicate that broad, categorical statements regarding corrosion are suspect. This is true. There is essentially no statement regarding corrosion or the use of a material that does not have an exception. Stainless steel, for instance, is not necessarily ''better'' than steel. Thus, providing the reader with an appreciation of the factors involved in identifying the corrosion resistance of a material is another objective of this book.

Material Selection

For a specific application, a certain alloy may be superior to another, yet many compete for the same application because of factors other than corrosion. Cost is a major factor and has already been mentioned. Other factors may include one or a combination of the following.

0. Safety.

TABLE 1.1 — The Elements

PERIOD	GROUP 0	GROUP I	GROUP II	GROUP III	GROUP IV	GROUP V	GROUP VI	GROUP VII	GROUP VIII
I		1[1] HYDROGEN H[2] 1.0080[3]							
II	2 HELIUM He 4.003	3 LITHIUM Li 6.939	4 BERYLLIUM Be 9.012	5 BORON B 10.81	6 CARBON C 12.011	7 NITROGEN N 14.007	8 OXYGEN O 16.00	9 FLUORINE F 19.00	
III	10 NEON Ne 20.183	11 SODIUM Na 22.989	12 MAGNESIUM Mg 24.31	13 ALUMINUM Al 26.98	14 SILICON Si 28.09	15 PHOSPHORUS P 30.974	16 SULFUR S 32.064	17 CHLORINE Cl 35.453	
IV	18 ARGON Ar 39.948	19 POTASSIUM K 39.102	20 CALCIUM Ca 40.08	21 SCANDIUM Sc 44.96	22 TITANIUM Ti 47.90	23 VANADIUM V 50.94	24 CHROMIUM Cr 52.00	25 MANGANESE Mn 54.94	26 IRON Fe 55.85 27 COBALT Co 58.93 28 NICKEL Ni 58.71
		29 COPPER Cu 63.54	30 ZINC Zn 65.37	31 GALLIUM Ga 69.72	32 GERMANIUM Ge 72.59	33 ARSENIC As 74.92	34 SELENIUM Se 78.96	35 BROMINE Br 79.91	
V	36 KRYPTON Kr 83.80	37 RUBIDIUM Rb 85.47	38 STRONTIUM Sr 87.62	39 YTTRIUM Y 89.91	40 ZIRCONIUM Zr 91.22	41 COLUMBIUM Cb 92.91	42 MOLYBDENUM Mo 95.94	43 TECHNETIUM Tc [99]	44 RUTHENIUM Ru 101.07 45 RHODIUM Rh 102.91 46 PALLADIUM Pd 106.4
		47 SILVER Ag 107.87	48 CADMIUM Cd 112.40	49 INDIUM In 114.82	50 TIN Sn 118.69	51 ANTIMONY Sb 121.75	52 TELLURIUM Te 127.60	53 IODINE I 126.90	
VI	54 XENON Xe 131.30	55 CESIUM Cs 132.91	56 BARIUM Ba 137.34	57-71 (LANTHANIDE RARE EARTHS)	72 HAFNIUM Hf 178.49	73 TANTALUM Ta 180.95	74 TUNGSTEN W 183.85	75 RHENIUM Re 186.22	76 OSMIUM Os 190.2 77 IRIDIUM Ir 192.2 78 PLATINUM Pt 195.09
		79 GOLD Au 196.97	80 MERCURY Hg 200.59	81 THALLIUM Tl 204.37	82 LEAD Pb 207.19	83 BISMUTH Bi 208.98	84 POLONIUM Po [210]	85 ASTATINE At [211]	
VII	86 RADON Rn [222]	87 FRANCIUM Fr [223]	88 RADIUM Ra 226.05	89 ACTINIUM Ac [227]	90 THORIUM Th 232.04	91 PROTACTINIUM Pa [231]	92 URANIUM U 238.03	→	MAN-MADE TRANS-URANIUM ELEMENTS 93-105

RARE EARTHS

PERIOD	SERIES										
VI	LANTHANIDE SERIES •	57 LANTHANUM La 138.91	58 CERIUM Ce 140.12	59 PRASEODYNIUM Pr 140.91	60 NEODYMIUM Nd 144.24	61 PROMETHIUM Pm [145]	62 SAMARIUM Sa 150.35	63 EUROPIUM Eu 151.96	64 GADOLINIUM Gd 157.25		
			65 TERBIUM Tb 158.92	66 DYSPROSIUM Dy 162.50	67 HOLMIUM Ho 164.93	68 ERBIUM Er 162.26	69 THULIUM Tm 168.93	70 YTTERBIUM Yb 173.04	71 LUTETIUM Lu 174.97		
VII	ACTINIDE SERIES	89 ACTINIUM Ac [227]	90 THORIUM Th 232.04	91 PROTACTINIUM Pa [231]	92 URANIUM U 238.03	93 NEPTUNIUM Np [237]	94 PLUTONIUM Pu [242]	95 AMERICIUM Am [243]	96 CURIUM Cm [245]	97 BERKELIUM Bk [249]	98 CALIFORNIUM Cf [249]
					99 EINSTEINIUM Es [254]	100 FERMIUM Fm [252]	101 MENDELEVIUM Md [256]	102 NOBELIUM No [254]	103 LAWRENCIUM Lw [258]	104 ?	105 ?

[1]Atomic number.
[2]Symbol.
[3]Atomic weight.

1. Cost.

2. Corrosion behavior.

3. Solderability or weldability.

4. Forming characteristics (bending, stretching, etc.).

5. Suitable mechanical properties (tensile strength, impact resistance, fatigue, etc.).

6. High or low temperature strength (ductility).

7. Availability of material and in proper form.

8. Compatibility with other materials in the system.

9. Thermal or electrical characteristics.

10. Unique characteristics such as low density, magnetism, or a nuclear requirement.

Notice that no number is given for the requirement of safety when selecting materials. If the safety of an individual, his or her associates, or the public is endangered in some manner by the selection of a material or the use of a corrosion control procedure, that approach must be abandoned. After assurance is given that such a criterion has been met, the other factors can then be weighed to arrive at the optimum solution.

Corrosion is often insidious, working unobserved in the internals of some machine, vessel, pipe, or structure until the device fails, sometimes in a catastrophic manner resulting in injury or death to operators and bystanders. Boiler explosions in the past were often triggered by corrosion processes and were responsible for the death or injury of thousands of people. In petrochemical plants, many process streams are at temperatures higher than those at which they will spontaneously ignite if in contact with the atmosphere. In these cases, great care must be taken to evaluate the corrosion resistance of the piping systems conveying these process streams, and rigorous inspection programs must be devised to ensure their safe operation. It is very difficult to put an economic value on human life, but safety considerations transcend all economic considerations.

Cost is the most important of the remaining factors in materials selection. However, the term *cost* in this sense does not refer solely to the initial cost or the installed cost, but involves these factors plus the expected life of the item, tax structure (investment or maintenance money), depreciation of the item, and the value of money. On this true cost basis, steel may be "better" than stainless steel, even though the rate of corrosion is higher in the intended environment. (Chapter 15 further comments on corrosion economics.)

To assess the safety of the installation or to define the economics of material selection, it is necessary to have an adequate grasp of the corrosion characteristics of a wide variety of materials. Knowing what substance to select is not a simple matter. In some applications, wood is still the best choice; others may use a plastic-lined metal pipe or tank, or ceramic-lined pipe, while still others often use a special alloy.

The mechanical and corrosion properties of these materials vary widely and can also be strongly affected by heat treatment. This aspect of corrosion control is discussed in a later chapter (Chapter 3).

Measuring Corrosion

When corrosion occurs, a weight gain or loss is often experienced. This is most commonly used as the measure of the extent of corrosion. Of course, if a given weight change occurs on a specimen under certain circumstances, a piece of twice the area will have twice the weight change. This must therefore be taken into consideration in addition to the time to develop such a weight change, since weight change generally will increase for longer exposure times. The units usually used are as follows.

1. *Weight Change:* Weight loss or gain per unit area per unit time. A common measure is *milligrams per square decimeter per day,* often abbreviated mdd. (A milligram is a thousandth of a gram; a decimeter is 10 centimeters or almost four inches.) The term is used primarily in laboratories where repetitive tests of a single metal are conducted.

2. *Dimension Change:* Loss of metal thickness per unit time. Commonly used units of this type are μm/y (microns per year; a micron is one-millionth of a meter), mm/y (millimeters per year; a millimeter is one-thousandth of a meter), or mpy (mils per years; a mil is one-thousandth of an inch or 0.001 in.). Other units also have been used on occasion and can easily be converted to these units (Chapter 14).

3. *Mechanical Property Change:* Percent loss in tensile strength, yield strength, ductility, or other mechanical property. These changes are the best indicators of the extent of corrosion when some localized form of attack (intergranular, stress corrosion cracking, etc.) has occurred on metals. The change in mechanical properties (hardness, tensile, ductility, etc.) of a plastic material of construction is most often used as the criterion of attack on the material.

Who Is Doing What About Corrosion Control

Organized studies have been devoted to corrosion for a longer time than most people realize. The British Association for the Advancement of Science appropriated a sum of money for a series of corrosion experiments on the corrosion of cast and wrought iron in the 1830s. Experiments were conducted by Robert Mallet and reported by him in 1838, '40, and '43.[6] Mallet's work was done in an era in which gifted scientists were investigating the properties of matter. Useful guidelines in electrochemical theory were developed during this time by Volta, Hall, Faraday, de la Rive, and others.

Sir Humphrey Davy published results of his work on cathodic protection of copper bottoms for British naval vessels in 1824.[7] These early experiments established a practical base for the application of cathodic protection which led to the development of galvanized iron.

In 1906, Committee U of the American Society for Testing Materials was formed to make corrosion tests. Shortly afterward, other organizations began to pay attention to corrosion and its control.

Among others, W. R. Whitney published his classic article[8] in the American Chemical Society's journal in 1903 on the electrochemical nature of corrosion, disproving a commonly accepted theory involving the presence of carbonic acid gas as necessary for the corrosion of iron.

Among pioneers in studying the effects of corrosion was the American Committee on Electrolysis, which noted in 1921[9] that its preliminary report had been published in October, 1916. This committee, composed of representatives of the American Institute of Electrical Engineers, American Electric Railway Association, American Railway Engineering Association, National Bureau of Standards, and others, concerned itself with the then serious problem of stray current damage to underground metal structures, especially the protection of communication cable from electrified street and interurban railways.

In England, the Corrosion Committee of the Iron and Steel Institute issued its first report in 1931 and its sixth in 1959.[10] An American Coordinating Committee on Corrosion was organized with representatives from 17 technical societies in 1938. This group, which aimed at coordinating the activities of societies to prevent duplicated work was absorbed by the National Association of Corrosion Engineers in 1948 and was renamed the Inter-Society Committee on Corrosion Control. It functioned in a semi-autonomous manner until it was finally disbanded about ten years later. This was largely because the growth of abstract publications and numerous other periodicals permitted easy interchange of most information. However, coordinating committees exist throughout the USA today to assimilate local data on cathodic protection works.

International convocations of those interested in corrosion are also now held biennially, and corrosion societies in the various countries have established a good liaison.

Although Germany had a corrosion journal (*Korrosion und Metallschutz*) prior to World War II, which was interrupted during the war and reissued after its end under a new title (*Werkstoffe und Korrosion*), it was not until after the National Association of Corrosion Engineers (NACE) began publishing the magazine *Corrosion* in 1945 that journals on corrosion control were started in other countries. Since then, one or more magazines about corrosion control have been started and are being published in Australia, Belgium, England, France, Holland, Italy, Japan, Portugal, Russia, and Sweden. Comprehensive abstract journals are published in the United States by NACE and in Sweden by a government agency.

In addition to NACE, whose sole activity centers around corrosion control, many other scientific engineering, governmental, and trade organizations are active in corrosion control work. Leaders among the scientific and engineering groups are American Society for Metals, American Society for Testing and Materials, American Chemical Society, and The Electrochemical Society. In government, the National Bureau of Standards has been working on corrosion control since 1922. NBS Circular C450 on Underground Corrosion was published in November, 1945. The data were updated in Circular 579 which was published in April, 1957.

Other governmental organizations, notably the armed services, Nuclear Regulatory Commission, Federal Housing Administration, National Aeronautics and Space Administration, and various subdivisions of the Department of Interior, among others, have conducted corrosion control programs of a greater or lesser degree.

Technical societies and trade organizations, especially those concerned with metals, have been and continue to be active in a wide variety of pursuits related to corrosion control.

What Causes Corrosion?

There is now little or no controversy about what factors cause most forms of corrosion on metals. Current thinking in the case of usual corrosion is firmly grounded in electrochemical theory, and various formulas and equations have been devised that describe the chemical reactions which make up most corrosion processes. In essence, electrochemical corrosion requires four primary factors: (1) an anode, (2) a cathode, (3) an electrolyte, and (4) an electronic circuit. (These will be discussed more fully in later chapters.) Thus, corrosion theorists are obliged to take into account considerations of the infinitely small and necessarily complex activities on the molecular and at the ionic, electronic, and atomic levels. It is at this point that three basic kinds of corrosion can be listed.

1. Chemical
2. Electrochemical
3. Physical

These differ according to the degree of involvement of the ions, electrons, and atoms.

For example, a common representation of the corrosion reaction may be expressed with respect to iron, water, and oxygen in this chemical reaction:

$$\underset{\text{Iron}}{Fe} + \underset{\text{Water}}{H_2O} + \underset{\text{Oxygen}}{1/2O_2} \rightarrow \underset{\text{Ferrous Hydroxide}}{Fe(OH)_2} \quad (1.1)$$

This indicates that the initial reaction of iron with oxygen in pure water is to form ferrous hydroxide. If any one of these elements were absent entirely, corrosion (rusting) would not occur. (The reasons for this will also be explained in later chapters.)

When other than pure water is contacted, different corrosion products are likely to form. In looking back at Figure 1.3, it appears safe to assume that just about any environment can cause corrosion, and this is correct. The factors which influence corrosion listed on the right-hand side of Figure 1.3 can make a tremendous difference.

Temperature Effects

Increasing the temperature of a corrosive system will normally have the effect of increasing corrosion rates. The *kinetics* (rate of motion or reaction) of the action are said to have increased. Even in water solutions at near room temperatures, that part of a piece of material which has a higher temperature than another part may be anodic to the other. For example, when iron is immersed in dilute aerated NaCl solution, the hot electrode is anodic to colder metal of the same composition. Temperature can also have many other effects on the environment or on the material exposed.

Potential (Emf) Difference

When there is a difference in potential between metals exposed in the same environment, as between zinc and steel in salt water (Chapter 2, Galvanic Series in Seawater), the metal higher in the series (zinc, in this case) will corrode and protect the one lower (steel) in the series.

Heat Treatment

As will be apparent from the discussion in Chapter 3, the corrosion behavior of many an alloy can be strongly influenced by its thermal history (heat treatment).

Surface Condition

The cleanliness of the surface, existence of surface films, and presence of foreign matter can exert a very strong influence on the initiation and rate of corrosion.

Effect of Erosion

Erosion itself is not corrosion. However, even mildly abrasive conditions may remove a corrosion film from a surface which is protective of a substrate, thus exposing a fresh metal to corrode and thereby accelerating damage.

Radiation

Not much is known about the effects of metals in corrosive environments subject to irradiation. The few known tests reveal only a slight additional increase in corrosion.

Environmental Impurities

Environmental impurities are extremely important factors and are therefore given considerable attention throughout the text.

Time

The extent of corrosion naturally increases with increased time. In some cases, the relationship is linear; in many instances, the rate of corrosion diminishes; in several well-documented instances, discussed in Chapters 11 and 13, the rate increases with time.

Effects of Stress

Under the proper conditions, a material will show a somewhat higher overall corrosion rate when under tensile stress. However, the major concern is the cracking of a metal or plastic under the combined effects of tensile stress and corrosion to produce a brittle failure of the material.

Pressure

Variations in pressure encountered in most liquid systems have little if any effect on the rate of corrosion, unless the pressure simply retains a corrosive species in the environment, *e.g.,* oxygen in heated water. However, there are environments primarly controlled by the partial pressure of the compounds in the system, such as in high-temperature, gaseous corrosion reactions (Chapter 13).

Other

Other factors are bound to develop as processes, conditions, etc., become more complex; a few can be listed.

Differential Aeration. A surface, one part of which is exposed to an aerated liquid while another part is exposed to a liquid with less aeration, will corrode if there is an electrical path through the liquid.

Concentration Difference. When there are differences in concentration or pH of corrosives in a liquid in contact with a metal surface, a corrosion cell usually will develop between the zones exposed to the differing solutions.

Biological Effects. Macro and microscopic organisms influence corrosion in two principal ways: (1) by creating mats or obstructions on the surface which produce differential aeration cells, or (2) by absorbing hydrogen from the surface of steel and thus removing the hydrogen as a resistance factor in the corrosion cell. Certain sulfate reducing bacteria operate this way, creating sulfurous acid in the vicinity of cathodic areas of steel, thereby accelerating its corrosion.

For those who wish to investigate in more detail the chemical and electrochemical reactions of the corrosion processes, the books cited in the bibliography at the end of this chapter will be useful.

These subjects and many others are often discussed in detail in *Corrosion,* the NACE magazine concerned with corrosion research. The other NACE monthly magazine, *Materials Performance,* provides engineering data relating to the testing and use of materials on a practical basis.

Industries With Control Programs

Some industries, because they are involved in handling material inherently poisonous or corrosive (as in the chemical industry), explosive or combustible (as in petroleum refining), or hot and under high pressure (as in steam power generating plants) have given close attention to corrosion problems. Many plants in these industries cannot operate at all unless corrosion is controlled.

Many industries pay attention to corrosion because they must maintain the integrity of their plant (as in the case of telephone companies) or preserve purity (as in the case of the food and soap industry, where minor contamination from corrosion will spoil products).

Other industries pay attention to corrosion in varying degrees of intensity, often as the result of the personality, initiative, and influence of one person who may be assigned responsibility for a company's programs.

In any event, all industry, as well as the consuming individual, should increase their awareness of the economic savings to be gained by exercising proper corrosion control. Such a program not only returns dollars to the individual, but also improves the economy of the country and conserves our natural resources.

References

1. NACE Glossary of Corrosion Terms, Materials Protection, Vol. 4, No. 1, pp. 79-80 (1965).
2. How One Company Saves Thousands of Dollars by Using Plastic Materials, Materials Protection, Vol. 2, No. 4, pp. 70-75 (1963).
3. Economic Effects of Metallic Corrosion in the United States, National Bureau of Standards Special Pub. 511-1.
4. Webster, Harold A., Automobile Body Corrosion Problems, Corrosion, Vol. 17, No. 2, pp. 9-12 (1961).
5. Staff Feature, Corrosion Damage on Automobile Exhaust Systems, Corrosion, Vol. 17, No. 10, pp. 18-29 (1961).
6. Mallet, Robert, Pioneer Corrosion Engineer, Wilson Lynes, Corrosion, Vol. 10, No. 2, pp. 59-62 (1954).
7. Davy, Sir Humphrey, Beginnings of Cathodic Protection, A collection of his papers, NACE, Houston, TX.
8. Whitney, W. R., The Corrosion of Iron, Corrosion, Vol. 3, No. 1, pp. 331-340 (1947).
9. American Committee on Electrolysis, Report, 1921.
10. Lund, Percy, Sixth Report of the Corrosion Committee, Special Report No. 66, Iron and Steel Institute, Humphries & Co., Ltd., London, 1959.

Bibliography of General Books on Corrosion and Corrosion Control

Alloy Digest. Engineering Alloy Digest, Inc., Upper Montclair, NJ.

Bakhvalov, G. T. and A. V. Turkovskaya. Corrosion and Protection of Metals, Pergamon Press, New York, N.Y., 1965.

Brasunas, Anton deS. and E. E. Stansbury, Eds. Symposium on Corrosion Fundamentals. University of Tennessee Press, Knoxville, TN, 1956.

Evans, Ulick R. An Introduction to Metallic Corrosion. Edward Arnold & Co., London, 1948.

Evans, Ulick R. Metallic Corrosion, Passivity and Protection. Longmans, Green & Co., New York, N.Y., 1948.

Fontana, M. G. and Norbert D. Greene. Corrosion Engineering. McGraw-Hill Book Co., Inc., New York, N.Y., 1978.

Henthorne, M. Fundamentals of Corrosion. Chem-Eng., May 1971-April 1972, or Carpenter Technology Corp.

Jastrzebski, Z. D. Nature and Properties of Engineering Materials. John Wiley & Sons, Inc., New York, N.Y., 1975.

Klinov, I. Ya., Ed. Corrosion and Protection of Materials Used in Industrial Equipment. Consultants Bureau, New York, N.Y., 1962.

Menzies, I. A. Corrosion and Protection of Metals. American Elsevier Publishing Co., Inc., New York, N.Y., 1965.

Process Industries Corrosion, NACE, Houston, TX, 1975.

Shreir, L. L., Ed. Corrosion. Vols. 1 and 2. John Wiley & Sons, Inc., New York, N.Y., 1976.

Speller, F. N. Corrosion Causes and Prevention. 3rd Ed. McGraw-Hill Book Co., Inc., New York, N.Y., 1951.

Tomashov, N. D. Theory of Corrosion and Protection of Metals. The MacMillan Company, New York, N.Y., 1966.

Uhlig, Herbert H. Corrosion and Corrosion Control. John Wiley & Sons, Inc., New York, N.Y., 1971.

Uhlig, H. H., Ed. Corrosion Handbook. John Wiley & Sons, Inc., New York, N.Y., 1958.

Woldman, N. Engineering Alloys. Reinhold Publishing Co., 1962.

Wronglen, G. An Introduction to Corrosion and Protection of Metals. Butler and Tanner Ltd., London, 1972.

Additional References on Corrosion Fundamentals

Among the many sources of information for the elementary facts on corrosion control, the following are recommended.

Short Course Proceedings

1. Appalachian Underground Corrosion Control Short course, held annually at University of West Virginia, Morgantown. Available from the university.
2. Liberty Bell Corrosion Course, held annually at Philadelphia, Pennsylvania. Available from NACE.
3. Oklahoma University Corrosion Control Course, held annually at Norman, Oklahoma. Available from NACE.
4. Western States Corrosion Seminar, held annually in California. Available from NACE.

Appendix A—Glossary of Corrosion-Related Terms

Active—A state in which a metal tends to corrode (opposite of passive).

Active Metal—A metal ready to corrode, or being corroded.

Active Potential—The potential of a corroding metal.

Additive—A substance added in a small amount, usually to a fluid, for a special purpose, such as to reduce friction, corrosion, etc.

Aeration Cell—An oxygen concentration cell; an electrolytic cell resulting from differences in dissolved oxygen at two points.

Anaerobic—An absence of unreacted or free oxygen. [Oxygen as H_2O, Na_2SO_4 (reacted) is not "free."]

Anion—An ion or radical which is attracted to the anode because of the negative charge on the ion or radical (*e.g.*, Cl^- or OH^-).

Anode—(Opposite of cathode) The electrode at which oxidation or corrosion occurs. A common anode reaction is:

$$Zn \rightarrow Zn^{++} + 2 \text{ electrons.}$$

Anode Corrosion Efficiency—Ratio of actual corrosion to theoretical corrosion based on the total current flow.

Anodic Polarization—Polarization of anode; *i.e.*, the decrease in the initial anode potential resulting from current flow effects at or near the anode surface. Potential becomes more noble (more positive) because of anodic polarization.

Anodic Protection—An appreciable reduction in corrosion by making a metal an anode and maintaining this highly polarized condition with very little current flow.

Anolyte—The electrolyte of an electrolytic cell adjacent to the anode.

Anti-Fouling—Refers to the prevention of marine organism attachment or growth on a submerged metal surface, generally through chemical toxicity caused by the composition of the metal or coating layer.

Aqueous—Pertaining to water; an aqueous solution is a water solution.

Austenitic—The name given to the face-centered cubic crystal structure (FCC) of ferrous metals. Ordinary iron and steel has this structure at elevated temperatures; also certain stainless steels (300 Series) have this structure at room temperature.

Auxiliary Electrode—An electrode commonly used in polarization studies to pass current to or from a test electrode. It is usually made of a noncorroding material.

Bimetallic Corrosion—Corrosion resulting from dissimilar metal contact; galvanic corrosion.

Carburizing—The absorption of carbon into a metal surface; this may or may not be desirable.

Cathode—(Opposite of anode) The electrode where reduction (and practically no corrosion) occurs. A typical cathode reaction is:

$$4 \text{ electrons} + O_2 + 2H_2O \rightarrow 4OH^-.$$

Cathodic Corrosion—An unusual condition (especially with Al, Zn, Pb) in which corrosion is accelerated at the cathode because cathodic reaction creates an alkaline condition which is corrosive to certain metals.

Cathodic Inhibitor—A chemical substance or combination of substances that prevent or reduce the rate of cathodic or reduction reaction by a physical, physico-chemical or chemical action.

Cathodic Polarization—Polarization of the cathode; a reduction from the initial potential resulting from current flow effects at or near the cathode surface. Potential becomes more active (negative) because of cathodic polarization.

Cathodic Protection—Reduction or elimination of corrosion by making the metal a cathode by means of an impressed d-c current or attachment to a sacrificial anode (usually Mg, Al, or Zn).

Catholyte—That portion of the electrolyte of an electrolytic cell adjacent to the cathode. Antonym: anolyte.

Cation—A positively charged ion (*e.g.*, H^+ or Zn^{++}) or radical (*e.g.*, NH_4^+) which migrates toward the cathode.

Caustic Embrittlement—Cracking as a result of the combined action of tensile stresses and corrosion in alkaline solutions (as at riveted joints in boilers).

Cavitation—Formation and sudden collapse of vapor bubbles in a liquid, usually resulting from local low pressures, as on the trailing edge of a propeller; this develops momentary high local pressure which can mechanically destroy a portion of a surface on which the bubbles collapse.

Cavitation Corrosion—Corrosion damage resulting from cavitation and corrosion: metal corrodes, pressure develops from collapse of cavity and removes corrosion product, exposing bare metal to repeated corrosion.

Cavitation Damage—Deterioration of a surface caused by cavitation (sudden formation and collapse of cavities in a liquid).

Cavitation Erosion—See *Cavitation Damage,* the preferred term.

Cell—A circuit consisting of an anode and a cathode in electrical contact in a solid or liquid electrolyte. Corrosion generally occurs only at anodic areas.

Cementation Coating—A coating developed on a metal surface by a high-temperature diffusion process (*e.g.*, as carburization, calorizing, or chromizing).

Chalking—Development of a loose, chalky, removable powder on or beneath a coating layer.

Checking—Surface cracking in a checkerboard-like pattern; this may be in a surface layer (coating) or on the metal surface itself.

Chemical Conversion Coating—A metal surface layer intentionally developed by chemical reaction for the purpose of protection or looks.

Concentration Cell—A cell involving an electrolyte and two identical electrodes, with the potential resulting from differences in the chemistry of the environments adjacent to the two electrodes.

Concentration Polarization—Polarization of an electrode caused by concentration changes in the environment adjacent to the metal surface.

Conversion Coating—An adherent reaction product layer on a metal surface formed by reaction with a suitable chemical; such as an iron phosphate film on iron developed by H_3PO_4.

Corrosion—The destruction of a substance; usually a metal, or its properties because of a reaction with its surroundings (environment).

Corrosion-Erosion—Corrosion which is increased because of the abrasive action of a moving stream; the presence of suspended particles greatly accelerates abrasive action.

Corrosion Fatigue—The combined action of corrosion and fatigue (cyclic stressing) in causing metal fracture.

Corrosion Potential—The potential that a corroding metal exhibits under specific conditions of concentration, time, temperature, aeration, velocity, etc.

Corrosion Rate—The speed (usually an average) with which corrosion progresses (it may be linear for awhile); often expressed as though it were linear, in units of mdd (milligrams per square decimeter per day) for weight change or mpy (mils per year) or μm/y (microns per year) for thickness changes.

Corrosion Fatigue Limit—The maximum cyclic stress value that a metal can withstand for a specified number of cycles or length of time in a given corrosive environment.

Couple—A cell developed in an electrolyte resulting from electrical contact between two dissimilar metals.

Cracking—Fracture of a metal in a brittle manner along a single or branched path.

Crazing—Development of a network of fine surface cracks.

Crevice Corrosion—Localized corrosion resulting from the formation of a concentration cell in a crevice formed between a metal and a nonmetal, or between two metal surfaces.

Critical Humidity—A humidity level above which corrosion in air increases sharply.

Current Density—The current per unit area; generally expressed as amps per sq ft or milliamperes per sq ft (also milliamps per sq cm, etc.).

Deactivation—The process of removing active constituents from a corroding medium, *e.g.,* removal of dissolved oxygen from water.

Dealloying—The selective corrosion (removal) of a metallic constituent from an alloy, usually in the form of ions.

Dealuminization, Denickelification, Demolybdenization, Dezincification, etc.—The selective leaching or corrosion of a specific constituent (Al, Ni, Mo, Zn) from an alloy; the terms *Parting, Dealloying,* or *Selective Corrosion* are preferred.

Decomposition Potential (or Voltage)—The potential of metal surface necessary to decompose the electrolyte of a cell or a substance thereof.

Demineralization—Removal of dissolved mineral matter, generally from water.

Depolarization—The elimination or reduction of polarization by physical or chemical means; depolarization results in increased corrosion.

Deposit—A foreign substance, which comes from the environment, adhering to a surface of a material.

Deposit Attack (Deposition Corrosion)—Pitting corrosion resulting from deposits on a metal surface which cause concentration cells.

Dezincification—The parting of zinc from an alloy. (In some brasses zinc is lost, leaving a weak, brittle, porous, copper-rich residue behind.)

Differential Aeration Cell—An oxygen concentration cell (a cell resulting from a potential difference caused by different amounts of oxygen dissolved at two locations).

Double Layer—The interface between the electrode and the electrolyte where charge separation takes place. The simplest model is represented by a parallel plate condenser of 2×10^{-8} cm in thickness. In general the electrode will be positively charged with respect to the solution.

Electrical Current—An electric current is caused by the flow of electrons. However, the electric current flows in a direction opposite to the flow of electrons. (This is the *positive* current concept.)

Electrochemical Equivalent—The weight of an element oxidized or reduced by a specific (unit) quantity of electricity, usually a *coulomb* (an ampere flowing for one second).

Electrochemical Potential, Electrochemical Tension—The partial derivative of the total electrochemical free energy of a constituent with respect to the number of moles of this constituent where all factors are kept constant. It is analogous to the chemical potential of a constituent except that it includes the electric as well as chemical contributions to the free energy.

Electrode—A metal in contact with an electrolyte which serves as a site where an electrical current enters the metal or leaves the metal to enter the solution.

Electrode Potential—The potential of an electrode as measured against a reference electrode. See text for sign convention. The electrode potential does not include any resistance loss in potential in the solution due to current passing to or from the electrodes; *i.e.,* it represents the reversible work to move a unit charge from the electrode surface through the solution to the reference electrode.

Electrokinetic Potential—This potential, often called zeta potential, is a potential difference in the solution caused by residual, unbalanced charge on an adjacent surface. This charge causes a counter charge distribution in the adjoining solution producing a double layer. The electrokinetic potential causes the effects of electrophoresis, electro-osmosis, streaming potential, and sedimentation potential. The electrokinetic potential is different from the electrode potential in that it occurs exclusively in the solution phase; *i.e.,* it represents the reversible work necessary to bring a unit charge from infinity in the solution up to the interface in question but not through the interface.

Electrolysis—Chemical changes in an electrolyte caused by an electrical current. The use of this term to mean corrosion by stray currents is discouraged.

Electrolyte—An ionic conductor (usually in aqueous solution).

Electrolytic Cleaning—A process of cleaning, degreasing, or descaling a metal by making it an electrode in a suitable bath.

Electromotive Force Series (Emf Series)—An orderly listing of elements according to their standard electrode potentials. (Hydrogen electrode is a reference point and given the value of zero.)

Electronegative Potential—A potential representing the active or anodic end of the Emf series; *anodic potential* is the preferred term.

Electropositive Potential—A potential representing the noble or cathodic end of the Emf series; *cathodic potential* is the preferred term.

Embrittlement—Severe loss of ductility of a metal (or alloy).

Endurance Limit—The maximum cyclic stress level a metal can withstand without a fatigue failure.

Environment—The surroundings or conditions (physical, chemical, mechanical) in which a material exists.

Equilibrium Potential—The electrode potential at equilibrium.

Erosion—Deterioration of a surface by the abrasive action of moving fluids. This is accelerated by the presence of solid particles or gas bubbles in suspension. When deterioration is further increased by corrosion, the term *erosion-corrosion* is often used.

Erosion-Corrosion—A corrosion reaction accelerated by velocity and air abrasion.

Exchange Current—When an electrode reaches dynamic equilibrium in a solution, the rate of anodic dissolution just balances the rate of cathodic plating. The rate at which either positive or negative charges are entering or leaving the surface at this point is known as the exchange current.

Exfoliation—A thick layer-like growth of loose corrosion products (observed in some cases on steel and aluminum alloys).

Fatigue—A process leading to fracture resulting from repeated stress cycles well below the normal tensile strength. Such failures start as tiny cracks which grow to cause total failure.

Ferritic—Pertaining to the body-centered cubic crystal structure (BCC) of many ferrous (iron-base) metals.

Filiform Corrosion—Random small threads of corrosion that develop beneath thin lacquers and other semipermeable films.

Film—A thin surface layer that may or may not be visible.

Fogged Metal—A metal whose luster has been reduced because of a surface film, usually a corrosion product layer.

Fouling—A term used to describe the submerged surfaces covered by marine growths such as barnacles.

Fretting Corrosion—Fretting refers to metal deterioration caused by repetitive slip at the interface between two surfaces. When metal loss is increased by corrosion, the term *fretting corrosion* is used.

Galvanic—Pertaining to an effect caused by a cell; often dissimilar metal contact which results in electrolytic potential.

Galvanic Cell—A cell consisting of two dissimilar metals in contact with each other and with a common electrolyte (sometimes refers to two similar metals in contact with each other but with dissimilar electrolytes; differences can be small and more specifically defined as a concentration cell).

Galvanic Corrosion—Corrosion that is increased because of the current caused by a galvanic cell (sometimes called *couple action*).

Galvanic Series—A list of metals arranged according to their relative corrosion potentials in some specific environment; seawater is often used.

Galvanostatic—Refers to the constant current technique of applying current to a specimen in an electrolyte. Synonym: intentiostatic.

General Corrosion—Corrosion in a uniform manner.

Grain—A portion of a solid metal (usually a fraction of an inch in size) in which the atoms are arranged in an orderly pattern. The irregular junction of two adjacent grains is known as a grain boundary (also a unit of weight, 1/7000th of a pound).

Graphitization (Graphitic Corrosion)—Corrosion of gray cast iron in which the metallic constituents are converted to corrosion products, leaving the graphite flakes intact. Graphitization is also used in a metallurgical sense to mean the decomposition of iron carbide to form iron and graphite.

Green Rot—A form of high-temperature corrosion of chromium-bearing alloys in which green chromium oxide (Cr_2O_3) forms, but certain other alloy constituents remain metallic; some simultaneous carburization is sometimes observed.

Half-Cell—A pure metal in contact with a solution of known concentration of its own ion, at a specific temperature develops a potential which is characteristic and reproducible; when coupled with another half-cell, an overall potential develops which is the sum of both half-cells.

Heat-Affected Zone (HAZ)—Refers to area adjacent to a weld where the thermal cycle has caused microstructural changes which generally affect corrosion behavior.

Holiday—A discontinuity (hole or gap) in a protective coating.

Hydrogen Blistering—Formation of blister-like bulges on a ductile metal surface caused by internal hydrogen pressures.

Hydrogen Disintegration—Deep internal cracks in a metal caused by hydrogen.

Hydrogen Embrittlement—Embrittlement of a metal caused by hydrogen; sometimes observed in cathodically protected steel, electroplated parts, pickled steel, etc.

Hydrogen Overvoltage—Overvoltage caused by the liberation of hydrogen.

Immunity—A state of resistance to corrosion or anodic dissolution caused by the fact that the electrode potential of the surface in question is below the equilibrium potential for anodic dissolution.

Impingement Attack—Localized erosion-corrosion caused by turbulence or impinging flow at certain points.

Inhibitor—A substance which sharply reduces corrosion when added to water, acid, or other liquid in small amounts.

Interdendritic Corrosion—Corrosion which occurs preferentially along dendrite boundaries. This type of attack is caused by local differences in chemical composition, typical of cast metals; known as *coring*.

Intergranular Corrosion—Corrosion which occurs preferentially at grain boundaries.

Internal Oxidation—Formation of particles of corrosion product beneath a metal surface at high temperatures. This results from preferential reaction of certain alloy constituents by the inward diffusion of oxygen, sulfur, nitrogen, etc.; also known as selective or subsurface corrosion.

Ion—An electrically charged atom (Na^+, Al^{+3}, Cl^-, S^{-2}) or group of atoms known as *radicals* ($NH_4{}^+$, $SO_4{}^{-2}$, $PO_4{}^{-3}$).

Ion Erosion—Deterioration of material caused by ion impact.

Iron Rot—Deterioration of wood in contact with iron.

Isocorrosion—Refers to lines on a graph or chart which show constant corrosion behavior with changing composition.

Knife-Line Attack (KLA)—A form of weld decay sometimes observed on stabilized stainless steel; the zone of attack is very narrow and very close to or in the weld.

Langelier Index—A calculated saturation index for calcium carbonate that is useful in predicting scaling behavior of natural water.

Local Action—Corrosion due to action of local cells, *i.e.,* galvanic cells caused by nonuniformities between two adjacent areas at a metal surface exposed to an electrolyte.

Local Cell—A galvanic cell caused by small differences in composition in the metal or the electrolyte.

Long-Line Current—Current flowing through the earth from an anodic to a cathodic area which returns along an underground metallic structure (generally applied only where the areas are separated by considerable distance and where current results from concentration cell action).

Luggin Probe or Luggin Haber Capillary—A scheme for measuring the potential of a specimen with a significant current density imposed on its surface. The purpose of the probe is to minimize the IR drop that is included in the measurement without significantly disturbing the current distribution on the specimen.

Metal Dusting—A unique form of high-temperature corrosion which forms a dust-like corrosion product and sometimes develops hemispherical pits on a susceptible metal surface.

Metal Ion Concentration Cell—A galvanic cell caused by a difference in metal ion concentration at two locations on the same metal surface.

Metallizing—A process of coating a surface with a layer of metal; spraying, vacuum deposition, dipping, plasma jet, cementation, etc., are used.

Mill Scale—The heavy oxide layer formed during heat treatment or hot working of metals; often refers to steel forming magnetic oxide, Fe_3O_4 (magnetite).

Mixed Potential—A potential resulting from two or more electrochemical reactions occurring simultaneously on one metal surface.

Nernst Layer and Nernst Thickness—The diffusion layer or the thickness of this layer as given by the theory of Nernst. It is defined by:

$$i_d = n F D \frac{C^\circ - C}{\delta}$$

where, i_d = the diffusion limited current density, D = the diffusion coefficient, C° = the concentration at the electrode surface, and δ = the Nernst thickness. It is a hypothetical thickness which has been found to be 0.05 cm in many cases of unstirred aqueous electrolytes.

Noble—Referring to positive direction of electrode potential, thus resembling noble metals such as gold and platinum. Antonym: active.

Noble Metal—A metal which is not very reactive, *e.g.,* silver, gold, or copper, and may be found naturally in metallic form on earth.

Noble Potential—A potential in the general range of the noble metals, *i.e.,* strongly cathodic to the standard hydrogen potential.

Nitriding (Nitrogenation)—The absorption of nitrogen atoms by a metal; it may remain dissolved or it may form metal nitrides.

Open Circuit Potential—The measured potential of a cell in which no current flows.

Overvoltage—The difference in electrode potential when a current is flowing vs when there is no current flow; also known as polarization.

Oxidation—Loss of electrons; as when a metal goes from the metallic state to the corroded state (opposite of Reduction). Thus, when a metal reacts with oxygen, sulfur, etc., to form a compound as oxide, sulfide, etc., it is oxidized.

Oxygen Concentration Cell—A galvanic cell caused by a difference in oxygen concentration at two points on a metal surface.

Parting—The selective corrosion or leaching of a component from an alloy, such as the parting of zinc from brass, leaving a copper residue (sometimes known as dezincification; similar leaching also may lead to terms such as demolybdenization, deberyllization, detantalizing, depraseodymization, etc.).

Passivation—A reduction of the anodic reaction rate of an electrode involved in electrochemical action such as corrosion.

Passivator—An inhibitor which changes the potential of a metal appreciably to a more cathodic or noble value (as when chromate is added to water).

Passive-Active Cell—A cell composed of a metal in the passive state and the same metal in the active state.

Passivity—The phenomenon of an active metal becoming passive.

Patina—A green coating which slowly develops on copper and some copper alloys consisting mainly of copper sulfates, carbonates, and chlorides after long-term exposure to the atmosphere.

pH—A measure of the acidity or alkalinity of a solution. A value of seven is neutral; low numbers are acid, large numbers are alkaline. Strictly speaking, pH is the negative logarithm of the hydrogen ion concentration.

Pickle—A solution, usually acid, used to remove mill scale or other corrosion products from a metal.

Pitting—Highly localized corrosion resulting in deep penetration at only a few spots.

Pitting Factor—The depth of the deepest pit divided by the *average penetration* as calculated from weight loss.

Polarization—The shift in electrode potential resulting from the effects of current flow, measured with respect to the *zero-flow* (reversible) potential; *i.e.,* the counter-emf caused by the products formed or concentration changes in the electrolyte.

Potentiostat—An electronic device which maintains an electrode at a constant potential; used in anodic protection devices or to draw E log I curves.

Prime Coat—A first coat of paint applied to inhibit corrosion or improve adherence of the next coat.

Protective Potential—A term sometimes used in cathodic protection to define the minimum potential required to suppress corrosion. For steel in seawater, this is claimed to be about 0.85 volt as measured against a saturated calomel.

Reduction—Gain of electrons, as when copper is electroplated on steel from a copper sulfate solution (opposite of Oxidation).

Relative Humidity—The ratio (%) of the amount of moisture in the air compared to what it could hold if saturated at the temperature involved.

Rest Potential—See Corrosion Potential.

Ringworm Corrosion—Localized corrosion frequently observed in oil well tubing in which a circumferential attack is observed near a region of metal "upset."

Rusting—Corrosion of iron or an iron-base alloy to form a reddish-brown product which is primarily hydrated ferric oxide.

Sacrificial Protection—Reduction or prevention of corrosion of a metal in an electrolyte by galvanically coupling it to a more anodic metal.

Scaling—(1) High-temperature corrosion resulting in formation of thick corrosion product layers. (2) Deposition of insoluble materials on metal surfaces, usually inside water boilers or heat exchanger tubes.

Season Cracking—Cracking caused by the combined action of corrosion and internal tensile stresses; usually applied to the stress corrosion cracking of brass.

Selective Corrosion—The selective corrosion of certain alloying constituents *from* an alloy (as dezincification), or *in* an alloy (as internal oxidation).

Slushing Compound—A nondrying oil, grease, or wax applied to metals for temporary corrosion protection.

Spalling—The separation of a surface layer caused by thermal or mechanical stresses (*e.g.,* cooling, bending, etc.).

Stabilized Steel—Usually refers to a stainless steel which has been alloyed with a carbide-forming element (*e.g.,* Cb, Ti, or Ta) which makes it less or not susceptible to carbide precipitation.

Standard Electrode Potential (Standard Potential)—The reversible potential of an electrode process when all reactants and products are at unit activity on a scale in which the potential for the standard hydrogen half-cell is zero. (This term is important in theoretical considerations, but is of limited value, as is, to the practical corrosion engineer.)

Stray Current Corrosion—Corrosion that is caused by stray currents from some external source.

Stress Corrosion or Stress-Accelerated Corrosion—Corrosion which is accelerated by tensile stress.

Subscale Formation—*Subsurface Corrosion* or *Internal Oxidation* are the preferred terms.

Subsurface Corrosion—Formation of isolated products of corrosion beneath the metal surface (resulting from selective reaction with certain alloy constituents; same as Internal Oxidation or Subscale Formation).

Sulfidation—Oxidation by sulfur.

Sulfide Stress Cracking—Hydrogen-induced cracking (HIC) of a metal in an environment containing hydrogen sulfide.

Tafel Line, Tafel Slope, Tafel Diagrams—When an electrode is polarized, it frequently will yield a current potential relationship over a region which can be approximated by:

$$\eta = \pm B \log \frac{i}{i_o}$$

where η = change in open circuit potential, i = the current density, B and i_o = constants. The constant

(B) is also known as the Tafel slope. If this behavior is observed, a plot on semetogrithmic components is known as the Tafel line and the overall diagram is termed a Tafel diagram.

Tarnish—Surface discoloration of a metal surface caused by a thin film of corrosion product.

Thermogalvanic Corrosion—Galvanic corrosion resulting from temperature differences at two points.

Transpassive—The noble region of potential where an electrode exhibits a higher than passive current density.

Tuberculation—Localized corrosion at scattered locations resulting in knoblike mounds.

Underfilm Corrosion—Corrosion which occurs under coatings and other organic films at exposed edges or by filiform corrosion.

Upset—A metallurgical term meaning a hot deformation process to cause a dimensional change; in this case a thickening of metal by compressive forces.

Voids—A term generally applied to paints to describe holidays, holes, and skips in the film.

Volatilization—Vaporization due to high vapor pressure at a particular temperature. Some metals and oxides evaporate easily (*e.g.*, Zn, MoO_3, etc.).

Wash Primer—A thin inhibiting paint, usually chromate pigmented with a polyvinyl butyrate binder.

Weld Decay—A term applied to areas adjacent to welds of certain alloys which have been subjected to intergranular corrosion because of metallurgical changes in the alloy (commonly applied to certain grades of stainless steel).

Working Electrode—The test or specimen electrode in an electrochemical cell.

NOTES

Chapter 2

Basics of Corrosion

BASICS OF CORROSION

Introduction

This is the most important chapter in the book. It is imperative to learn the basic mechanisms of the corrosion process in order to properly analyze corrosion problems and arrive at effective solutions.

The information in this chapter is fundamental to an understanding of the remainder of the text. Thus, it is suggested that the reader proceed through this chapter carefully, concentrating particularly on learning the component parts of a corrosion cell and their interrelationships.

Why Metals Corrode

The driving force that causes metals to corrode is a natural consequence of their temporary existence in metallic form. To reach this metallic state from their occurrence in nature in the form of various chemical compounds (ores), it is necessary for them to absorb and store up for later return by corrosion, the energy required to release the metals from their original compounds. The amount of energy required and stored varies from metal to metal. It is relatively high for metals such as magnesium, aluminum, and iron, and relatively low for metals such as copper and silver. Table 2.1 lists some commonly used metals in order of diminishing amounts of energy required to convert them from their ores to metal.

A typical cycle is illustrated by iron. The most common iron ore, hematite, is an oxide of iron (Fe_2O_3). The most common product of the corrosion of iron, rust, has the same chemical composition. The energy required to convert iron ore to metallic iron is returned when the iron corrodes to form the same compound. Only the rate of energy change is different.

TABLE 2.1 — Positions of Some Metals in the Order of Energy Required to Convert Their Ores to Metal

Most Energy Required ↑	Potassium
	Magnesium
	Beryllium
	Aluminum
	Zinc
	Chromium
	Iron
	Nickel
	Tin
	Copper
	Silver
Least Energy Required ↓	Platinum
	Gold

The energy difference between metals and their ores can be expressed in electrical terms which are related to heats of formation of the compounds.

The difficulty of extracting metals from their ores in terms of the energy required, and the consequent tendency to release this energy by corrosion, is reflected by the relative positions of pure metals in a list, which is discussed later as the *electromotive series*.

Forms of Corrosion

Destruction by corrosion takes many forms, depending on the nature of the metal or alloy; the presence of inclusions or other foreign matter at the surface; the homogeneity of its structure; the nature of the corrosive medium; the incidental environmental factors such as the presence of oxygen and its uniformity, temperature, and velocity of movement; and other factors such as stress (residual or applied, steady or cyclic); oxide scales (continuous or broken); porous or semiporous deposits on surfaces, built-in crevices; galvanic effects between dissimilar metals; and the occasional presence of stray electrical currents from external sources.

Except in rare cases of a grossly improper choice of material for a particular service, or an unanticipated drastic change in the corrosive nature of the environment or complete misunderstanding of its nature, failures of metals by rapid general attack (wasting away) are not often encountered. Corrosion failures are more often localized in the form of pits, intergranular corrosion, attack within crevices, etc. These and several other forms of attack are discussed in Chapter 5.

Chemistry and Electrochemistry of Corrosion

The Atom

Matter, itself being made up of atoms, is also composed of those lesser particles which make up the atoms. These numerous particles arrange themselves so that those bearing positive charges or those which are neutral cluster together to form a *nucleus* around which negatively charged particles rotate in orbits much like the rotation of planets around the sun. In a normal atom, the *negative* particles which are called *electrons* exactly balance the positive charges on the nucleus. The electrons occupy "shells," which in the case of iron are "filled" when they contain up to eight electrons plus any additional electrons that are required to balance the positive charge on the nucleus. The electrons in the outermost shell can be "stripped" from the atom, changing its properties. Thus, the charge on the nucleus is unbalanced and the atom displays a positive charge. This charged atom is called an *ion* and the process is called *ionization*.

There is a chemical shorthand to denote this state of affairs. For example, Fe is the chemical shorthand for a neutral atom of iron, whereas Fe^{++} denotes an iron atom that has been stripped of two electrons and is called a ferrous ion. Similarly, Fe^{+++} denotes an iron atom stripped of three electrons and is called a ferric ion. The process of stripping electrons from atoms is referred to by electrochemists as *oxidation*.

An opposite process can also occur in which extra electrons are added to the neutral atom giving it a net negative charge. Any increase in negative charge (or decrease in positive charge) of an atom or ion is called *reduction*.

Many chemical compounds are made up of two or more ions of opposite charge. When these are dissolved in water, they spontaneously split into two or more separate ions which display equal but opposite charges. This process is also called ionization. It is these particles that are responsible for the conduction of electric currents in aqueous solutions.

Acidity and Alkalinity (pH)

The ions referred to above will exist in an electrically conductive medium, normally water. When discussing any such aqueous medium, the question often arises as to how acid (or alkaline) is the solution. Quite simply, this refers to whether there is an excess of H^+ (hydrogen) or OH^- (hydroxyl) ions present. The H^+ ion is acid. The hydroxyl ion is alkaline. The other portion of an acid or alkali added to water increases the conductivity or other property of the liquid, but does not increase or decrease the acidity. For instance, whether a given amount of H^+ ion is produced in water by introducing HCl, H_2SO_4, H_2S, or acetic acid is immaterial. The pH of the solution will be the same for the same number of dissolved hydrogen atoms. Other properties of the solution may differ, but the pH is simply a statement of the H^+ concentration in the solution.

The pH may be measured with a meter or calculated if certain parameters are established (Figure 2.1). Water itself dissociates to a small extent to produce equal quantities of H^+ and OH^- ions. That is shown as:

$$HOH \leftrightarrow H^+ + OH^-. \quad (2.1)$$

Since there are equal quantities of H^+ (acid) and OH^- (alkali) ions, the solution is said to be neutral. By a manipulation of the number of H^+ ions present under these conditions, the solution is said to

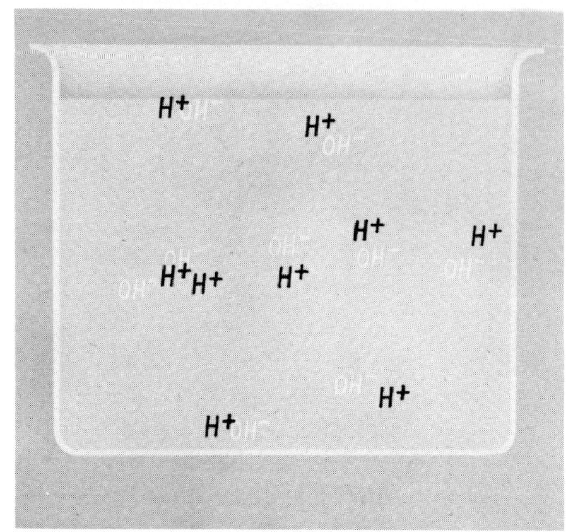

FIGURE 2.1 — Hydrogen (H^+) and hydroxyl (OH^-) ions in water.

have a pH of 7 (neutral). If the number decreases (<7), there are more H^+ ions than OH^- ions, and the solution is acidic. If the number increases (>7), there are more OH^- ions than H^+ ions, and the solution is alkaline. The greater the variation of this number from 7, the greater the acidity or alkalinity. Thus, a pH of 2 is very acid and a pH of 12 is very alkaline.

Many salts added to an aqueous system also have some effect on the pH of that mixture.

Corrosion as a Chemical Reaction

Corrosion in Acids

One of the common ways of generating hydrogen in a laboratory is to place zinc into a dilute acid, such as hydrochloric or sulfuric. When this is done, there is a rapid reaction in which the zinc is attacked and hydrogen is evolved as a gas. This is shown in Equations (2.2) and (2.3),

$$\underset{\text{Zinc}}{Zn} + \underset{\text{Hydrogen Chloride}}{2HCl} \rightarrow \underset{\text{Zinc Chloride}}{ZnCl_2} + \underset{\text{Hydrogen}}{H_2 \uparrow} \quad (2.2)$$

which is chemical shorthand for the statement: One zinc atom + two hydrochloric acid molecules becomes one molecule of zinc chloride (a salt) + one molecule of hydrogen gas which is given off as indicated by the vertical arrow.

Similarly, zinc combines with sulfuric acid to form zinc sulfate (a salt) and hydrogen gas as shown in Equation (2.3).

$$\underset{\text{Zinc}}{Zn} + \underset{\text{Sulfuric Acid}}{H_2SO_4} \rightarrow \underset{\text{Zinc Sulfate}}{Zn\,SO_4} + \underset{\text{Hydrogen.}}{H_2 \uparrow} \quad (2.3)$$

Note that each atom of a substance that appears on the left-hand side of these equations must also appear on the right-hand side. There are also some rules that denote in what proportion different atoms combine with each other, if at all.

Other metals are also corroded or "dissolved" by acids and they, too, yield a soluble salt and hydrogen gas as shown in Equations (2.4) and (2.5).

$$\underset{\text{Iron}}{Fe} + 2HCl \rightarrow \underset{\text{Ferrous Chloride}}{FeCl_2} + H_2\uparrow \qquad (2.4)$$

$$\underset{\text{Aluminum}}{2Al} + 6HCl \rightarrow \underset{\text{Aluminum Chloride}}{2AlCl_3} + 3H_2\uparrow \qquad (2.5)$$

Equations (2.4) and (2.5) show that both iron and aluminum are also corroded by hydrochloric acid solutions.

Note that zinc and iron combined with two Cl^- ions, whereas aluminum combined with three. This is due to the fact that both zinc and iron, when corroding, each lose two electrons and display two positive charges in their ionic form. They are said to have a valence of 2, whereas aluminum loses three electrons when leaving an anodic surface and hence displays three positive charges and is said to have a valence of 3. Some metals have several common valences, others only one.

Corrosion in Neutral and Alkaline Solutions

The corrosion of metals can also occur in fresh water, seawater, salt solutions, and alkaline or basic media. In almost all of these systems, corrosion only occurs if dissolved oxygen is also present. Water solutions rapidly dissolve oxygen from the air, and this is the source of the oxygen required in the corrosion process. The most familiar corrosion of this type is the rusting of iron when exposed to a moist atmosphere or water.

$$4Fe + 6H_2O + \underset{\text{Oxygen}}{3O_2} \rightarrow \underset{\text{Ferric Hydroxide}}{4Fe(OH)_3\downarrow} \qquad (2.6)$$

In Equation (2.6), we see that iron will combine with water and oxygen to produce an insoluble reddish-brown corrosion product that falls out of the solution, as shown by the downward pointing arrow.

During rusting in the atmosphere, there is an opportunity for drying, and this ferric hydroxide dehydrates and forms the familiar red-brown iron oxide (rust), as shown below.

$$2Fe(OH)_3 \rightarrow \underset{\text{Ferric Oxide}}{Fe_2O_3} + 3H_2O \qquad (2.7)$$

Similar reactions occur when zinc is exposed to water or moist air.

$$2Zn + 2H_2O + O_2 \rightarrow \underset{\text{Zinc Hydroxide}}{2Zn(OH)_2\downarrow} \qquad (2.8)$$

$$Zn(OH)_2 \rightarrow \underset{\text{Zinc Oxide}}{ZnO} + H_2O \qquad (2.9)$$

The resulting zinc oxide is the whitish deposit seen on galvanized pails, rain gutters, and imperfectly chrome-plated bathroom faucets.

As discussed previously, the iron that took part in the reaction with hydrochloric acid in Equation (2.4) had a valence of 2, whereas the iron that takes part in the reaction shown in Equation (2.6) has a valence of 3. The clue to this lies in the examination of the equation for the corrosion product $Fe(OH)_3$. Note that water ionized into H^+ and OH^-. It is further known that hydrogen ion has a valence of 1 (it has only one electron to lose). It would require three hydrogen ions with the corresponding three positive charges to combine with the three OH^- ions held by the iron. It can thus be concluded that the iron ion must have been Fe^{+++} or a ferric ion.

This roundabout method of determining valence is very useful; for example, in Equation (2.4) note that one hydrogen (valence of 1) combines with one chlorine atom in HCl. The valence of chlorine is now known to be 1. With this knowledge it is quickly discovered that the iron in $FeCl_2$ has a valence of 2. Fe^{++} is called a *ferrous* ion.

For purposes of comparison, consider a reaction which is not electrochemical. If a solution of silver nitrate is added to a solution of sodium chloride, a white precipitate of silver chloride precipitates from solution. The overall reaction is:

$$\underset{\text{Silver Nitrate}}{AgNO_3} + \underset{\text{Sodium Chloride}}{NaCl} \rightarrow \underset{\text{Silver Chloride}}{AgCl\downarrow} + \underset{\text{Sodium Nitrate}}{NaNO_3}. \qquad (2.10)$$

Recognizing that some of the substances in the above equation exist as separate ions in solution, Equation (2.10) can be rewritten in this fashion:

$$Ag^+ + NO_3^- + Na^+ + Cl^- \rightarrow AgCl\downarrow + Na^+ + NO_3^-. \qquad (2.11)$$

Examining this equation shows that both nitrate and sodium ions appear on both sides of the reaction. Therefore, they are not directly involved and can be disregarded.

$$Ag^+ + Cl^- \rightarrow AgCl\downarrow \qquad (2.12)$$

Thus, Equation (2.12) is a simplification of the reaction shown in Equation (2.11) (Figure 2.2). Note that there is no oxidation or reduction (electron transfer) during this reaction. The valences of both

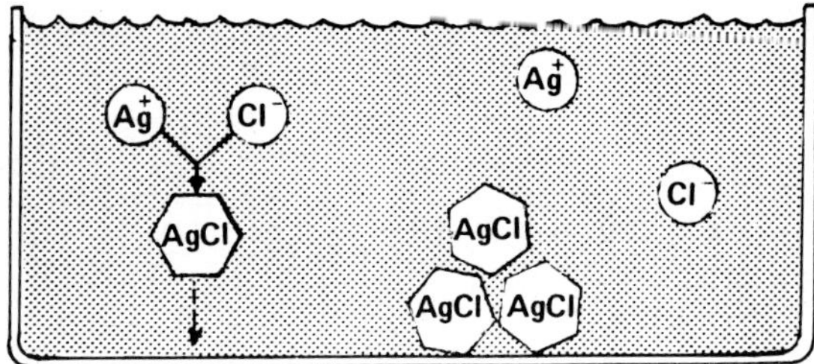

FIGURE 2.2 — A chemical reaction which is not electrochemical in nature (precipitation of AgCl).

silver and chlorine remain unchanged throughout the course of this reaction, and it is consequently not possible to divide this reaction into individual oxidation and reduction reactions. Corrosion reactions are usually electrochemical processes which involve electron transfer.

To summarize, corrosion reactions are electrochemical in nature. Because of this, it is possible to divide corrosion into anodic and cathodic reactions (oxidation and reduction). This has the advantage of simplifying the presentation of most corrosion processes.

Corrosion in Other Systems

Metals can also be corrosively attacked in solutions containing neither oxygen nor acids. The most typical types of such solutions are oxidizing salts such as ferric and cupric compounds. Corrosion reactions of this type are indicated by:

$$\underset{}{Zn} + \underset{\text{Ferric Chloride}}{2FeCl_3} \rightarrow ZnCl_2 + \underset{\text{Ferrous Chloride}}{2FeCl_2} \quad (2.13)$$

$$Zn + \underset{\text{Copper Sulfate}}{CuSO_4} \rightarrow \underset{\text{Zinc Sulfate}}{ZnSO_4} + \underset{\text{Copper}}{Cu\downarrow} \quad (2.14)$$

Note that in one case *ferric* chloride is changed to *ferrous* chloride as it corrodes the zinc. In the other case, zinc reacts with copper sulfate to yield soluble zinc sulfate plus a spongy mass of metallic copper deposited on the surface of the zinc. Equation (2.14) is often called a metal replacement reaction.

Corrosion Products

The term *corrosion products* refers to the substances produced during a corrosion reaction. These can be soluble, such as zinc chloride or zinc sulfate in the examples cited earlier, or insoluble compounds such as iron oxide or hydroxide. The presence of corrosion products is one way corrosion is detected (*e.g.*, rust). However, it should be noted that insoluble corrosion products are not always visible. Upon exposure to air, aluminum forms an almost invisible oxide film which protects it from extensive atmospheric corrosion. It is invisible because it is so thin. This explains the widespread use of aluminum in storm windows, gutters, and automobile trim.

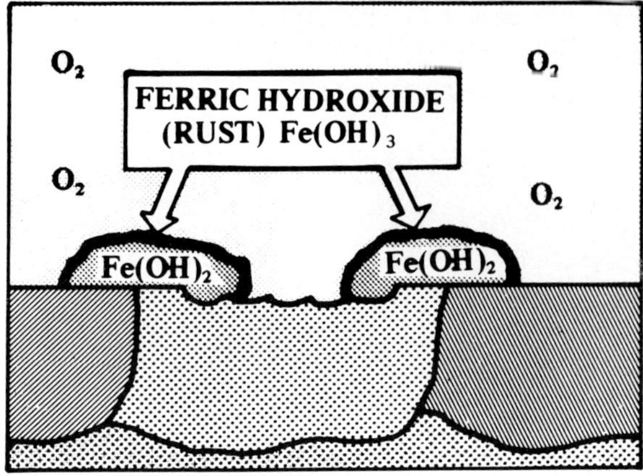

FIGURE 2.3 — Formation of ferrous and ferric hydroxides by interaction of products of anodic and cathodic reactions.

The products of the anodic and cathodic processes frequently migrate through the solution and meet to enter into further reactions that yield many of our common visible corrosion products. For example, with iron in water, the hydroxyl ions from the cathodic reaction, in their migration through the electrolyte towards the anodic surfaces, encounter ferrous ions moving in the opposite direction. These ions combine to form ferrous hydroxide which subsequently reacts further with oxygen in solution to form ferric hydroxide. This is illustrated in Figure 2.3 and represents a form of iron rust with which we are all quite familiar.

Electrochemistry of Corrosion

While corrosion can take any one of the several forms that have been mentioned, the mechanism of attack in aqueous solutions will involve some aspect of electrochemistry. There will be a flow of electricity from certain areas of a metal surface to other areas through a solution capable of conducting electricity, such as seawater or hard water.

The term *anode* is used to describe that portion of the metal surface that is corroded and from which current leaves the metal to enter the solution. On the other hand, the term *cathode* is used to describe the metal surface from which current leaves the solution and returns to the metal.

The *circuit* is completed outside the solution through the metal or through a conductor joining two pieces of metal. The essential components are shown in Figure 2.4. The dots represent electricity (not electrons) flowing in the solution from the anode (−) to the cathode (+) and returning from the cathode to the anode through the metal wires.

A solution capable of conducting electricity is called an *electrolyte.* Its ability to conduct electricity is due to the presence of ions. These are positively or negatively charged atoms or groups of atoms in solution. Pure water, depicted in Figure 2.1, contains positively charged hydrogen ions (H^+) and negatively charged hydroxyl ions (OH^-) in equal concentration. The electrolyte forming a corrosive environment may be any solution, rain, or even

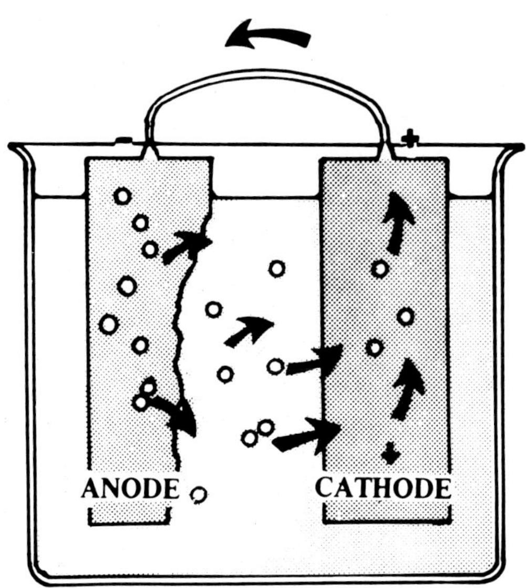

FIGURE 2.4 — Sketch showing flow of current between an anode and a cathode in a corrosion cell.

moisture condensed from the air. It can range from fresh water or salt water to the strongest alkali or acid.

The anodes and cathodes involved in a corrosion reaction are called *electrodes*. The electrodes may consist of two different kinds of metal or they may be different areas on the same piece of metal. The *negative* electrode (anode) is where corrosion occurs.

Electrochemical Reactions

Definition and Terminology

An *electrochemical reaction* is defined as a chemical reaction involving the transfer of electrons. It is also a chemical reaction which involves oxidation and reduction. Since metallic corrosion is almost always an electrochemical process, it is very important to understand the basic nature of electrochemical reactions. The above definition of electrochemical reactions can be most simply understood by looking at a typical corrosion reaction in detail. The most common (and beneficial) corrosion reactions available to us, wherein all the electrochemistry just described occurs, is the dry cell battery.

The typical flashlight battery, shown in Figure 2.5, depends on galvanic corrosion to generate electrical power. As illustrated, zinc (anode) is electrically connected to graphite or carbon (cathode) in the presence of a corrosive electrolyte. When these are connected through a flashlight bulb or buzzer, electrical current flows between these two electrodes. This causes accelerated corrosion of zinc and produces a cathodic reaction at the graphite electrode. The battery is completely depleted of power when the zinc is completely corroded. This sometimes causes trouble, since perforation of the zinc cup allows the corrosive electrolyte to leak into the flashlight. This is solved by encasing the battery in a steel container.

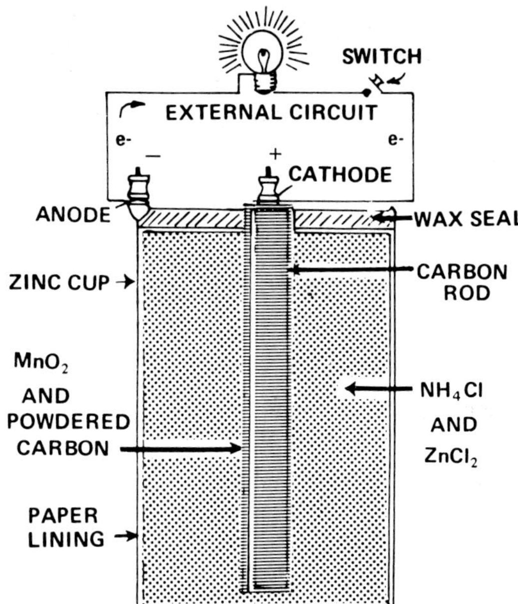

FIGURE 2.5 — Cross-sectional view of a typical dry cell.

Note that there are four essential elements required for this cell: (1) an anode (Zn); (2) a cathode (C); (3) an electrolyte (NH_4Cl and $ZnCl_2$); and (4) an external circuit. All corrosion cells must have these four elements. In this case, a significant potential (voltage) is developed between the highly cathodic carbon electrode and the zinc anode. The reactions involved can be clearly understood by considering the corrosion of zinc by hydrochloric acid discussed earlier:

$$Zn + 2HCl \rightarrow ZnCl_2 + H_2\uparrow \qquad (2.2)$$

Remembering that hydrochloric acid and zinc chloride are ionized in water solutions, the above equation can be rewritten, as shown in Equation (2.11), as:

$$Zn + 2H^+ + 2Cl^- \rightarrow Zn^{+2} + 2Cl^- + H_2\uparrow \qquad (2.15)$$

When written in this form, it becomes obvious that the chloride ion does not directly participate in the reaction. That is, chloride appears on both sides of the equation, but is not altered by the corrosion reaction (*i.e.,* the valence of the chloride ion remains unchanged). Thus, we can further simplify Equation (2.15) by omitting the nonreacting chloride.

$$Zn + 2H^+ \rightarrow Zn^{+2} + H_2\uparrow \qquad (2.16)$$

As shown in Equation (2.16), the corrosion of zinc by hydrochloric acid simply consists of the reaction between zinc and hydrogen ions which yield zinc ions and hydrogen gas. During this reaction, zinc is oxidized to zinc ions. It can also be said that the valence of zinc is increased by the reaction. Simultaneously, hydrogen ions are reduced (valence decreased) to hydrogen gas during the corrosion process.

The reaction shown in Equation (2.16) can be further simplified by dividing it into a separate oxidation reaction and a separate reduction reaction.

$$Zn \rightarrow Zn^{+2} + 2e \text{ oxidation (anodic reaction)} \quad (2.17)$$

$$2H^{+} + 2e \rightarrow H_2\uparrow \text{ reduction (cathodic reaction)} \quad (2.18)$$

$$Zn + 2H^{+} \rightarrow Zn^{+2} + H_2\uparrow \quad (2.16)$$

An *oxidation reaction,* such as Equation (2.17), is indicated by an increase in valence or a production of electrons. In a similar fashion, a *reduction reaction* is indicated by a decrease in valence or the consumption of electrons, as shown in Equation (2.18). Note that the summation of Equations (2.17) and (2.18) yields the overall reaction shown in Equation (2.16). In corrosion terminology, an *oxidation* reaction is often called an *anodic* reaction, while *reduction* reactions are usually termed *cathodic* reactions. These terms are used interchangeably throughout this text.

Corrosion reactions actually proceed as shown in Equations (2.17) and (2.18). That is, the corrosion consists of at least one oxidation and one reduction reaction. This is illustrated schematically in Figure 2.6. In this figure, a piece of zinc immersed in hydrochloric acid solution is undergoing corrosion. At some point on the surface, zinc is transformed to zinc ions, according to Equation (2.17). This reaction produces electrons and these pass through the solid conducting metal to other sites on the metal surface where hydrogen ions are reduced to hydrogen gas according to Equation (2.18).

Equations (2.17) and (2.18) and Figure 2.6 illustrate the nature of an electrochemical reaction. During such a reaction, electrons are transferred, or, viewing it another way, an oxidation process occurs together with a reduction process.

Briefly then, for corrosion to occur there must be a formation of ions and release of electrons at an anodic surface where oxidation or deterioration of the metal occurs. There must be a simultaneous acceptance at the cathodic surface of the electrons generated at the anode. This acceptance of electrons can take the form of neutralization of positive hydrogen ions, or the formation of negative ions. *The anodic and cathodic reactions must go on at the same time and at equivalent rates.* However, corrosion occurs only at the areas that serve as anodes.

Anodic Processes

Let us consider in greater detail what takes place at the anode when corrosion occurs. Positively charged atoms of metal leave the solid surface and enter into solution as ions. They leave their corresponding negative charges in the form of electrons which are able to flow through the metal or any external electronic conductor. The ionized atoms can bear one or more positive charges. In the corrosion of iron, each iron atom becomes an iron ion carrying two positive charges and generates two electrons (Figure 2.7). These electrons travel through the metal or an external electronic conductor to complete the circuit at the cathode, where a corresponding reaction consumes these electrons. During corrosive attack, the anodic reaction always is the oxidation of a metal to a higher valence state (usually from zero to some positive value).

For instance, reconsider Equations (2.2), (2.3), (2.4), and (2.5) discussed earlier.

$$Zn + 2HCl \rightarrow ZnCl_2 + H_2\uparrow \quad (2.2)$$

$$Zn + H_2SO_4 \rightarrow ZnSO_4 + H_2\uparrow \quad (2.3)$$

$$Fe + 2HCl \rightarrow FeCl_2 + H_2\uparrow \quad (2.4)$$

$$2Al + 6HCl \rightarrow 2AlCl_3 + 3H_2\uparrow \quad (2.5)$$

All of these reactions involve the reduction of hydrogen ions to hydrogen gas, according to Equation (2.18), and the only difference between them is the nature of their anodic or oxidation processes. Thus, understanding corrosion by acids is greatly simplified, since in every case the cathodic reaction is

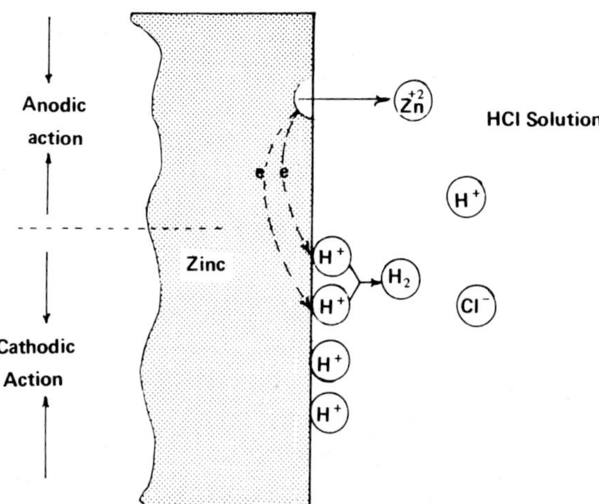

FIGURE 2.6 — Electrochemical reactions occurring during the corrosion of zinc in air-free hydrochloric acid.

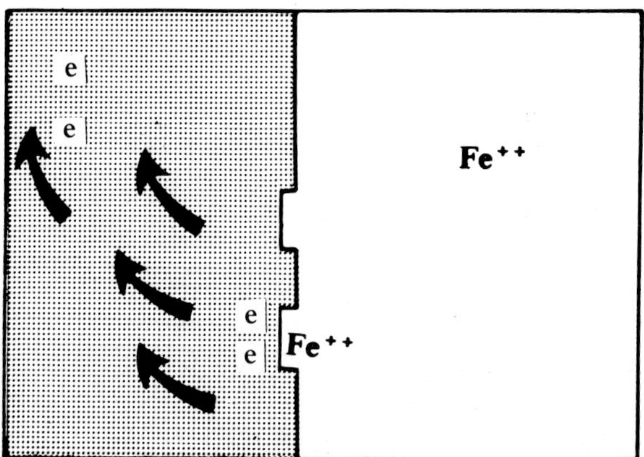

FIGURE 2.7 — Formation of ferrous ions and release of electrons in the corrosion of iron.

simply the evolution of hydrogen as gas, as was previously shown in Equation (2.18). This hydrogen evolution reaction occurs with a wide variety of metals and acids, including hydrochloric, sulfuric, perchloric, hydrofluoric, formic, and other strong acids.

By separating Equations (2.2) through (2.5) into anodic and cathodic reactions, the only difference that is found is in the oxidation reaction. Equations (2.2) and (2.3) involve the oxidation of zinc to its ions, while Equations (2.4) and (2.5) involve the oxidation of iron and aluminum to their ions. These individual anodic reactions are listed as follows.

$$Zn \rightarrow Zn^{+2} + 2e \quad (2.17)$$

$$Fe \rightarrow Fe^{+2} + 2e \quad (2.19)$$

$$Al \rightarrow Al^{+3} + 3e \quad (2.20)$$

Examining the above equations shows that the anodic reaction occurring during corrosion can be written in the general form:

$$M \rightarrow M^{+n} + ne. \quad (2.21)$$

That is, the corrosion of metal M results in the oxidation of metal M to an ion with a valence charge of +n and the release of n electrons. The value of n, of course, depends primarily on the nature of the metal. Some metals, such as silver, are univalent, while others such as iron, titanium, and uranium are multivalent and possess positive charges as high as 6. Equation (2.21) is general and applies to all corrosion reactions.

Just remember, the **A**node is where the **A**ction is.

Cathodic Processes

What takes place at the cathode that parallels what goes on at the anode? The electrons generated by the formation of metallic ions at the anode have passed through the metal to the surface of the cathodic areas immersed in the electrolyte. Here, they restore the electrical balance of the system by reacting with the neutralizing positive ions, such as hydrogen ions, in the electrolyte. Hydrogen ions can be reduced to atoms, and these often combine to form hydrogen gas through reaction with electrons at a cathodic surface. This reduction of hydrogen ions at the cathodic surface will disturb the balance between the acidic hydrogen H^+ ions and the alkaline hydroxyl $(OH)^-$ ions and make the solution less acid or more alkaline in this region (Figure 2.8).

This change in the concentration of hydrogen ions can be shown by the use of chemical indicators which change color with changes in hydrogen ion concentration and thus can serve to demonstrate and locate the existence of surfaces on which the cathodic reactions in corrosion are taking place.

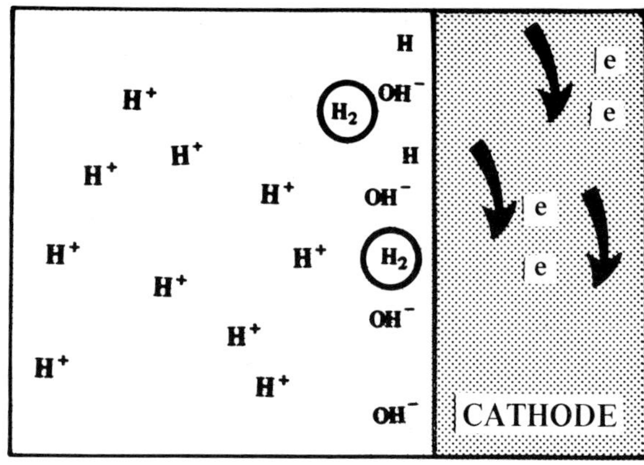

FIGURE 2.8 — Reduction of hydrogen ions at the cathode to form hydrogen atoms and subsequently hydrogen molecules (gas). Hydroxyl ions also accumulate.

There are several other cathodic reactions encountered during the corrosion of metals. These are listed below.

Oxygen Reduction (acid solutions)	$O_2 + 4H^+ + 4e \rightarrow 2H_2O$	(2.22)
Oxygen Reduction (neutral and alkaline solutions)	$O_2 + 2H_2O + 4e \rightarrow 4OH^-$	(2.23)
Hydrogen Evolution	$2H^+ + 2e \rightarrow H_2\uparrow$	(2.18)
Metal Ion Reduction	$Fe^{+3} + e \rightarrow Fe^{+2}$	(2.24)
Metal Deposition	$Cu^{+2} + 2e \rightarrow Cu$ Copper	(2.25)

As a mnemonic device, remember 2OHM2 which indicates the two oxygen, one hydrogen, and two metal reactions to be considered at the cathode. Hydrogen ion reduction, or hydrogen evolution, has already been discussed. This is the cathodic reaction that occurs during corrosion in acids. Oxygen reduction [Equations (2.22) and (2.23)] is a very common cathodic reaction, since oxygen is present in the atmosphere and in solutions exposed to the atmosphere. Metal ion reduction and metal deposition, although less common, cause severe corrosion problems.

Note that all of the above reactions are similar in one respect; they consume electrons. All corrosion reactions are simply combinations of one or more of the above cathodic reactions, together with an anodic reaction similar to Equation (2.21). Thus, almost every case of metallic corrosion can be reduced to these six equations, either singly or in combination. We have already seen how acid corrosion can be reduced to the oxidation of a metal and the reduction of hydrogen according to Equations (2.18) and (2.21) and the oxygen reduction shown in Equation (2.23)

Consider the corrosion of zinc by water or moist air. By multiplying the zinc oxidation reaction by 2 and summing this with the oxygen reduction reaction, the overall equation is a simplified form of that shown previously in Equation (2.8).

$$2Zn \rightarrow 2Zn^{+2} + 4e \text{ (oxidation)} \quad (2.26)$$

$$+ O_2 + 2H_2O + 4e \rightarrow 4OH^- \text{ (reduction)} \quad (2.23)$$

$$2Zn + 2H_2O + O_2 \rightarrow 2Zn^{+2} + 4OH^- \rightarrow$$

$$2Zn(OH)_2\downarrow \quad (2.27)$$

The products of this reaction are Zn^{+2} and OH^-, which immediately react to form insoluble $Zn(OH)_2$. Likewise, the corrosion of zinc by copper sulfate [Equation (2.14)] is merely the summation of the oxidation reaction for zinc and the metal deposition reaction involving cupric ions [Equation (2.25)].

$$Zn \rightarrow Zn^{+2} + 2e \quad (2.17)$$

$$+ Cu^{+2} + 2e \rightarrow Cu\downarrow \quad (2.25)$$

$$Zn + Cu^{+2} \rightarrow Zn^{+2} + Cu\downarrow \quad (2.28)$$

A comparison of Equations (2.28) and (2.14) shows that they are essentially identical.

During corrosion, more than one oxidation and one reduction reaction may occur. For example, during the attack on an alloy, its component metal atoms go into solution as their respective ions. Thus, during the corrosion of a chromium-iron alloy, both chromium and iron are oxidized. Also, more than one cathodic reaction can occur on the surface of a metal.

Consider the corrosion of zinc in a hydrochloric acid solution containing dissolved oxygen. Two cathodic reactions are possible: the evolution of hydrogen and the reduction of oxygen (Figure 2.9). Since there are two cathodic reactions or processes which consume electrons, the overall corrosion rate of zinc is increased. Thus, acid solutions which either contain dissolved oxygen or are exposed to the air are generally more corrosive than air-free acids.

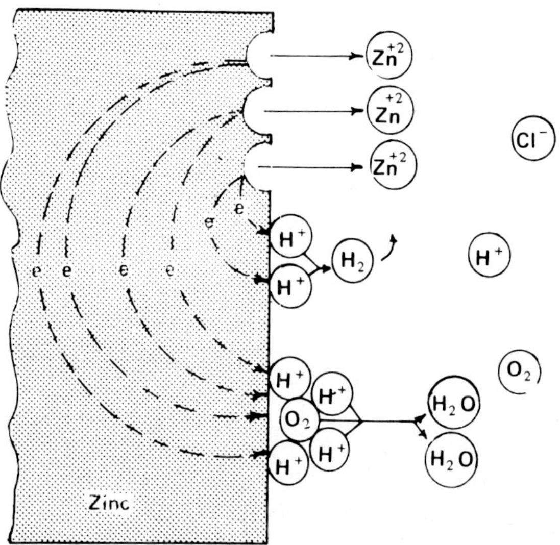

FIGURE 2.9 — Electrochemical reactions occurring during the corrosion of zinc in aerated hydrochloric acid.

Therefore, removing oxygen from acid solutions will render them less corrosive. This is a common method for reducing the corrosivity of many environments. Oxygen removal may be accomplished by either chemical or mechanical means.

If a piece of mild steel is placed in a solution of hydrochloric acid, a vigorous formation of hydrogen bubbles is observed. Under such conditions, the metal corrodes very quickly. The dissolution of the metal occurs only at anodic surfaces. The hydrogen bubbles form only at the cathodic surfaces, even though it may appear they come from the entire surface of the metal rather than at well-defined cathodic areas. The anodic and cathodic areas may shift from time to time so as to give the appearance of uniform corrosion. If this action could be seen through a suitable microscope, many tiny anodic and cathodic areas would be observed shifting around on the surface of the metal. These areas, however, are often so small as to be invisible and so numerous as to be almost inseparable.

Combined Anodic and Cathodic Processes

In summary then, if just one anode and one cathode could be seen in a magnified view of a piece of iron in an acid solution, electrons generated by the formation of ferrous ions would be observed flowing through the metal from an anodic area to a cathodic area (Figure 2.10). At the cathodic surface, the electrons would meet hydrogen ions from the solution. One hydrogen ion would accept one electron and be converted into a hydrogen atom which could enter the metal and lead to hydrogen embrittlement, or, as in most cases, it could combine with another hydrogen atom and become molecular hydrogen gas which would either cling to or be released as a bubble from the cathodic surface. As this process continues, oxidation (corrosion) of the iron occurs at the anodic surfaces and reduction of hydrogen ions occurs at the cathodes. Note that the term *oxidation* is not necessarily associated with oxygen.

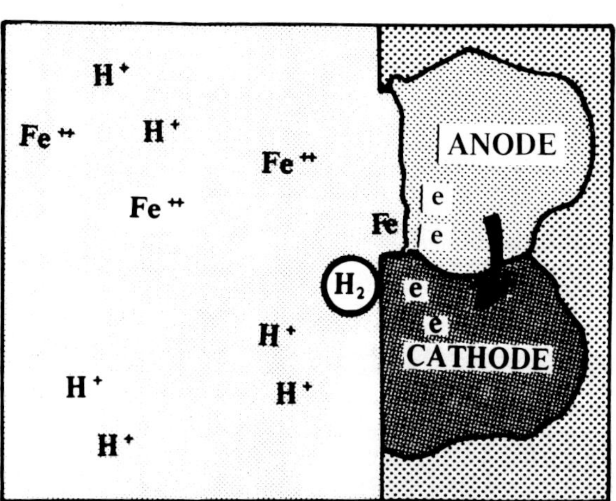

FIGURE 2.10 — Formation of ions at an anodic area and release of hydrogen at a cathodic area in a local cell on an iron surface.

Polarization

As is the case with other chemical reactions that tend to reach some equilibrium rate lower than the initial rate, corrosion action tends to slow down as a result of the effects of the products of anodic and cathodic reactions. The cathodic reaction, and with it the overall corrosion reaction, would slow down if the hydrogen product of the cathodic reaction were not removed by evolution as gas or some reaction involving oxygen. This slowing down is said to be the result of *cathodic polarization.*

It is possible to measure this effect in terms of the potential of the metal on which the reaction is occurring. For example, if the potential of the surface of the more noble metal, the cathode, were to be measured before the flow of any galvanic current and subsequently after current flow had occurred for some time, it would be found that the potential measured would have changed to a value closer to that of the less noble metal in the couple.

Similarly, measurements of the potential of the anodic member of the couple would show a drift in potential closer to that of the cathodic member of the couple. This could be the result of an increase in the concentration of the ions of the anodic metal in the immediate vicinity of the corroding metal surface.

There are two different types of polarization or ways that electrochemical reactions are retarded. These are activation polarization and concentration polarization.

The term *activation polarization* is used to indicate retarding factors which are inherent in the reaction itself. For example, consider the evolution of hydrogen gas illustrated previously in Equation (2.18) and Figure 2.6. The rate at which hydrogen ions are reduced to hydrogen gas will be a function of several factors, including the speed of electron transfer to the hydrogen ion at the metal surface. Thus, there is an inherent rate for this reaction depending on the particular metal, hydrogen ion concentration, and the temperature of the system. In fact, there are wide variations in the ability of various metals to transfer electrons to hydrogen ions and, as a result, the rate of hydrogen evolution from different metal surfaces is observed to be quite different.

In contrast, *concentration polarization* refers to the retardation of an electrochemical reaction as a result of concentration changes in the solution adjacent to the metal surface (Figure 2.11). Here, we are looking at the evolution of hydrogen on a rapidly corroding metal surface. In order to remain simplistic, the corresponding metal oxidation reaction is not shown.

If this reaction is proceeding at a fairly rapid rate, and the concentration of hydrogen ions in solution is relatively low, it can be seen that the region very close to the metal surface will become depleted of hydrogen ions because these are being consumed by the cathodic reaction. Under these

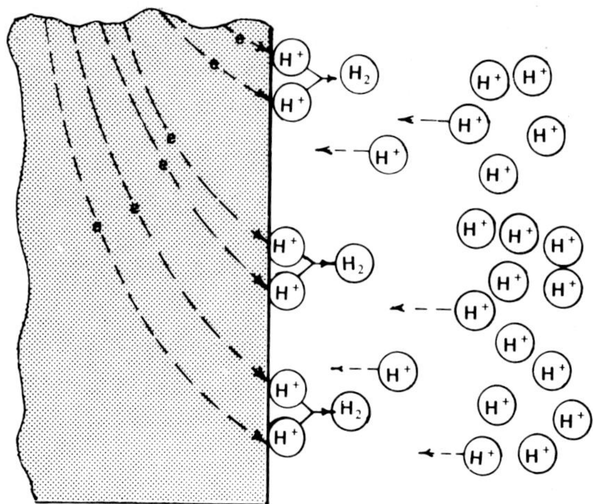

FIGURE 2.11 — Concentration polarization during the cathodic reduction of hydrogen ions.

conditions, the reaction is controlled by the diffusion rate of the hydrogen ions to the metal surface.

Activation polarization is usually the controlling factor during corrosion in strong acids. Concentration polarization usually predominates when the concentration of the active species is low; for example, in dilute acids and in aerated water and salt solutions (O_2 solubility is low in water and aqueous solutions). Knowing the kind of polarization which is occurring is very helpful, since it allows the prediction of characteristics of the corroding system.

For example, if corrosion is controlled by concentration polarization, then any change which increases the diffusion rate of the active species (*e.g.*, H^+) will increase the corrosion rate. In such a system, it would therefore be expected that agitating the liquid or stirring it would tend to increase the corrosion rate of the metal.

However, if the cathodic reaction is activation controlled, then stirring or agitation will have no effect on the corrosion rate. Knowing the kind of polarization which is controlling the corrosion reaction therefore allows us to make very useful predictions concerning the relative effect on corrosion rate that would be produced by, say, increasing the flow of liquid in a pipeline. Polarization will be more thoroughly discussed later in this chapter. Meanwhile, it should be obvious that the *polarization* occurring at the anode and cathode *determine the corrosion rate* generated by most electrochemical cells.

Area Effects

As stated previously, the corrosion effects of current flow on polarization phenomena are related not just to the total amount of current flow, but also to the *current density* or current flow per unit area. It is easy to understand that the effect of a certain amount of current concentrated on a small area of metal surface will be much greater than when the effect of the same amount of current is dissipated over a much larger area.

This area effect in terms of current density is illustrated by combinations of steel and copper as either plates or the fastenings used to join them and immersed in a corrosive solution. If steel rivets are used to join copper plates, the current density on the relatively large cathodic copper plates will be low, cathodic polarization of the copper will be slight, and the voltage of the galvanic couple will maintain a value close to the open circuit potential. At the same time, the current density on the small anodic steel rivets will be high and the consequent corrosion quite severe.

With the opposite arrangement of copper rivets joining steel plates, the current density on the copper cathodes will be high, with consequently considerable cathodic polarization of the copper reducing the open circuit potential below its initial value. The diminished anodic current will be spread over the relatively large steel plates and the undesirable galvanic effect will hardly be noticeable.

Open circuit potential measurements are grossly inadequate for predicting the magnitude of galvanic effects since they do not take into account area and polarization effects. They are reliable only for predicting the direction of such effects.

Importance of Oxygen

Oxygen is the most common of the cathodic depolarizers. The action of oxygen in increasing corrosion is easily demonstrated by placing iron in two flasks filled with water. Oxygen is allowed to bubble through the water in one flask to supply it with oxygen. The water in the second flask is saturated with nitrogen to help eliminate dissolved oxygen. After the gases have bubbled for several hours, the iron in the oxygen-free solution remains bright, but the iron in the water saturated with oxygen already begins to rust.

The oxygen content of any solution ranks high on the list of factors influencing the corrosion of iron and numerous other metals. Elimination of oxygen by deaeration is a potent means of preventing corrosion, as in the case of steam boilers which are operated with completely deaerated feed water.

Oxygen Concentration Cells

The role of oxygen in enabling a corrosion reaction to occur forms the basis for the fact that oxygen can not only maintain a cathodic reaction, but can promote one.

$$1/2O_2 + 2e^- + H_2O \rightarrow 2OH^- \qquad (2.23)$$

This occurs where there is a difference in the concentration of dissolved oxygen at one point on a metal surface as compared with another point.

Since, here again, the direction of the reaction is towards equilibrium, the only way that equilibrium can be approached by corrosion will be to reduce the concentration of oxygen where it is highest. Such reduction can be done by consuming the oxygen. The result is that where there is a difference in the concentration of dissolved oxygen at

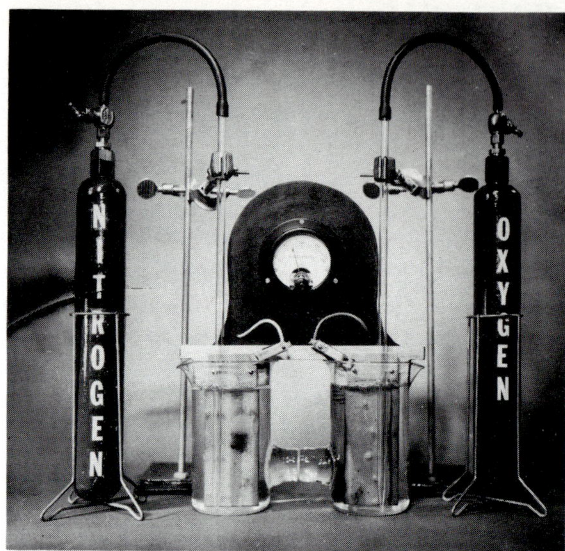

FIGURE 2.12 — Experiment to demonstrate generation of a corrosion current by an oxygen concentration cell.

two points on a metal surface, the surfaces in contact with the solution containing the higher concentration of dissolved oxygen will become cathodic to the surfaces in contact with the solution containing the lower concentration of dissolved oxygen. These surfaces exposed to the lower O_2 concentration will suffer accelerated corrosion as anodes in an oxygen concentration cell.

It is easy to demonstrate an oxygen concentration cell with an experimental setup using a two-compartment vessel similar to that used to demonstrate a metal ion concentration cell (Figure 2.12). In this experiment, pieces of steel are connected and immersed in a sodium chloride solution in the two compartments. The solution in one compartment is saturated with oxygen and the solution in the other compartment is saturated with nitrogen. This establishes a large difference in the concentration of dissolved oxygen in contact with the two pieces of steel. The high concentration of dissolved oxygen in the solution in contact with the steel in one compartment makes this steel surface strongly cathodic to the steel in the other compartment. The potential that is measured is determined by the difference in oxygen concentration and the magnitude of the current by the areas of the metal surfaces and the resistance of the circuit.

Dissolved oxygen concentration differences can be established by velocity gradients and by crevices, but the location of anodes and cathodes from these sources is just the opposite of that to be described for the metal ion concentration cells. More oxygen is brought to the surfaces moving at the highest velocity so that these surfaces become cathodic to the anodic surfaces with lower oxygen availability because they are moving at the lower velocity. Those surfaces nearer the center of a rotating disc will suffer accelerated corrosion as a result, as illustrated by Figure 2.13.

Similarly, because of the difference in the direction of oxygen concentration cells compared

FIGURE 2.13 — Corrosion by an oxygen concentration cell near the center of an iron disc rotating in seawater.

with metal ion concentration cells, corrosion accelerated by an oxygen concentration cell will occur within a crevice or under a deposit rather than outside of it. This difference between the two types of concentration cells complicates the prediction of the intensity and location of corrosion resulting from concentration cell action. It does, however, facilitate explanation after the fact.

As a general rule, those metals towards the top of the electromotive series in Table 2.2, *e.g.*, iron, are likely to be more susceptible to acceleration of attack by oxygen concentration cells, while those towards the bottom, *e.g.*, copper, are more vulnerable to metal ion concentration cell action. Metals and alloys in the middle of the range, *e.g.*, copper-nickel alloys, benefit from the opposing effects of the two types of cells.

Alloys made by combining metals near the top of the electromotive series, *e.g.*, iron and chromium (stainless steels) which exhibit a more noble potential than that of their constituents as a result of the passivating effect of a film based on a reaction with oxygen, will be particularly sensitive to oxygen availability and will, therefore, be particularly vulnerable to oxygen concentration cell action. (Passivity phenomena will be discussed later in greater detail.)

Metal Ion Concentration Cells

In the coming discussion of the basis for the electromotive series in Table 2.2, it is pointed out that in measuring the potential used to establish the position of a metal in this series, it is necessary to place the metal in a solution containing a specified concentration of the ions of that metal. The reason for this is that when a metal is in contact with a solution of its ions, an equilibrium becomes established between a tendency for the metal to go into solution to increase the concentration of its ions and an opposing tendency for the ions to plate out on the

TABLE 2.2 — Position of Some Metals in The Standard Electromotive Series

Metal		Standard Potential Volt
Potassium	Active End	−2.922
Magnesium		−2.34
Aluminum		−1.67
Zinc		−0.762
Chromium		−0.710
Iron		−0.440
Nickel		−0.250
Hydrogen		0.000
Copper		+0.345
Silver		+0.800
Platinum	Noble or	+1.20
Gold	Passive End	+1.68

metal and thereby reduce the concentration in solution. From this it follows that the tendency of a metal to go into solution, as reflected by its measured potential, will be greater in a solution containing a low concentration of its ions than in one in which the concentration of metal ions is greater.

Under circumstances where there is a relatively low concentration of metal ions at one point on a metal surface and a higher concentration at another point, a difference in a potential between the two points will be established. For the reason that has been cited, the surface in contact with the lower concentration of metal ions will go into solution more easily, have the more negative potential, and will thus act as the corroding anode in what is called a *metal ion concentration cell.*

Such cells, like other chemical reactions, operate in a direction that will restore equilibrium. Corrosion occurs at the anodic surface where the metal ion concentration is relatively low so that this concentration will increase. At the same time, metallic ions will plate out on the cathodic surface from the solution containing the higher concentration so as to decrease this concentration towards that of the originally lower concentration around the anode.

The action of a metal ion concentration cell can be demonstrated by an experiment in which pieces of copper are immersed in two solutions of copper sulfate separated by a porous membrane which prevents the solutions from mixing, but which provides ionic conductance between the two solutions (Figure 2.14). The electrolytic cell established in this way generates a current at a voltage which depends on the difference in the concentrations of the copper ions in the two solutions. The magnitude of the current will be determined by the areas of the metal specimens and the resistance of the circuit.

In practice, differences in metal ion concentrations that can give rise to corrosion cells of this type can also be aided by velocity gradients over a metal surface. Metal ions in corrosion products will be washed away faster where the velocity and turbulence are greatest and corrosion will be accelerated by the cell established in this way.

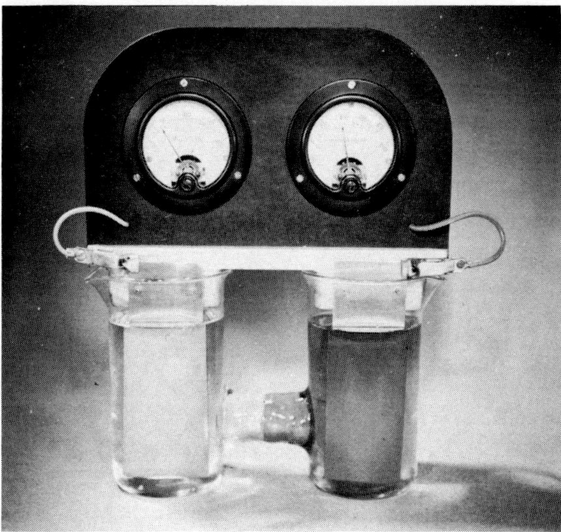

FIGURE 2.14 — Experiment to demonstrate generation of a corrosion current by differences in concentration of a metal ion; a potential of 40 millivolts and a current of 380 microamperes is obtained.

A velocity gradient over a metal surface can be established by spinning a disc immersed in a solution. The surfaces towards the periphery will move progressively faster than the surfaces nearer the center of the disc. A metal ion concentration cell set up in this way on a copper alloy will cause accelerated corrosion of the faster moving surfaces near the outer edge of the disc, as illustrated by Figure 2.15.

Metal ions can accumulate within crevices or under loosely adherent deposits so that the surfaces within such crevices can become cathodic to surfaces just outside of the crevices which will suffer accelerated attack as a result of the metal ion concentration cell action.

Galvanic Action

The term *galvanic action* is generally restricted to the changes in normal corrosion behavior that result from the current generated when one metal is in contact with a different one, and the two metals are in a corrosive solution (Figure 2.16).

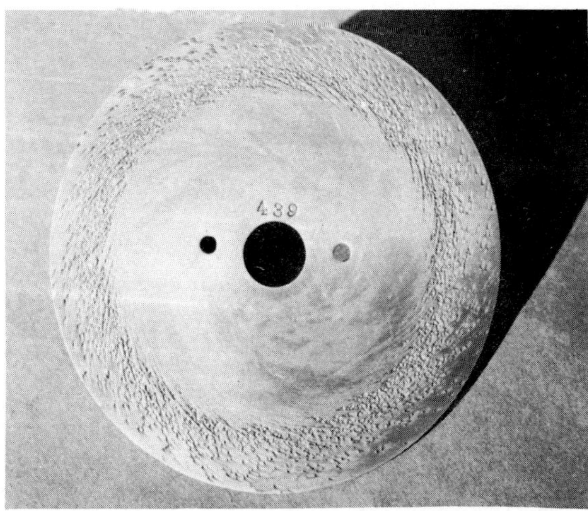

FIGURE 2.15 — Corrosion by a metal ion concentration cell near the periphery of a brass disc in seawater.

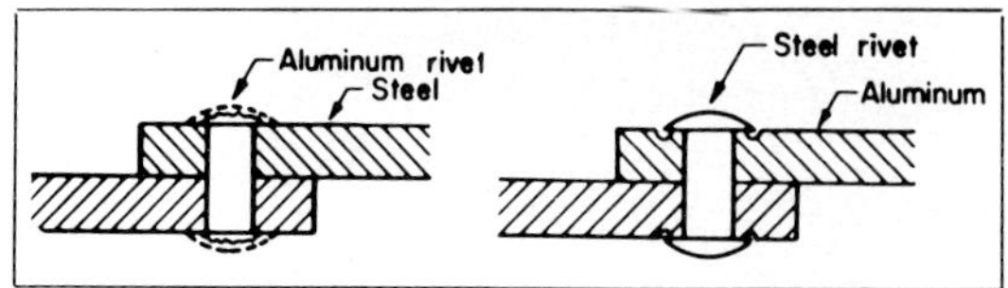

FIGURE 2.16 — Galvanic couple between steel and aluminum.

In such a galvanic *couple,* the corrosion of one of the metals (aluminum) will be accelerated and the corrosion of the other (steel) will be reduced or stopped. Therefore, the first thing to be established is which of the metals will be affected in the one way and which will be affected in the other. The answer to this is provided by the direction in which an electric current will flow from the one metal (the anode) to the other (the cathode) in the corrosive solution as a result of the galvanic effect. This direction in turn will be determined by the difference in the potentials of the two metals in the solution in question. In the case of an aluminum-steel couple, the aluminum acts as the anode.

Corrosion Potentials and Direction of Galvanic Effects

The potential of a metal in a solution is related to the energy that is released when the metal corrodes, as discussed previously. Such corrosion potentials are capable of measurement in at least a relative sense; for example, by placing a more reactive metal such as zinc and a less reactive metal such as copper in a solution of sodium chloride and measuring the direction of the current that is generated by their galvanic action. Such an experiment could be repeated with all the possible combinations of metals in any corrosive solution.

Examination of the results would make it possible to arrange the metals in what could be called a *galvanic series.* If the experiments were to be repeated using a different solution, a different concentration of sodium chloride, a different degree of aeration, a different velocity of movement, or a different temperature, the values recorded could be different, and the positions of some of the metals relative to each other in the new galvanic series may change.

Galvanic Series

There is no absolute value of the electropotential of a metal independent of the factors that influence the corrosive characteristic of the solution in which the potential is measured. Values of potential can change from one solution to another or in any solution when influenced by such factors as temperature, aeration, and velocity of movement. Consequently, there is no way, other than by potential measurements in the exact environment of interest, to predict the potentials of the metals and the consequent direction of a galvanic effect in that environment.

As an example, zinc is normally very negative or anodic to steel at ambient temperatures, as indicated in Table 2.3. However, the potential difference decreases with an increase in temperature until the potential difference may be zero or actually be reversed at 60 C (140 F).

However, the situation is not quite as bad as the preceding statement implies. The relative tendencies of metals to corrode remain about the same in many of the environments in which they are likely to be used. Consequently, their relative positions in a galvanic series may be about the same in many environments. Since more observations of potentials and galvanic behavior have been made in seawater than in any other single environment, an arrangement of metals in a galvanic series based on such observations is frequently used as first approximation of the probable direction of the galvanic effects in other environments in the absence of data more directly applicable to such environments.

Such a galvanic series is shown in Table 2.3. In a galvanic couple involving any two metals in this list, the normal corrosion of the metal higher in the list is likely to be accelerated, while the corrosion of the metal lower in the list is likely to be reduced or completely stopped. Metals with more positive corrosion potentials are called noble or cathodic, and those with more negative corrosion potentials are referred to as active or anodic metals and alloys.

Note that several metals in Table 2.3 are grouped. The potential differences within a group are not likely to be great and the metals can be combined without substantial galvanic effects under many circumstances.

Magnitude of Galvanic Effects

Up to now we have been concerned only with the direction of galvanic action as determined by the relative potentials of the metals in a galvanic couple. What we are most concerned with in a practical way is the intensity of whatever galvanic action occurs. This intensity is determined by the amount of current or, more specifically, the current density (current per unit area).

Since, according to Ohm's law, the amount of current that can flow is directly proportional to the voltage for any given value of resistance, galvanic couples in which the difference in potential is high, *e.g.,* zinc and copper (700 millivolts in seawater), can generate more current (and therefore, corrosion) than couples having a lower potential difference, *e.g.,* naval brass and copper (40 millivolts in seawater).

The potentials that have been cited for illustration are the potentials that would be measured before the flow of any current between the two metals. This is sometimes referred to as the *open circuit potential.*

Area Effect

Another important factor in galvanic corrosion is the area effect or the ratio of cathodic to anodic area which was discussed previously. As the ratio of the cathodic to anodic area increases (that is, the size of the cathode increases in relation to the anode), the corrosion rate of the more anodic metal is rapidly accelerated, as shown in Figure 2.17.

From the standpoint of practical corrosion resistance, the most unfavorable ratio is a very large cathode connected to a very small anode. This effect is illustrated in Figures 2.18 and 2.19. Table 2.3 in-

TABLE 2.3 — Galvanic Series of Some Commercial Metals and Alloys in Seawater

Active or Anodic (−)	Magnesium Magnesium Alloys Zinc Galvanized Steel
	Aluminum 1100
	Aluminum 2024 (4.5 Cu, 1.5 Mg, 0.6 Mn)
	Mild Steel Wrought Iron Cast Iron
	13% Chromium Stainless Steel Type 410 (Active) 18-8 Stainless Steel Type 304 (Active)
	Lead-Tin Solders Lead Tin
	Muntz Metal Manganese Bronze Naval Brass
	Nickel (Active) 76 Ni-16 Cr-7 Fe alloy (Active)
	60 Ni-30 Mo-6 Fe-1 Mn
	Yellow Brass Admiralty Brass
	Red Brass Copper Silicon Bronze
	70:30 Cupro Nickel G-Bronze Silver Solder Nickel (Passive) 76 Ni-16 Cr-7 Fe Alloy (Passive)
	13% Chromium Stainless Steel Type 410 (Passive) Titanium
	18-8 Stainless Steel Type 304 (Passive)
Noble or Cathodic (+)	Silver Graphite Gold Platinum

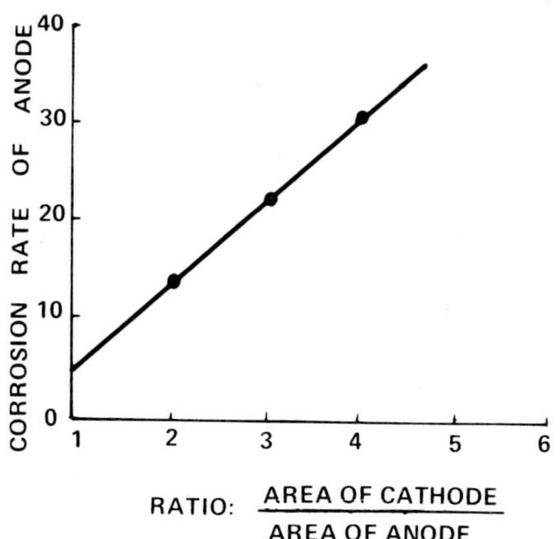

FIGURE 2.17 — Area effect during galvanic corrosion.

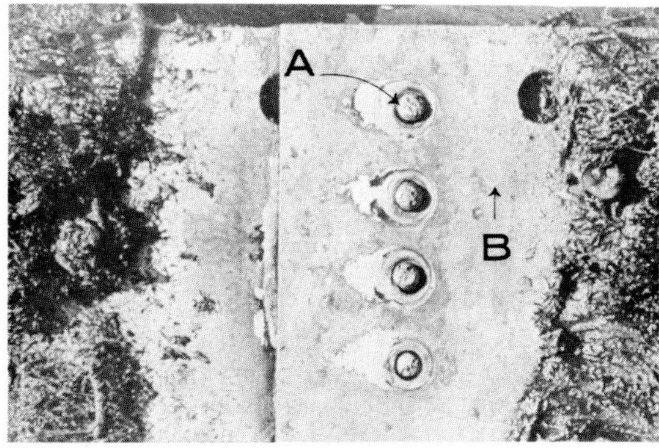

FIGURE 2.18 — Representation of reactions encountered when copper plates are connected by steel rivets after seawater exposure. Intense attack on small anodes (steel): (A) Steel rivets heavily corroded; (B) Copper, very slight corrosion.

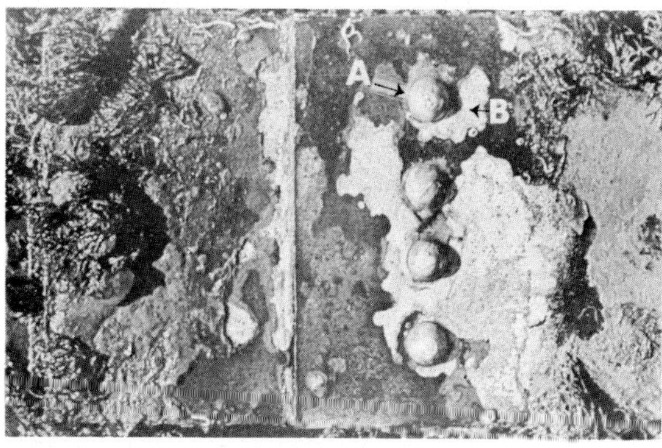

FIGURE 2.19 — Representation of reactions encountered when steel plates are connected by copper rivets after seawater exposure (exposure duration identical to test shown in Figure 2.16). Large anode (steel) and small cathode (Cu). Results in negligible galvanic corrosion: (A) Copper rivets, very slight corrosion; (B) Steel, mild corrosion.

dicates that iron is anodic with respect to copper and therefore is more rapidly corroded when placed in contact with it. This effect is greatly accelerated if the area of the iron is small in comparison to the area of the copper, as shown in Figure 2.18. However, under the reverse conditions; namely, when the area of the iron is very large compared to the copper, the corrosion of the iron is only slightly accelerated.

Recognizing Galvanic Corrosion

Before discussing ways of preventing galvanic corrosion, it is necessary first to be sure that galvanic corrosion is occurring. For galvanic corrosion to occur, three conditions are generally necessary: (1) electrochemically dissimilar metals must be present; (2) these metals must be in electrical contact; and (3) the metals must be exposed to an electrolyte. All of these conditions must be present for galvanic corrosion to occur.

Consider, for instance, that 18-8 stainless steel (Type 304; S30400) in electrical contact with 18-8 Mo stainless steel (Type 316; S31600) is observed to be rapidly corroding. Table 2.3 indicates that this is not the result of galvanic corrosion (grouping). Therefore, separating these two metals would not improve the corrosion resistance of the 18-8SS.

Consider also that a piece of aluminum connected to a cast iron motor block immersed in hot motor oil was rapidly attacked. This is not a case of galvanic corrosion, since motor oil and most organic liquids are not electrolytes. Again, separating these two metals would not improve the resistance of the aluminum.

In addition to the three previously listed conditions necessary for galvanic corrosion, another way of recognizing this kind of attack is to look for localized corrosion near the junction between the two dissimilar metals. Galvanic corrosion is usually most intense immediately adjacent to the cathodic material, *e.g.,* Figure 2.19 where it was noted that the corrosion of the iron plate was most intense near the copper rivets.

Preventing Galvanic Corrosion

There are a number of ways that galvanic corrosion can be prevented. These can be used singly or in combination. All of these preventive measures follow directly from the basic mechanism of galvanic corrosion.

1. Avoid the use of dissimilar metals wherever possible. If this is not practical, try to use metals which are close together in the galvanic series (Table 2.3).

2. Avoid an unfavorable area ratio whenever possible. Under no circumstances should a small anode be connected to a large cathode.

3. If dissimilar metals are used, insulate these electrically from one another. An example of this technique is shown in Figure 2.20, which illustrates

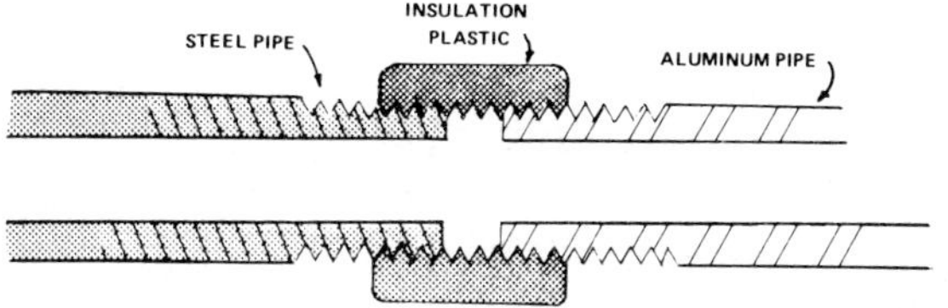

FIGURE 2.20 — Prevention of galvanic corrosion of aluminum pipe by an insulating pipe coupling.

the insulation of aluminum and steel pipes carrying water. It is very important to make sure that there is no electrical contact. In the example shown in Figure 2.20, if two pipes are screwed together too far, so as to make contact, galvanic corrosion will occur.

4. If it is necessary to use dissimilar metals, and these cannot be insulated, then the more anodic part should be designed for easy replacement or should be constructed of thick materials to longer absorb the effects of corrosion.

5. Coat the cathode (or both anode and cathode) near the junction to reduce the effective cathodic area. Never coat the anode alone.

Standard Potentials

Since it is just as inconvenient to relate the potentials of different metals to each other by measuring all sorts of combinations as it would be to describe the relative heights of mountains by a similar procedure, the practice has developed of providing what might be called *bench marks* for potential measurements. These may be related to any measured potential just as land elevations are related to sea level as a basis of height measurements.

There are several potential bench marks in common use, but all of them are related to a basic standard in which one-half of the cell which generates the potential that is measured is represented by a platinized platinum electrode over which hydrogen gas is bubbled while immersed in a solution having a definite concentration of hydrogen ions (expressed as an activity of 1). Using such an electrode as one-half of a galvanic cell and immersing pure metals in solutions having a concentration of their ions at an activity of 1 (1 molal concentration), a series of voltage measurements can be made.

If it is arbitrarily agreed that the potential of the platinized platinum electrode covered with hydrogen in its standard solution is zero on a scale of potentials, then the potentials of all the other metals in their appropriate solutions can be described in terms of the voltage that is generated in the several cells that have been examined as just described. With some of the combinations of metal half-cells with the hydrogen half-cell, the measured voltage would be of one polarity, while with others, it would be of opposite polarity.

Since by definition the direction of flow of the cell current is from the anode to the cathode in the solution, and from the cathode to the anode in the metallic path outside the solution, a more or less arbitrary decision has to be made as to which of the electrodes is to be said to have the positive potential and which the negative one when the values are recorded in a table of potentials.

Unfortunately, there are two recognized and opposite conventions for the sign of potential, which is symbolized by the letter E. Without going into a debate to justify the choice that is made, it will suffice to state that the convention used most widely in corrosion circles in this country and abroad, and that accepted by NACE, shows a metal like zinc to have a negative potential and a metal like gold to have a positive potential, relative to the hydrogen half-cell which is assigned a zero potential in the series of standard potentials shown in Table 2.2.

$$Zn \rightarrow Zn^{+2} + 2 \text{ electrons (E is } -)$$

$$Au \rightarrow Au^{+} + \text{electron (E is } +)$$

$$H \rightarrow H^{+} + \text{electron (E is 0)}$$

Corrosion Potential

The potential of a corroding metal is most useful in corrosion studies, and fortunately, it can be readily measured in the laboratory or under field conditions. The corrosion potential is measured by determining the voltage difference between a metal immersed in a corrosive and an appropriate reference electrode. Examples of such reference electrodes are the saturated calomel electrode, the copper-copper sulfate electrode, and the platinum-hydrogen electrode.

Figure 2.21 illustrates the experimental technique for measuring the corrosion potential of a metal M immersed in an electrolyte. This is accomplished by measuring the voltage difference between the reference electrode and the metal using a potentiometer. A potentiometer is used because it is capable of accurately measuring small voltages without drawing any appreciable current. Note that in Figure 2.21 a salt bridge is used between the reference electrode and the corrosive solution. This is to prevent contamination of the reference electrode by the corrosive liquid.

In measuring and reporting corrosion potentials, it is neccssary to indicate the *magnitude* of the voltage *and* its *sign*. In the example shown in Figure 2.21, the corrosion potential of metal M is −0.175 volt. The minus sign indicates that the metal is negative with respect to the reference electrode. A negative sign also indicates that the metal was connected to the negative terminal of the potentiometer.

There is no need to worry about mixing up these connections, since the potentiometer cannot

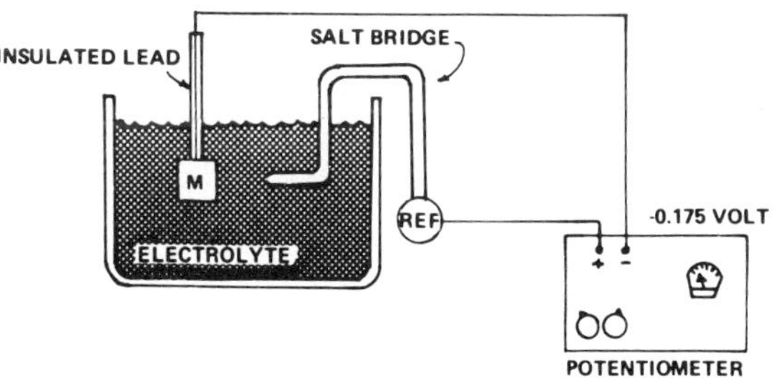

FIGURE 2.21 — Experimental measurement of corrosion potential.

be balanced unless it is properly connected to the reference electrode and metal. Thus, in making a corrosion potential measurement, it is first necessary to experiment by connecting the metal to either the positive or negative terminal of the potentiometer, and finding which connection allows the potentiometer to be balanced.

In addition to recording the voltage and the plus or minus sign, it is also necessary to specify the kind of reference electrode used in making the corrosion potential measurement. For example, if a saturated calomel electrode is used, the experiment shown in Figure 2.21 would be reported as "−0.175 volt vs saturated calomel electrode."

The magnitude and sign of the corrosion potential is a function of the metal, the composition of the electrolyte, and the temperature and agitation of the electrolyte.

If electrometers are used instead of the potentiometer referred to above, greater care must be taken to accurately denote polarities. In an analogue-type meter, such as the D'Arsonval, an upscale deflection will be observed if the external circuit has the polarities shown on the instrument terminals. A downscale deflection denotes that the polarities of the cell being measured are opposite to that marked on the meter terminals. On digital meters, a reading with no polarity indicator on the readout panel denotes that the polarities are those indicated on the instrument terminals, whereas a negative sign means that the opposite pertains.

Reference Electrodes (Half-Cells)

The standard hydrogen half-cell is rather awkward to use under many circumstances in which potential measurements are to be made. Any other combination of a metal electrode and a solution containing a specific concentration of its ions could be used if, first, the potential of such a half-cell is given a reproducible value by measurement in a cell in which the other half of the cell is the standard hydrogen electrode. When this has been done, a measurement made with any stable half-cell can be related to the standard hydrogen half-cell by simple arithmetic.

The other half-cells most frequently used in corrosion studies, along with their potentials relative to the standard hydrogen half-cell, are listed in Table 2.4.

To illustrate conversion of values of potential measured with any one of these half-cells to values on the hydrogen electrode scale, we can take the case of a measurement of the potential of a steel pipe buried in the ground, using a copper-copper sulfate reference half-cell. This might show a potential of −0.700 volt measured in this way. To convert this potential to a value on the scale in which the hydrogen electrode has a potential of zero, it is necessary to add 0.316 volt to the potential that was measured, making it −0.384 volt vs hydrogen.

TABLE 2.4 — Potential Values of Reference Electrodes (Half-Cells) Referred to Standard Hydrogen Electrode

Half-Cell	Potential Volt
Saturated Calomel	+0.2415
Normal Calomel	+0.2800
Tenth Normal Calomel	+0.3337
Silver-Silver Chloride	+0.2222
Copper-Copper Sulfate (saturated)	+0.3160
Hydrogen	0.0000

The calomel half-cells are, in fact, mercury electrodes in contact with specific concentrations of mercury ions controlled by the concentration of potassium chloride (one-tenth normal to saturated) in a saturated solution of mercurous chloride. Calomel half-cells are used most frequently in laboratory experiments. The calomel cells in which there is an easily controlled saturated solution of potassium chloride are most common.

The Silver

Silver chloride half-cells, which are more rugged than the calomel half-cells, are employed frequently for measurements in seawater.

The Copper

Copper sulfate half-cells are favored on a traditional basis for measurements of potentials of underground steel pipes. What is often referred to as a *pipe-to-soil* potential is actually the potential measured between the pipe and the half-cell (reference electrode) used to make the measurement. The soil itself has no value of potential against which the potential of a pipe can be measured independently of the potential of whatever reference electrode is used in making the measurement. Use and misinterpretation of the term *pipe-to-soil potential* can easily lead to confusion and should, therefore, be avoided in favor of defining the reference electrode to which the stated potential is referred.

Oxidation-Reduction Potentials

Definition

Oxidation-reduction potential refers to the relative potential of an electrochemical reaction under equilibrium or nonreacting (zero current flow) conditions. These potentials are measured by special electrochemical techniques under carefully controlled equilibrium conditions. Table 2.5 lists some of the values for various electrochemical reactions. Since these potentials refer to the equilibrium state, the reactions are shown to proceed at equal rates in both directions. These potentials are also called by other terms such as redox potentials, half-cell potentials, and the electromotive force or EMF series.

Criterion for Corrosion

Oxidation-reduction potentials are very useful since they can be used to predict whether or not a

TABLE 2.5 — Standard Oxidation-Reduction (Redox) Potentials at 25 C[(1)]

$K \rightleftharpoons K^+ + e$	-2.92	Active
$Na \rightleftharpoons Na^+ + e$	-2.71	
$Mg \rightleftharpoons Mg^{+2} + 2e$	-2.38	
$Al \rightleftharpoons Al^{+3} + 3e$	-1.66	
$Zn \rightleftharpoons Zn^{+2} + 2e$	-0.763	
$Cr \rightleftharpoons Cr^{+3} + 3e$	-0.71	
$Fe \rightleftharpoons Fe^{+2} + 2e$	-0.44	
$Cd \rightleftharpoons Cd^{+2} + 2e$	-0.402	
$Co \rightleftharpoons Co^{+2} + 2e$	-0.27	
$Ni \rightleftharpoons Ni^{+2} + 2e$	-0.23	
$Sn \rightleftharpoons Sn^{+2} + 2e$	-0.140	
$Pb \rightleftharpoons Pb^{+2} + 2e$	-0.126	
$2H^+ + 2e \rightleftharpoons H_2$	0.000	Reference
$Sn^{+4} + 2e \rightleftharpoons Sn^{+2}$	0.154	
$Cu \rightleftharpoons Cu^{+2} + 2e$	0.34	
$O_2 + 2H_2O + 4e \rightleftharpoons 4OH^-$	0.401	
$Fe^{+3} + e \rightleftharpoons Fe^{+2}$	0.771	
$2Hg \rightleftharpoons Hg_2^{++} + 2e$	0.798	
$Ag \rightleftharpoons Ag^+ + e$	0.799	
$Pd \rightleftharpoons Pd^{++} + 2e$	0.83	
$O_2 + 4H^+ + 4e \rightleftharpoons 2H_2O$	1.23	
$Pt \rightleftharpoons Pt^{+2} + 2e$	1.2	
$Au \rightleftharpoons Au^{+3} + 3e$	1.42	Noble

[(1)]Data from: Kortum, G. and Bockris, J. O'M., Textbook of Electrochemistry, Vol. 2, Elsevier, New York, N.Y., 1951, pp. 745-755, and Latimer, W. M., Oxidation Potentials, Prentice-Hall, Englewood Cliffs, New Jersey, 1952, p. 39.

metal will be corroded in a given environment. This is accomplished quite simply by following the generalized rule: *In any electrochemical reaction, the most negative half-cell tends to be oxidized and the most positive half-cell tends to be reduced.*

For example, assumc thal it is not commonly known whether or not zinc tends to react with an acid. In looking at Table 2.5, it can be seen that the potential of the zinc-zinc ion half-cell is more negative than that of the hydrogen ion-hydrogen gas half-cell. Thus, an application of the preceding rule indicates that zinc does tend to be corroded by acid solutions. In fact, it can be seen that all metals which have redox potentials more negative than the hydrogen gas-hydrogen ion potential tend to be corroded by acid solutions. These include lead, tin, nickel, iron, chromium, and aluminum together with the other metals with negative potentials.

The oxidation-reduction potentials listed in Table 2.5 are for standard conditions and therefore must be corrected for other conditions (temperature, concentration, etc.). However, these corrections are usually small and can be neglected in corrosion calculations.

In a similar fashion, it can be noted that copper, mercury, silver, palladium, and the other metals with potentials more positive than the hydrogen-hydrogen ion electrode will not be corroded by acid solutions. Thus, copper would be predicted to be a good container material for acid media, a prediction which has been proven acurate by corrosion tests. However, copper will tend to be corroded by acids or any medium which contains dissolved oxygen, since the redox potential of copper is more negative (less positive) than the two oxygen reduction reactions shown in Table 2.5. Platinum and gold, however, would not be expected to corrode, even by oxygenated acids because of their relatively high positive potentials compared to the oxygen half-cell.

The above examples show the utility of oxidation-reduction potentials in predicting corrosion. As a result, such potentials are widely used to make initial selections of possible corrosion-resistant alloys for different media.

Deposition Corrosion

The listing of metals in Table 2.5 indicates also the approximate tendency of one metal to plate out on another metal when one metal is placed in a solution containing ions of another metal. A common example of this is the plating out of silver on a piece of copper immersed in a solution of silver nitrate (Figure 2.22). The bubbling, or "plating out of hydrogen" on zinc (Figure 2.23) is analogous.

This plating out action, or *deposition corrosion,* may be an important factor in the corrosion of the more reactive metals near the top of the series, *e.g.*, magnesium, zinc, and aluminum, when these latter metals come into contact with solutions containing ions of metals (particularly copper) lower in the series. Copper ions in concentrations less than one part per million have been observed to have a sig-

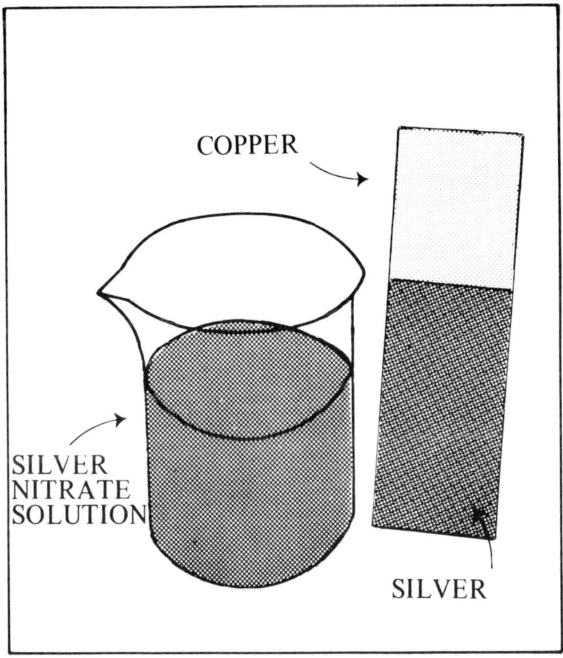

FIGURE 2.22 — Silver film formed by deposition on a copper strip immersed in a solution of silver nitrate.

FIGURE 2.23 — Hydrogen bubbles formed by deposition on a strip of zinc immersed in hydrochloric acid.

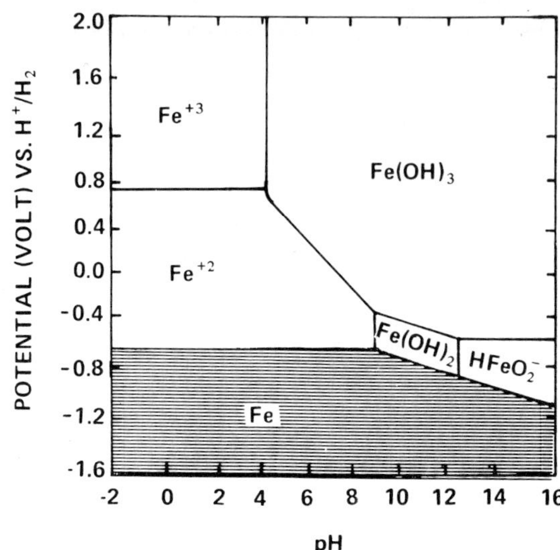

FIGURE 2.24 — Simplified potential-pH diagram for the Fe-H_2O system. [SOURCE: Pourbaix, M., Atlas of Electrochemical Equilibrium in Aqueous Solutions, Pergamon, Press, New York, NY, 1966.]

nificant effect on the corrosion of aluminum by water. Metals, such as copper, that can aggravate corrosion of aluminum are sometimes referred to as "heavy metals in solution." The fact that they are heavier than aluminum is less significant than that they occupy a position lower than aluminum in the electromotive or galvanic series.

Corrosion initiated by the plating out action just described is frequently aggravated and continued by galvanic action between the more noble metal that is plated out and the less noble (more anodic) metal on which it deposits.

Oxidation-Reduction Potentials vs Galvanic Series

There has been some confusion in the literature regarding oxidation-reduction (EMF) potentials and the galvanic series. Examination of Tables 2.3 and 2.5 shows that these two tabulations are quite similar. The differences between them, however, can be made clear by re-examining the previous discussion.

The oxidation-reduction table is used to predict whether or not corrosion of a given single metal will occur. In contrast, the galvanic series is used to predict whether or not galvanic corrosion will occur, and if so, which of the two coupled metals will exhibit increased corrosion. Thus, these two tabulations have entirely different uses and should therefore not be confused.

Potential-pH Diagram

The use of oxidation-reduction potentials can be further extended by plotting these potentials as a function of solution pH. Such diagrams, often called *Pourbaix diagrams,* are constructed using electrochemical calculations, solubility data, and equilibrium constants. To refresh our memory, pH is simply the negative logarithm of the hydrogen ion concentration. For example, a pH of 7 indicates that there are 10^{-7} gram atoms of hydrogen ion per liter of solution. A pH of 7 indicates a neutral solution while a pH of 0 represents a very acidic media, and a pH of 14 or above denotes a highly alkaline solution.

Figure 2.24 illustrates the potential-pH diagram for iron exposed to water. The various regions indicate the compounds which are stable under those conditions. For example, at potentials more positive than -0.6 and at pH values below about 9, ferrous ion is the stable substance. This indicates that iron will be corroded under these conditions, yielding Fe^{+2} [Equation (2.19)]. In other regions of this diagram, it can be seen that the corrosion of iron produces ferric ions, ferric hydroxide, ferrous hydroxide, and at very alkaline conditions, complex iron ions.

The major uses of such diagrams, which can be constructed for all metals, are: (1) predicting whether or not corrosion can occur; (2) estimating the composition of the corrosion products formed; and (3) predicting environmental changes which will prevent or reduce corrosion attack.

For example, the large region in Figure 2.24 labeled Fe indicates that iron will not corrode under these potential and pH conditions. Thus, if the corrosion potential of iron is made sufficiently negative (below approximately -1.2 volt), iron will not be corroded in *any* system, ranging from very acidic to very basic. One way of causing this change in potential is by application of an external voltage, sometimes called cathodic protection. This is discussed briefly in the following sections, and in more detail in Chapter 9.

Passivity and Protective Films

Although only briefly mentioned in previous discussions, corrosion products and other surface films can have profound effects on the corrosion behavior of metals. Oxide films which form naturally upon most metals when they are exposed to the air can provide substantial protection against further attack by many environments. If it were not for such films, many of the common metals near the top of the electromotive series would corrode rapidly in ordinary air and water. This is the case, for example, with magnesium and aluminum.

Other corrosion product films or scales are also protective. For example, insoluble films of lead sulfate are responsible for the resistance of lead to corrosion by sulfuric acid. The films that form on copper alloys in seawater contribute greatly to their durability. The extent to which these films are able to adhere, resist removal by turbulence effects, or be restored rapidly if broken, largely determines the relative merits of the copper alloys in resisting velocity effects.

The effect of oxygen and other oxidizing agents on corrosion is variable and complex. Oxygen can accelerate corrosion by participation in cathodic reactions; oxygen and other oxidizing agents can sometimes retard corrosion by forming protective films. Metals, like iron, may carry very thin, invisible oxygen or oxide films; if so, they are said to be rendered *passive* by such films. Passivity is exhibited by iron, stainless steel, and other metals if its measured potential resembles that of a noble metal (*e.g.*, platinum or gold) rather than the potential of the unfilmed metal. It can be demonstrated also by a resistance to corrosion orders of magnitude greater than that of the unfilmed or unpassivated "active" metal.

Definition and Nature

Passivity can be defined as the loss of chemical reactivity exhibited by certain metals and alloys under specific environmental conditions. That is, metals and alloys such as chromium, iron, nickel, titanium, and alloys containing these elements, become essentially inert and act as if they were noble metals. Although the oxidation-reduction potentials of these metals as shown in Table 2.5 indicate that they should be corroded by acid solution, this is not always the case.

Although passivity phenomena have been studied for more than 100 years, the exact nature or cause of these effects is still not completely understood. It is generally agreed, however, that these effects are due to the formation of a surface film which acts as a barrier to further corrosion.

What is not known is the nature or composition of this surface film. Some scientists believe that it is a very thin oxide layer which tends to shield the metal from the electrolyte (like the film described for aluminum), while other investigators believe that it is an adsorbed layer. An adsorbed layer is simply a

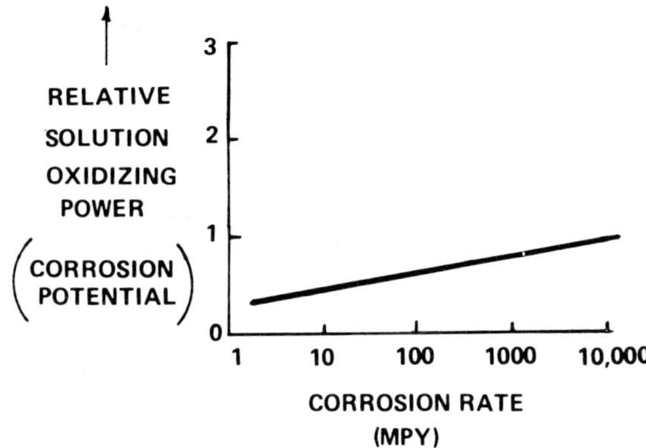

FIGURE 2.25 — Corrosion rate of a nonpassivating metal as a function of solution oxidizing power (corrosion potential).

monomolecular film of a substance, such as oxygen or some ionic species from solution. The reason for the confusion regarding the nature of the passive film is that it is extremely thin and fragile. The film is 30 Å[1] or less in thickness and contains considerable amounts of water. Thus, when removing or isolating this film from the metal surface for purposes of study, it often dehydrates and suffers mechanical and physical damage.

Effect of Oxidizers

One of the easiest ways of showing the unusual characteristics of a metal which demonstrates passivity is to compare it with a metal which does not show this effect. Figure 2.25 shows the corrosion behavior of a nonpassivating metal or alloy as a function of solution oxidizing power. Such data can be obtained by making a series of corrosion tests in solutions containing increasing amounts of oxidizing agents, such as cupric ions, etc. If air-free acid was used to begin a series of such experiments, a plot similar to Figure 2.25 would be obtained.

Note that corrosion rate rapidly increases with increasing oxidizing power or concentration of oxidizer in the solution. As described previously, this is simply the result of the cathodic reaction associated with oxidizers which are capable of consuming electrons, and thus increasing overall corrosion rate. Figure 2.25 clearly indicates that the presence of oxidizers detrimentally affects the corrosion resistance of nonpassivating metals. Examples of this kind of behavior include zinc, lead, and copper exposed to oxidizing acid mixtures. Note that corrosion potential is a measure of oxidizing power.

However, if a metal showing passivity effects is exposed to an electrolyte and increasing amounts of oxidizing agents are added to it, the results shown in Figure 2.26 are obtained. Initially, slight increases in the oxidizing power of the solution cause the corrosion rate to increase. As more oxidizing agent is added, the corrosion rate shows a sudden decrease,

[1]Å is the Angstrom, a fine measure of length equal to 0.00000001 cm or 10^{-8} cm, or about 0.004 millionths of an inch.

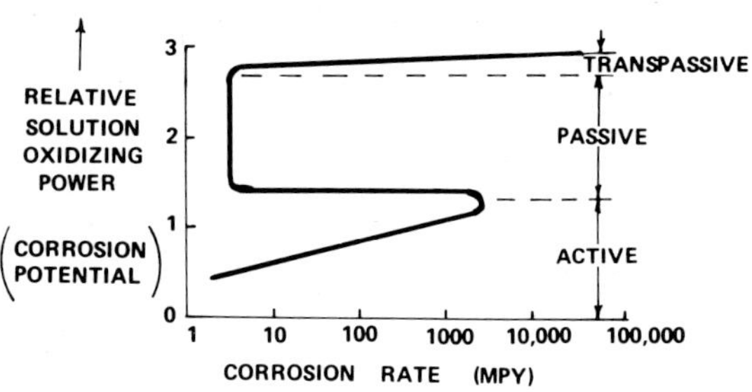

FIGURE 2.26 — Corrosion rate of a passivating metal as a function of solution oxidizing power (corrosion potential).

and then remains essentially constant as further oxidizing agent is added. Finally, at very high concentrations of oxidizers, corrosion rate again increases. As shown in Figure 2.26, three reactions can be identified: (1) active; (2) passive; and (3) transpassive.

In the passive region, the corrosion rate is frequently 10,000 times less than in the active region. Passivity is a very useful phenomenon from a practical standpoint, as it can be used to achieve a high degree of corrosion resistance. The chromium and chromium-nickel stainless steels, which find widespread application in corrosion engineering, owe their resistance to a passivity effect.

Examination of Figure 2.26 shows that the achievement of the passive state requires the proper amount of oxidizing power. If too little or too much oxidizer is added to the solution, corrosion rate will be very high. Thus, the stainless steels and other metals showing passivity are usually attacked rapidly in air-free acid solutions, but are resistant to aerated acids and acids containing oxidizers such as cupric and ferric ions.

On the other hand, extremely powerful oxidizer solutions such as boiling, fuming nitric acid (100%) tend to corrode the stainless steels. Thus, the selection of stainless steels and other passive metals must be done with caution to ensure that they will be used under optimum conditions.

Under circumstances where there may be limited access of oxygen to a stainless steel surface, passivity may be destroyed on such surfaces while the remainder of the stainless steel surface will remain passive. A difference in potential between the active and passive stainless steel surfaces can be as large as 700 mv. This will set up a powerful galvanic cell between the active and passive stainless steel surfaces and result in serious corrosion of anodic areas (pitting).

The difference in potential between the active and passive states of stainless steels, and some other metals and alloys that develop passivity, will account for the dual location of these materials in positions representing either their active or passive states in the galvanic series shown in Table 2.3.

Active-passive cells are principally responsible for the severity of pitting and crevice corrosion of stainless steels under circumstances in which such attack occurs.

Restoration of passivity within pits or crevices is frequently prevented by the acid nature of corrosion products when they accumulate within the pits or crevices. Consequently, circumstances such as stagnation and gravity effects that favor accumulation of corrosion products within pits or crevices will promote the most severe localized corrosion.

Because of the action of such active-passive cells, it is necessary in the use of stainless steel to eliminate designed-in crevices. It is also necessary to avoid any opportunities for the accumulation of loosely adherent deposits of any kind which could permit passivity to be destroyed within such crevices or under such deposits.

Effect of Alloying

As was previously explained, passivity can be achieved by controlling the amount of oxidizer in the corrosive medium. Also, by appropriate alloy additions, it is possible to improve the corrosion characteristics of a metal. This is illustrated in Figure 2.27 which schematically compares the corrosion behavior of iron and chromium stainless steel in dilute sulfuric acid solutions as a function of oxidizing power. The addition of 18% chromium to iron appreciably reduces the amount of oxidizer necessary to achieve passivity. The addition of chromium to iron also decreases the corrosion rate in the passive state.

As revealed in Figure 2.27, ferritic stainless steel, containing primarily iron and chromium, corrodes more rapidly than iron under nonoxidizing conditions (*e.g.*, air-free acids). It is this unusual fact that retarded the application of stainless steels for more than 30 years because no one thought to test these alloys in oxidizing media. Alloying is one of the more important ways of improving corrosion resistance, and is further discussed in Chapter 3.

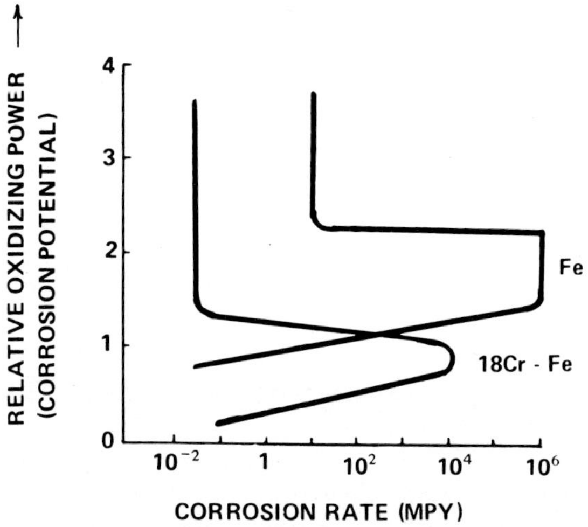

FIGURE 2.27 — Corrosion characteristics of iron and 18% Cr stainless steel in dilute sulfuric acid as a function of solution oxidizing power (corrosion potential).

Corrosion Prevention by Electrochemical Methods

Anodic Protection

The term *anodic protection* refers to corrosion protection achieved by maintaining an active-passive metal or alloy in the passive region by an externally applied anodic current. The basis for this type of protection is shown in Figure 2.26. Solution oxidizing power and corrosion potential are equivalent, and therefore it is possible to achieve passivity by altering the potential of the metal by an appropriate external power supply.

Since the potential must be maintained within the passive region, it is necessary to use a special device called a *potentiostat* which is capable of maintaining a constant electrode potential by controlling the anodic current (Figure 2.28). The potentiostat has three terminals, and these must be connected to the proper electrodes.

The main advantages of anodic protection are: (1) low current requirements; (2) large reductions in corrosion rate (typically 10,000-fold or more); and (3) applicability to certain strong, hot acids and other highly corrosive media.

It is important to emphasize that anodic protection can only be applied to metals and alloys possessing active-passive characteristics such as titanium, stainless steels, steel, and nickel-base alloys. Furthermore, it can only be utilized in certain environments since electrolyte composition influences passivity.

The corrosion rate of a nonpassive metal is markedly accelerated if its potential is increased, as shown in Figure 2.25. Thus, anodic protection techniques must be used with caution.

Anodic protection has been successfully used to reduce the corrosion rate of steel, 18-8 stainless steels, and other alloys in such media as sulfuric acid, phosphoric acid, sodium hydroxide, and corrosive salts such as aluminum sulfate and ammonium nitrate.

Cathodic Protection

A second electrochemical method of protecting metals is more widely used. Since it has been demonstrated that electrochemical corrosion results from, or is accompanied by, a flow of current between anodic and cathodic surfaces, it should be possible to prevent corrosion by controlling the flow of corrosion currents. The ultimate objective is to suppress all current flowing from the anode in a corrosion cell. This can often be accomplished by applying current from an external source so that current will be made to flow to, instead of away from, the original anodic surface. This will result in a cathodic rather than an anodic reaction on these surfaces.

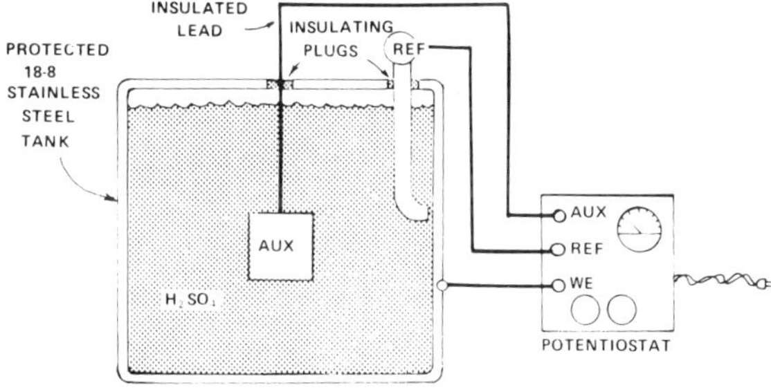

FIGURE 2.28 — Anodic protection of an 18Cr-8Ni stainless steel tank containing H_2SO_4. Aux = Auxiliary electrode; Ref = Reference electrode; WE = Working electrode.

To accomplish this, the source of the protective current must be at a higher potential than that of the anodic surface to be protected. The amount of current that will be needed will depend on the requirement to support a cathodic reaction over the whole of the surface to be protected.

Cathodic protection can be illustrated by a simple experiment using two iron nails and a piece of zinc. If one iron nail is immersed in water, but in contact with the zinc, it will not corrode. The nail by itself will corrode. This is why galvanized iron (zinc-coated steel) is so widely used.

A similar experiment can be conducted using iron nails with half their surfaces plated with copper and with one of these being connected to a piece of zinc. As would be expected, the unplated half of the first iron nail would become anodic to the copper-plated half and quickly corrode. In the case of the partially plated nail connected to the zinc, the galvanic corrosion of the iron half would be suppressed and a cathodic reaction would be made to occur along both the bare iron and copper-plated surfaces.

In normal corrosion, the amount of current (quantity of electrons) required by the cathodic reaction that is occurring is supplied by the electrons generated by corrosion of the anodic surfaces. By means of "artificial" cathodic protection, this quantity of electrons can also be provided by an external source. Chapter 9 will provide a more detailed discussion of this subject.

Corrosion Rate Measurements by Electrochemical Techniques

Apparatus

The corrosion potential of a metal can be altered using a simple external power supply, as shown in Figure 2.29. Here, a variable voltage, DC power supply is used to pass current through the sample, or working electrode (WE), and an auxiliary electrode (AUX) immersed in solution.

The potential change of the working electrode which occurs as a result of this external current is measured by means of a reference electrode and a voltmeter. The voltmeter for this purpose is usually a high-resistance-type instrument such as a vacuum tube voltmeter or potentiometer, as shown.

If the change of voltage of a corroding metal is plotted against the applied current, a graph similar

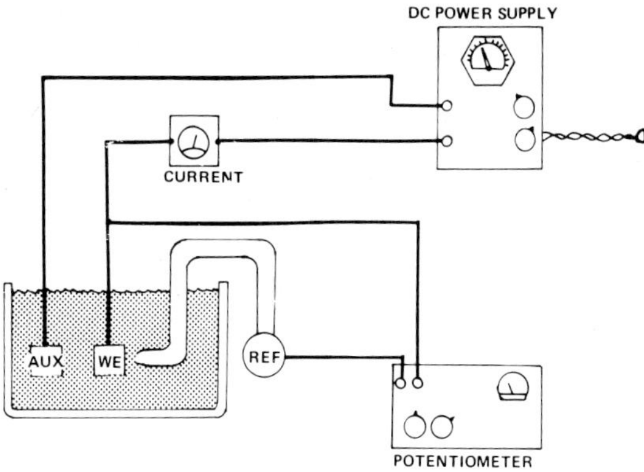

FIGURE 2.29 — Circuit for conducting linear polarization measurements.

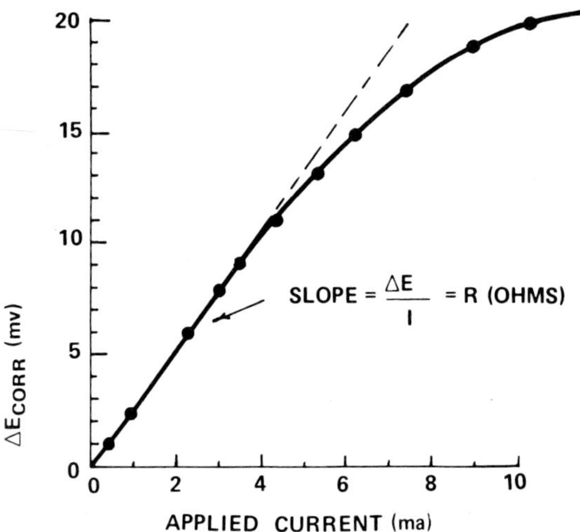

FIGURE 2.30 — Typical data obtained during linear polarization experiment.

to that shown in Figure 2.30 is obtained. Here the potential change of the corroding metal is expressed in terms of millivolts and the applied current is expressed in milliamperes. For potential changes of 10 millivolts or less, there is a straight line relationship between the voltage change and the applied current. Beyond 10 millivolts, curvature becomes evident. This initial portion is termed the *linear polarization* region.

Calculation of Corrosion Rates

The corrosion rate of a metal can be calculated from linear polarization data such as shown in Figure 2.30. Note that the slope of the line is voltage over current, or resistance in ohms. It can be shown by electrochemical calculations that this slope is related to the corrosion rate by the following equation:

$$\text{mpy} = \frac{K}{RA} \tag{2.29}$$

where mpy is the corrosion rate in mils penetration per year, K is an electrochemical constant depending on the metal and corrosive, R is the resistance in ohms read from the linear polarization graph, and A is the total area of the corroding specimen in square inches.

For iron, cobalt, nickel, and alloys containing these elements, the constant K is approximately equal in most environments and can be substituted into this equation:

$$\text{mpy} = \frac{2000}{RA} \tag{2.30}$$

Thus, by conducting the linear polarization measurement and by obtaining the slope of the linear portion of the line, this value can be substituted into Equation (2.30) to obtain the corrosion rate of the metal.

Applications

Although most corrosion testing is performed by immersing the specimen into the corrosive medium and observing it or measuring its weight change after a given period of time (Chapter 14), electrochemical measurement of corrosion rate has several unique advantages and can be applied to cases where conventional tests are not applicable.

Generally, electrochemical measurements require only a short period of time. Note that in Figure 2.30, only a few points are needed to define the linear region, and since each of these points takes approximately 1 to 2 minutes, the entire measurement can usually be performed in less than 10 minutes. Thus, the technique provides a way of measuring instantaneous corrosion rates. That is, it can be repeatedly applied to a corroding metal and the corrosion rate calculated and plotted as a function of time.

Another advantage of this technique is that it can be performed remotely, which is evident from Figure 2.29. The actual corroding specimen can be considerably removed from the measuring instrument and need only be connected by electrical wires. Because of this, electrochemical methods can be used to measure the corrosion rate of a specimen without removing it from solution or where it is inconvenient to periodically remove the specimen and examine it.

Electrochemical techniques such as those described above have been used to measure the corrosion rate of metals under unusual conditions. Examples include the internal surfaces of food cans, corrosion occurring in or near the reactor portion of nuclear power plants, and the corrosion of metals embedded in living tissue. This is very important in many medical applications, such as bone pins and plates.

Electrochemical methods can be readily adapted to such systems, as shown in Figure 2.31. In this figure, the corroding specimen is a bone plate which has been implanted in an experimental animal

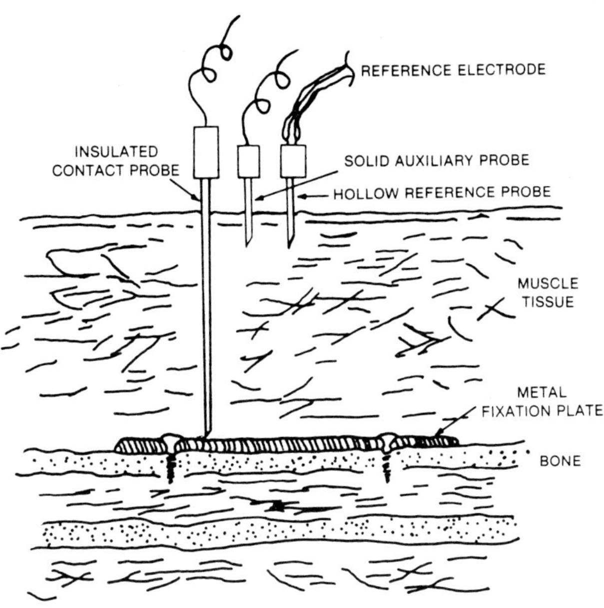

FIGURE 2.31 — Apparatus for performing *in vivo* corrosion tests by linear polarization. [SOURCE: Greene, N. D. and Jones, D. A., J. Materials, Vol. 1, p. 345 (1966).]

(dog). The current is passed through this implant by means of the insulated solid probe needle, which is pressed against the implant, making electrical contact. Another solid, uncoated probe serves as an auxiliary electrode, while a hollow needle connected to a salt bridge is used to measure the potential of the fixation device.

The connections to these three electrodes are identical to those shown in Figure 2.29. In this application, it is possible to measure the corrosion rate of the fixation device as a function of time by periodically inserting the three probes and conducting repeated measurements. It is quite obvious that conventional corrosion tests could not be conveniently applied to this system because of the difficulty of removing and replacing the specimen without severely damaging the living system.

Summary

This chapter has examined the fundamental reactions of corrosion and found these to be primarily electrochemical reactions. It has explained how most corrosion processes involve electron transfer and can be greatly simplified by considering them as a series of anodic and cathodic reactions.

Electrochemical principles, namely oxidation-reduction potentials, allow us to predict whether or not a pure metal will be corroded by a given environment. Furthermore, this chapter has shown that the electrochemical nature of corrosion often results in galvanic attack which can be both understood and controlled by recognizing and understanding basic electrochemical principles. An understanding of electrochemistry has been shown to be useful because it can be utilized to reduce or prevent corrosion, and it can also be used to measure the corrosion rate of a metal.

Bibliography

Brubaker and Phipps. Corrosion Chemistry, 1979.

Fontana, M. G. and Greene, N. D. Corrosion Engineering. Electrochemistry and Corrosion. McGraw-Hill, New York, N.Y., 1967.

Henthorne, M. Fundamentals of Corrosion. Chem. Eng., May, 1971-April, 1972.

NACE. Electrochemical Techniques for Corrosion, 1977.

Pourbaix, M. Atlas of Electrochemical Equilibrium in Aqueous Solutions. Potential-pH Diagrams. Pergamon Press, New York, N.Y., 1966.

Schaum, D. College Chemistry. Schaum Outline Series. McGraw-Hill, New York, N.Y., 1966.

Shrier, L. L. Corrosion. George Newnes Ltd., London, England, 1963.

Uhlig, H. H. Corrosion and Corrosion Control. John Wiley & Sons, New York, N.Y., 1963.

Uhlig, H. H. Corrosion Handbook. John Wiley & Sons, New York, N.Y., 1948.

NOTES

Chapter 3

Metallurgy

METALLURGY

Since metals are the principal materials which suffer corrosive deterioration, it is important to develop a background in the principles of metallurgy in order to fully understand corrosion.

General Characteristics of Metals

Nearly all metals and alloys exhibit a crystalline structure. The atoms which make up a crystal exist in an orderly three-dimensional array. There are solid materials, principally glass, that exist in an amorphous state. However, only crystals have the unique condition in which atoms are geometrically and uniformly arranged in all three dimensions.

Figure 3.1 is a schematic representation of the unit cells of the most common crystal structures found in metals and alloys. The unit cell is the smallest portion of the crystal structure which contains all of the geometric characteristics of the crystal. The crystals, or grains, of a metal are made up of these unit cells repeated in a three-dimensional array.

The crystalline nature of metals is not readily obvious because the metal surface usually conforms to the shape in which it has been cast or formed. Therefore, the crystalline nature of metals is difficult to understand since the usual concept of a crystal is a geometrically shaped object. In some rare instances, this crystallinity can be observed naturally, *i.e.,* brass door knobs are normally bright and shiny, however, after a time, the corrosive perspiration from hands etches the crystalline features of the alloy on the surface (Figure 3.2).

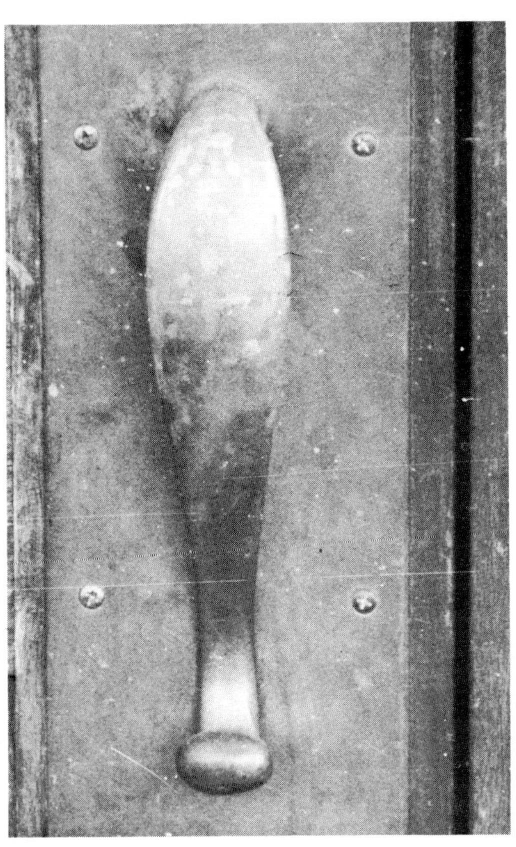

FIGURE 3.2 — Brass door knob with evidence of crystal structure etched by corrosive perspiration.

Controlled etching with selected electrolytes will normally show the granular characteristics of metals and alloys. Figure 3.3 is a photograph of an aluminum strip which has been specially treated to

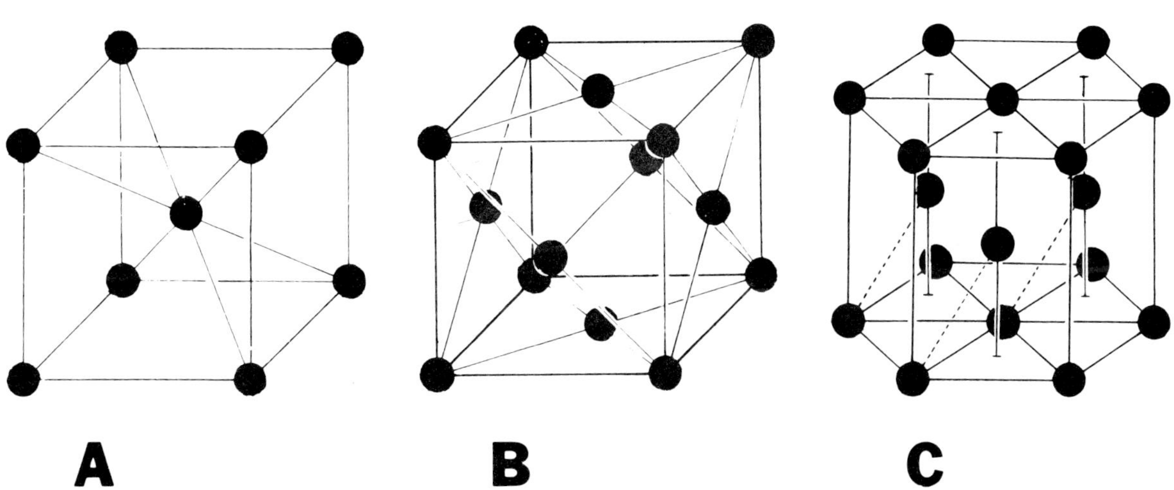

FIGURE 3.1 — Schematic representation of the unit cells of the most common crystal structures found in metals: (a) body-centered cubic; (b) face-centered cubic; and (c) hexagonal close packed.

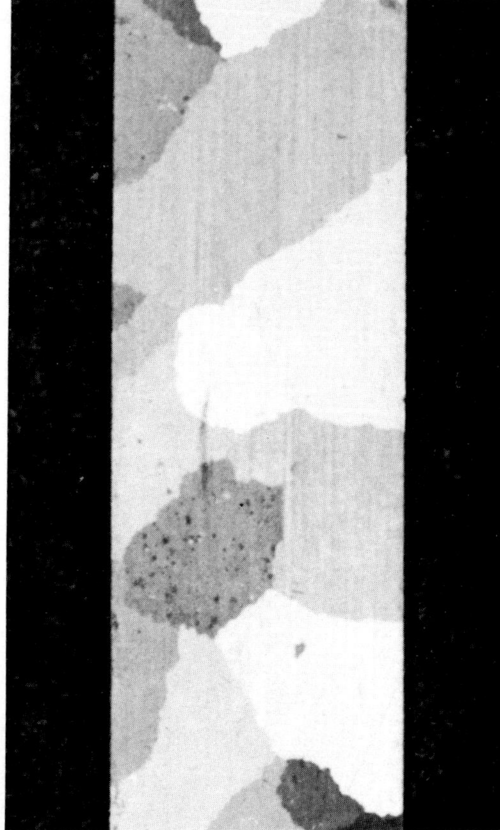

FIGURE 3.3 — Aluminum strip specially treated to grow exceptionally large grains.

grow exceptionally large grains. The grains have been clearly defined by the etching solution which promotes a controlled electrolytic reaction; actually, it could be called a controlled corrosion process. The grains shown in these photographs are extremely large for most metal crystals. Normally, metal grains are so small they can only be satisfactorily observed under a microscope. The general range of grain size usually runs from 0.25 to 0.025 mm (0.01 to 0.001 inch) in diameter.

To determine the grain size or *microstructure* (structure under the microscope) of a metal or alloy, it is usually necessary to prepare a sample for microscopic study by special grinding and polishing techniques of the specimen surface. The polished surface is then reacted with a suitable etching reagent. This reveals the grain boundaries and the distinguishing features of the microstructure.

Figure 3.4 shows *photomicrographs* (photographs taken at high magnification through a microscope) of annealed pure iron and pure copper at a magnification of 100 times. Even though grains are very small, each is still made up of thousands upon thousands of unit cells. The contrast in the grains in these photographs is due to the difference in orientation of the crystal with respect to the surface being examined.

Many features in the microstructure of metals can be related to the properties of the material. For example, Figure 3.5a is a relatively pure iron with sulfide inclusions. Sulfide inclusions have a marked tendency to react in even mild corrosive environments. Figure 3.5b, also of iron, shows a sample in which the grains have been heavily deformed by *plastic deformation* (permanent deformation from the original shape). Metals are frequently plastically deformed in fabricating. Terms often used for metals deformed at room temperature are cold-worked, cold-rolled, cold-drawn, etc., depending upon the method of fabrication.

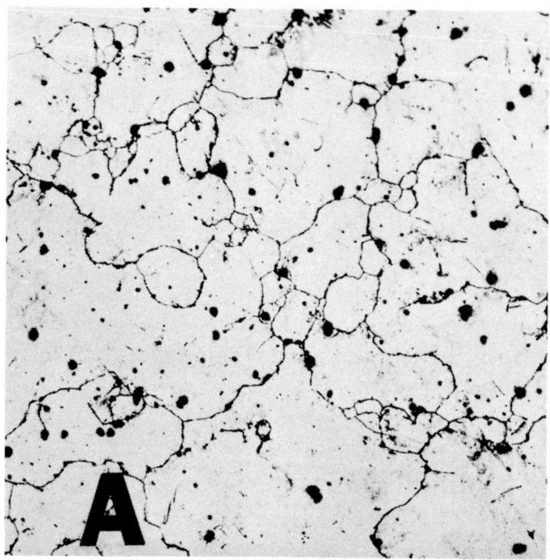

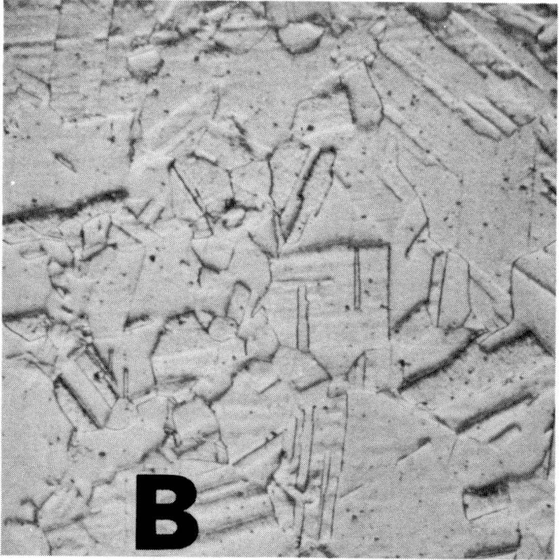

FIGURE 3.4 — Photomicrographs of: (a) annealed pure iron; (b) pure copper. 100 X.

In highly deformed metals, the grains are deformed and the grain structure is completely disrupted. Normally, in this condition, the material is somewhat more reactive in electrochemical environments. Relatively pure metals may show definite increases in electrochemical reactivity due to impurities or mechanical deformation.

In addition to the influence of impurities, inclusions, and cold work, grain boundaries and differences in grain orientation also may result in significantly different electrochemical reactivities in many metals and alloys. For example, the grain boundaries of the samples shown in Figures 3.3 and 3.4 have been selectively attacked by the etching reagent. In Figures 3.4a and b, a marked difference

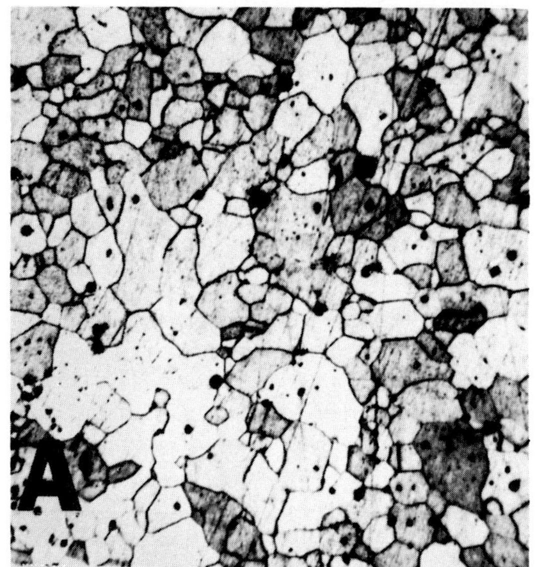

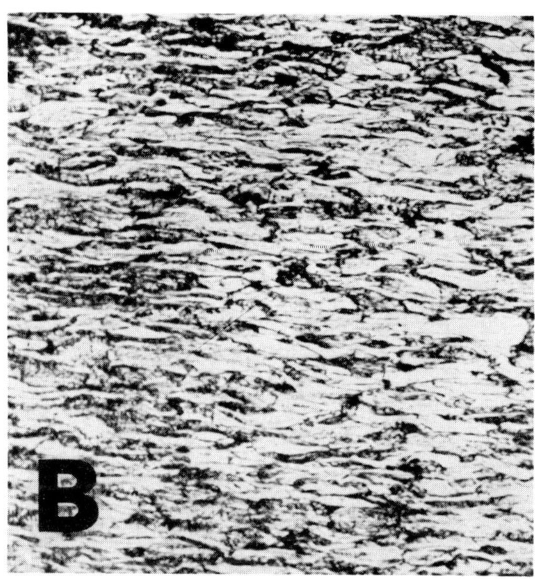

FIGURE 3.5 — (a) Relatively pure iron with sulfide inclusions; (b) iron in which grains have been heavily deformed.

in grain contrast has occurred because of the variation in etching characteristics of different grain orientations.

Imperfections and Defects

On a smaller scale, differences in submicroscopic characteristics of a metal also must be considered. Previously, the crystal structure has been represented as a perfect three-dimensional array, but in reality there are variations in the structure caused by crystalline defects. These defects may be vacancies caused by the absence of atoms in the crystal, impurity atoms of different sizes, interstitial atoms (small atoms in the space between large atoms) and large lattice disturbances, called dislocations (Figure 3.6).

Each of the illustrated imperfections can produce highly localized differences in electrochemical behavior of the metal. However, the vacancy, the impurity atom, and the interstitial atom are point

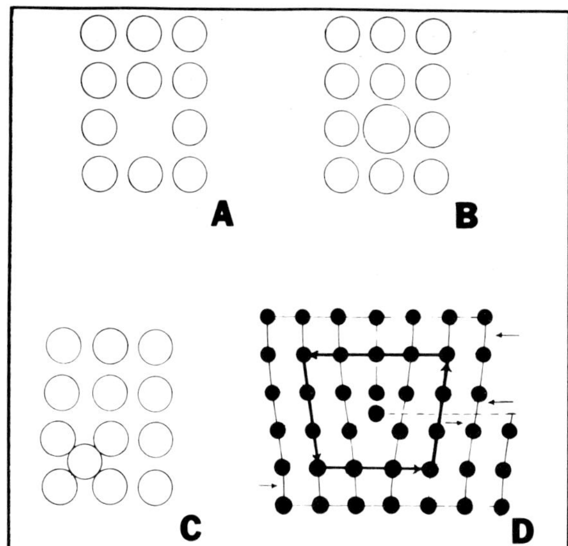

FIGURE 3.6 — Schematic diagram of crystal imperfections caused by: (a) vacancies (absences of atoms in the crystal); (b) impurity atoms of a different size; (c) interstitial atoms; and (d) large lattice disturbance (dislocations).

defects affecting a much greater volume of the crystal. In etching or corroding environments, these areas are usually more anodic than the surrounding matrix.

Examples of pits formed at intersections of dislocations with the surface of a copper sample are shown in Figure 3.7. The large number of triangular etch pits are a result of selective electrochemical attack due to the stress field around a dislocation. The shape of the etch pit is related to the orientation of the grain to the etched surface.

Metals listed as pure or commercially pure actually contain a variety of impurities and imperfec-

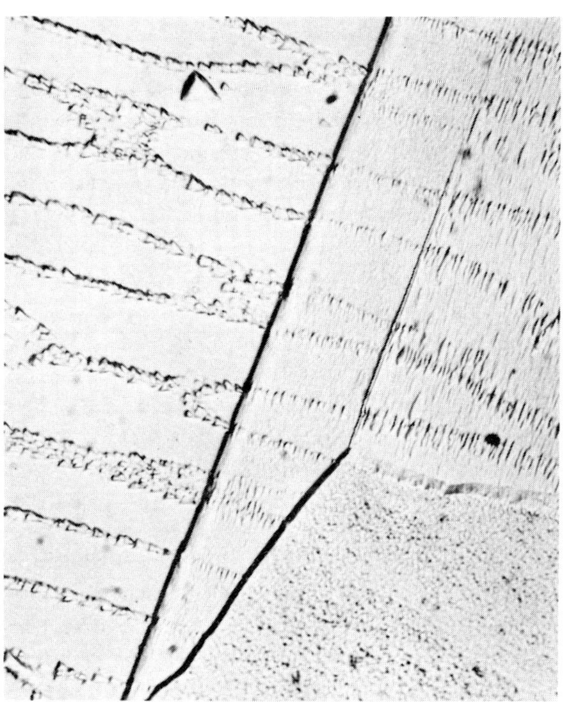

FIGURE 3.7 — Pits formed at the intersection of dislocations with the surface of a copper sample.

tions. These impurities and imperfections are inherent causes of corrosion in an aggressive environment. It has been shown that as purity increases, the tendency for a metal to react in electrochemical environments reduces proportionately. Very high-purity metals can be produced by a special technique called *zone refining.* Final purities of 99.99999% or better are possible, depending on the metal and nature of impurity. Such materials often exhibit corrosion resistance much better than their commercial counterparts which are only 99 to 99.9% pure. Commercially pure metals may contain as many as 1 impurity atom for every 100 in the material.

However, high-purity metals generally have low mechanical strength; hence, they are rarely used in engineering applications. It is necessary, therefore, to work with metallic materials which are stronger and are usually formed from a combination of several elemental metals. One metal usually serves as a major or base metal to which other metals or nonmetallic constituents are added. These metallic mixtures are properly called *alloys* and are a major class of engineering materials.

While alloys can exist in an almost unlimited number of combinations, only a portion are useful. The common alloys are those which have a good combination of mechanical, physical, and fabrication qualities that tend to make them structurally, as well as economically, useful. Although an alloy's resistance to corrosion is also very important, this factor is often sadly neglected in the aim to improve mechanical strength.

Alloying by itself has produced a number of materials that have improved corrosion resistance compared to high-purity metals. Designations of many of these so-called corrosion-resistant alloys may be misleading because they are not truly corrosion-resistant under all conditions. In fact, they may be subject to catastrophic failure as, for example, intergranular attack of sensitized stainless steel and stress corrosion cracking of brass, stainless steel, and high-strength steel. To understand the corrosion characteristics of alloys, it is important to examine some of the fundamental characteristics of alloying.

Alloying

When an alloying element is added to a base metal, it is possible that the crystal structure will remain essentially stable and produce a simple solid solution. For example, copper is face-centered cubic. The addition of nickel, which is also face-centered cubic, does not alter the crystal structure. This ability of nickel to dissolve in copper, or copper to dissolve in nickel, without a change of crystal structure, results in what is termed a *solid solution* or a *single phase.*

An analogy can be made with a liquid solution also. For example, water will dissolve salt up to a given concentration at a specific temperature. However, the solubility of salt in water changes with temperature. Many alloy systems have similar characteristics in that the solubility limit of one metal in another, in the solid state, changes with temperature. Thus, the liquid solution of salt dissolved in water is analogous to the face-centered cubic solid solution of zinc in copper where zinc dissolves in copper up to approximately 40 weight percent (Wt%) before another phase begins to precipitate.

There are many alloy systems which exhibit more than one solid solution or phase, depending on the concentration and temperature. To understand this more thoroughly, it is important to consider how one metal dissolves in another in the solid state.

Basically, there are two common solid solutions. In the *substitutional solid solution,* the atoms of the alloying element take a position on the lattice of the base metal. For substitutional solid solutions to exist, a number of limiting conditions related to atomic size and electronic structure must be met. Figure 3.8a schematically represents a substitutional solid solution of this type.

Another type of solid solution is possible in which small atoms actually fit into void (interstitial) areas between the atoms in the crystal lattice of a metal with a larger atomic size. Such alloy phases are called *interstitial solid solutions.* The most notable of these is carbon which dissolves interstitially in iron to produce steel. A schematic representation of an interstitial solid solution in a body-centered cubic structure is schematically represented in Figure 3.8b.

Although there is a wide range of solid solutions, in many cases it is impossible to dissolve a large amount of one metal into another. When this occurs in an alloy, it results in the formation of two or more phases, depending on the number and type of components in the alloy. Plain carbon steel, for example, is a mixture of an interstitial solid solution of carbon in body-centered cubic iron (ferrite) and an intermetallic compound called cementite (Fe_3C). Figure 3.9 is the typical two-phase microstructure of plain carbon steel.

The properties of steel, or any multiphase material, depend greatly on the relative physical and structural characteristics (amount, distribution,

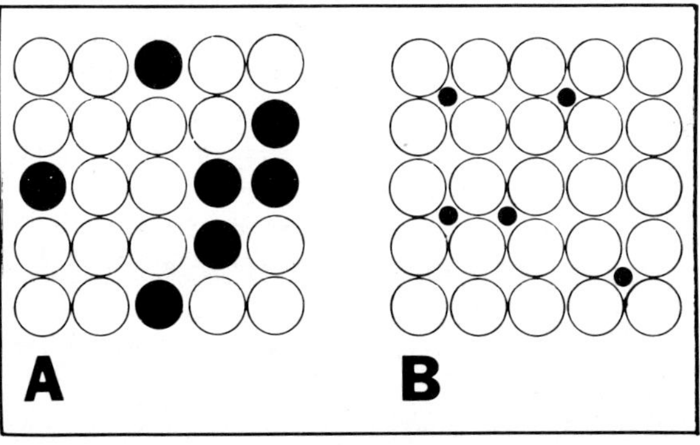

FIGURE 3.8 — Schematic diagrams of: (a) substitutional solid solution of face-centered cubic type; (b) interstitial solid solution in a body-centered cubic structure.

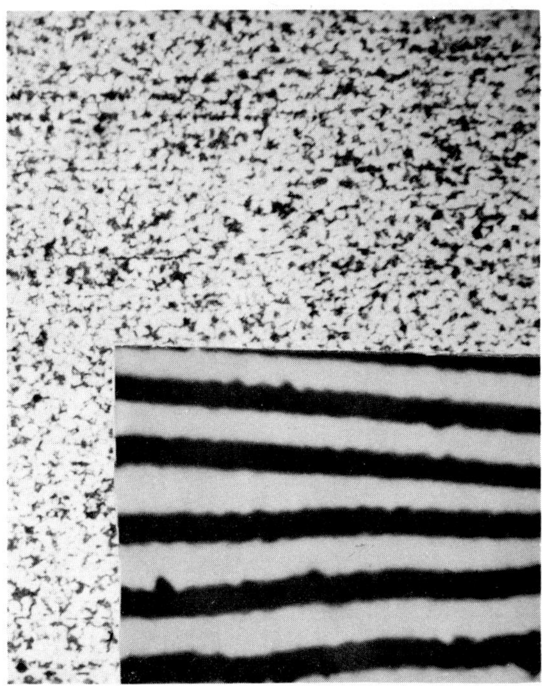

FIGURE 3.9 — Two-phase microstructure of carbon steel. Insert shows typical iron-Fe_3C plates which make up the dark (pearlite) microconstituent in the large photomicrograph.

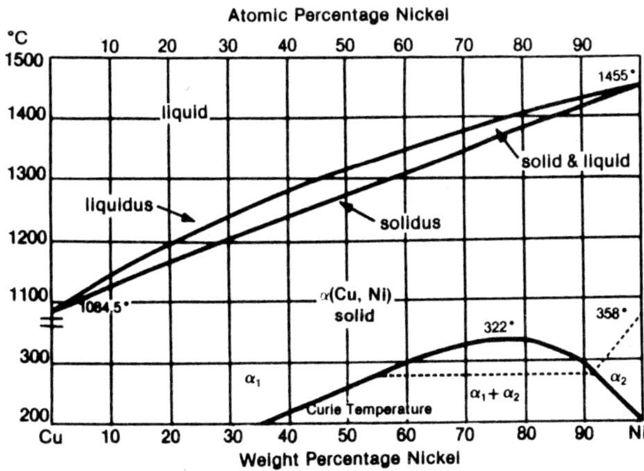

FIGURE 3.10 — Copper-nickel phase diagram. (SOURCE: ASM Metals Handbook, Vol. 8, p. 294, 1973.)

size, shape, and strength) of the various phases in the alloy. In many cases, multiphase materials present a problem from a corrosion standpoint because the two phases may have marked differences in electrochemical characteristics. It is possible that one phase will be selectively attacked in a corrosive environment. However, some two-phase alloys have small electrochemical differences and excellent corrosion resistance, and others develop a protective film which results in improved corrosion properties for both phases. The existence of more than one phase in an alloy usually results in poorer corrosion resistance than for equivalent single-phase materials. There are, however, many notable exceptions.

Phase Diagrams

Graphs can be plotted showing the equilibrium relation between various compositions of one metal in another as a function of temperature. While pressure also can be considered a variable, as a rule it has little influence on the metallurgy of metals. These graphs of phase stability as a function of temperature and composition are called *phase diagrams.* The phase diagram often is called the *equilibrium diagram,* because it is based on equilibrium conditions in the alloy, with the most stable phases represented at each temperature and composition on the diagram.

A diagram consisting of two metallic components is called a *binary phase diagram.* Binary systems can range from the very simple to the very complex. The copper-nickel system (Figure 3.10), which was briefly discussed previously, is one of the simplest since there is complete solubility of the metallic components in all compositions in the liquid and solid state. The vapor state is ignored for most metals because it normally exists well above the temperatures of exposure. However, in simple binary systems, rates of heating and cooling can greatly influence the characteristics of certain alloys. For example, rapid cooling may not allow diffusion of atoms to occur, and a nonequilibrium phase, or phases, may result at room temperature.

One of the most basic binary phase diagrams is that of the iron-carbon system (Figure 3.11). This complex binary system is the key to the broad range of properties available in steels.

Figure 3.12 is an example of a phase diagram which is extremely important to stainless steel technology, the iron-chromium system. Although this diagram is for the Fe-Cr binary system, the influence of a third important alloying element, carbon, will cause an enlargement of the gamma (γ, face-centered cubic) region with increasing carbon content. The sigma (σ) phase is also noted in this diagram. This phase is important because its formation usually results in a reduction in the mechanical properties and corrosion resistance of the alloy.

A ratio of 1 chromium atom to 7 iron atoms results in a naturally occurring surface oxide which greatly improves corrosion resistance. On a weight basis, this represents slightly more than 10% chromium, and this portion of the iron-chromium system is the basis for the 400 Series of stainless steels (S40000).

The 400 Series of stainless steels (S40000) listed in Chapter 4 is divided into two types: (1) the *ferritic* type, represented by alloys which remain alpha (α, body-centered cubic iron) at room temperature, and (2) the *martensitic* type which result from the formation of a nonequilibrium phase by quenching from the gamma (γ) or *austenite* region. The latter usually have higher carbon content to enlarge the gamma region. Quenching from the gamma region (face-centered cubic iron) produces the extremely hard, strong, and often brittle, nonequilibrium phase called *martensite.*

Although changes, such as increasing the size of the gamma region with increasing carbon content

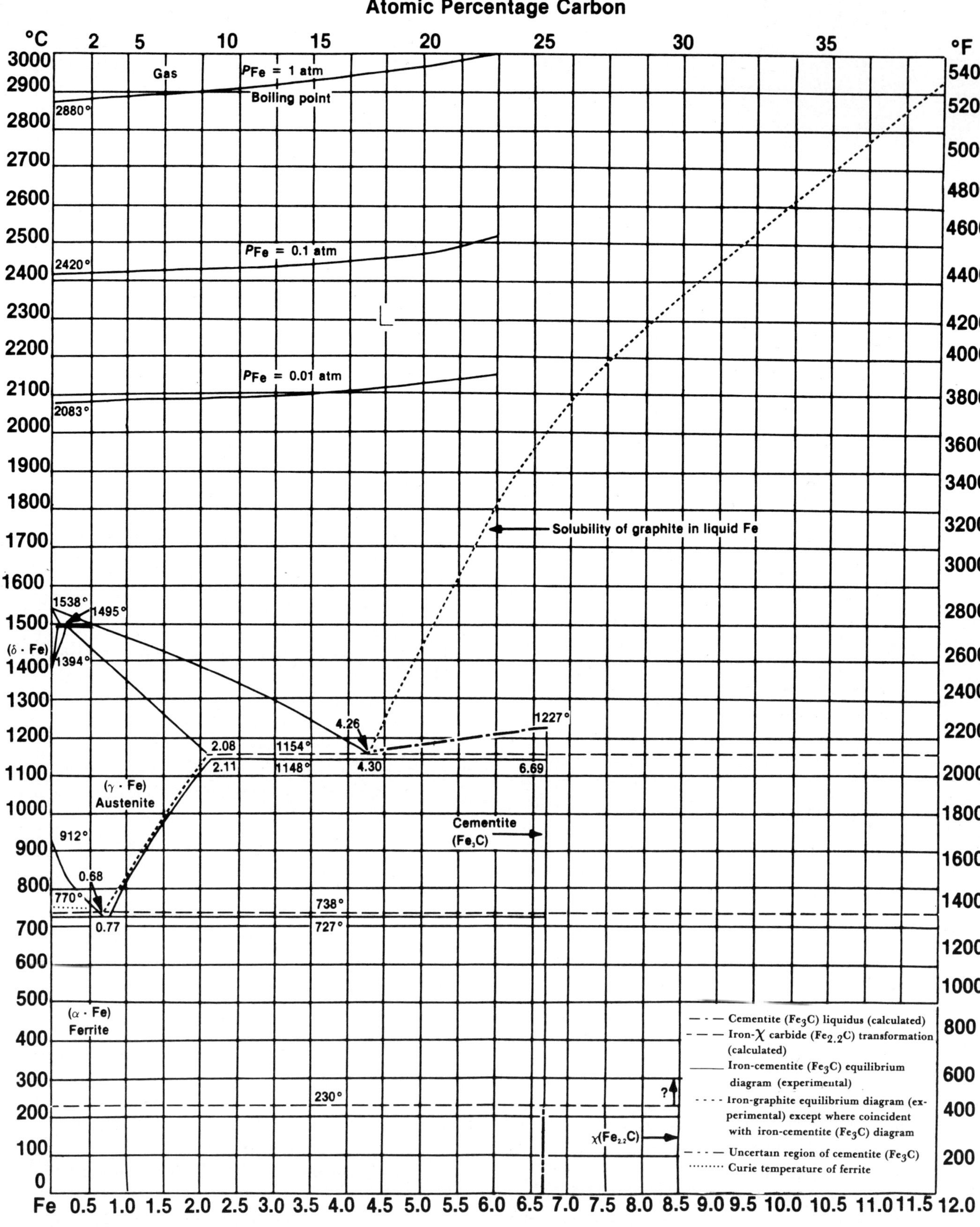

FIGURE 3.11 — Iron-carbon phase diagram. (SOURCE: ASM Metals Handbook, Vol. 8, p. 275, 1973.)

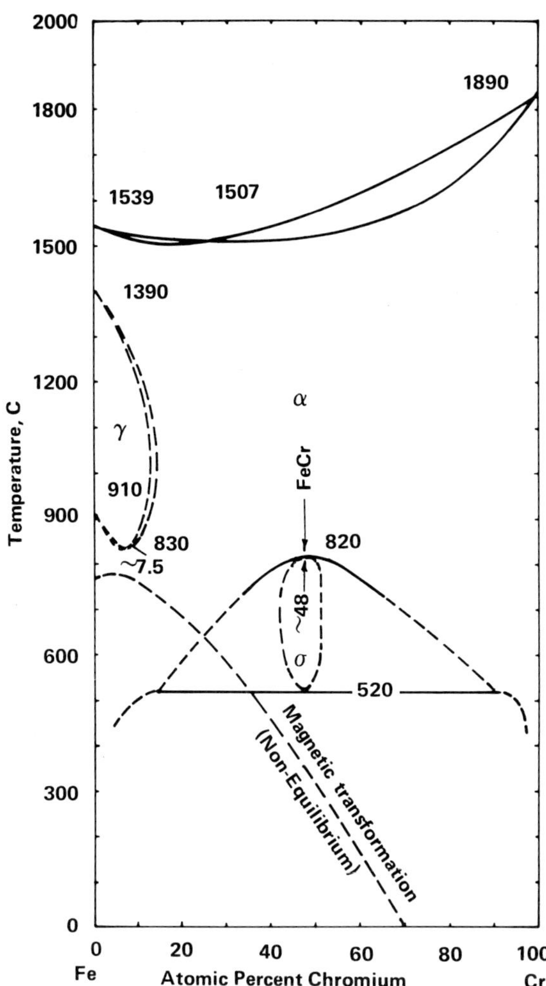

FIGURE 3.12 — Iron-chromium phase diagram.

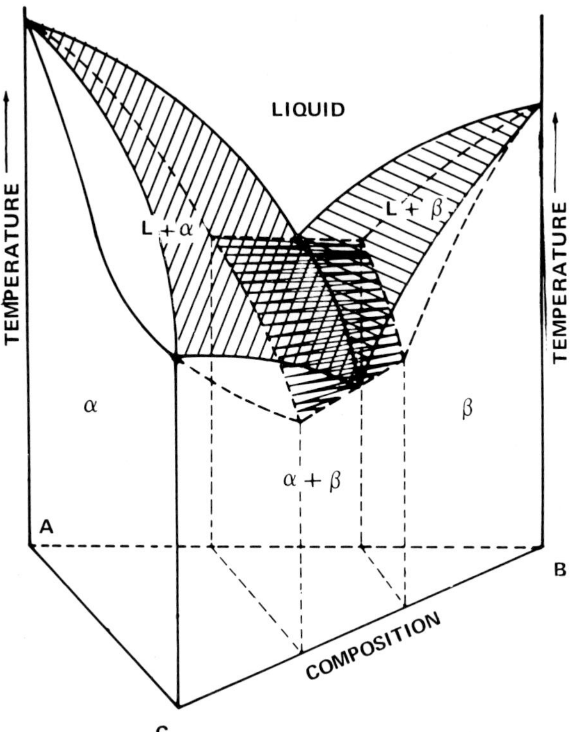

FIGURE 3.13 — Schematic diagram of ternary liquidus and solidus relationships among elements A, B, and C.

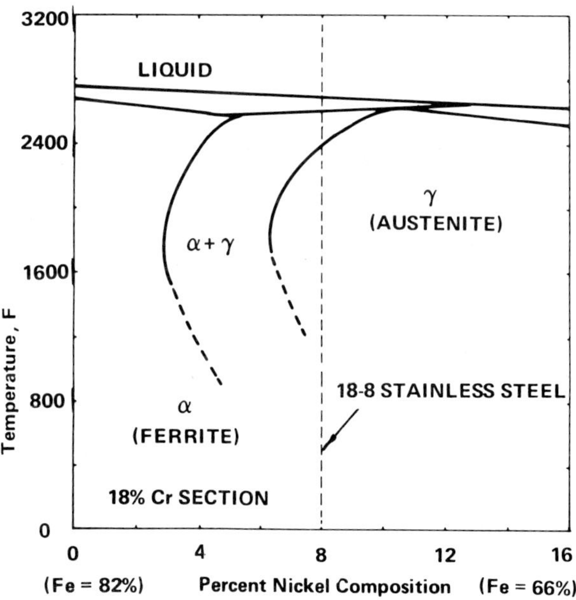

FIGURE 3.14 — Vertical section of a ternary phase diagram with constant (18%) chromium, but with nickel and iron variable.

(Figure 3.12), can be depicted on binary diagrams, in order to truly represent what happens in more complex systems, *ternary* (three-component) and *quarternary* (four-component) diagrams are necessary.

Figure 3.13 is a schematic diagram showing the ternary relationship between three elements designated as A, B, and C. The binary of these elements is shown on the face of each side of the three-dimensional system. Even with simple reactions, the internal portion of a ternary diagram becomes quite complex. In practice, it is often easier to represent a two-dimensional portion of the diagram showing the important features at a particular temperature. These are called *isothermal* (constant temperature) sections. They represent a horizontal plane through a ternary diagram showing the equilibrium phases existing at a particular temperature.

Another way of representing important areas in a ternary diagram is through the use of what is called the *vertical section*. Here, the percentage of one component may be held constant and the other two varied in such a way that a constant ratio is maintained between them.

Figure 3.14 represents the general features of the iron-rich end for a constant 18% chromium in the iron-nickel chromium system. A dotted line represents the composition, 18Cr-8Ni, which is the most common composition for austenitic stainless steels. Austenitic Fe-Cr-Ni, 300 Series stainless steels (S30000), are generally more corrosion-resistant than the Fe-Cr, 400 Series (S40000). Of course, there are some exceptions due to environment, particularly in environments which contain the chloride (Cl^-) ion. It is obvious from the diagram that nickel has a strong influence in changing the structure of the alloy from body-centered-cubic alpha (ferrite), to face-centered-cubic gamma (austenite). This section shows that austenite is essentially "stable" at room temperature for composition above 8% nickel.

However, this stability depends somewhat on the extreme sluggishness of the two transformations in these steels. A dotted line schematically represents a temperature-composition area where ferrite may exist, or if the alloy is deformed, a meta-stable martensite may form. This is the reason some austenitic stainless steels become magnetic after being subjected to a large amount of plastic deformation. In general though, these alloys retain their nonmagnetic austenitic structure.

From the foregoing discussion it should be obvious that while phase diagrams are complex, they also are extremely important to the understanding of the application, fabrication, and even prediction of nonequilibrium states of the most commonly used alloys. Reference to equilibrium diagrams for heat treatment, annealing, and fabrication information will be made later in this chapter.

Castings

Probably the best example of how the phase diagram can be applied to predicting a nonequilibrium condition is found in the prediction of inhomogeneous chemical composition as a result of nonequilibrium solidification. This inhomogeneity is very important in determining the corrosion characteristics of many cast alloys.

When an alloy cools through the solid-liquid, two-phase region, solid and liquid phases will exist with widely different compositions (Figure 3.15). Under slow equilibrium cooling, these variations disappear in the solid state due to rapid diffusion of the atoms in the solid near its melting point. The result is a homogeneous solid once the alloy has completely solidified. But what happens if cooling is rapid? Under these conditions, sufficient diffusion cannot occur and nonequilibrium variations result.

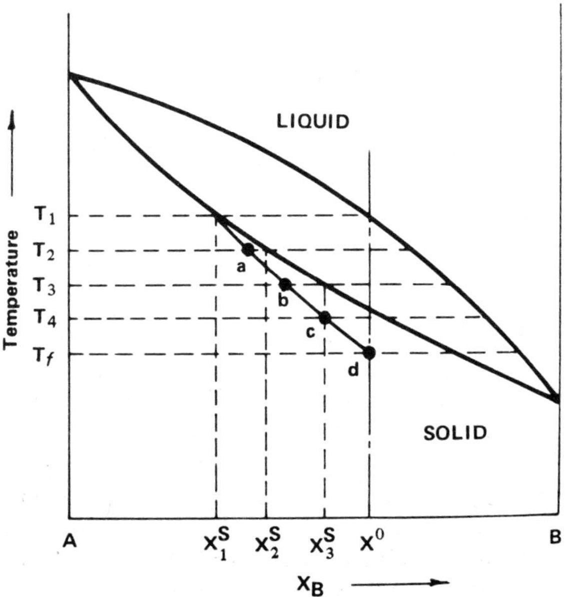

FIGURE 3.15 — Schematic diagram of nonequilibrium cooling through a two-phase region.

Consider a single grain of material rapidly solidified from a liquid. A variation in composition will exist from the interior to the surface of the grain because diffusion is not fast enough to chemically homogenize the grain as it solidifies and grows. When grains of a rapidly cooled alloy grow together, the grain boundaries have a definitely different composition than their interiors. This variation in chemical composition from the interior to the exterior of a grain is referred to as *coring*.

In some cases, it is possible to reduce major chemical differences in a cored structure by reheating the alloy and holding it for a relatively long time at a temperature just below the solidus line. This allows diffusion to occur more rapidly and assists in homogenization of the alloy. Cored structures generally have widely different corrosion characteristics and actually have a built-in electrochemical cell.

The possibility of severe coring in a cast alloy can be predicted by examining the phase diagram. It is possible also to determine the approximate temperatures for homogenization annealing. The importance of chemical homogeneity in effecting corrosion resistance for a cast, corrosion-resistant alloy cannot be overstressed. It is important also to note that rapid cooling through two-phase regions in the solid state also can result in inhomogeneous phases. Many cases of intergranular corrosion result from this type of chemical inhomogeneity.

These are examples of the important metallurgical and corrosion data contained in alloy equilibrium diagrams. This information, if used wisely, can be a tremendous help in obtaining the maximum corrosion resistance from alloys.

Mechanical Properties of Metals and Alloys

Chapter 6 of this text, entitled "Environmental Cracking," gives a detailed account of the relation between environment and stress. Many unexpected catastrophic failures have occurred due to stress-environment interactions. Knowledge of the relation between stress, mechanical properties, and stress corrosion resistance is critical to the proper application of many materials in engineering design. An understanding, or at least an awareness of this problem cannot be overemphasized.

A *mechanical property* can be defined best as a measure of the ability of a material to withstand the mechanical forces applied to it. These properties are often wrongly referred to as physical properties; the term, *physical property,* is correctly applied to electrical conductivity, magnetic properties, thermal conductivity, etc., as opposed to a mechanical property which deals strictly with the behavior of a material under mechanical loading.

Unfortunately, the ability of a metal or alloy to withstand mechanical loading often is used as the sole criterion in material selection. Only if the structure is protected from all environmental effects is the mechanical property the most important consideration. Too often, however, the relation of the metal to stress plus environment is neglected. It is necessary to relate environmental effects to the loads on a structure and to knowledge of the mechanical properties of the metal.

This section will serve only to summarize the general terminology and basic definitions of mechanical properties. Appendix A, which lists some common mechanical properties, briefly defines each term, and, where applicable, relates the properties to the corrosion characteristics of the material.

Heat Treatment of Metals

Many of the final mechanical and corrosion resistance properties of a material can be related to its heat treatment. The following section will define and discuss the most common heat treatments and their relation to mechanical strength and corrosion resistance of many metals and alloys.

Annealing

The annealing of metals is usually done to produce one of two effects:

1. Homogenization of cast alloys which may exhibit chemical inhomogeneity (coring); and
2. Remove residual stress and cold work in deformed metals.

Variations in chemical homogeneity of alloys usually are the result of rapid cooling of the cast metal. (The way in which this occurs was covered in the preceding section on phase diagrams.) The homogenization anneal is designed to eliminate or reduce this type of chemical inhomogeneity.

A metal which has been *cold worked* (plastically deformed to a different permanent shape) has experienced distortion and fragmentation of its crystalline grain structure that has produced high residual stresses. This often can be detrimental if the material is subject to stress corrosion. The second form of annealing reduces or eliminates this cold-worked condition by stress relief and recrystallization of the metal.

Homogenization Anneal

The importance of chemical homogeneity of an alloy in relation to corrosion resistance cannot be overstressed. In the as-cast condition, coring or chemical segregation often exists. Appreciable reduction of the mechanical properties, as well as reduced corrosion resistance, often occur due to this inhomogeneity.

The major purpose of this type of anneal is to produce chemical homogeneity by *diffusion* (movement of atoms in the solid state). To obtain the maximum diffusion rates, temperatures as close as possible to the melting point are usually used. For example, for most nonferrous alloys (alloys not based on iron), temperatures not far below the solidus temperature (Figure 3.10) are usually used for annealing. In ferrous alloys (based on iron), homogenization may be realized by heating to some intermediate temperature above the gamma to alpha transformation.

Length of time for homogenization anneal can vary from a few minutes to several days, depending upon the material and the inhomogeneity of the structure, as well as the economic factors involved. Alloys that have been previously cold-worked usually homogenize more rapidly than castings. Rearrangement of the disrupted crystal lattice at elevated temperatures aids in the redistribution of the various elements in an alloy. (Specific information on homogenization of particular alloys can be obtained from the general references listed at the end of this chapter.)

Annealing of Cold-Worked Material

Because of the *plasticity* of metals (their ability to deform without failure), it is possible to produce large changes in shape and cross section. Thus, hot and cold forging, rolling, drawing, etc., results in many useful fabricated shapes. Figure 3.16 (a and b) shows the typical changes in grain structure that occur under the effect of cold rolling mild (low-carbon) steel. The structure after cold working shows elongated and highly deformed grains. Figure 3.15c illustrates the grain structure of a metal after hot rolling where the grains have rapidly reformed by diffusion, are stress-free, and have uniform size.

When a metal or alloy is deformed by cold working, hardness and strength increase, but the ability to withstand more deformation decreases. Figure 3.17 shows the effect of cold working and the resulting change in properties. It is often important to be able to remove the effects of cold work. This can be done by annealing at a temperature at which the cold-worked structure can be completely eliminated by reforming of stress-free grains by diffusion.

Annealing of deformed material involves three stages: (1) recovery, (2) recrystallization, and (3) grain growth. These stages are noted in Figure 3.18, which is a series of photomicrographs showing the typical microstructure changes that occur in each stage.

In the first stage, *recovery,* the cold-worked metal recovers a slight amount of its lost ductility while losing little or none of its strength characteristics. Because higher strength properties in the cold-worked condition often are desired, a heat treatment in the recovery portion of the annealing curve is often used. As a result, the hardness and tensile properties remain relatively high, and improvement often is noted in fatigue life, toughness, and corrosion resistance. This type of partial or preanneal is called *stress relief.*

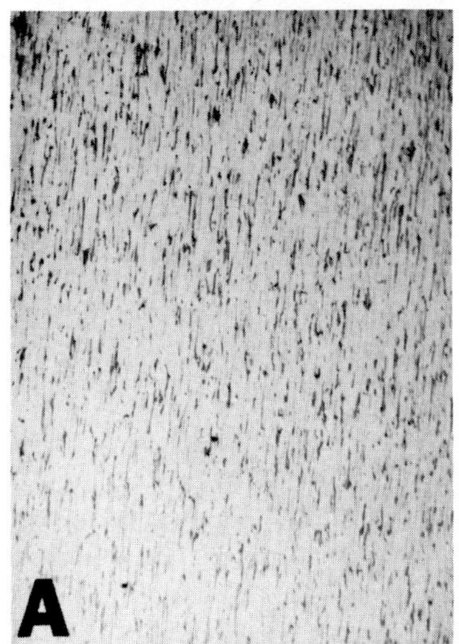

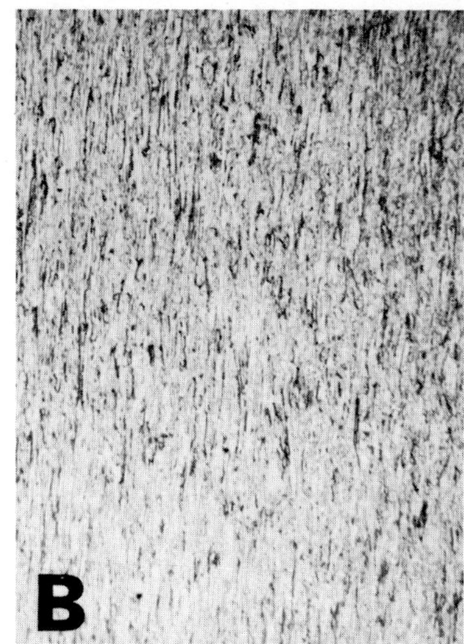

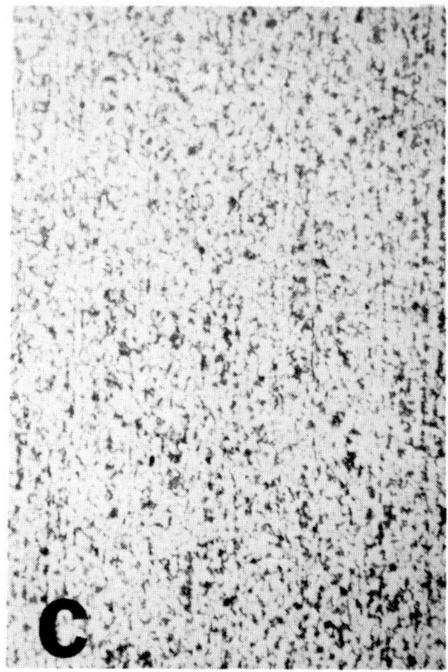

FIGURE 3.16 — Typical structure of mild (low-carbon) steel after: (a) 30% cold rolling; (b) 60% cold rolling; and (c) hot rolling.

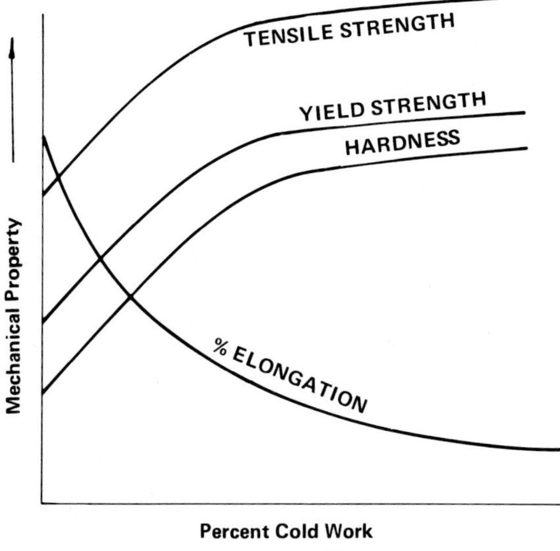

FIGURE 3.17 — Schematic curves showing typical effect of cold work on the mechanical properties of a metal.

The second stage of annealing of a cold-worked material is *recrystallization.* In this stage, new stress-free grains are initiated at the grain boundaries of the cold-worked metal. These grains grow by diffusion until new stress-free grains replace all the heavily deformed metal. Upon recrystallization, large changes in mechanical properties occur in which ductility, hardness, and other strength properties return to their pre-cold-worked level.

The last step in an annealing heat treatment results in *grain growth.* In this step, some recrystallized grains grow at the expense of others. This further increases ductility and lowers the strength and hardness of the metal. It is important to control grain size because a large grain size usually leads to difficulties in further fabrication processes and fatigue life.

Relation of Cold Work to Corrosion Resistance

In general, cold working can reduce resistance to general corrosion. In many alloys, it results in a condition more prone to stress corrosion cracking. However, evidence in recent years indicates that cold working of ultra high-purity metals has very little effect on their overall corrosion resistance. It appears that impurities and alloying elements, in addition to cold work, are responsible for diminished corrosion resistance in plastically deformed metals. However, the cost and low-strength properties of very high-purity metals rule them out for most applications.

Deformation decreases corrosion resistance basically because it increases dislocations in the cold-deformed metal. As shown in Figure 3.7, these areas are usually subject to pitting. Impurities or atoms of alloying metals often migrate to these imperfections causing an even greater change in the electrochemical characteristics of these defects. Thus, annealing a cold-worked metal may result in a decrease in dislocations and important improvements in corrosion resistance. Just as a homogenization anneal is required to produce a more uniform chemical composition, full annealing tends to produce a more uniform crystal structure with fewer defects.

Hardening Heat Treatments

Certain alloys may be subjected to heat treatments which result in improved mechanical properties. This hardening may be produced in some alloys by what is termed *age hardening* or *precipitation hardening.* In other alloys, usually steels, quenching can produce martensite, a nonequilibrium phase with extremely high mechanical strength. The next two sections will cover the essential features of both types of these hardening reactions.

In many alloy systems exhibiting a definite change of solubility of one element in another as a function of temperature, it is possible to produce a supersaturated condition by rapid cooling. With an alloy in this supersaturated condition, it is often possible to control the precipitation from the supersaturated phase by controlled reheating to improve

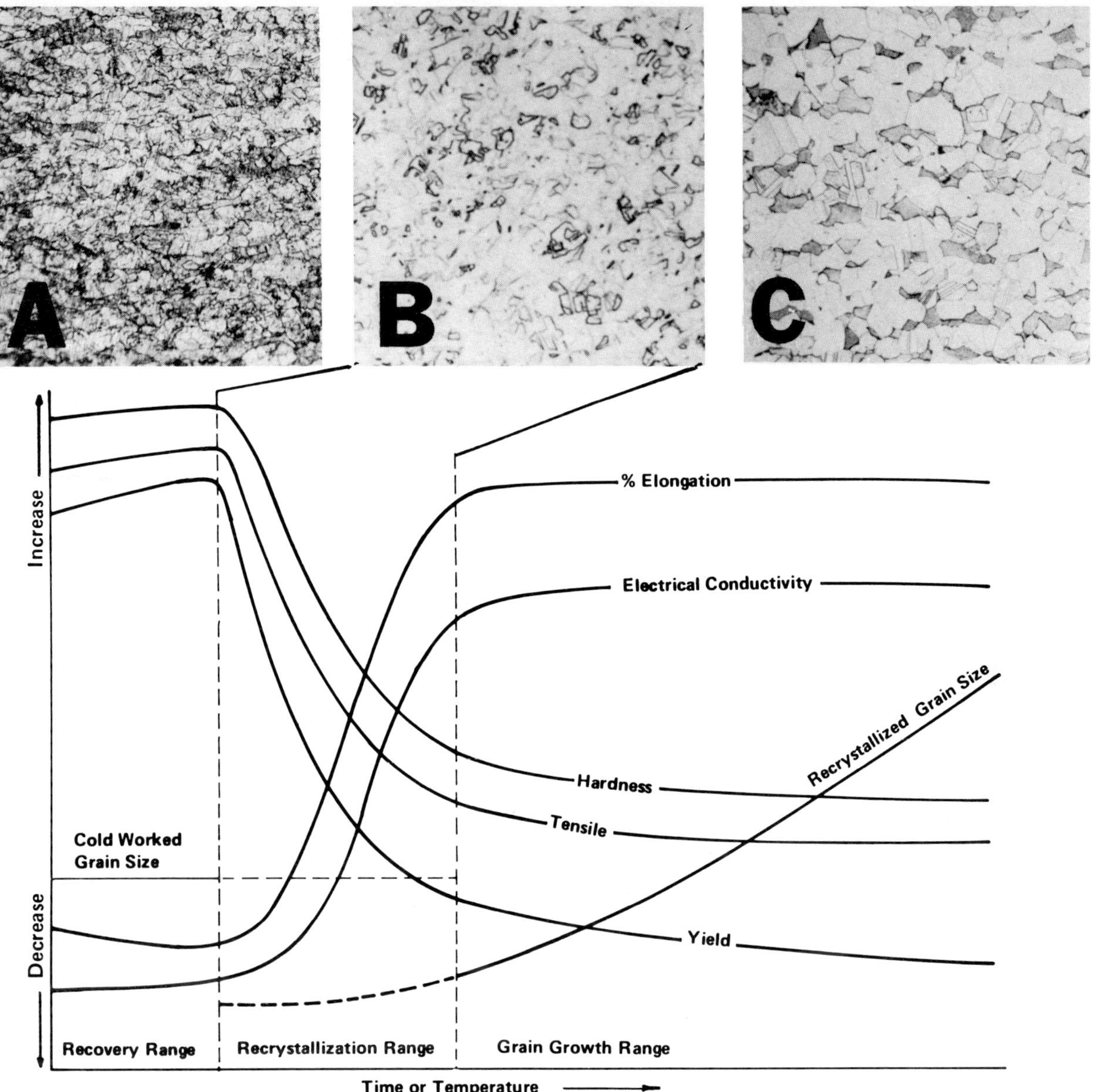

FIGURE 3.18 — Diagram showing variations in stages in annealing of cold-worked brass as functions of time and temperature. Insert micrographs of grain structures are typical of those in the: (a) recovery; (b) recrystallization; and (c) grain growth zones.

the strength. Many alloys with only average mechanical strength in their annealed condition have become extremely important because they have the ability to be strengthened by controlled precipitation from a supersaturated phase. Heat treatments which result in this type of strengthening are usually called age hardening or precipitation hardening.

Aluminum, magnesium, nickel, copper, and some forms of stainless steel, when alloyed with certain elements, can be strengthened in this manner. A number of these alloys have achieved a high degree of importance in alloy technology. Aluminum and magnesium age-hardenable alloys have become particularly important as structural components in aircraft because of their combination of lightness and strength.

Historically, Duralumin was the first precipitation hardening alloy. Duralumin is the trade name for the original German alloy developed and patented early in the 20th century. This alloy contains approximately 4% copper, which is the important element in the age hardening process of this particular aluminum alloy. Strengthening is produced by the controlled precipitation of fine, highly dispersed $CuAl_2$ precipitates. The increase in the yield strength as a result of this treatment is more than double that of the alloy in the equilibrium (slow-cooled) state.

Aluminum-copper alloys (2000 Series) have poor corrosion resistance. Basically, they are more noble than pure aluminum. However, it is well known that pure aluminum forms a thin film of aluminum oxide which provides a very effective barrier to environmental attack. This explains the extremely good corrosion resistance of aluminum, which is otherwise a rather active element.

The protective oxide that forms on pure aluminum in contact with the atmosphere also forms on aluminum alloys. On the 2000 and 7000 series age hardening alloys, the oxide is not complete in that it does not fully cover intermetallic constituents such as $CuAl_2$. As a result, the corrosion resistance of these aluminum alloys is not as good as that of pure aluminum and other aluminum alloys.

The corrosion resistance of these high-strength, age hardening aluminum alloys can be improved by using a composite aluminum product known as Alclad aluminum. The composite consists of a thin layer (usually 5 to 10% of the total composite thickness) of either pure aluminum or a special aluminum alloy that is metallurgically bonded to one or both sides of the high-strength aluminum alloy core. The cladding component is selected to be anodic to the high-strength core. As a result, the cladding will provide cathodic protection to the core.

It is possible that deterioration of the resistance to corrosion of clad materials can occur due to improper heat treatment. If the solution anneal is overextended, or several reanneals are used, diffusion of copper into the aluminum cladding will result. This occurs rapidly at the grain boundaries and reduces the ability of clad material to retain its original oxide protection where the grain boundaries intersect the surface. Thus, with an overextended annealing time or too high a solution-anneal temperature, it is possible to reduce the corrosion resistance of a clad alloy.

In the precipitation hardening aluminum alloys, the precipitating phase may cause what is referred to as a denuded zone adjacent to the grain boundaries which is more subject to corrosion attack. In various corrosive environments, this results in preferential corrosion adjacent to and at the grain boundaries. This is termed *intergranular corrosion.* An intergranular type of corrosion that occurs in lamellar-structured, age-hardened, high-strength aluminum alloys is *exfoliation.* Exfoliation proceeds inward from cut or sheared edges and forces the aluminum apart in layers or leaves. An example of exfoliation attack along the elongated grain structure of an age-hardened 2024-T6 aluminum alloy is shown in Figure 3.19.

The Martensite Transformation

In contrast to the quenching of age-hardenable alloys, which results in a soft material in the as-quenched condition, steels quenched properly from the gamma region will form a hard nonequilibrium phase before reaching room temperature. This phase is termed *martensite,* after Martens, an early German metallurgist who recognized the basic characteristics of the reaction. The extent of this transformation depends upon the composition of the steel, the size and shape of the work piece, and the quenching ability of the cooling medium.

Quenching from the gamma (face-centered cubic phase) region results in a highly supersaturated condition, since solubility of carbon in alpha iron may be more than a 100 times less than it is in gamma iron. Thus, the intermediate martensite phase is formed as the result of the enormously supersaturated condition. The martensite that forms has a highly stressed lattice which results in exceptional hardening of the steel.

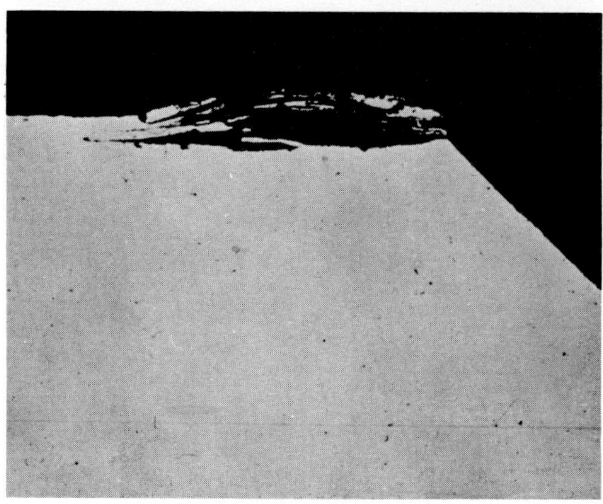

FIGURE 3.19 — Photograph of exfoliation corrosion in a 2024-T6 aluminum alloy.

In general, an as-quenched martensitic steel is extremely brittle. Many steels may crack or fracture during rapid quenching as the result of martensite formation. However, while the martensite structure is hard and brittle at room temperature, it can be softened by reheating to some intermediate temperature. This causes the metastable martensite to begin to decompose into a more stable structure. This process, called *tempering,* decreases brittleness, and although the total strength is lowered, ductility is increased and the best combination of strength and toughness results. Thus, steels are nearly always tempered following quenching.

Generally, an increase in strength and hardness due to heat treatment is accompanied by decreased corrosion resistance. If possible, hardened steels are protected in a corrosive environment by some form of surface treatment ranging from simply painting or lubricating to special plating or coating procedures.

A good example of the problem encountered when using a heat-treated, unprotected, high-strength steel is the so-called sucker rods used in oil well pumping. When salt water or hydrogen sulfide solutions are encountered, stress corrosion cracking is very likely to occur. Consequently, special alloys and extreme care in heat treatment must be used to prevent these failures.

Sensitization of Austenitic Stainless Steel

Stainless steels, particularly the 300 Series (S30000) are subject to a heat-treating effect called *sensitization.* These steels, when heated in the 427 to 760 C (800 to 1400 F) range, form chromium carbide. During short exposures at these temperatures, the chromium carbide forms only at the grain boundary. (This is analogous to the $CuAl_2$ precipitates referred to in aluminum alloy aging treat-

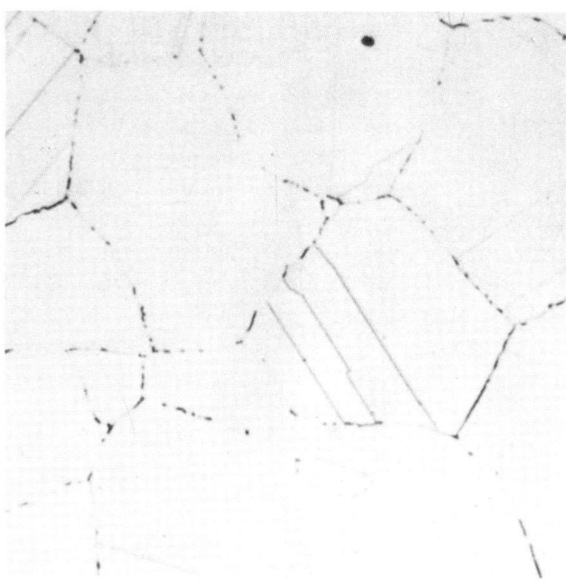

FIGURE 3.20 — Microstructure of a sensitized Type S30400 stainless steel.

ments.) Thus, the chromium near the grain boundaries is tied up as carbide and no longer can act as a deterrent to corrosion. The grain boundaries are susceptible to intergranular attack and are anodic to the surrounding grains.

Figure 3.20 shows the microstructure of a highly sensitized Type 304 (S30400) stainless steel. In this state, the most highly resistant of the 300 (S30000) series stainless steels can fail rapidly by corrosion attack in certain media. Sensitized stainless steels can deteriorate completely in the space of hours in strongly acidic solutions.

The importance of care in the heat treatment not only of stainless steels, but also other metals and alloys cannot be overstressed. The principles of heat treatment applicable to each material must be considered to produce the most effective combination of mechanical properties and environmental stability.

Welding

In this section, only a few of the many ways welding may affect the corrosion resistance properties of metals and alloys can be discussed. To fabricate complex equipment and structures for modern industry, it is necessary to produce structurally sound joints using various welding procedures.

Through the application of heat, welding may:

1. Induce phase transformations,
2. Cause secondary precipitation, and
3. Produce high stresses in and adjacent to welds which can greatly reduce corrosion resistance in these zones.

For example, welding can cause intergranular sensitization of stainless steels, as was discussed in the previous section. In the heat-affected zone adjacent to welds, sensitization can occur in austenitic stainless steels which can lead to rapid deterioration and destruction.

During welding, a large variation of thermal expansion occurs between the solidifying molten metal pool and the base metal. Upon solidification, this often can result in high residual tensile stresses. Under these circumstances, the highly stressed areas can be susceptible to stress corrosion cracking in a corrosive environment. It is important to stress relieve welds exposed to environments that could cause stress corrosion cracking. Design changes can often eliminate a weld zone which might induce stresses that could lead to stress corrosion failure.

The other basic metallurgical considerations discussed in this chapter should be considered in welding. For example, for optimum corrosion resistance, it is important to maintain homogeneity between the weld and base metals. Weld filler-metals should be used which are chemically and electrochemically (neither strongly anodic nor cathodic to the base metal) similar to the base metals. In addition, any phase transformations which could occur in welding, especially in the heat-affected zone (HAZ), must be considered in estimating the corrosion stability of welded structures.

Corrosion Prevention By Application of Metallurgical Principles

It is possible to reduce or prevent corrosion by application of the metallurgical principles discussed in this chapter, *i.e.:*

1. Use of high-purity material,
2. Use of alloying additions,
3. Effective heat treatment,
4. Surface coatings, and
5. Knowledge of the metallurgical history of the material, etc.

The following brief sections will illustrate the importance of metallurgical principles in corrosion control.

High-Purity Metals

A high-purity metal has better general corrosion resistance and a reduced tendency to pitting compared to its commercial counterpart. Of course, the limits on their mechanical properties greatly reduce their application possibilities. One of the best examples of improvement of corrosion characteristics of a material is found in results achieved by reducing the sulfur content of plain carbon steels. Corrosion attack on a steel may be greatly reduced when its sulfur level is low.

Another example where an impurity plays an extremely large role in the corrosion characteristics of a material is the effect of iron in magnesium. The solid solubility of iron in magnesium is 0.003%. If the iron is below this amount, it has little or no effect on the corrosion tendencies, but above this amount (0.003% Fe), an iron-rich phase precipitates. Because of the extremely large potential difference between the magnesium and the iron phase, the magnesium corrodes rapidly. This results in severe pitting in areas containing the iron-rich precipitate.

Lead in zinc die-cast alloys has a marked effect on the corrosion characteristics of the material. If the lead concentration exceeds 0.002%, it precipitates at the grain boundaries and produces an increased tendency for intergranular corrosion of the casting alloy.

A special series of 300 (S30000) stainless steels, 300L (S30003), has a maximum limit of 0.03% carbon. This reduces the possibility of sensitization as a result of welding and heat treatment. This gives rise to a whole series of low-carbon, or "L" grades of stainless steels which have very little or no tendency toward sensitization. There are many other similar examples of instances when minor concentrations of alloying elements have a great effect upon corrosion resistance of a metal.

The importance of keeping elements of widely different electrochemical potentials in solid solution cannot be minimized. In fact, very small amounts of impurities in solid solution can be detrimental to corrosion resistance. The amount of carbon and nitrogen in the conventional stainless steel is very small. Yet when these are reduced further in quantity by vacuum melting, an increased corrosion resistance is noted.

Alloy Additions

It is possible to improve the corrosion resistance of certain alloys by specific alloying additions. The most notable, discussed previously, is the addition of chromium to iron in amounts of 12% or more. At or above this concentration, a passive film of chromium-iron oxide, which is the basis of stainless steel's corrosion resistance, is formed on the surface.

It is also possible by controlled additions of much smaller amounts of selected alloying additions to greatly improve stability of the surface films which form on an alloy surface. The best example of this is the weathering of certain low-alloy architectural steels which form a stable protective rust after a short exposure.

Corrosion resistance also may be improved by changing the electrochemical potential of the second phase in an alloy. For aluminum alloys, this is accomplished by adding chromium or manganese. In these alloys, the intermetallic compound, $FeAl_3$, greatly affects pitting and corrosion resistance. The addition of manganese or chromium changes the $FeAl_3$ to a complex Al-Fe-Mn or Al-Fe-Cr compound whose potential now approaches that of the aluminum itself. Consequently, the pitting tendency is greatly reduced.

The intergranular sensitization of stainless steels due to the formation of chromium carbide at grain boundaries was discussed previously, but alloying for prevention of this effect was not. In these alloys, it is possible to add columbium (S34700) or titanium (S32100) which selectively tie up the carbon as carbides, eliminating chromium carbide formation and the sensitization of the alloy.

A number of alloying elements, most notably chromium and aluminum, improve the corrosion resistance of alloys by forming, or assisting in the formation of passive surface films which are resistant to corroding environments. It is possible to add aluminum or manganese to copper alloys to get passive surface films which resist corrosive attack. The addition of magnesium to aluminum, even though it may lower the potential of the alloy, improves corrosion resistance because magnesium stabilizes and thickens the natural oxide film on aluminum. The addition of chromium, particularly to nickel and iron-base materials increases their high-temperature oxidation resistance.

Additional information on alloying for corrosion resistance is discussed in Chapter 13 and is also contained in several of the references cited at the end of this chapter.

Heat Treatment

It has been shown that it is possible to modify the structure of metals and alloys in many ways through heat treatment. The improvement of cast alloys by homogenization annealing has been discussed already. Age hardening heat treatments also have a great effect on the corrosion resistance of alloys. Many alloys have a temperature range for aging in which they are not susceptible to intergranular corrosion. However, the aging temperature for optimum mechanical properties of an alloy may produce susceptibility to intergranular corrosion.

Often, it is important to sacrifice some mechanical properties in order to improve corrosion resistance. This is particularly true of alloys susceptible to stress corrosion. It has been shown that stress relief greatly improves the resistance of 300 Series (S30000) stainless steels to stress corrosion. High-zinc brasses, susceptible to stress corrosion in the cold-worked state, also have been protected effectively by stress relief.

Heat treatment of the surface to increase surface hardness or improve the stability of surface films is important in resisting all types of corrosion, but is particularly useful in improving fretting and erosion-corrosion resistance.

Usually, the producers or users of a specific alloy have carefully evaluated the various heat treating procedures to produce the maximum stability for a particular corrosion environment. This information should be examined carefully before heat treating an alloy for a specific application.

Surface Coatings for Corrosion Control

Because the corrosion reaction occurs at the metal-environment interface, it is logical that the interposition of barriers between the substrate and the environment would influence the corrosion rate. Various types of barriers are commonly used for corrosion control, including a wide range of metals and inorganic and organic materials. (Chapter 12 deals with this subject in more detail.)

As reported elsewhere in the text, metallic coatings may be of two main types: anodic or cathodic. Their function may be to confer reflectivity, minimize adhesion, provide sacrificial protection (as from zinc on steel), and in the case of certain high-temperature gaseous environments, limit the transfer of metal outward and corrosive gases inward. An example of the latter type would be the metal-ceramic coatings applied to the throats of rocket nozzles to resist the effects of high-velocity hot exhaust gases.

Various nonmetallics, usually organics, function as dielectrics, as electrical barriers between the environment and substrate, and/or to carry special inhibitive or passivating materials which tend to reduce the corrosion rate at the coating-base interface. An example of a coating that tends to inhibit corrosion of the base metal would be one containing chromates. The chromates are effective, when present in sufficient concentration, in limiting corrosive attack. (More about the mechanism of inhibitors and passivators can be found in Chapter 7.)

Relation of Metallurgical History to Corrosion

Nearly all forms of metal deterioration are dependent upon the metallurgical history of the material. Impurities retained from the original extractive processes, the inclusions and imperfections introduced in casting and forming, plus structure variations due to heat treatment, all alter the corrosion stability of the metal or alloy. Thus, a thorough knowledge of the background of the material is important. For example:

1. When steels were made in the acid open hearth, higher sulfur resulted, and, of course, corrosion resistance was lower.

2. Hot rolling a steel results in scale formation which, if not properly removed by pickling, serves to produce corrosion pit initiation sites.

3. Slow cooling of regular 300 Series (S30000) stainless steel through the 760 to 427 C (1400 to 800 F) temperature range sensitizes it to intergranular corrosion susceptibility. Rapid cooling or quenching through this range prevents sensitization.

4. Residual stress due to cold working of an alloy or unequal cooling of welded structures can lead to stress corrosion, etc.

Thus, a knowledge of the basic principles of metallurgy and an understanding of the metallurgical history of the material is extremely important in predicting the final behavior of the metal in service.

It has been beyond the scope of this particular chapter to go into details concerning the design of materials for a particular corrosion application from a metallurgical standpoint. A number of important factors relating both basic metallurgy and metallurgical history to behavior in corrosion environments have been discussed. Because corrosion of metal is so deeply involved in the basic principles of metallurgy, a few pertinent references which give additional details on the various metallurgical phenomena are cited in the bibliography.

Metallurgical References

The bibliography below lists only a few of the many comprehensive works on the various aspects of metallurgy. However, this combination of handbooks and general texts provides more than a sufficient background for the metallurgical concepts introduced in this section. Other modern metallurgical textbooks may serve as well.

No effort has been made to include corrosion texts with specific information related to metallurgy because many of these have been referenced in other chapters. The list here is intended only to provide further sources on the various aspects of metallurgy introduced in this chapter. They can be found in most city and industrial libraries.

Bibliography

Amer. Soc. for Metals. Metals Handbook. Vol. 1. Properties and Selection. Metals Park, OH, 1961.

Amer. Soc. for Metals. Metals Handbook. Vol. 2. Heat Treating, Cleaning, and Finishing. Metals Park, OH, 1964. (Volumes 7, 8, 11, and 12 are also of prime interest.)

Barrett, C. S. and T. S. Massalski. Structure of Metals. 3rd Ed. McGraw-Hill, New York, NY, 1966.

Brick, R. M., R. B. Gordon, and A. Phillips. Structure and Properties of Alloys. McGraw-Hill, New York, NY, 1965.

Chalmers, B. Physical Metallurgy. John Wiley & Sons, New York, NY, 1959.

Grossmann, M. A. and E. D. Bain. Principles of Heat Treatment. 5th Ed. Amer. Soc. for Metals, Metals Park, OH, 1964.

Guy, A. G. Elements of Physical Metallurgy. Addison Wesley, Reading, MA, 1959.

Samans, C. H. Metallic Materials in Engineering. MacMillian, New York, NY, 1963.

Smithells, C. J. Metals Reference Book. Vols. 1 and 2. 3rd Ed. Butterworths, Washington, DC, 1962.

APPENDIX A—Definitions of Metallurgical and Mechanical Property Terms

Acid Embrittlement—A form of hydrogen embrittlement which may be induced into some metals by acid cleaning treatment.

Brinell Hardness Test—A test for determining the hardness of a material by forcing a hard ball of a specified diameter into the metal under a specified load. Such hardness tests may provide some measure of mechanical properties.

Brittle Fracture—Fracture with little or no plastic deformation.

Brittleness—The property of a material which leads to the propagation of a fracture without appreciable deformation.

Cleavage Fracture—A type of fracture observed in crystalline materials which is transgranular and results in bright, reflecting facets from the cleaved surface.

Compressive Strength—The maximum compression stress a material is capable of withstanding without fracture or excessive flattening based on the load applied to its original cross-sectional area.

Creep Strength—That stress which, when applied to a material at a specific elevated temperature, will cause a specified amount of elongation.

Ductile Crack Propagation—Slow crack propagation that is accompanied by noticeable plastic deformation and requires energy to be continuously supplied from the outside to continue crack growth.

Ductility—The ability of a material to withstand significant plastic deformation prior to fracture. It is often measured by the elongation or reduction in the cross-sectional area of a tensile test specimen.

Elastic Deformation—Changes of dimensions of a material upon the application of a stress within the elastic range. Following the release of an elastic stress, the material will return to its original dimensions without any permanent deformation.

Elasticity—The property of a material which allows it to recover its original dimensions following deformation by a stress below its elastic limit.

Elastic Limit—The maximum stress to which a material may be subjected without retention of any permanent deformation after the stress is removed.

Elongation in Percent—In tensile testing, the percent increase in gage length of a specimen after fracture has occurred.

Endurance Limit—The stress below which a material can presumably endure an infinite number of cycles of stress without failure.

Fatigue Strength—The maximum stress that can be sustained for a specific number of stress cycles without failure under fatigue loading. Corrosive environments have deleterious effects on fatigue life.

Hardness—The property of a material which can be defined best by measurement of resistance to indentation or scratching.

Hydrogen Embrittlement—A condition of low ductility in certain materials as a result of the absorption of hydrogen.

Impact Strength—The amount of energy required to fracture a material under an impact load. The type of specimen, the test conditions, and particularly temperature affect the values and therefore should be specified in impact tests.

Modulus of Elasticity—A measure of the stiffness or rigidity of a material. It is actually the ratio of stress to strain in the elastic region of a material. If determined by a tension or compression test, it is also called Young's Modulus or the coefficient of the elasticity.

Plasticity—The ability of a material to deform permanently (nonelastically) without fracturing.

Residual Stress—Stress present in a material which is free from external forces. These stresses may be due to some prior mechanical deformation, phase transformation, or to nonuniform cooling.

Reduction in Area—The percent difference in the minimum cross-sectional area of a tensile specimen compared to its original cross-sectional area after fracture.

Rockwell Hardness Test—A common test for determining the hardness of a material based on the depth of penetration of a shaped indentor under a specified load.

Strain—A measure of the change in dimensions of a material when loaded compared to its original size or shape. For example, linear strain would be the change in length of part compared to original length. It is usually expressed as a percentage or in inches per inch or centimeters per centimeter.

Stress—Load or force per unit area of the cross section through which the load is acting. In equation form it is: Stress Load/Area (S = L/A).

Stress Raisers—Changes in contour or discontinuities in structure that cause local increases in stress.

Toughness—The ability of material to absorb energy and deform plastically before fracturing.

Ultimate Strength—The maximum stress that a material can sustain.

Yield Point—The stress on a material at which the first significant permanent or plastic deformation occurs without an increase in stress. In some materials, particularly annealed low-carbon steels, there is a well-defined yield point from the straight line defining the modulus of elasticity.

Yield Strength—The stress which causes a specified (usually 0.2%) deviation from elastic dimensions.

NOTES

Chapter 4

Materials

MATERIALS[1]

A large variety of materials, ranging from platinum to concrete, is used by the engineer to construct bridges, automobiles, process plant equipment, pipelines, power plants, etc. The corrosion engineer is primarily interested in the chemical properties (*i.e.,* corrosion resistance) of materials, but it is also necessary to have a knowledge of mechanical, physical, and other properties to ensure desired materials performance. The properties of engineering materials depend upon their physical structure as well as their basic chemical composition.

Mechanical Properties

Mechanical properties are related to behavior under load or stress in tension, compression, or shear. Properties are determined by engineering tests under appropriate conditions. Commonly determined mechanical properties are tensile strength, stress rupture, fatigue, elongation (ductility), impact strength (toughness and brittleness), hardness, and modulus of elasticity (ratio of stress to elastic strain—rigidity). Strain may be elastic (present only during stressing) or plastic (permanent) deformation. These properties are helpful in determining whether or not a part can be produced in the desired shape and also resist anticipated mechanical forces.

Other Properties

The corrosion engineer is often required to consider one or more properties in addition to corrosion resistance and strength when selecting a material. These include density or specific gravity (needed to calculate corrosion rates); fluidity or castability; formability; thermal, electrical, optical, acoustical, and magnetic properties; and resistance to atomic radiation. (Radiation sometimes enhances properties of a material, *e.g.,* the strength of polyethylene can be increased by controlled radiation.) The requirements for a particular part may, for example, include the ability to be cast into an intricate shape, good heat transfer characteristics, and nondegradation by atomic radiation. In another case, equipment requirements may include being a good insulator, heat reflector, and of low unit weight.

Although *cost* is not a property of a material, keep in mind that it may be the overriding factor in selection of a material for engineering use based on economic considerations.

Specifying Materials

When specifying a metal to be purchased, use the existing ASTM (American Society for Testing Materials) or other available specifications. For most metals, there exists a standard which normally identifies the analysis, form, properties, and other details of their characteristics. The use of such specifications provides a recognizable basis for the supplier to furnish the specific material desired. The various fabrication and use codes are based, with few exceptions, on these standardized materials.

Many of the nonmetallic materials may also be identified by existing standards. Use these consensus standards whenever possible. Prepare proprietary specifications only when it is known that some new material or property is required. Even in these cases, use as much standardized data as possible in the preparation of the new specification.

Most metals can be generally identified by a common name (*e.g.,* yellow brass), by a group number (*e.g.,* Copper Development Association No. 268), by the newer UNS (Unified Numbering System) number (*e.g.,* C26800), or by the ASTM specification (*e.g.,* B36 or B587). The ASTM designation is more precise and generally the most desirable. The UNS is a newer, coordinated effort devised by the ASTM and the SAE (Society of Automotive Engineers) to provide a consistent mode of identification for all alloys. A description of the alloy identifiers is given in ASTM E527. Throughout this text, the UNS number is used in parentheses following the older designator for an alloy to familiarize the reader with the newer five-place UNS system.

Metals and Alloys

Cast Irons

Cast iron is a generic term that applies to high carbon-iron alloys containing silicon. The common ones are designated as gray cast iron, white cast iron, malleable cast iron, and ductile or nodular cast iron. Ordinary gray irons contain about 2 to 4% carbon and 1 to 3% silicon. These are the least expensive of the engineering metals. The dull or grayish fracture is due to the free graphite flakes in the microstructure. Gray cast irons can be readily cast into intricate shapes because of their excellent fluidity and relatively low melting points. They can also be alloyed for improvement of corrosion resistance and strength.

ASTM specification A48 classifies gray irons as listed in Table 4.1. Their high compressive strength is indicated in the table. These materials are brittle and exhibit practically no ductility. While they do

TABLE 4.1 — Typical Mechanical Properties of Gray Iron Test Bars[1]

ASTM Class	Tensile[1] Strength, psi	Compressive Strength, psi	Fatigue Limit, psi	Brinell Hardness No.
20	22,000	83,000	10,000	156
25	26,000	97,000	11,500	174
30	31,000	109,000	14,000	201
35	36,500	124,000	16,000	212
40	42,500	140,000	18,500	235
50	52,500	164,000	21,500	262
60	62,500	187,500	24,500	302

Source: ASM Metals Handbooks, Vol. 1, p. 354 (1961).
[1]Can be improved substantially by appropriate heat treatment.

not show a clearly defined yield point, the yield strength is about 85% of the tensile strength. The modulus of elasticity in tension varies from 65 to 160 GPa (9.6 to 23.5 million psi) for these classes, with Class 20 exhibiting the lowest value and Class 60 the highest. (Theoretically, gray irons do not have a modulus of elasticity because the stress-strain curve is not a straight line.) Impact strength is generally low, but is best for material with the highest ratio of tensile strength to hardness. These irons show high damping capacity (vibration damping). Specific gravity varies with carbon content and is in the range of 7 to 7.35. Thermal and electrical conductivities are lower than pure iron.

Malleable Cast Irons

Malleable cast irons are produced by high-temperature heat treatment of white irons of suitable composition. The graphite forms as rosettes or clusters instead of flakes, and the material shows good ductility (hence the name *malleable*).

White Cast Irons

White cast irons have practically all of their carbon in the form of iron carbide. These are extremely hard and brittle. Silicon content is low because this element promotes graphitization. Graphite formation is related to rate of cooling from the melt, so chilling can produce a white iron from one that would be normally gray.

Ductile Cast Irons

These materials, also known as nodular cast irons, exhibit ductility in the as-cast form. The graphite is present as nodules or spheroids as a result of special ladle additions to the molten metal. The mechanical properties of ductile cast irons can be altered by heat treatment similar to that used for ordinary steels.

High-Silicon Cast Irons

When the silicon content of gray cast iron is increased to over 14%, it becomes extremely corrosion-resistant to many environments. The notable exception is hydrofluoric acid. In fact, these high-silicon irons are the most universally resistant of the commercial (nonprecious) metals and alloys. Their inherent hardness makes them resistant to erosion-corrosion. A straight high-silicon iron, such as Duriron,[1] contains about 14.5% silicon and 0.95% carbon. This composition must be closely controlled within narrow limits to provide the best combination of corrosion resistance and mechanical strength.

Durichlor 51[1] exhibits increased resistance to hydrochloric acid, chlorides, bleaches, chlorinated compounds, and pitting and has replaced the former molybdenum-containing alloy, Durichlor. Durichlor 51 contains chromium in addition to some molybdenum, which imparts improved resistance to oxidizing conditions. For instance, the presence of ferric or cupric chlorides in hydrochloric acid inhibits corrosion instead of causing severe selective attack as it does with most metals and alloys.

Duriron and Durichlor 51 exhibit tensile strengths around 138 MPa (20,000 psi) and hardness of Rockwell C53. The specific gravity is 7.0. Both alloys are brittle and can be machined only by grinding. Welding of the alloy is very difficult, and while simple shapes like pipe can be welded when proper precautions are observed, it is not practical to weld complex shapes.

These alloys are available only in cast form for drain lines, pumps, valves, and other process equipment (Figure 4.1). They have also found extensive use as anodes for impressed current cathodic protection (Figure 4.2).

[1]Trademark Duriron Co., Inc., Dayton, OH.

FIGURE 4.1 — High-silicon iron pump.

FIGURE 4.2 — High-silicon iron anode with an epoxy encapsulation of cable-to-anode connection.

TABLE 4.2 — Composition of Steels and Other Iron-Base Alloys

Carbon and Low-Alloy Steels[1]

Industrial carbon and low-alloy steels are generally designated by four numerical digits.[2] The first two refer to the alloy type and the next indicate the carbon content. Thus:

10XX are plain carbon steels.
11XX are carbon steels with higher than normal sulfur content for easier machining.
13XX contain 1 3/4% Mn (Manganese).
23XX contain 3% Ni (Nickel).
25XX contain 5% Ni.
31XX have 1% Ni and some chromium.
33XX have 3% Ni and some chromium.
40XX contain 1/4% Mo (Molybdenum).
41XX contain 1/4% Mo and 1% Cr.
43XX contain 1/4% molybdenum, 3/4% chromium, and 1 3/4% nickel.

50XX contains 0.3%Cr.
51XX contains 1.0% Cr.
52XX contains 1.5% Cr.
61XX contains 1% Cr and 0.15% V (Vanadium)
86XX contains 1/2% Cr, 1/2% Ni, and 0.2% Mo.
92XX contains 2% Si.
XX refers to percent carbon; thus a 4340 steel contains:

C	Mo	Cr	Ni
0.40	0.25	0.8	1.8

as well as the usual minor elements:

Mn	S	P	Si
0.7	0.040 max.	0.040 max.	0.3

[1]These metals have a prefix of G in the UNS (*e.g.,* AISI 1020 is G10200).
[2]This system, promoted by American Iron & Steel Institute and the Society for Automotive Engineers, is in use throughout the U.S.A.

The excellent corrosion resistance of high-silicon irons is due to the formation of an inert SiO_2 surface layer which forms during exposure to the environment.

Other Alloy Cast Irons

In addition to silicon and molybdenum, nickel, chromium, and copper are added to cast irons for improved corrosion, abrasion, and heat resistance and improved mechanical properties. Copper additions impart better resistance to sulfuric acid and atmospheric corrosion.

The high nickel-chromium cast irons, with and without copper (up to 7%), are the most widely used of this group. These austenitic alloys, known as Ni-Resist,[2] are the toughest of the gray cast irons. Seven varieties of Ni-Resist contain from 14 to 32% nickel and from 1.75 to 5.5% chromium and possess tensile strength from 170 to 310 MPa (25 to 45,000 psi). They are also produced as ductile cast irons with tensile strengths up to 480 MPa (70,000 psi) and elongations up to 40%. One variety contains 35% nickel and is used where low thermal expansion is required.

Ni-Hard[2] is a white cast iron containing about 4% nickel and 2% chromium. It is very hard, with a Rockwell hardness of C55 to C65. Ni-Hard has found wide application where erosion-corrosion resistance is needed in near-neutral and alkaline solutions or slurries.

Carbon Steels and Irons

The structure of steel, the principles of heat treatment, and the effect of cold working have been described in Chapter 3. A wide range of carbon steels is available to take advantage of the great diversity of properties obtainable from this most common material of construction (Table 4.2).

Commercially pure irons are ingot iron and Armco iron.[3] These are soft, weak, and not used where strength is a major requirement. However, the softness of the metal can be used to advantage in gaskets and other applications where deformation is required.

The lower carbon steels (G10050-G10100) may be severely cold-worked without cracking. Consequently, shipping cans and drums, or other items requiring extensive cold bending operations, are made from such low-strength alloys.

The "garden variety" of construction steel contains about 0.2% carbon. However, the carbon steels can vary widely in properties. Hardness and strength depend largely on their carbon content and heat treatment. Plain carbon steels exhibit mechanical properties in approximately the following ranges: tensile 275 to 1400 MPa (40 to 200 ksi); hardness, B55 to C52 Rockwell; and elongation, 5 to 50%.

Corrosion resistance of the various carbon steels is essentially the same. If any difference were noted, it would be a somewhat higher corrosion rate for the higher carbon materials in certain media (*e.g.,* strong sulfuric acid at ambient temperatures). Phase changes developed by welding or heat treatment produce greater changes in corrosion resistance than any difference in the annealed metals.

Low-Alloy Steels

Carbon steel is alloyed, singly or in combination, with chromium, nickel, copper, molybdenum, phosphorus, and vanadium in the range of a few

[2]Trademark International Nickel Co., Inc., New York, N.Y.

[3]Armco Steel Corporation, Middletown, Ohio.

TABLE 4.3, Part 1 — Wrought Stainless Steels[1]

Type	C	Cr	Ni	Other
	Composition-Percent			
Martensitic				
403	0.15 max.	12.5	0	1.0 Mn max., 0.5 Si max.
410	0.15 max.	12.0	—	—
414	0.15 max.	12.5	2.0	—
416	0.15 max.	13.0	—	—
420	0.15 max.	13.0	—	—
422	0.20 max.	13.0	0.7	1 Mo, 1 W, 0.3 V
431	0.20 max.	16.0	2.0	—
440A	0.70 max.	17.0	—	0.75 Mo max.
440B	0.85 max.	17.0	—	0.75 Mo max.
440C	1.10 max.	17.0	—	0.75 Mo max.
Ferritic				
405	0.08 max.	12.5	—	0.20 Al
430	0.12 max.	17.0	—	—
442	0.25 max.	21.0	—	—
443	0.20 max.	21.0	0.5 max.	1.1 Cu, 0.75 Si max.
446	0.20 max.	26.0	—	0.25 N max.
501	0.10 max.	5.0	0	1 Si max., 0.5 Mo
502	0.10 max.	5.0	0	1 Si max., 0.5 Mo

Type	C	Cr	Ni	Other
	Composition-Percent			
Austenitic				
201	0.15 max.	17.0	4.5	6.5 Mn, 0.25 N max.
202	0.15 max.	18.0	5.0	8 Mn, 0.25 N max.
301	0.15 max.	17.0	7.0	—
302	0.15 max.	18.0	9.0	—
303	0.15 max.	18.0	9.0	0.15 min. S or Se
304	0.08 max.	18.5	9.5	—
304L	0.03 max.	18.5	10.0	—
305	0.12 max.	18.0	12.0	—
308	0.08 max.	20.0	11.0	—
309	0.20 max.	23.0	13.5	—
310	0.25 max.	25.0	20.0	—
314	0.25 max.	25.0	20.0	2.5 Si
316	0.08 max.	17.0	12.0	2.25 Mo
316L	0.03 max.	17.0	13.0	2.25 Mo
317	0.08 max.	19.0	14.0	3.25 Mo
321	0.08 max.	18.0	11.0	(5 × C)Ti min.
322	0.12 max.	17.0	7.0	1.0 Ti, 1.0 Al
325	0.25 max.	9.0	21.0	1.3 Cu
330	0.025 max.	35.0	15.0	—
347	0.08 max.	18.0	11.0	(10 × C)Cb + Ta min.
A286	0.05	15.0	26.0	4 Mo, 1.6 Ti, 0.1 Al, 1.5 Mn

Precipitation Hardening Stainless Steels

Type	C	Cr	Ni	Other	TM[2]
		Composition-Percent			
17-4 PH	—	16.5	4	4 Cu, 0.3 Cb + Ta	AS
17-7 PH	—	17	7	1.1 Al	AS
17-10 P	0.12	17	10	0.25 P, 0.65 Si	AS
PH 15-7 Mo	—	15	7	2.5 Mo, 1.2 Al	AS
Stainless W	0.07	17	7	0.7 Ti, 0.2 Al	USS
AM 350	0.08	17	4	3.0 Mo	AL
AM 355	0.12	15.5	4.5	3.0 Mo, 0.1 N	AL
HNM	0.30	18.5	9.5	3.5 Mn, 0.25 P	Cru
Maraging (18-250)	0.03 max.	0	18	4.8 Mo, 7.8 Co, 0.4 Ti, 0.1 Al	—

[1] These stainless steels have a prefix S in the UNS (*e.g.*, AISI 304 is S30400).
[2] Trademarks: AS-Armco Steel Co.; USS-U.S. Steel Co.; AL-Allegheny Ludlum Steel Co.; Cru-Crucible Steel Co.

percent or less to produce low-alloy steels. The higher alloy additions are usually for better mechanical properties and hardenability. The lower range of about 2% total maximum is of greater interest from the corrosion standpoint. Strengths are appreciably higher than those of plain carbon steel, but the most important attribute is a better resistance to atmospheric corrosion when freely exposed (Chapter 11).

Steels with very high strengths are of particular interest for aerospace applications where a high strength-weight ratio is the overriding factor. A good example is H-11 steel (5 Cr, 1.5 Mo, 0.4 V, and 0.35 C) which can be heat-treated to tensile strengths over 2060 MPa (300,000 psi).

Stainless Steels

The main reason for the existence of the stainless steels is their resistance to corrosion. Chromium is the main alloying element, and the steel should contain at least 11%. Chromium is a reactive element, but it and its alloys passivate and exhibit excellent resistance to many environments. A large number of stainless steels are available. Their corrosion resistance, mechanical properties, and cost vary over a broad range. For this reason, it is important to specify the exact stainless steel desired for a given application.

Table 4.3, Part 1 lists the compositions of most of the common stainless steels and the four groups or classes of these materials. Group III steels are the most widely utilized, with Groups II, I, and IV following in order. The American Iron and Steel Institute (AISI) type numbers which are shown designate wrought compositions.

Group I: Martensitic

Group I materials are termed *martensitic* stainless steels because they can be hardened by heat treatment like ordinary carbon steel. Strength increases and ductility decreases with increasing hardness. Corrosion resistance is usually less than in Groups II and III. Martensitic steels can be heat-treated to obtain high tensile strengths. Corrosion resistance is generally better in the hardened condition than in the annealed or soft condition. They are used in applications requiring moderate corrosion resistance plus high strength or hardness.

Several examples are indicated in Table 4.3, Part 2. Others are valve parts; ball bearings (440A, S44002); and surgical instruments (420, S42000). These steels are not normally used in process equipment such as tanks and pipelines.

Type 416 (S41600) is easier to cut and is used for valve stems, nuts, bolts, and other parts to reduce machining costs.

TABLE 4.3, Part 2 — Nominal Mechanical Properties of Some Stainless Steels

Material	Condition	Tensile Strength, psi	Yield Point, psi 0.2% Offset	Elongation, % in 2 in.	Hardness Rockwell	Hardness Brinell
Type 410	Annealed	75,000	40,000	30	B-82	155
Type 410	Hardened, tempered at 600 F	180,000	140,000	15	C-39	375
Type 410	Hardened, tempered at 1000 F	145,000	115,000	20	C-31	300
Type 420	Annealed	95,000	50,000	25	B-92	195
Type 420	Hardened, tempered at 600 F	230,000	195,000	25	C-50	500
Type 440A	Annealed	105,000	60,000	20	B-95	215
Type 440A	Hardened, tempered at 600 F	260,000	240,000	5	C-51	510
Type 430	Annealed	75,000	45,000	30	B-82	155
Type 446	Annealed	80,000	50,000	23	B-86	170
Type 301	Annealed	110,000	40,000	60	B-85	165
Type 301	Cold-worked, 1/2 hard	150,000	110,000	15	C-32	320
Type 304	Annealed	85,000	35,000	55	B-80	150
CF-8	Annealed (15% ferrite)	87,000	47,000	52	B-80	150
Type 304L	Annealed	80,000	30,000	55	B-76	140
Type 310	Annealed	95,000	40,000	45	B-87	170
Type 347	Annealed	92,000	35,000	50	B-84	160
17-7PH	Annealed	130,000	40,000	35	B-85	165
17-7PH	Aged at 950 F	235,000	220,000	6	C-48	480
17-4PH	Aged at 900 F	200,000	178,000	12	C-44	420
14-8MoPH	Annealed	130,000	50,000	30	B-85	162
14-8MoPH	Cold-rolled, aged at 900 F	280,000	270,000	2	C-52	520
AM350	Annealed	160,000	55,000	40	B-95	215
AM350	Aged at 850 F	220,000	190,000	13	C-45	450
CD4MCu	Annealed	105,000	85,000	20	C-25	240
CD4MCu	Aged at 950 F	140,000	120,000	15	C-31	310

Group II: Ferritic

Group II is designated as *ferritic* nonhardenable steels because they do not undergo a high-temperature phase transformation, as was discussed in Chapter 3, and therefore cannot be hardened by heat treatment.

Type 430 (S43000) can be formed readily and has good corrosion resistance to the atmosphere. This is one reason why its most extensive use is for automobile trim. It is used also in ammonia oxidation plants for making nitric acid and in tank cars and tanks for the storage of nitric acid. The first chemical plant application of stainless steel was a Type 430 tank car for shipping nitric acid. However, in these chemical applications, it has been largely displaced by 18-8 because of its ease of welding and better ductility, plus better corrosion resistance.

Type 442 and 446 (S44200 and S44600) find application where heat resistance is required, such as in furnace parts and heat treating equipment. They possess good resistance to high-temperature oxidation and sulfur gas attack because of their high chromium content. These materials do not possess very good structural stability or high-temperature strength and should be selected with care.

One of the most interesting aspects of the Group II steels is their resistance to stress corrosion. They do quite well in many cases where the 18-8 types fail, particularly in chloride-containing waters.

Group III: Austenitic

Group III *austenitic* stainless steels are essentially nonmagnetic and cannot be hardened by heat treatment. Like the ferritic steels, they are hardenable only by cold working. Most of these steels contain nickel as the principal austenite former, but the relatively new ones like Types 201 and 202 (S20100 and S20200) contain less nickel and therefore more substantial amounts of manganese.

The austenitic steels possess better corrosion resistance than the straight chromium steels (Groups I and II) and generally the best resistance of any of the four groups. For this reason, austenitic steels are widely specified for the more severe corrosion conditions such as those encountered in the process industries. They are rust-resistant in the atmosphere and find wide use for architectural purposes, in the kitchen, in food manufacture and dispensing, and for applications where contamination (rust) is undesirable.

Types 201 and 202 steels show about the same corrosion resistance as the Type 302 grade. The "workhorses" for the process industries are Types 304, 304L, and 316. The molybdenum-bearing steel, Type 316, is considerably better in many applications than Type 304. Type 316 exhibits much better resistance to pitting, sulfuric acid, and hot organic acids. Corrosion resistance and heat resistance generally increase with nickel and chromium contents. For instance, Type 310 (S31000), also called 25-20, is one of the better heat-resistant alloys.

Group IV: Age- or Precipitation-Hardened

Group IV steels are hardened and strengthened by solution-quenching, followed by heating for substantial times at temperature in the approximate range of 427 to 538 C (800 to 1000 F). Some of these steels are listed in Table 4.3. Tensile strengths up to around 1375 MPa (200,000 psi) can be obtained. The first five listed find their widest application in the aircraft and missile industry under relatively mild corrosion conditions. Corrosion resistance to severe environments is generally less than that of 18-8, except for CD-4MCu which is better than the other five listed. It is also superior to the 18Cr-8Ni steels.

The higher hardness of Group IV steels reduces the tendency for seizing and galling of rubbing parts such as valve seats and disks.

Cast

Cast stainless steels and alloys usually have somewhat different chemical compositions than their wrought counterparts. One difference is higher silicon content, which increases castability. Silicon content is usually 1% maximum in wrought stainless steels. This is necessary particularly for casting thin sections or small parts.

A cast part is not rolled or deformed after casting into shape, so the microstructure can be varied considerably with regard to ferrite content for "duplex" alloys. High ferrite content in an austenitic matrix substantially increases strength. The amount and extent of ferrite phase present can be controlled by raising the percentage of ferrite formers (Cr and Mo) and by decreasing the percentage of austenite formers and stabilizers. (Ni, N, and C). (Ordinary 18-8 has enough nickel added to balance the alloy, *i.e.,* make it completely austenitic.) Producers of wrought products cannot take advantage of this situation because mixed austenite-ferrite structures cause difficulty in rolling. For these and other reasons, the ACI (Alloy Casting Institute) designations for cast grades are shown in Table 4.4.

C indicates grades used primarily for aqueous environments and *H* for heat-resistant applications. Alloy 20, a superior stainless steel, falls into the CN-7M (J95150) grade. For most environments, the corrosion resistance of cast and wrought structures are considered equivalent.

Table 4.3, Part 2 lists mechanical properties of several stainless steels in each of the four groups. Note the wide range of properties available from these materials. The high-strength materials exhibit good strength-weight ratios for aircraft and missile applications. High hardness is desirable for wear and some applications where resistance to erosion-corrosion is required (*e.g.,* trim for high-pressure steam valves).

The 200 and 300 series stainless steels all exhibit roughly the same mechanical properties in the annealed condition. The major exception involves cast alloys with duplex microstructures. Note the higher yield strength (and accordingly higher design strength) for CF-8 (J92600) containing 15% ferrite as compared with Type 304 (S30400) which has zero ferrite.

The austenitic steels retain good ductility and impact resistance at very low temperatures and are used for handling liquid oxygen and nitrogen. The only method available for hardening the austenitic stainless steels is cold working. This usually decreases corrosion resistance only slightly, but in certain critical environments, a galvanic cell could form between cold-worked and annealed material. Type 301 (S30100) is utilized mostly in the cold-worked condition for such applications as train and truck bodies. The austenitic steels can be cold-rolled to strengths in the neighborhood of 300,000 psi in wire form. Types 301 and 302 (S30100 and S30200) are not used for severe corrosion applications, the former because of its lower Cr and Ni, and the latter because of higher carbon.

One of the main uses for stainless steels in the chemical industry is for handling nitric acid. Figure 4.3 is an isocorrosion chart showing corrosion of 18-8 (Type 304 or CF-8) by nitric acid as a function of temperature and concentration. The lines in Figure 4.3 are constant corrosion rate lines. Many data are needed to delineate these lines. Passivity

TABLE 4.4 — Cast Stainless Steels[1]

Corrosion (C) and Heat (H) Resistant Alloys

Cast Alloy Designation[2]	Composition-Percent (Balance Fe) Cr	Ni	Other Elements	Cast Alloy Designation[2]	Composition Percent (Balance Fe) Cr	Ni	Other Elements
CA-15	11.5-14	1 max.	—	CK-20	23-27	19-22	—
CA-40	11.5-14	1 max.	—	CN-7M	18-22	21-31	Mo-Cu
CB-30	18-22	2 max.	—	CY-40	14-17	Bal.	Fe 11.0 max.
CB-7Cu	15.5-17	3.6-4.6	Cu 2.3-3.3				
CC-50	26-30	4 max.	—	HA	8-10	—	M0 0.90-1.20
CD-4MCu	25-27	4.75-6.00	2 Mo, 3 Cu	HC	26-30	4 max.	—
CE-30	26-30	8.11	—	HD	26-30	4-7	—
CF-3	17-21	8-12	—	HE	26-30	4-7	—
CF-8	18-21	8-11	—	HF	19-23	9-12	—
CF-20	18-21	8-11	—	HH	24-28	11-14	N o.2 max.
CF-3M	17-21	9-13	Mo 2.0-3.0	HI	26-30	14-18	—
CF-8M	18-21	9-12	Mo 2.2-3.0	HK	24-28	14-18	
CF-8C	18-21	9-12	Cb(8 × C) min., 1.0 max.	HL	28-32	18-22	—
CF-16F	18-21	9-12	Mo 1.5 max., Se 0.20-0.35	HT	13-17	33-37	—
CG-8M	18-21	9-13	Mo 3.04-4.0	HU	17-21	37-41	—
CH-20	22-26	12-15	—	HW	10-14	58-62	—
				HX	15-19	64-68	

(1)These alloys have a prefix J in the UNS (*e.g.,* ACI CA-115 is J91150

(2)Numbers greatly refer to "pints" of carbon (really hundreds of percent).

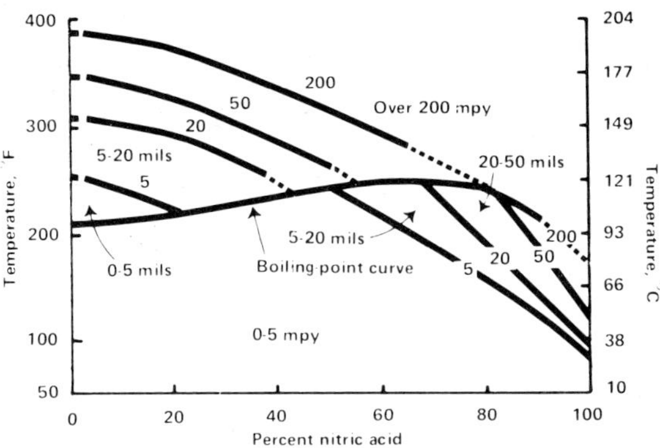

FIGURE 4.3 — Isocorrosion chart showing variations in corrosive attack on Type 304 stainless steel (18Cr-8Ni) as influenced by temperature and changing nitric acid concentration.

breaks down at high temperatures and high acid concentrations.

Aluminum and Its Alloys

Aluminum is a reactive metal, but it develops an aluminum oxide coating or film that protects it from corrosion in many environments. This film is quite stable in neutral and many acid solutions, and it can be artificially increased by passage of electric current. This process is called *anodizing*.

The high-copper alloys are utilized mainly for structural purposes. When using these alloys (>2.5% Cu), considerations must be given to their protection even in the atmosphere. The copper-free or low-copper alloys are used in the process industries or where better corrosion resistance is required.

In addition to corrosion resistance, other properties contributing to its widespread application are colorless and nontoxic corrosion products, appearance, electrical and thermal conductivity, reflectivity, and lightness or good strength-weight ratio.

Pure aluminum is soft and weak, but it can be alloyed and heat-treated to a broad range of mechanical properties, as was discussed earlier in Chapter 3. Strengthening usually decreases corrosion resistance, particularly resistance to stress corrosion, and for this reason the structural alloys are Alclad[4] or covered with a thin skin of pure aluminum.

Table 4.5 lists compositions of several wrought and cast alloys. The 3000, 5000, and 6000 series alloys are widely used in the process industries. Alloys are sandcast, die-cast, or cast into permanent molds. This table also lists mechanical properties of some aluminum alloys. This table illustrates the wide range of properties available. For example, the tensile strength of annealed commercially pure aluminum is 90 MPa (13,000 psi) and 600 MPa (88,000 psi) for heat-treated 7178 (A97178) alloy. Alloy 5052 (A95052) exhibits the highest strength of non-heat-treatable alloys. No. 7178 is one of the highest strength heat-treatable aluminum alloys utilized in aircraft and aerospace applications.

Aluminum alloys lose strength rapidly when exposed to temperatures of 177 C (350 F) and higher. Aluminum shows excellent mechanical properties at sub-zero temperatures.

[4]Trademark Aluminum Co. of America, Pittsburgh, PA.

TABLE 4.5 — Aluminum Alloys[1]

System for Numbering Wrought Alloys		No.
Aluminum, Al	99 + %	1xxx
Copper, Cu	as major alloying element	2xxx
Manganese, Mn	as major alloying element	3xxx
Silicon, Si	as major alloying element	4xxx
Magnesium, Mg	as major alloying element	5xxx
Mg and Si	as major alloying element	6xxx
Zinc, Zn	as major alloying element	7xxx
Other	as major alloying element	8xxx

F-as fabricated	W-solution heat treated
O-annealed	T-heat treated other than H,
H-strain hardened	O, or W(T-1 to T-10 are different degrees of temper)

Thus some of the more common aluminum alloys are as follows:

No. Designation	Wrought Alloys[1] Composition-Percent (Balance Aluminum) Cu	Mg	Si	Other
1100	0	0	0	99 + Aluminum
2011	5.5	0	0	0.8 Mn
2017	4.0	0.5	0	0.5 Mn
2024	4.5	1.5	0	0.6 Mn
2217	6.3	0.02	0.2	
3003	0.2	0	0	1.3 Mn
4043	0.3	0.05	5.5	
5052	0	2.5	0	0.25 Cr
5183	0.1	4.8	0	0.2 Cr, 0.7 Mn
6061	0.25	1.0	0.6	0.25 Cr
7075	1.6	2.5	0	5.6 Zn, 0.3 Cr
7178	2.0	2.8	0	6.8 Zn, 0.2 Ti, 0.3 Cr

No. Designation	Cast Alloys[1] Composition-Percent (Balance Aluminum) Cu	Mg	Si	Other
12	7.0	0	2.5	2.0 Zn, 0.2 Ti
13	0	0	12.0	
43	0	0	5.0	
A-132	0.8	1.2	12.0	2.5 Ni
142	4.0	1.5	0	2.0 Ni
195	4.5	0	0.8	
B-195	4.5	0	2.5	
214	0	3.8	0	
218	0	8.0	0	
220	0	10.0	0	
333	4.0	0	9.0	
380	3.5	0	9.0	0.8 Fe

[1]The aluminum alloys have the prefix A in the UNS (*e.g.*, 1100 is A91100 and cast 12 is A02012).

Magnesium and Its Alloys

Magnesium is one of the lightest commercial metals, with a specific gravity of 1.74. It is utilized in trucks, automobile engines, ladders, portable saws, luggage, aircraft, and missiles because of its light weight and also its good strength when alloyed. However, it is one of the least corrosion-resistant and is accordingly used as sacrificial anodes for cathodic protection (Chapter 9) and dry cell batteries. It is generally anodic to most other metals and alloys and must be insulated from them.

Magnesium exhibits good resistance to ordinary atmospheres due to the formation of a protective oxide film. This protection tends to break down (pit) in air contaminated with salt, so protective measures are required. These include coatings and "chrome" pickling, which also provides a good base for a coating. Corrosion resistance generally decreases with impurities and alloying. Alloys are quite susceptible to stress corrosion and must be protected. Presence of dissolved oxygen in water has no significant effect on corrosion. The metal is susceptible to erosion-corrosion. Magnesium is much more resistant than aluminum to alkalies. It is attacked by most acids except chromic and hydrofluoric. The corrosion product in HF acts as a protective film.

Magnesium and its alloys are available in a variety of wrought forms and die castings. Tensile strengths in the approximate range of 100 to 340 MPa (15,000 to 50,000 psi) are obtainable.[5]

Lead and Its Alloys

Lead is one of the oldest metals used by man. It was used for water piping during the time of the Roman Empire, and some of it is still in operation. Lead ornaments and coins were utilized several thousand years ago. Lead forms protective films consisting of corrosion products such as sulfates, oxides, and phosphates. Lead can be used in sulfuric acid service (Figure 4.4). Lead and its alloy may be used as piping, sheet linings, solders (Pb-Sn), type metals, storage batteries, radiation shields, cable sheath, terneplate (steel coated with Pb-Sn alloy), bearings, roofing, and ammunition. Lead is soft, easily formed, and has a low melting point. Lead-lined steel is often made by "burning on" the lead. It is subject to erosion-corrosion because of its softness.

When protection against corrosion is required for process equipment, chemical lead containing about 0.06% copper is specified, particularly for sulfuric acid. This lead is resistant to sulfuric, chromic, hydrofluoric, and phosphoric acids in certain strengths, neutral solutions, seawater, and soils. It is rapidly attacked by acetic acid and generally is not used in nitric, hydrochloric, or organic acids.

Chemical lead exhibits a tensile strength of about 16 MPa (2300 psi) at room temperature. Hard leads, containing 3 to 18% antimony double this strength. However, the strength of both materials drops rapidly as temperature increases, and they show about the same strength around 110 C (230 F). Design strength at appreciably higher temperatures drops to zero.

[5]For a comprehensive discussion of magnesium and its alloys, refer to Metals Handbook, 8th ed., Vol. 1, pp. 1067-1112, American Society for Metals, 1961.

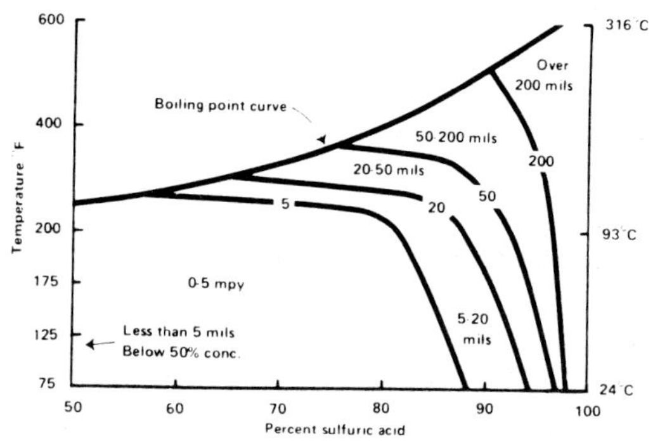

FIGURE 4.4 — Isocorrosion chart for chemical lead in stagnant sulfuric acid showing effects of temperature and concentration.

Copper and Its Alloys

Copper is different from other metals in that it combines corrosion resistance with high electrical and heat conductivity, formability, machinability, and strength for use in urban, marine, and industrial atmospheres and waters. Copper is a relatively noble metal, and hydrogen evolution is not usually part of the corrosion process. For this reason it is not corroded by acids unless oxygen or other oxidizing agents (*e.g.*, HNO_3) are present. For example, reaction between copper and sulfuric acid is not thermodynamically possible, but corrosion proceeds in the presence of oxygen, and the products are copper sulfate and water. Reduction of oxygen to form hydroxide ions is the predominant cathodic reaction for copper and its alloys.

Copper-base alloys are resistant to neutral and slightly alkaline solutions with the exception of those containing ammonia, which cause stress corrosion and sometimes rapid general attack. In strongly reducing conditions at 300 to 400 C (575 to 750 F), copper alloys are often superior to stainless steels and stainless alloys.

Table 4.6 lists chemical compositions and mechanical properties of some common copper-base materials. Hundreds of compositions with a wide variety of mechanical properties are commercially

TABLE 4.6 — Nominal Composition of Some Common Copper Alloys[1]

Name	Composition-Percent Cu	Zn	Sn	Other
Admiralty Metal	71	28	1.0	0.06 Sb or As
Aluminum Bronze[2]	90	0	0	7 Al, 2.5 Fe
Architectural Bronze	57	40	0	3 Pb
Beryllium Bronze	98	0	0	1.9 Be, 0.25 Co
Cartridge Brass	70	30	0	—
Commercial Bronze	90	10	0	—
Cupronickel[2]	69	0	0	30 Ni, 1 Fe, 0.5 Mn
Gilding Metal	95	5	0	—
Muntz Metal	61	39	0	—
Naval Brass	60.5	38.7	0.75	—
Nickel Silver	65	17	0	18 Ni
Phosphor Bronze[2]	Balance	0	5.0	0.2 P
Red Brass	85	15	0	—
Silicon Bronze[2]	97	0	0	3 Si
Yellow Brass	65	35	0	—

[1]The UNS numerical designations for copper alloys eliminate some of the ambiguities inherent in the traditional nomenclature given in this table and used in this book. The UNS prefix is C (*e.g.*, wrought yellow brass is C26800).
[2]There are numerous variations to this and some other alloys.

available. For example, a range of tensile strength from about 205 to 1350 MPa (30,000 to 200,000 psi) is exhibited by pure copper and copper alloyed with 2% beryllium. The most common alloys are brasses (Cu-Zn), bronzes (Sn, Al, or Si additions to Cu), and cupronickels (Cu-Ni). Nickel silver (an alloy of Cu, Ni, and Zn) does not contain silver, but resembles it in appearance. Everdur,[6] a silicon bronze, possesses relatively high strength, and this property is utilized in hardware such as nuts, bolts, and valve stems for copper-base equipment, thus minimizing two-metal corrosion.

Copper and brasses are subject to erosion-corrosion or impingement attack. The bronzes and aluminum brass are much better in this respect. The bronzes are stronger and harder. The cupronickels with small iron additions are also superior in erosion-corrosion resistance.

Copper and copper alloys are available as duplex tubing or plate (one metal inside, another outside) in combination with steel, aluminum, and stainless steels. This construction solves many heat-exchanger materials problems, *e.g.,* steel tubing carrying ammonia and immersed in brackish water (admiralty metal).

Copper and its alloys find extensive application as water piping, pumps, valves, heat-exchanger tubes and tube sheets, hardware, wire, screens, shafts, roofing, bearings, stills, tanks, and other vessels.

Nickel and Its Alloys

An important group of materials for corrosion applications is based on nickel. Nickel is resistant to many corrosives and is a natural for alkaline solutions. Most tough corrosion problems involving caustic and caustic solutions are handled with nickel. In fact, the corrosion resistance of alloys to sodium hydroxide is roughly proportional to their nickel content. For example, 2% Ni cast iron is much superior to unalloyed cast iron.

Another important attribute is the large and rapid increase in stress corrosion resistance as the nickel content of stainless alloys exceeds 30%. For example, Inconel[7] shows excellent stress corrosion resistance, and many tons of it are used for this reason. Nickel generally shows good resistance to neutral and slightly acid solutions. It is widely used in the food industry. It is not resistant to strongly oxidizing solutions, *e.g.,* nitric acid and ammonia. Nickel forms a good base for alloys requiring strength at high temperatures. However, nickel and its alloys are attacked and embrittled by sulfur-bearing gases at elevated temperatures.

Table 4.7 lists some common nickels and nickel-base alloys. A wide range of compositions and mechanical properties is available. Aged Duranickel[7] possesses very good mechanical properties and also good resistance to many environments. Monel is a natural for hydrofluoric acid. Chlorimet 3[8] and Hastelloy C[9] are two of the most generally corrosion-resistant alloys commerically available. Chlorimet 2 and Hastelloy B are very good in many cases where oxidizing conditions do not exist. Nichrome[7] is used for electrical resistors (heating elements). Nickel, high-nickel alloys, and alloys

[6]Trademark Anaconda American Brass Co., Waterbury Connecticut.

[7]Trademark International Nickel Co., Inc., New York, N.Y.

[8]Trademark Duriron Co., Inc., Dayton, Ohio.

[9]Trademark Cabot Corporation, Kokomo, Indiana.

TABLE 4.7 — Nominal Composition and Mechanical Properties of Some Nickel and Other Alloys

Material	Ni	C	Cr	Percent Mo	Cu	Fe	Other	Tensile Strength psi	Yield Point psi	% Elongation	Rockwell Hardness
Nickel 200	99.5	0.06	—	—	0.05	0.15		60,000	20,000	40	B-50
Nickel 210 (cast)	95.6	0.80	—	—	0.05	0.50	1.6 Si, 0.9 Mn	50,000	25,000	20	B-55
Duranickel 301	94	0.15	—	—	0.05	0.15	4.5 Al, 0.5 Ti	170,000[1]	130,000	15	C-34
Monel 400	66	0.15	—	—	31	1.4		75,000	30,000	40	B-65
Monel K-500	66	0.15	—	—	29	0.9	3 Al	150,000[1]	110,000	25	C-28
Monel 505 (cast)	63	0.10	—	—	30	2	4 Si	120,000[1]	90,000	2	C-31
Inconel 600[2]	76	0.08	16	—	0.2	8		85,000	35,000	40	B-70
Chlorimet 2 (cast)[5]	62	0.07	—	32	—	3	1 Si	80,000	55,000	20	B-90
Hastelloy B (wrought)[6]	62	0.10	—	28	—	5	1 Si	130,000	56,000	50	B-92
Chlorimet 3 (cast)[5]	60	0.07	18	18	—	3	0.6 Si	75,000	50,000	25	B-90
Hastelloy C (wrought)[6]	56	0.08	15	17	—	5	1 Si, 4W	120,000	52,000	49	B-94
Incoloy 825[2]	42	—	22	3	1.8	30	1 Ti	90,000	35,000	50	B-91
Ilium G (cast)[3]	56	0.07	22.5	6.5	6.5	6		68,000	38,000	7	B-70
Nichrome V[2]	80	—	20	—	—	—		95,000	35,000	30	B-85
Carpenter 20Cb3[4]	34	0.07 max.	20	2.5	3.5	Bal.	1 Si				
Durichlor D-51	—	0.9	4	0.5	—	Bal.	14.5 Si				
Duriron[5]	—	0.9	—	—	—	Bal.	14.5 Si				
Hastelloy X[6]	Bal.	0.2 max.	22	9	—	18	1.5 Co, 1 W				
Ni Resist (No. 1)[2]	15.5	3.0 max.	2	—	6.5	Bal.	2 Si, 1.25 Mn				
Multiphase MP 35N[7]	3.5	—	20	10	—	Bal.	35 Co				
Zircaloy -2	0.05	—	0.1	—	—	0.12	1.5 Sn, Bal. Zr				

[1]Age hardened.
[2]Trademarks of the International Nickel Co., Inc., New York, N.Y.
[3]Trademark of Stainless Foundry and Eng. Co., Inc., Milwaukee, WI.
[4]Trademark of Carpenter Steel Co.
[5]Trademarks of Duriron Co., Inc., Dayton, OH.
[6]Trademarks of Cabot Corporation, Kokomo, IN.
[7]Trademarks of Standard Pressed Steel Co.

containing substantial amounts of nickel (8%) are the most common materials used for most of the more severe corrosion problems.

Zinc and Its Alloys

Zinc is not a corrosion-resistant metal, but it is utilized as a sacrificial metal for cathodic protection of steel. Its chief use is in galvanized (zinc-coated) steel for piping, fencing, nails, etc. It is also utilized in the form of bars or slabs as sacrificial anodes to protect ship hulls, pipelines, and other structures.

Zinc alloy parts are made by die-casting because of their low melting points. Many automobile components such as grilles and door handles are die-cast, but are usually plated with corrosion-resistant metals.

Tin and Tinplate

Over half of the tin produced goes into coating other metals, primarily steel to make the "tin" can. In addition to offering corrosion resistance, tin-coated steel is easily formed and soldered, provides a good base for organic coatings, is nontoxic, and has a pleasant appearance.

Tin coating can be applied by dipping, but most of it is electroplated. Two advantages of the latter method are controlled thinness and the fact that steel can be plated with a different thickness on each side of a sheet, usually thicker on the inside of the can than on the outside. Tin cans are used for food products, beverages, petroleum products, and paints. Alloys of tin with zinc, nickel, cadmium, or copper can be deposited as electroplates.

Tin is cathodic to iron in most common environments, but the potential reverses in most sealed cans containing food products and tin acts as a sacrificial coating, thus protecting the steel. Complex ion formation apparently causes this reversal. Tin is relatively inert, but in the presence of oxygen or other oxidizing agents, it is attacked.

Tin shows excellent resistance to relatively pure water. Solid tin pipe and sheet and also tinned copper are utilized for producing and handling distilled water. Collapsible tubes are used for dentifrices and medicants. Solders are another important outlet for tin in which tin and lead contents vary from 30 to 70%. Tin babbits are used for bearings.

Tin shows good resistance to atmospheric corrosion, dilute mineral acids in the absence of air, and many organic acids, but is corroded by strong mineral acids. It is generally not used for handling alkalies.

Tin is a weak, soft, and ductile metal. Tensile strength at room temperature is around 17 MPa (2500 psi), and this drops rapidly with increasing temperature.

Cadmium

Except for low melting alloys and some bearings, cadmium is used almost exclusively as an electroplated coating. Its bright appearance and ease of soldering accounts for its use in electronic equipment and hardware. Cadmium is less electronegative than zinc and not as effective as a sacrificial anode in many cases. It is somewhat less protective than zinc for steel in industrial atmospheres. Cadmium is more expensive than zinc and its salts are toxic. Unlike zinc, cadmium has some measure of resistance to alkalies.

Cadmium plating is utilized on high-strength steels in aircraft because of improved resistance to corrosion fatigue. However, hydrogen embrittlement is a problem. Proper plating techniques and baking out the hydrogen after plating have solved this problem. At temperatures in the neighborhood of the melting point (321 C), steel can be attacked by cadmium.

Cadmium is a relatively weak metal with a tensile strength around 70 MPa (10,000 psi) and elongation of 50%.

Titanium and Its Alloys

Titanium is a relative newcomer in that it was first used as a structural metal in 1952. It is strong and has a specific gravity of 4.5, about halfway between aluminum and steel. It has a high strength-to-weight ratio and accordingly was first adapted for aircraft and ordnance. In addition, it is now utilized in the space and chemical process industries. Titanium is a reactive metal and depends on a protective film (TiO_2) for corrosion resistance. Melting and welding must be done in inert environments or the metal becomes brittle due to absorbed gases.

Titanium has three outstanding characteristics which account for many of its applications in corrosive services. These are resistance to: (1) seawater and other chloride salt solutions; (2) hypochlorites and wet chlorine; and (3) nitric acid, including fuming acids. Salts such as $FeCl_3$ and $CuCl_2$, which tend to pit most other metals and alloys, actually inhibit corrosion of titanium. It is not resistant to relatively pure sulfuric and hydrochloric acids, but does a good job in many of these acids when they are heavily contaminated with heavy metal ions such as ferric and cupric. Titanium shows surprisingly low galvanic effects because it readily passivates.

Titanium is catastrophically attacked in red fuming nitric acid with high NO_2 and low water content and also in dry halogen gases.

Alloying with about 30% Mo greatly increases resistance to hydrochloric acid. Small amounts of tin reduce scaling losses during hot rolling. Small additions of palladium, platinum, and other noble metals increase resistance to moderately reducing acids. One such commercial titanium alloy contains about 0.15% palladium. Other commercial alloys contain aluminum, chromium, iron, manganese, molybdenum, tin, vanadium, and zirconium.

Table 4.8 lists mechanical properties of titanium and a few commercial alloys. The alloys fall into three metallurgical types, as indicated in the

TABLE 4.8 — Nominal Composition and Mechanical Properties of Titanium and Titanium Alloys

Material	Type Phase	Condition	Tensile Strength, psi	Yield Strength, psi	Elongation, %
Commercial Ti	a	Annealed	85,000	70,000	26
5 Al, 2.5 Sn	ab	Annealed	125,000	120,000	18
8Mn	ab	Annealed	140,000	125,000	15
4 Al, 3 Mo, 1 V	ab	Heat-treated	195,000	165,000	6
6 Al, 4 V	ab	Annealed	135,000	120,000	11
6 Al, 4 V		Heat-treated	170,000	150,000	7
3 Al, 13 V, 11 Cr	b	Heat-treated	180,000	170,000	6

table. Titanium has a low modulus of elasticity (115 GPa-16,800,00 psi) as compared with iron (200 GPa-28,500,000 psi).

Castings of commercial titanium are available as pumps, valves, etc., and the utilization of castings is increasing as manufacturing techniques improve and costs are lowered.

Refractory Metals

Refractory metals are characterized by very high melting points, as compared with iron and steel. The jet engine and outer-space programs have provided the impetus in making these metals commercially available. Unfortunately, they show poor resistance to high-temperature oxidation, and protective coatings are needed (Chapter 13).

Columbium, molybdenum, tungsten, and zirconium are relative newcomers in the field of corrosion by aqueous solutions. Many alloys of these metals are commercially available. Tantalum has long been used for severely corrosive conditions. Melting points of these metals are shown in Table 4.9.

Columbium

Columbium exhibits good corrosion resistance to organic and inorganic acids except hydrofluoric and hot concentrated sulfuric and hydrochloric acids. Apparently the formation of a Cb_2O_5 film results in protection. Columbium is poor in alkaline solutions.

TABLE 4.9 — Melting Points of Refractory Metals

Metal	Melting Point F	C
Chromium	3434	1890
Columbium	4474	2468
Molybdenum	4710	2600
Rhenium	5755	3180
Tantalum	5396	2980
Tungsten	6170	3410
Zirconium	3366	1860
Iron (for comparison)	2798	1537

Molybdenum

Molybdenum shows good resistance to hydrofluoric, hydrochloric, and sulfuric acids, but oxidizing agents such as nitric acid cause rapid attack. It is good in aqueous alkaline solutions. This metal forms a volatile oxide (MoO_3) in air at temperatures above about 700 C (1300 F).

One great advantage of molybdenum from the mechanical standpoint is its high modulus of elasticity, which is 345 GPa (50,000,000 psi). This means it is much stiffer and will deflect much less than steel under a given load and cross section.

Tantalum

Tantalum has been used for many years because of its superior resistance to most environments. A few exceptions include alkalies and hydrofluoric and hot concentrated sulfuric acid. It is used in handling chemically pure solutions of such corrosives as hydrochloric acid. Because of tantalum's wide spectrum of corrosion resistance, it is utilized in repair of glass-lined equipment. Any evolution of hydrogen (corrosion reaction or otherwise) near tantalum will result in absorption and embrittlement of this metal. Desorption of hydrogen to restore ductility is not practical because of the high temperature and high vacuum required. Tantalum sheet is strong, so cost is reduced through the use of thin sections. Tantalum is used for surgical implants.

Tungsten

As indicated in Table 4.9, tungsten has the highest melting point of the metals. Its chief use involves strength at high temperatures. A common example is filaments in light bulbs. Tungsten shows good resistance to acids and alkalies, but it is not commonly used for aqueous solutions. Tungsten exhibits a tensile strength of 138 MPa (20,000 psi) at 1649 C (3000 F).

Zirconium

Zirconium has found increased use following the advent of nuclear energy, primarily because it has a low thermal neutron cross section and resists high-temperature water and steam. Its excellent corrosion resistance is due to a protective oxide film.

This metal exhibits good corrosion resistance to alkalies and acids (including hydriodic and hydrobromic acids), except for hydrofluoric acid and hot concentrated hydrochloric and sulfuric acids. Cupric and ferric chlorides cause pitting. Zirconium has found some application in hydrochloric acid service. Its corrosion resistance is affected by impurities such as nitrogen, aluminum, iron, and carbon.

Zirconium alloyed with small amounts of tin, iron, chromium, and nickel (Zircaloys) shows improved resistance to high-temperature water. Zirconium and its alloys in high-temperature water first show a decreasing corrosion rate that may be followed by a rapid linear rate of attack (termed *breakaway*). They also tend to pick up hydrogen from the corrosion reaction and become brittle. Zirconium possesses a tensile strength of about 110 MPa (16,000 psi) at 427 C (800 F) and 550 MPa (80,000 psi) at room temperature. Its modulus of elasticity is 95 GPa (13,700,000 psi).

Noble Metals

Noble materials are characterized by highly positive potentials relative to the hydrogen electrode, excellent corrosion resistance, and high cost. The latter accounts for the term *precious* metals. In most cases, formation of protective films (passivity) is not required. The noble metals are gold, silver, platinum, and the other five "platinum" metals: iridium, osmium, palladium, rhodium, and ruthenium. Gold, silver, platinum, and palladium are available in most commercial forms. The first three are used extensively in industry. The other noble metals are utilized chiefly as alloying elements, *e.g.*, Pt-Rh thermocouple wire.

The use of noble metals for jewelry is well known. In spite of their high cost, the noble metals are the most economical for numerous corrosion applications. Their high scrap value is an advantageous characteristic. Cladding, linings, and coatings with inexpensive substrates providing strength are common combinations.

Rhodium and osmium electroplate well and are used for critical valve parts and other applications requiring total resistance to an aggressive environment.

Gold

Gold is one of the oldest metals utilized by man because it is found in nature in the pure state. Jewelry and coinage represent its first applications, with jewelry accounting for most of its modern consumption. For jewelry applications, gold is alloyed with copper because of the extreme softness of pure gold. A karat is 1/24 part, so pure gold is 24 karats and 12 karat would be 50% gold. In addition, gold and its alloys are used for dental inlays, electrical contacts, plating, tableware, process equipment (*e.g.*, condensers and stills), printed circuits, gold leaf, in surgical and body implants, and decorative purposes in signs and displays. A thin gold underplate (topped with cadmium) prevents hydrogen embrittlement of high-strength steels. Gold leaf and foil are readily produced because of the softness and excellent malleability of pure gold.

Gold is very good in dilute nitric acid and hot strong sulfuric acid. It is attacked by aqua regia, concentrated nitric acid, chlorine and bromine, mercury, and alkaline cyanides.

Platinum

Platinum is used because it is resistant to many oxidizing environments and particularly to air at high temperatures. Platinum is also used for thermocouples (Pt-PtRh), tanks, spinnerets for molten glass, crucibles for chemical analytical work, windings for electrical resistance furnaces (alloys up to 1760 C), and combustion or reaction chambers for extremely corrosive environments at temperatures over 982 C (1800 F). Sheet linings on a base alloy substrate are used for the latter. An applied oxide layer (*e.g.*, Al_2O_3) prevents alloying or reaction with the substrate material. It has replaced fused quartz in several chemical applications of this type. Incidentally, contact with silica at very high temperature embrittles the platinum. The inertness of platinum is attested to by its extensive service as a catalyst.

Other uses for platinum and its alloys are spinnerets (70 Au, 30 Pt) for rayon, safety or frangible disks up to 482 C (900 F) [silver is good only up to 149 C (300 F) and gold only up to 71 C (160 F)], sulfuric acid absorbers, electroplating anodes, impressed-current anodes for cathodic protection, other chemical equipment, and high-quality jewelry. It is not attacked by mercury.

Platinum exhibits good mechanical properties at high temperatures, as shown in Table 4.10. Gold and silver, however, are poor in this respect.

Platinum is attacked by aqua regia, hydriodic and hydrobromic acids, ferric chloride, and chlorine and bromine.

Silver

Silver is best known for its use as coinage and tableware in solid and plated form. Sterling silver contains not more than 7.5% copper for hardness. Silver loses its "nobility" in contact with sulfur compounds and is evident by tarnishing. It serves as electrical contacts, electrical bus bars (even at red heat), brazes, solders, and dental alloys with mercury. Silver is widely used in the chemical industry as solid silver and also as loose, clad, or brazed linings. Applications include stills, heating coils and condensers for pure hydrofluoric acid, evaporating pans for production of chemically pure anhydrous sodium hydroxide, autoclaves for production of urea, and all kinds of equipment for production of foods and drugs where purity of the product is paramount. It is highly resistant to organic acids.

TABLE 4.10 — Some Mechanical and Physical Properties of Noble Metals

Material	Temperature F	Temperature C	Tensile Strength, psi	Yield Strength, psi	Elongation, %	Brinell Hardness No.	Modulus of Elasticity	Melting Point, F
Silver	Room	Room	18,000	8,000	55	26	11,000,000	1761
Gold	Room	Room	19,000	nil	70	25	11,600,000	1945
70 Au-30 Pt	Room	Room	29,000	3,500	—	130	16,500,000	2640
Platinum	Room	Room	21,000	< 2,000	40	40	21,000,000	3217
Platinum	750	400	13,000					
Platinum	1830	1000	4,000					
Platinum	2190	1200	2,400					

Silver is attacked by nitric acid, hot hydrochloric acid, hydriodic and hydrobromic acids, mercury, and alkaline cyanides, and may be corroded by reducing acids if oxidizing agents are present.

Nonmetallics

Natural and Synthetic Rubbers

The outstanding characteristic of rubber and other elastomers is resilience, or low modulus of elasticity. Flexibility accounts for most applications such as tubing, belting, gasketing, and automobile tires. However, chemical and abrasion resistance and good insulating qualities result in many corrosion applications. Rubber and hydrochloric acid form a natural combination in that rubber-lined steel pipes and tanks have been used for this service for many years.

Generally speaking, the natural rubbers have better mechanical properties (resiliency and resistance to cuts and their propagation) than the synthetic or artificial rubbers, but the synthetics have better corrosion resistance.

Natural Rubber

Organically speaking, rubber is a long-chain molecule of isoprene (polyisoprene). It comes from trees as a liquid latex. The coil-like structure of these molecules accounts for the high elasticity (100 to 1000% elongation). Soft rubber has a temperature limitation of around 70 C (160 F). This temperature limit can be raised to about 82 C (180 F) by hardening through alloying (compounding). Adding sulfur and heating makes the rubber harder and more brittle. Charles Goodyear first discovered this process, called *vulcanization,* in 1839.

About 50% sulfur results in a hard product, known as ebonite, which is used to make bowling balls. Semihard and hard rubbers are used for tires, tank linings, and many other items. Corrosion resistance usually increases with hardness. In the case of lining pipes and tanks, the rubber is usually applied soft and then cured in place (for large items) or in autoclaves. Modulus of elasticity varies roughly from 3 to 3500 MPa (500 to 500,000 psi) for soft and hard rubber, respectively.

Synthetic Rubbers

At the beginning of World War II, when the main sources of natural rubber were taken over by the enemy, there was great interest in developing substitutes. Neoprene was developed in the early thirties by Du Pont, but it was in small supply in 1941. Neoprene was one of the five strategic materials in World War II—the other four were metals.

A wide variety of synthetic rubbers is available, including combinations with plastics. Plasticizers, fillers, and hardeners are compounded to obtain a large range of properties, including elasticity, and temperature and corrosion resistance. Table 4.11 illustrates these points. Note the variations obtainable in hardness, elongation, tensile strength, elasticity, and temperature, tear, and corrosion resistance. Neoprene and nitrile rubber possess resistance to oils and gasoline. One of the first extensive applications of neoprene was, and is, gasoline hoses.

An outstanding characteristic of butyl rubber is impermeability to gases. This accounts for its use as inner tubes and process equipment such as seals for floating-top storage tanks. Butyl rubber exhibits better resistance to oxidizing environments such as air and dilute nitric acid. The temperature resistance shown in Table 4.11 is for air. Note 304 C (580 F) for silicone rubber. Imide and perfluoro elastomers exceed this service temperature. Temperatures listed are considerably reduced in some corrosives—to room temperature for natural rubber in 70% sulfuric acid, for example. Neoprene and nitrile-rubber-lined vessels handle pure and strong sodium hydroxide.

Soft rubbers are best for abrasion resistance. A common mistake, because of metals-oriented thinking, is to use hard rubber for erosion-corrosion conditions. Linings also may consist of hard and soft layers.

One of the newer elastomers is a perfluoro compound known as Kalrez.[10] Other fluorinated products were often considered to be comparable to the TFE Teflon plastic in chemical stability when first introduced. This was certainly not true. How-

[10]Trademark E.I. du Pont de Nemours & Co., Wilmington, Delaware.

TABLE 4.11 — Property Comparisons of Some Natural and Synthetic Rubbers

Property	Natural Rubber	Butyl (GR-I)	Buna S (GR-S)	Neoprene	Nitrile (Buna N)	Polyacrylic Rubber	Silicone Rubber	Urethane
Hardness range (Shore "A")(1)	40-100	40-90	40-100	30-90	45-100	50-90	40-80	35-100
Tensile strength, psi(2)	4500	3000	3500	3500	4000	1500	900	4000
Max. elongation, %	900	900	600	1000	700	200	250	600
Abrasion resistance(3)	Excellent	Good	Excellent	Very good	Excellent	Fair	Poor	Excellent
Resistance to compression set at 158 F (70 C)(3)	Good	Fair	Excellent	Good	Excellent	Good	Excellent	—
Resistance to compression set up to 250 F (121 C)(3)	Poor	Poor	Excellent	Fair	Excellent	Good	Excellent	—
Aging resistance (normal temp.)	Good	Excellent	Excellent	Excellent	Excellent	Excellent	Excellent	—
Max. ambient temp. allowable, °F in air	160	275	275	225	300	400	580	240
°C in air	71	135	135	107	149	204	304	115
Resistance to weather and ozone(3)	Fair	Very good	Fair	Excellent	Fair	Excellent	Excellent	Good
Resistance to flexing	Excellent	Excellent	Good	Excellent	Fair	Excellent	Poor	—
Resistance to diffusion of gases	Fair	Excellent	Fair	Very good	Fair	Good	Poor	
Resilience	Excellent	Poor at low temp. Good at high temp.	Fair	Very good	Fair	Poor	Good	Poor
Resistance to petroleum oils and greases	Poor	Poor	Poor	Good	Excellent	Very good	Good	Excellent
Resistance to vegetable oils	Good	Excellent	Fair	Excellent	Excellent	Good	Excellent	Excellent
Resistance to nonaromatic fuels and solvents	Poor	Poor	Poor	Fair to good	Very good	Excellent	Fair	Excellent
Resistance to aromatic fuels and solvents	Poor	Poor	Poor	Fair	Good	Fair	Poor	Excellent
Resistance to water and antifreezes(3)	Good	Good	Good	Fair	Excellent	Poor	Fair	Poor
Resistance to dilute acids	Good	Good	Good	Good	Good	Fair	Good	Fair
Resistance to oxidizing agents	Poor	Fair	Poor	Poor	Poor	Excellent	Excellent	Good
Resistance to alkali	Fair	Fair	Fair	Good	Fair	—	—	Poor
Dielectric strength(3)	Excellent	Good	Excellent	Fair	Fair	Fair	Excellent	Fair
Flame resistance	Poor	Poor	Poor	Good	Poor	Poor	Good	Good
Processing characterics	Excellent	Good	Good	Good	Good	Fair	Poor	Good
Low temp. resistance(3)	Very good	Fair	Good	Fair	Good	Poor	Excellent	Fair
Tear resistance(3)	Excellent	Excellent	Good	Good	Good	Fair	Poor	Fair

(1)100 Durometer reading is bone hard and indicates that ebonite or hard rubber can be made.
(2)Indicates soft rubber type. Hard rubber types run higher in value.
(3)These properties available in specific compounds.

ever, the Kalrez product approaches this ideal and is the first elastomer to have such a wide range of resistance.

The chemistry and compounding of all types of rubbers are quite complex and information is not readily available to consumers. The best procedure is to discuss the situation at hand with suppliers in order to obtain the proper material for a given problem. If corrosion testing is involved, it is imperative that the specimens be representative of the actual installation. Furthermore, if lining is involved, the test specimens should be covered, as bond failure is sometimes the primary cause of unsatisfactory performance. (All of the statements in this paragraph also apply to plastics.)

Plastics

Production and utilization of plastics have increased tremendously during the past 30 years. One of the early incentives for development of these materials was the need for an inexpensive substitute for ivory billiard balls. Plastics are now available in a wide variety of forms, including billiard balls, pumps, valves, pipe, fans, nosecones, airplane canopies, telephones, hosiery, radio cabinets, insulation, bushings, drawer rollers, waxes, pot handles, heart valves, and other body implants. Plastics are produced by casting, molding, extrusion, and calendering. They are available as solid parts, linings, coatings, foams, fibers, and films.

The American Society for Testing and Materials states: "A plastic is a material that contains as an essential ingredient an organic substance of large molecular weight, is solid in its finished state and at some stage in its manufacture or in its processing into finished articles, can be shaped by flow." In other words, they are high-molecular-weight organic materials that can be formed into useful articles.

Some plastics occur in nature, but most are produced synthetically. In general, plastics, as compared with metals, are much weaker, softer, more resistant to chloride ions and hydrochloric acid, less resistant to concentrated sulfuric and oxidizing acids such as nitric, less resistant to solvents, and most have definitely lower temperature limitations. Cold flow, or creep, at ambient temperatures is a problem, particularly with the thermoplastics.

Plastics are readily divided into two classes, *thermoplastics* and *thermosets.* The former soften with increasing temperature and return to their original hardness when cooled. Most are meltable; for example, nylon is extruded into fibers or filaments from the molten state. The thermosets harden when heated and retain hardness when cooled. They "set" into permanent shape by catalysis or when heated under pressure. Generally, they cannot be reworked by scrap.

Tables 4.12 and 4.13 list some properties and the corrosion resistance of several well-known plas-

TABLE 4.12 — Mechanical and Physical Properties of Some Plastics at Room Temperature

Material	Tensile Strength, psi	Elongation %	Hardness, Rockwell	Impact, Izod, ft.-lb.	Modulus of Elasticity, psi × 10^3	Specific Gravity	Heat Distortion Temp. °F/264 psi[(1)]	Coeff. of Expansion 10^{-5} in/in°C
Thermoplastics								
ABS	7,000	5-30	100 R	2.0-8	300	1.04	215	10.0
Acetals	9,000	25-75	120 R	1.2	500	1.42	240	8.2
Cellulosics	5,000	50-100	80 R	0.4-10	220	1.22	150	15.0
Fluorocarbons	2,500	100-350	70 R	4.0	60	2.13	270	10.0
Methyl methacrylate	8,000	5	220 R	0.5	420	1.19	200	7.0
Nylon	10,000	45	110 R	1.5	400	1.14	325	8.2
Polycarbonates	9,000	100-130	75 M	16.0	325	1.20	275	7.0
Polyethylene (low density)	2,000	90-800	10 R	16.0	25	0.92	100	10.0-22
Polyethylene (high density)	4,000	15-100	40 R	1.0-12	120	0.95	120	12.0
Polyimides	11,000	6	50 E	1.0	450	1.43	680	4.0
Polyphenylene oxide	8,500	50	116 R	5.0	360	1.06	240	5.2
Polyphenylene sulfide	10,000	3	125 R	0.5	480	1.36	278	5.5
Polypropylene	5,000	10-700	90 R	1.0-11	200	0.91	150	9.0
Polystyrene	7,000	1-2	75 R	0.3	450	1.05	180	7.0
Polysulfones	11,000	50-100	120 R	—	360	1.24	345	5.4
Rigid Polyvinyl Chloride	6,000	2-30	110 R	1	400	1.4	150	8.0
Vinyls (flexible)	2,500	100-450	80 R	good	low	1.18	145	—
Thermosets								
Epoxy	10,000	nil	90 R	0.8	1,000	1.1	350	1.0
Phenolics	7,500	nil	125 R	0.3	1,000	1.4	300	1.5
Polyesters	4,000	nil	100 R	0.4	1,000	1.1	350	1.5
Silicones	3,500	nil	89 R	0.3	1,200	1.75	350-900	1.0
Ureas	7,000	nil	115 R	0.3	1,500	1.48	265	—
Furanes	11,000	nil	110 R	—	650	1.45	420	1.0

(1)According to ASTM Specification D 648-56, *e.g.*, the heat distortion temperature of fluorocarbons is 270 F when loaded at 264 psi. The 264 psi is a constant applied against all materials.

TABLE 4.13 — Suitability of Various Plastics Versus Environmental Factors

Material	Acids Weak	Acids Strong	Alkalies Weak	Alkalies Strong	Organic Solvents	Water Absorp. %/24h	Oxygen and Ozone	Ionizing Radiation	Temperature Resistance, F High	Temperature Resistance, F Low[(1)]
Thermoplastic Materials										
ABS	A	A-O	R	R	P	0.25	F	—	150	—
Acetals	F	A	F	A	F	0.25	—	—	200	—
Cellulosics	A	A	A	A	P	1.2	—	—	150	—
Fluorocarbons (TFE)	inert	inert	inert	inert	inert	0.0	inert	P	550	-275
Methyl methacrylate	R	A-O	R	A	A	0.2	R	P	180	—
Nylon 6	G	A	R	R	R	1.5	SA	F	300	-70
Polycarbonates	R	A	A	A	F	0.16	—	—	250	—
Polyethylene (low density)	R	A-O	R	R	G	0.15	A	F	140	-80
Polyethylene (high density)	R	A-O	R	R	G	0.1	A	G	160	-100
Polyimides	G	A	A	A	G	0.32	—	R	550	-400
Polyphenylene oxide	R	G	R	R	P	0.10	—	—	200	—
Polyphenylene sulfide	R	F	G	G	R	0.01	—	—	450	—
Polypropylene	R	A-O	R	R	R	<0.01	A	G	300	P
Polystyrene	R	A-O	R	R	A	0.04	SA	G	160	P
Polysulfones	G	G	G	G	F	0.22	—	—	330	-150
Rigid polyvinyl chloride	R	R	R	R	A	0.10	R	P	150	P
Vinyls (chloride)	R	R	R	R	A	0.45	R	P	160	—
Thermoset Materials										
Epoxy (cast)	R	SA	R	R	G	0.1	SA	G	400	L
Phenolics	SA	0.6 A	SA	A	SA	0.6	—	G	400	L
Polyesters	R	SA	A	A	G	0.2	A	G	350	L
Silicones	SA	SA	SA	SA	A	0.15	R	F	550	L
Ureas	A	A	A	A	R	0.6	A	P	170	L
Furanes	R	R	R	G	R	0.1	A	—	300	L

NOTE: R = resistant, A = attacked, SA = slight attack, A-0 = attacked by oxidizing acids, G = good, F = fair, P = poor, L = little change.
(1)Numbers in this column refer to − °F at which the indicated reaction occurs, *e.g.*, high-density polyethylene has good resistance at temperatures as low as −100 F.

tics. A wide range of properties is available. These properties can be changed considerably through plasticizers, fillers, and hardeners.

Plastics do not generally dissolve like metals. Degradation or corrosion occurs from three basic sources. Some active species from an environment can be absorbed into the plastic to swell or react internally with the polymer chains. Softening and distortion normally develop, although actual loss of weight from the plastic can occur. Oxidation of the resinous molecule can occur in the atmosphere or other oxidizing conditions. This often results in hardening and cracking of the plastic. Continued polymerization of the resin can also occur with certain resinous components resulting in hardening, shrinkage, and cracking of the material.

The important aspect of the corrosion mechanism is that degradation is not a surface effect, but occurs internally in a plastic. Also, be aware of the fact that a plastic material will absorb even parts per million of an aggressive agent from an otherwise innocuous stream. The total loss of mechanical properties can result from such attack.

Plastic materials will react the same as metals under many circumstances. They will stress corrosion crack, fatigue, be eroded, cavitate, and creep as do the metals. (Further comment regarding corrosion of the plastic is given in Chapter 14.)

Thermoplastics

ABS (acrylonitrile-butadiene-styrene)

ABS resin is used extensively in the production of pipe for the distribution of gas and water. Many irrigation and sprinkling systems are installed in ABS. The material is available in a high impact grade that is used widely for telephones, tool housings, automotive parts, and toys.

Poor resistance to sunlight has been a characteristic of past polymers, but new alloys are said to have overcome this problem.

The ABS and PVC resins have made possible the low cost distribution of water to outlying districts throughout the country. Many attractive chrome-plated parts of cars and home furnishings are also made from the plastic.

Acetals

Acetals are tough, strong polymers used for many pipe and plumbing parts. The good stability of the resins in cold and hot water recommends their use for many household and appliance parts. Gears, cams, and bushings are often made of the tough material for appliance and automotive parts.

The material is not resistant to continued ultraviolet exposure. However, steaming of the plastic for long periods of time has no effect.

Acrylics

Methyl methacrylate (Lucite[11] and Plexiglas[12]) is the best known of this group. It is used for brush and purse handles, transparent displays, working models, and airplane canopies and tailights. It is attractive and transmits and conducts light. The acrylics are soft, easily scratched, and not very temperature resistant.

Cellulosics

Cellulosics are best represented by the cellulose acetate-butyrate product which has long been used for piping to convey natural gas, oils, brines, and sour crudes. It is unaffected by exposure in the atmosphere over long periods of time.

Care must be taken to assure that streams of crude products do not contain aromatic hydrocarbons or oxygenated solvents which will readily attack the plastic.

Fluorocarbons

The fluorocarbons are composed of polymer chains composed of carbon and fluorine, although some chlorine atoms are present in certain polymers. The first polytetrafluoroethylene (TFE) was produced by Du Pont and designated Teflon.[12] It is the "noble metal" of the plastics field with corrosion resistance to practically all environments up to 285 C (550 F).

The very inertness of TFE made it difficult to fabricate. Now there are a number of Teflons available, each with differing mechanical and corrosion resistance properties. Thus, the appropriate fluorocarbon plastic must be identified for a specific application. A number of companies now produce fluorocarbon plastics.

In addition to corrosion resistance, the low coefficient of friction of Teflon is the basis for a very successful application of the plastic illustrated in Figure 4.5. A confined Teflon sleeve separates the plug and body of this plug valve and acts as a lubricant, thereby preventing sticking or freezing of the metal parts. Ordinary petroleum greases that might be used to lubricate moving parts such as these are attacked by many environments, and consequently cannot be used in corrosive environments. Other uses are seals and gaskets, wire insulation, expansion joints, linings for pipe, tape, tubing, valve diaphragms, coatings, and even heat exchangers (thin-walled tubing).

Nylon

Nylon is a group of materials best known for its use in hosiery, but also finds much use for strictly corrosion-resistant applications. Its main applications depend on strength, low coefficient of friction, and wear resistance. These are gears, drawer and shelf rollers, fishline, sutures, tennis racket strings, dentures, automobile door catches, pipe fittings, and brush bristles. It is also utilized for electrical insulation and permits higher temperature operation than rubber.

FIGURE 4.5 — Nonlubricated, coated plug valve with TFE (Teflon) sleeve.

[11]Trademark E.I. du Pont de Nemours & Co., Wilmington, Delaware.

[12]Trademark Rohm and Haas Co., Philadelphia, Pennsylvania.

Polycarbonates

Like the acetal and polysulfone resins, the polycarbonates have good dimensional stability with temperature changes and a high modulus for a thermoplastic. Thus, the material is widely used in automobile and aircraft parts, lighting fixtures, tool housings, and electrical parts where properties and good impact strength are desired.

The plastic in the stressed condition (residual from molding stresses or applied) can stress crack in certain very mild chemical exposures such as ketone vapors. Consequently, the stress relief of fabricated parts may be desirable for many applications.

Polyethylenes

Polyethylene materials represent the greatest tonnage of all the plastics and are marketed by many producers. They are used for packaging film, sheet, squeeze bottles, pans, tumblers, ice trays, and many other household items. Piping for water and other chemicals is common. Certain solvents produce stress corrosion cracking of polyethylene. Considerable latitude in properties is possible, as indicated by Tables 4.12 and 4.13.

Polypropylenes

Polypropylene originated in Italy. It exhibits better heat and corrosion resistance and is stiffer than polyethylene. It is made into valves and bottles to be sterilized by heat, pipe, rope, fittings, and scooter parts.

Polyimides

Polyimide materials have the highest heat stability of any of the thermoplastics. Continuous use of the materials up to temperatures of 285 C (550 F) is possible, with shorter excursions to temperatures as high as 480 C (900 F) permissible. Some 90% of the tensile strength is retained after 1000 hours exposure at air temperatures of 300 C (575 F).

The material also has exceptional resistance to low-temperature embrittlement. Use of the plastic in cryogenic systems operating at temperatures as low as −250 C (−420 F) has been successful. No other thermoplastic has such a range of temperature stability.

The plastic also has an unusual resistance to ionizing radiation where most thermoplastics become embrittled. Evaluation of the material for use around radiation sources should be considered. Compressor parts, valve parts, bearings, and parts for electrical insulation have also been made from the material. The material is resistant to fungus attack, but loses strength in sunlight exposures.

Polyphenylene Sulfide

Polyphenylene sulfide has unique corrosion resistance (Table 4.13). Only strong oxidizing agents appear to attack the material. The outstanding corrosion resistance combines with good strength, high modulus, and good heat stability to provide a useful product for many chemical exposures.

The material has been used for pumps, valves, piping, coatings, insulators, and other applications in severe industrial exposures.

Polystyrene

Polystyrene is used for wall tile, battery cases, flowmeters, radio cabinets, bottle closures, and refrigeration equipment. It possesses good chemical resistance, but is too brittle for many structural applications. Polystyrene shows good resistance to hydrofluoric acid.

Polysulfones

The good heat stability, toughness at high and low temperatures, and good electrical insulating properties brought the polysulfones quickly into substantial use both in home and industry. Electric coffee pots, housing for many electric and electronic devices, sight glasses, microwave oven internals, and chemical processing equipment are but some of the past applications for this plastic.

The material can be continuously steamed without difficulty, but weathering in sunlight will produce some change in properties. Like polyethylene, polycarbonate, and some other thermoplastics, the polysulfones can be stress cracked when in the stressed condition. Chlorinated or oxygenated chemicals are known cracking agents when sufficient stress is present.

Polyvinyl Chlorides (PVC)

The most extensive use of plastic piping, particularly for water service, involves the PVC materials. A "high impact" grade is normally used. The lightness, strength, resistance to attack in soils and waters, good acid and alkali resistance, and ease of joining have made the material a standard for many utility operations and industrial applications.

The product is resistant to essentially all inorganic acid, alkali, and salt exposures, oils, and alcohols. Most aliphatic and aromatic chemicals will cause rapid attack. The material has a relatively low modulus and weakens quickly with increasing temperatures.

The vinylidene chlorides (Saran)[13] are related to the PVC materials, but contain a higher percentage of chlorine in the polymer chains. There are also pliable, soft, vinyl copolymers (usually polyvinyl chloride-acetate) which are used for tubing, tile, packaging, phonograph records, garden hose, etc.

Thermosets

Epoxies

Epoxies represent perhaps the best combination of corrosion resistance and mechanical proper-

[13]Trademark Dow Chemical Co., Midland, Michigan.

FIGURE 4.6 — Epoxy sink and laboratory bench top.

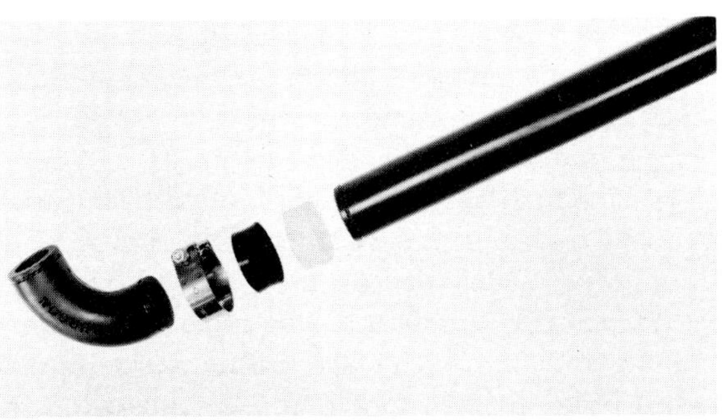

FIGURE 4.7 — Epoxy pipe and fitting, TFE (Teflon) inner sleeve, Neoprene outer sleeve, and stainless steel clamp.

ties. Epoxies are available as castings, extrusions, sheet, adhesives, and coatings. They are used as sinks, bench tops, pipe, valves, pumps, small tanks, potting compounds, adhesives, linings, protective coatings, dies for forming metal, printed circuits, insulation, and containers (Figures 4.6 and 4.7). Liquid epoxy resins can be cast into molds (as in a foundry) and then set by heating.

Phenolics

Phenolic materials are among the earliest and best known plastics. They are mostly based on phenol formaldehyde. Bakelite[14] was developed by Leo Baekeland shortly after the turn of the century. Applications include radio cabinets, telephones, electrical sockets and plugs, piping, pumps, valves, trays, auto distributors, rollers, and coatings.

Polyesters

Polyester plastics are best known to the general public for their use in the "space mirror" satellite (Echo). The satellite was made of thin, reflective Mylar film. Corrosion resistance of the polyesters compares favorably with many other plastics, as shown in Table 4.13. One of the main uses of polyesters involves reinforced material. A popular application is luggage. The body of the Corvette automobile is also made of fiberglass-reinforced polyester.

Silicones

Silicones offer outstanding heat resistance. Mechanical properties change little with variations in temperature. These plastics differ from most plastics in that an important ingredient is inorganic silicon. Silicones are used for molding compounds, laminating resins, and insulation for electric motors and electronic equipment. Resistance to attack by chemicals is not outstanding.

Ureas

Ureas were the second important resins to be developed. They are based on urea and formaldehyde. Corrosion resistance is not good. Uses include kitchen dishware and utensils, electrical fixtures, radio cabinets, closures, and adhesives (*e.g.*, bonding plywood).

Laminates and Reinforced Plastics

These materials usually consist of thermosetting resins filled or laminated with cloth, mats, paper, chopped fabrics, or fibers such as fiberglass, which is commonly used. The main advantage is that tensile strength can be increased to as high as around 345 MPa (50,000 psi). This results in a high strength-weight ratio for these light materials. Availability and uses include tanks, pipe, ducts, sheets, rod, car bodies, boats, and missile and satellite parts.

Other Nonmetallics

Ceramics

Ceramic materials consist of compounds of metallic and nonmetallic elements. A simple example is MgO or magnesia. Other ceramics include alumina (Al_2O_3), thoria (ThO_2), glass, brick, stone, fused silica, stoneware, clay tile, porcelain, concrete, abrasives, mortars, and high-temperature refractories.

In general, compared with metals, ceramics resist higher temperatures, have better corrosion and abrasion resistance, including erosion-corrosion resistance, and are better insulators; but they are brittle, weaker in tension, and subject to thermal shock. Most ceramic materials exhibit good resistance to chemicals, with the main exceptions of hydrofluoric acid and caustic. Parts are formed by pressing, extrusion, or slip-casting.

Acid Brick

This material is made from fireclay with a silica content about 10% greater than ordinary firebrick. A common application is lining of tanks and other vessels to resist corrosion by hot acids or erosion-corrosion. A brick-lined steel tank usually contains

[14]Trademark Union Carbide Corporation, Danbury, Connecticut.

an intermediate lining of lead, rubber, or a plastic. Acid-resistant cements and mortars join the brick. Floors subject to acid spillage are made of acid brick.

Stoneware and Porcelain

Both of these find many applications because of their good corrosion resistance. Porcelain parts are usually smaller in size than stoneware and the porcelain is less porous. Both can be glazed for easy cleaning. Stoneware and porcelain show tensile strengths of about 14 and 35 MPa (2000 and 5000 psi), respectively. Stoneware sinks, crocks and other vessels, absorption towers, pipes, valves, and pumps are available. Porcelain can be made into similar equipment (*e.g.*, acid nozzles) and is widely used as insulators and spark plugs.

Structural Clay

These clay products include building, fire, sewer, and paving brick; terra-cottas; pipe and roofing; and wall tile. Hot acids sometimes attack these materials.

Glass

Glass is an amorphous inorganic oxide, mostly silica, cooled to a rigid condition without crystallization. Glass laboratory ware, such as Pyrex,[(15)] and containers are well known. Piping and pumps are available. Transparency is utilized for equipment such as flowmeters. Glass fibers are widely used for air filters, insulation, and reinforced plastics. Hydrofluoric acid and caustic attack glass, and it shows slight attack in hot water. Glass-lined vessels are also available.

Vitreous Silica

Vitreous silica, also called fused quartz (almost pure silica), has better thermal properties than most ceramics and excellent corrosion resistance at high temperatures. It is used for furnace muffles, burners, reaction chambers, absorbers, piping, etc., particularly where contamination of the product is undesirable.

Concrete

Vast quantities of concrete are used throughout the world because of the good strength, durability, and low cost of the product. The material is used to construct large tanks, large diameter pipe, vessel linings, pump bases, and flooring for industrial work areas; and as linings on the interior of steel pipe, a ballast coating on the exterior of steel pipe, and foundations for many steel and plastic tanks, in addition to the more common paving and structural forms observed each day. It is important to know the properties of the concretes.

The term *concrete* identifies a mass of inert aggregate surrounded by a binder. The word is not much more descriptive than the word *stainless steel.*

Most of the binders used in concrete are based on portland cement. There are five basic types of portland cement as defined by ASTM C150. These are: (1) a general purpose cement, (2) a cement with somewhat greater resistance to sulfate attack, (3) a rapid hardening material, (4) a low heat-of-reaction cement, and (5) a more sulfate-resistant cement.

The first three types are also available as air-entraining cements (Types IA, IIA, and IIIA). The cements contain lime, gypsum, silica, alumina, magnesia, some iron, and a little free alkali. Upon reaction with water, interactions take place to produce various hydrated compounds, the most important of which are the silicates and aluminates (*e.g.*, tricalcium aluminate). Varying the ratio of these components, varying the amount of size of the void space, the rate at which the reactions take place, and the possible addition of other compounds, all have an influence on the mechanical, physical, and corrosion resistance properties of the cured cement.

An aggregate must also be chosen to resist the proposed environment. Naturally occurring rock, gravel, mill slags, and metal lumps have been used as aggregate for various reasons. With the aggregate constituting some three-quarters of the total concrete volume, it is important to select an inert, dense material compatible with the environment.

Unlike the metals, the strength of concrete is normally described in terms of compressive strength. Although great variations in strength can be obtained, a value of 40 MPa (6000 psi) should be achieved in a fully cured concrete.

The deterioration of concrete by freezing-thawing cycles, vibration, and attack by strong acids is well known. However, any environmental agent which can break the inorganic bonds of the cement binder will create corrosion of the concrete. Sulfates, ammonium salts, strong alkalies, weak acids, organic esters, carbon dioxide, magnesium salts, and many other agents can destroy the binder over a period of time. (Biczok's text is highly recommended for anyone concerned with the environmental compatibility of concrete.[2])

Much of the concrete deterioration which is experienced is caused by corrosion of the steel reinforcing bar. The expanding corrosion product (six times the volume of the metal from which it is formed) cracks the concrete, resulting in even greater corrosion since the steel is then exposed to the atmosphere. Epoxy or zinc-coated bar is often used to forestall this problem. Other means of lessening the problem are to use a heavier covering of denser concrete over the bar, using a richer cement mix, ensuring that salt-free water and aggregate are used, selecting a proper cement initially, and curing the product well before exposure. Exterior treatment or coating of the concrete to prevent the access of moisture to the steel is also effective. Such recommendations are particularly worthwhile if exposing the concrete to salt water.

[(15)]Trademark Corning Glass works, Corning, New York.

TABLE 4.14 — Some Properties of High-Temperature Refractories

Item	Magnesia[1] F	Magnesia[1] C	Mullite F	Mullite C	Silicon Carbide F	Silicon Carbide C	Stabilized Zirconia F	Stabilized Zirconia C	Bonded 99% Al_2O_3 F	Bonded 99% Al_2O_3 C
Fusion Point	4800	2650	3300	1815	—	—	4700	2600	3650	2010
Use Limit	4170 oxid.	2300	3000	1650	3000	1650	4400	2430	3300	1815
Modulus of rupture, psi	2500		1500		2000[3]		1900		2000	
Moh's Hardness[2]	6.0		6.5		9.6		7.0		9.0	
Thermal Shock Resist.	Poor		Good		Good		Fair		Fair	
Relative Cost	2.8		1.0		2.1		10		3.1	

[1]Basic refractories have poor resistance to hot acids.
[2]Scale 1 to 10. Talc = 1, low-carbon steel = 4, diamond = 10.
[3]At 2500 F (1371 C).

Concrete can also fatigue and is susceptible to impingement and cavitation damage.

A number of coating materials (bituminous, epoxy, and latex) are compatible with concrete and are used to seal the surface from water penetration (Chapter 12). Other surface treatments, such as magnesium fluosilicate washes, can be used to make the surface harder, more impervious, and more chemical-resistant. (Comments on the use of concrete underground are given in Chapter 10.)

Super Refractories

Table 4.14 lists some properties of a few very high-temperature refractories. These are also called super refractories. They are used to resist molten metals, slags, and hot gases. Alumina has also found application for pump and valve seats because of its high hardness and good wear and corrosion resistance. Silicon carbide performs well as spray nozzles for hot sulfuric acid and for many other severe applications.

Carbon and Graphite

These are unique nonmetallics in that they are good conductors of heat and electricity. High thermal conductivity results in excellent thermal shock resistance. They are used for heat exchangers, columns, pumps, and impressed-current anodes. Carbon and graphite are inert to many corrosive environments. They are weak and brittle compared to metals. Tensile strength varies between about 3 to 20 MPa (500 to 3000 psi) and impact resistance is nil. Abrasion resistance is poor. High-temperature stability is good, and they can be used at temperatures up to 2760 C (5000 F) if protected from oxidation (burning). Silicon-base coatings (silicides or silicon carbide) and iridium coatings are claimed to give protection up to around 1593 C (2900 F) in air.

Carbon exhibits good resistance to alkalies and most acids, but oxidizing acids such as nitric, concentrated sulfuric, and chromic acid attack it. Fluorine, iodine, bromine, chlorine, and chlorine dioxide are likely to attack carbon. Karbate,[16] a resin-bonded graphite, has found wide application in the chemical process industries. Graphite in nuclear reactors is quite well known.

Pyrolytic graphite, which is a dense, anisotropic material, has better strength and oxidation resistance than the more common types of carbon.

Wood

Cypress, pine, oak, and redwood are the main woods used for corrosion applications. There is no "better" wood. The choice depends on the exposure. Filter-press frames, structural members of buildings, barrels, and tanks are often made of wood. Containers must be kept wet or the staves will shrink, warp, and leak. Generally speaking, wood is limited to water and dilute chemicals. Strong acids and dilute alkalies attack woods. They are also subject to biological attack. Impregnation with waxes and plastic resins helps reduce chemical and biological attack (Table 4.15 and Chapter 8 figures).

References

1. This chapter is based largely on: Fontana, M. G. and Greene, N. D., Corrosion Engineering, Chapter 5, McGraw-Hill Book Co., New York, N.Y. (1967).
2. Biczok, I., Concrete Corrosion—Concrete Protection, Chemical Publishing Co., Inc., New York, N.Y. (1967).

Bibliography

Floyd, Don E. Polyamide Resins. 2nd Ed. Reinhold, New York, N.Y. (1967).

Harper, C. A. Handbook of Plastics and Elastomers McGraw-Hill, New York, N.Y. (1975).

[16]Trademark Union Carbide Corporation, Danbury, Connecticut.

TABLE 4.15 — Some Commonly Used Woods and Their General Corrosion Resistance to Moist Conditions

Good		Fair		Poor		
Black Locust	Redwood	Douglas Fir	Pine, South Yellow	Ash	Cottonwood	Red Oak
Cedar	Walnut	Honey Locust	Sassafras	Aspen	Fir	Spruce
Chestnut	Yew	Larch	White Oak	Beech	Hemlock	Willow
		Pine, East White		Birch	Maple	

Jastrzebski, Z. D. Nature and Properties of Engineering Materials. John Wiley, New York, N.Y. (1975).

Kinney, Gilbert Ford. Engineering Properties and Applications of Plastics. John Wiley, New York, N.Y. (1975).

Modern Plastics Encyclopedia, Vol. 58, No. 10A (Oct. 1981).

NACE. Process Industries Corrosion. Houston, Texas (1975).

Neville, Kris and Henry Lee. Handbook of Epoxy Resins. McGraw-Hill, New York, N.Y. (1967).

Rosato, D. V. and Robert T. Schwartz. Environmental Effects on Polymeric Materials, Vol. 1 and 2. Interscience, New York, N.Y. (1968).

Seymour, R. B. and J. M. Church. Plastics for Corrosion Resistant Applications. Reinhold, New York, N.Y. (1955)

Simonds, H. R. and J. M. Church. A Concise Guide to Plastics. 2nd Ed. Reinhold, New York, N.Y. (1963).

Troxell, G. E., et al. Composition and Properties of Concrete. McGraw-Hill, New York, N.Y. (1968).

NOTES

Chapter 5

Localized Corrosion

LOCALIZED CORROSION

Introduction

General (*i.e.*, substantially uniform) *corrosion* is the most desirable type of attack, allowing one to forecast with some degree of accuracy the probable life of equipment. Unfortunately, most corrosion encountered in engineering practice is of a more localized type.

Localized corrosion can be defined as selective attack on a metal by corrosion at small special areas or zones on a metal surface in contact with an environment. It usually occurs under conditions where the largest part of the original surface either is not attacked or is attacked to a much smaller degree than at the local sites.

The most common type of localized corrosion is pitting, in which small volumes of metal are removed by corrosion from certain areas on the surface to produce craters or pits. Pitting corrosion may occur on a metal surface in a stagnant or slow moving liquid. It also may be caused by crevice corrosion, poultice corrosion, deposition corrosion, cavitation, impingement, and fretting corrosion. All of these are forms of localized corrosion discussed in this chapter.

Another common type is intergranular corrosion (sometimes called intercrystalline corrosion). In this form, a small volume of metal is preferentially removed along paths that follow the grain boundaries to produce what might appear to be fissures or cracks. The same kind of subsurface fissures can be produced by transgranular corrosion (sometimes called transcrystalline corrosion). In this, a small volume of metal is removed in preferential paths that proceed across or through the grains. This occurs only under certain conditions and with certain alloys, as will be described later.

Intergranular and transgranular corrosion sometimes are accelerated by tensile stress. In extreme cases, the cracks proceed entirely through the metal, causing rupture or perforation. This condition is known as stress corrosion cracking, a subject that will be dealt with in Chapter 6. Intergranular and transgranular subsurface cracks also can be produced by hydrogen. Caustic embrittlement and corrosion fatigue are two other mechanisms of metal deterioration which form fissures at or beneath the surface. Hence, brief reference to these subjects is also made in this chapter.

In a completely different type of corrosion which may become localized, one of the metals in an alloy may be selectively leached out without producing visible pits or cracks and without changing the dimensions of the metal. At a casual glance the metal may appear to be intact. Under a microscope, however, it can be seen to be porous. The mechanical properties of the alloy are greatly reduced by the selective attack.

The most common example of this type is dezincification of brass in which the zinc is selectively dissolved out of the alloy. Another case is graphitic corrosion of cast iron, in which the iron is selectively dissolved or leached away, leaving a porous mass which appears intact, but in reality consists largely of graphite.

The several forms of corrosion (Figure 5.1) may be divided into three groups:

1. Those recognizable with the unaided eye.
2. Those which are more easily discerned with specific aids (*e.g.*, dye penetrants, magnetic particles, or low-power microscopy).
3. Those which can only be identified definitely by optical or electronic microscopy.

Local Cells

Pitting, intergranular and transgranular corrosion, and selective dissolution are caused by *local elements* or *local cells* which exist at the surface of a metal and have an electrochemical mechanism.

Remember that a local cell is a small surface area undergoing corrosion which contains both anodic and cathodic sites at separate nearby locations (Figure 5.2).

By definition, the *anode* (A) is the site at which *oxidation* occurs. Since corrosion is an oxidation reaction, it proceeds at the anode. The anode reaction is where metal atoms release electrons and are no longer in the metallic state. In the case of aluminum in water, for instance:

$$Al \rightarrow Al^{+3} + 3e(\text{oxidation}) \qquad (5.1)$$

The aluminum ions are not very soluble in water and combine with hydroxyl ions in the water to form aluminum hydroxide which is not soluble and therefore precipitates as corrosion product:

$$Al^{+3} + 3OH^- \rightarrow Al(OH)_3\downarrow \qquad (5.2)$$

This reaction lowers the hydroxyl ion concentration in the vicinity of the anodes so that the anodic areas tend to become acidic. Negatively charged ions such as chlorides or sulfates tend to

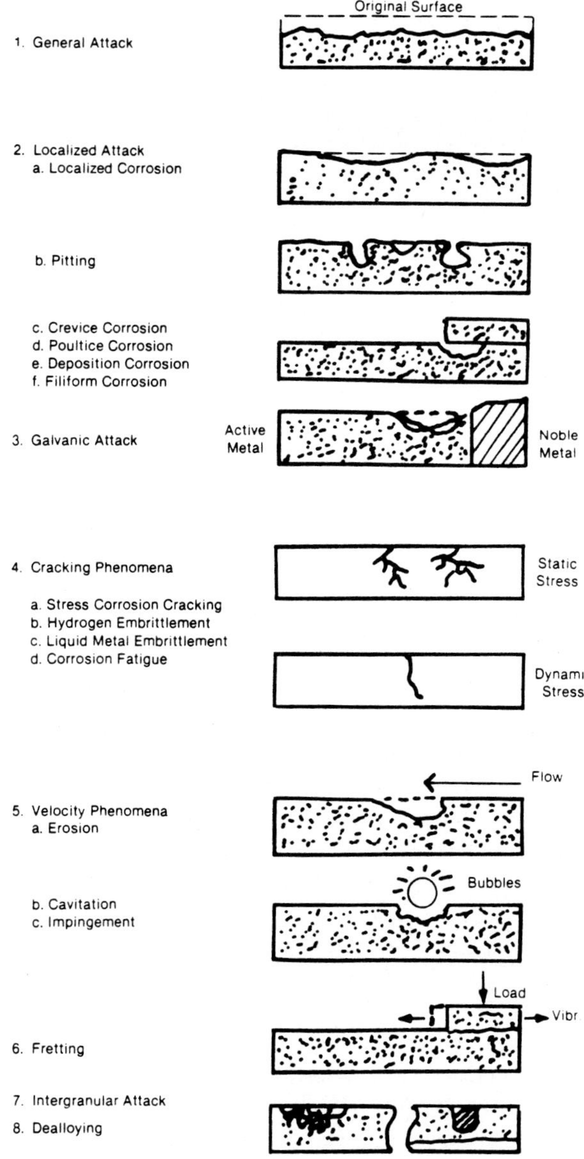

FIGURE 5.1 — Corrosion schematics.

migrate toward the anode and may indirectly further increase the acidity. The pH of active pits in a neutral solution may reach 2 to 4.

Also, by definition, the *cathode* (C) is the site at which *reduction* occurs, or where electrons are consumed. The three most common cathodic reactions are the reduction of hydrogen ions, of dissolved oxygen, and of metal ions:

$$2H^+ + 2e \rightarrow H_2 \quad (5.3)$$

(hydrogen evolution or hydrogen ion reduction)

$$1/2\,O_2 + H_2O + 2e \rightarrow 2OH^- \quad (5.4)$$

(oxygen reduction)

$$Cu^{+2} + 2e \rightarrow Cu \quad (5.5)$$

(metal ion reduction)

Within the metal, the electrons released by the oxidation reaction at the anode flow to the cathode where they take part in the reduction reaction at the cathode. As a result of the consumption of hydrogen ions and/or the production of hydroxyl ions at the cathode, the cathodic zone tends to become alkaline. However, the pH shift is less than that which occurs at the anode because in localized corrosion the cathode is usually larger than the anode, so that the cathode current density is less and the ion concentration changes less.

Local cells are produced by differences among small nearby areas on the metal surface. They may result from differences in the metal, or the environment, or from impressed currents. Metal variations may be the result of composition differences. A second phase constituent will have a different corrosion potential compared to that of an adjacent solid solution. The difference may be in the thickness of a surface film at adjacent sites (which in turn may reflect metal differences below). Diversities of the environment may be the result of differences in con-

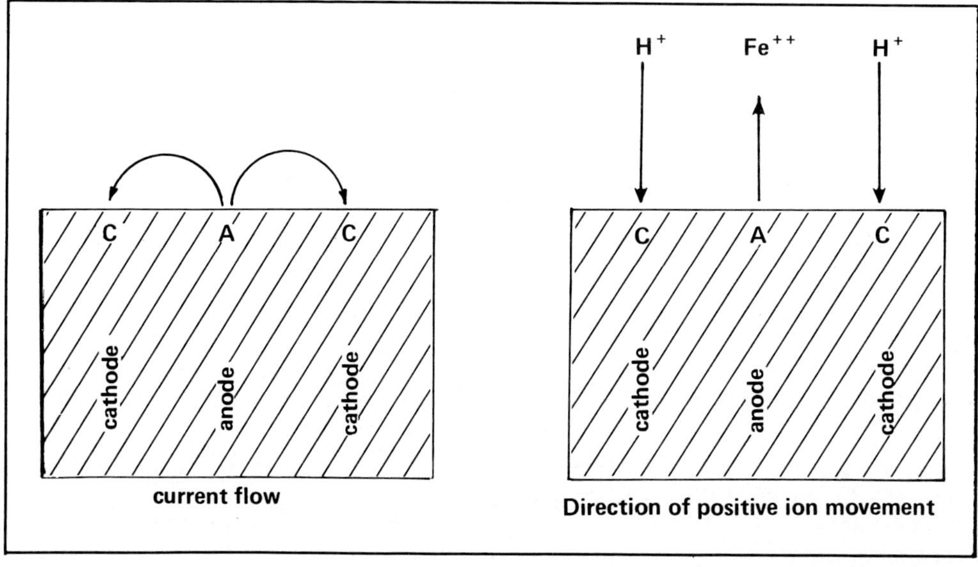

FIGURE 5.2 — Schematic of a local cell showing direction of current flow (left) and ion migration (right).

centration (called *concentration cells*), in temperatures, or in agitation.

A crevice or a local surface deposit may cause one part of the surface to be screened from contact with the bulk of the solution and thus from access to oxygen, while the remainder is in contact with liquid saturated in oxygen. This cause of local cells is known as *differential aeration,* which is the most common form of concentration cells. Differential aeration cells are also the cause of localized corrosion that occurs at the water line when a metal is partially immersed in a liquid.

The classic case is that of a zinc sheet partially immersed in KCl solution (Figure 5.3). The supply of oxygen in the solution nearest the water line is more plentiful than it is at greater depths, so this condition produces local cell with the illustrated pattern of attack.

Impressed currents may be caused by contact with a dissimilar metal (*galvanic corrosion*) or by application of an external potential from an "outside" electrical circuit. Impressed currents, regardless of their cause, frequently cause local cell action.

Open Pitting

Pitting is a form of localized corrosion that proceeds because of local cell action which produces cavities beginning at the surface. These cavities may or may not become filled with corrosion products. Corrosion products may form caps over pit cavities which are described as *nodules* or *tubercles.* While the shapes of pits vary widely, they usually are roughly saucer-shaped, conical, or hemispherical. Pit walls usually are irregular when viewed under a microscope. Their shape distinguishes pits from other forms of localized corrosion (Figure 5.4).

Pitting occurs when a film-protected metal is almost, but not completely, resistant to corrosion. Only a small amount of metal is corroded, but perforations can lead to costly repair of expensive equipment. It has been said that pitting has caused more unexpected corrosion losses than any other type of corrosion.

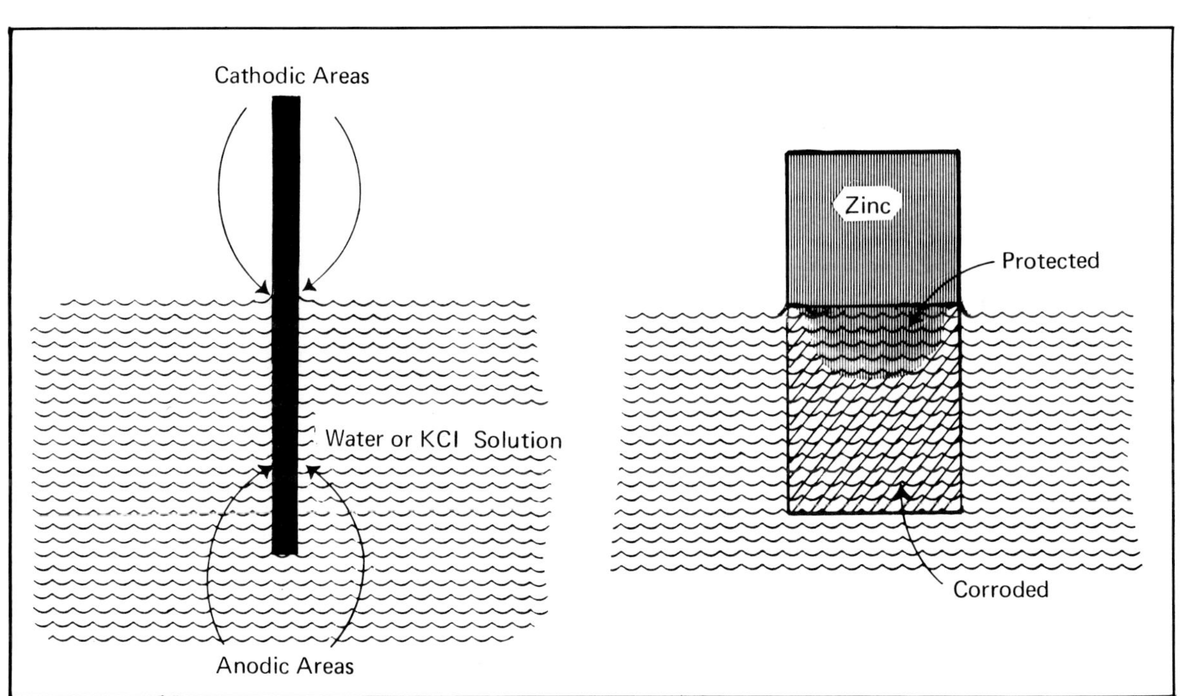

FIGURE 5.3 — Schematic of differential aeration cell involving zinc partially immersed in liquid.

FIGURE 5.4 — Attack under deposit in a condenser tube. [SOURCE: Corrosion Resistance of Metals, ACS Monograph No. 158, Reinhold, New York, N.Y. (1963).]

Pitting usually occurs on a metal surface immersed in a solution or moist environment (such as soil). It can occur on a surface exposed to the atmosphere if there are droplets of moisture or a condensed moisture film present on the metal surface. Pitting is sometimes associated with intergranular corrosion. For example, intergranular cracks may progress from the main pit cavity further into the metal. Pitting also may occur in crevices, in which case it is called crevice corrosion, a subject that will be discussed later.

Cavitation erosion also will produce a pitted surface, but a different mechanism is involved, as will also be discussed later.

Pitting usually occurs on metals that are covered with a very thin, often invisible, adherent protective surface film which may be formed during fabrication or be produced by reaction with the environment. Thus, pitting occurs on magnesium,[1] aluminum,[2] titanium,[3] stainless steels,[4] and copper in cases where surface films develop. It also may occur on iron, steel, lead, and other metals. Pits develop at weak spots in the surface film and at sites where the film is damaged mechanically under conditions where self-repair will not occur (Figure 5.5).

The practical importance of pitting depends on the thickness of the metal and on the penetration rate (which may be small). The rate usually decreases with time. Thus, on thin sections pitting may be serious, while on a thick section it may be almost unimportant. In general, the rate of penetration decreases if the number of pits increases. This is because adjacent pits have to share the available adjacent cathodic area, which controls the corrosion current that can flow. Movement of the solution over a metal surface often reduces and even may prevent pitting that otherwise would occur if the liquid were stagnant (*e.g.,* as with some austenitic stainless steels in seawater).

When an electric current leaves a metal surface and flows into the environment, pitting may occur. The current may be caused by contact of the surface with a dissimilar metal or by an external impressed current (*i.e.,* a *stray current*). Stray alternating currents in general do not cause very much corrosion of metals, except in the case of aluminum. In this case, the metal is unable to plate back onto the surface during the reverse part of the cycle. Most severe stray current corrosion is caused by direct current leakage (*e.g.,* from DC welding of reverse polarity or from DC power of electrified systems).

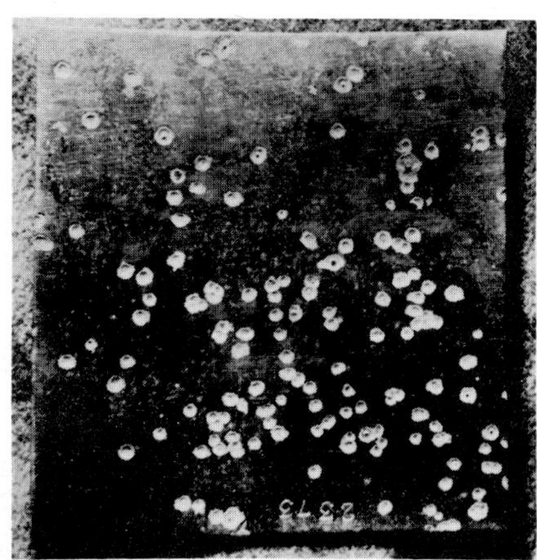

FIGURE 5.5 — Small corrosion pits distributed at random. [SOURCE: Corrosion Resistance of Metals, ACS Monograph No. 158, Reinhold, New York, N.Y. (1963).]

Theoretically, a local cell that leads to the initiation of a pit can be caused by an abnormal anodic site surrounded by normal surface which acts as a cathode, or by the presence of an abnormal cathodic site surrounded by a normal surface which acts as a cathode, or by the presence of an abnormal cathodic site surrounded by a normal surface in which a pit will have disappeared due to corrosion. In the second case, post-examination should reveal the local cathode, since it will remain unattacked. Most cases of pitting are believed to be caused by local cathodic sites in an otherwise normal surface.

A pit may go through four separate stages: (1) initiation, (2) propagation, (3) termination, and (4) reinitiation. The causes of initiation (*i.e.,* local cell action) have been discussed. In the propagation stage, the rate increases due to changes in the anodic and cathodic environment (these become more acidic and alkaline, respectively). A pit may terminate due to increased internal resistance of the local cell (caused by filling with corrosion products, filming of the cathode, etc.). If a pitted surface is dried out, the pits will, of course, terminate. When rewetted, some of the pits may reinitiate. This may be due to the reestablishment of the conditions or to differential aeration between the solution in the main pit cavity and solutions in some of the cracks which emanate from it deeper into the metal.

When evaluating the condition of pitted metal, most weight should be given the deepest pit, although this is not always done. Pit depths can be measured using a calibrated microscope focused first on the shoulder and then on the bottom of the pit, by using a dial gauge with a needle probe, by metallographic sectioning, or by using a metal cutting tool in a carpenter's router and cutting down until no corrosion is evident. This last method is the most accurate because even metallographic sectioning requires taking a section at the proper location, which may not be evident. A variation of the routing method that can be used to measure the maximum pit depth on the outside of a pipe is to place a sample length in a lathe and machine it until no further corrosion is evident.

Efforts have been made to treat the rate of penetration statistically. For example, equations have been developed to predict the perforation of buried steel pipelines.[5] It has been shown that the rate of penetration of pits in aluminum exposed to water follows a cube root curve.[6] Thus,

$$d = Kt^{1/3} \tag{5.6}$$

where

d = maximum pit depth,

K = a constant, depending on the alloy and the environment, and

t = time.

Such an equation predicts that doubling the wall thickness will increase the time to perforation by a factor of 8 and emphasizes the earlier remarks about the influence of metal thickness.

When considering the pitting of steel and other alloys, investigations have indicated the power in this equation can be 1/3, 1/2, 2/3, or other fractions, depending on the alloy composition, environmental conditions, density of the pitting, and time after initiations of the pit.[7]

Since pitting is electrochemical, it can be stopped by cathodic protection (Chapter 9). It can be prevented also by the use of inhibitors, which alter the electrode reactions of the local cell and remove their driving force.

Penetration by pitting often can be prevented by coating the surface of a metal with a sacrificial layer of another alloy (*e.g.*, zinc on steel or Alclad[(1)] aluminum) or by protective coatings. A zinc-rich paint is sacrificially active and will prevent the pitting of either steel or aluminum. In some cases, agitation of the environment will prevent environmental differences from developing and will prevent pitting that otherwise would occur. (For a more thorough treatment of the subject, refer to relevant works listed in this chapter's bibliography.)

Crevice Corrosion

Crevice corrosion occurs in cracks or crevices formed between mating surfaces of metal assemblies, and usually takes the form of pitting or etched patches (Figure 5.6 and 5.7). Both surfaces may be of the same metal or of dissimilar metals, or one surface may be a nonmetal. It can occur also under scale and surface deposits and under loose fitting washers and gaskets that do not prevent the entry of liquid between them and the metal surface. The crevice may proceed inward from a surface exposed to air, or may exist in an immersed structure.

Crevice corrosion is believed to initiate as the result of the differential aeration mechanism mentioned earlier. Oxygen in the liquid which is deep in the crevice is consumed by reaction with the metal. Oxygen content of liquid at the mouth of the crevice which is exposed to air (or of bulk liquid in the case of immersion) is greater, so a local cell develops in which the anode or area being attacked is the surface in contact with the oxygen-depleted liquid. It is also thought that subsequent pH changes at anodic and cathodic sites further stimulate local cell action, as in the case of pitting corrosion.

[(1)]An aluminum product consisting of a metal sandwich in which the upper and lower surfaces are anodic to the core alloy and thus provide built-in cathodic protection.

FIGURE 5.6 — Corrosion at crevice between a tube and a tube sheet. [SOURCE: Corrosion Resistance of Metals, ACS Monograph No. 158, Reinhold, New York, N.Y. (1963).]

FIGURE 5.7 — Crevice corrosion at intersections of wires in a screen. [SOURCE: Corrosion Resistance of Metals, ACS Monograph No. 158, Reinhold, New York, N.Y. (1963.]

Crevice corrosion occurs most commonly on film-protected metals such as aluminum, magnesium, stainless steels, and titanium.

Crevice corrosion often can be prevented at the design stage by avoiding crevices, or during construction by filling them with a durable jointing compound that will exclude moisture and remain resilient. Several synthetic types are available with much longer lives than the vegetable-based compounds that have been used for years.

Poultice Corrosion

Poultice corrosion is a special case of localized corrosion due to differential aeration, which usually takes the form of pitting when an absorptive material such as paper, wood, asbestos, sacking, cloth, etc., is in contact with a metal surface that becomes wetted periodically. No action occurs while the entire assembly is wet, but during the drying period, adjacent wet and dry areas develop. Near the edges of the wet zones, differential aeration develops which leads to pitting, as in the case of crevice corrosion.

Poultice corrosion is prevented by avoiding the contact of absorptive materials with a metal surface, by painting the surface that will contact such materials, or by designing to prevent such materials from becoming wet in service.

A form of poultice corrosion that has caused extensive damage to integral fuel tanks in aircraft, and which is found in other systems where mats of organic material may accumulate, is caused inside tanks by bacterial and fungal growths in jet aviation fuel. Aluminum surfaces under these mats suffer extensive damage, as shown in Figure 5.8.

Filiform Corrosion

Filiform corrosion is a special form of oxygen cell corrosion occuring beneath organic or metallic coatings on steel, zinc, aluminum, or magnesium. The attack results in a fine network of random "threads" of corrosion product developed beneath the coating material with a shallow grooving of the metal surface (Figure 5.9). Such attack develops beneath semipermeable films in a high-humidity (>60% RH) environment on such items as coated cans, office furniture, cameras, aircraft structures, auto interiors and exteriors, and a host of other common products.

In filiform corrosion, the head of the advancing filament (about 0.1 mm wide) becomes anodic, with a low pH and a lack of oxygen as compared with the cathodic area immediately behind the head where oxygen is available through the semipermeable film. Corrosion proceeds as the cathode follows behind

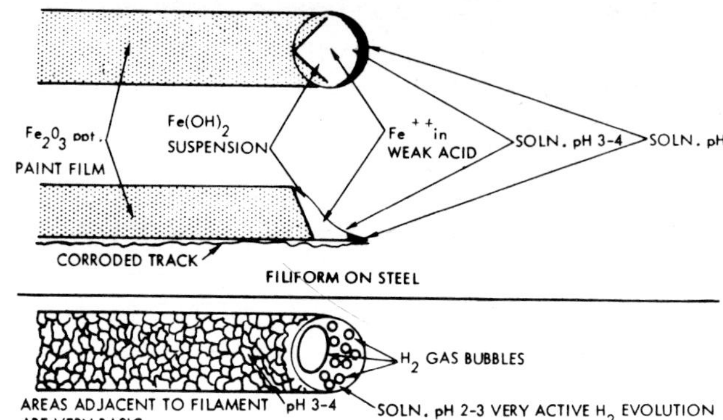

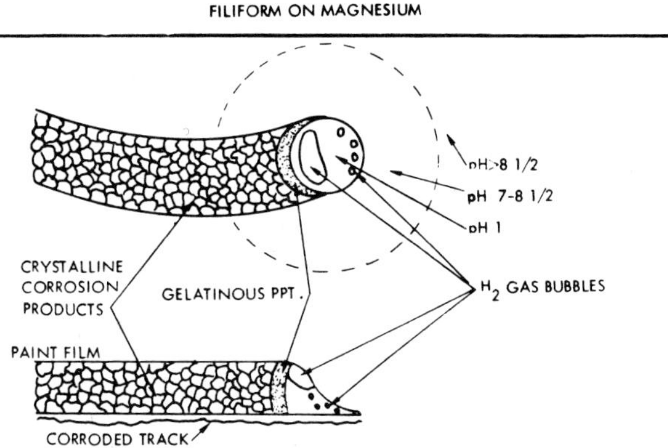

FIGURE 5.9 — Comparison of filaments on various metal structures. [SOURCE: Staehle, R. W., *et al.*, Localized Corrosion, NACE, Houston, Texas (1979).]

FIGURE 5.8 — Showing pits inside an aircraft fuel tank from sludge accumulations in jet fuel. [SOURCE: Figure 2, Corrosion Resulting From Microbial Fuel Tank Contamination, C. R. Ward, Materials Protection, Vol. 2, No. 6, pp. 10-18 (1963).]

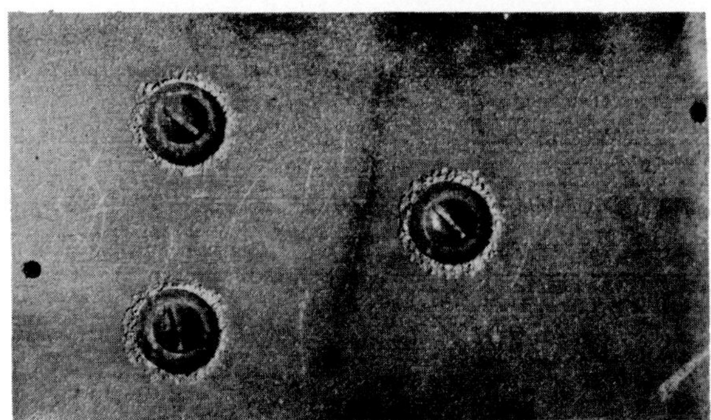

FIGURE 5.10 — Galvanic corrosion around fastener. [SOURCE: Corrosion Resistance of Metals, ACS Monograph No. 158, Reinhold, New York, N.Y. (1963).]

the anodic head. The water and oxygen present in the cathodic area convert the anodic products to the usual oxides of the metal. Intergranular attack on a sensitized metal also can occur by this process.

Galvanic Attack

Galvanic attack on the anodic member of two different alloy compositions can cause pitting in a highly localized manner or disperse corrosion over a larger area (Figure 5.10). (Galvanic action is thoroughly discussed in Chapters 2 and 15.)

Electrolysis, the discharge of direct current from a metal surface, will severly pit a metal. The amount of metal removed is proportional to the current discharged (Chapter 9).

Deposition Corrosion

Deposition corrosion is a form of pitting corrosion that can occur in a liquid environment when a more cathodic metal is plated out of solution onto a metal surface. It generally occurs with the more anodic metals such as magnesium, zinc, and aluminum. Common cathodic ''activators'' are mercury and copper ions in solution.

For example, soft water passing through a copper water pipe will accumulate some copper ions. If water is then admitted to a galvanized or aluminum vessel, particles of metallic copper will plate out, *i.e.,* deposit on the surface and stimulate pitting by local cell action (Figure 5.11).

Deposition corrosion can be avoided by preventing the pick-up of cathodic ions that will enter the equipment, or by scavenging them by passing the contaminated product through a tower packed with more anodic metal turnings on which the ions can deposit.

Velocity-Related Attack

Erosion

A metal surface over which a smooth, linear flow of liquid is maintained will corrode at essentially the same rate as the metal would in the stagnant solution. This is true because at the surface/liquid interface there is a static film of the liquid molecules. Unless the liquid is aggressive as a corrosive, the metal will form some semiprotective film over the surface and remain stable.

However, if the flow of liquid becomes turbulent, the random liquid motion impinges on the surface to remove the thin film. Additional oxidation then occurs by reaction with the liquid. This alternate oxidation and removal of the film will accelerate the rate of corrosion. The resulting erosive attack may be uniform, but quite often produces pitted areas over the surface (Figure 5.12).

Obviously, the presence of solid particles or gaseous bubbles in the liquid can accentuate the attack. Also, if the fluid dynamics are such that impingement or cavitation attack is developed, even more severe corrosion can occur. However, the term erosion or erosion-corrosion is normally reserved for that mechanism of accelerated attack associated with abrasion of the surface and further oxidation of the metal (Figure 5.13).

Cavitation

Cavitation damage (sometimes referred to as cavitation corrosion or cavitation erosion) is a form of localized corrosion combined with mechanical damage that occurs in turbulent or rapidly moving liquids and takes the form of areas or patches of pitted or roughened surface. It has been defined in Chapter 1 as the ''deterioration of a surface caused by the sudden formation and collapse of bubbles in

FIGURE 5.11 — Deposition corrosion. Copper deposited on the interior of an aluminum pipe caused the severe pitting.

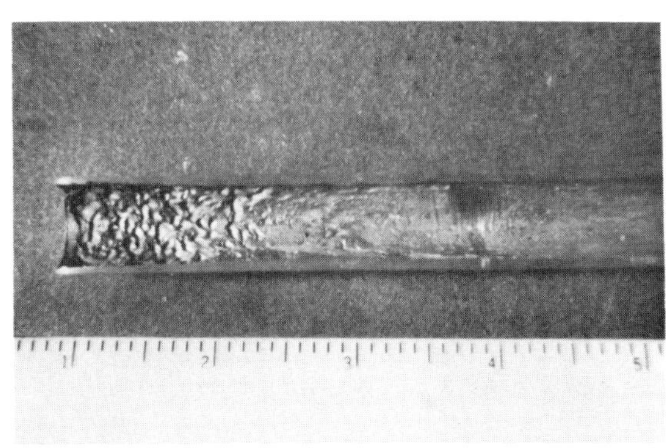

FIGURE 5.12 — Erosion on a copper tube.

FIGURE 5.13 — Erosion-corrosion of valve seat by escaping steam. [SOURCE: Corrosion Resistance of Metals, ACS Monograph No. 158, Reinhold, New York, N.Y. (1963).]

FIGURE 5.14 — Vibratory cavitation damage of diesel engine cylinder liner. [SOURCE: Corrosion Resistance of Metals, ACS Monograph No. 158, Reinhold, New York, N.Y. (1963).]

a liquid." It has been similarly defined as the "localized attack that results from the collapse of voids or cavities due to turbulence in a liquid at a metal surface" (Figure 5.14).

Cavitation is the formation of voids or cavities in a liquid due to turbulence or temperature which causes the pressure in local zones of the liquid to fall below the vapor pressure. This happens on the backside of ship propellers, water turbine blades in pumps, in high-velocity flow lines, and on the submerged planes of hydrofoil craft, depending on the design of equipment and degree of turbulence.

The voids are actually "holes" in the liquid which contain only water vapor. These usually exist for a very short time only because changes in liquid pressure in the turbulent local zones close the voids [in times as short as a microsecond (one millionth of a second)]. As the liquid surfaces meet during bubble collapse, considerable kinetic energy is released. This had been estimated to be as high as 687,000 kPa (100,000 psi). Depending on the speed of collapse, the energy released can produce a *clap,* similar to the familiar *water hammer* that occurs in water pipes when a valve is closed suddenly. The claps of energy impinge on the metal surface and may break a pro-

FIGURE 5.15 — Pump impeller perforated by cavitation attack. Note that the same area on each vane is attacked.

tective surface film and lead to the initiation of corrosion.

A great many cavities may form and collapse in a small zone (perhaps more than a million) in the course of one second. As a result, cavitation damage can occur very rapidly. For example, a cast Al + 10 Mg alloy propeller installed to drive a 13-foot boat at 31 knots developed visible cavitation damage areas on the back sides of the blades in only 15 minutes. Cavitation of a pump impeller is also quite common (Figure 5.15).

In some instances, cathodic protection has been successful in reducing or preventing cavitation damage, but because cavitation damage usually involves physical as well as electrochemical processes, it cannot always be prevented by this means. In some cases, inhibitors have been used successfully to limit cavitation corrosion, as in the water side of diesel engine cylinder liners.

It is believed that work hardening of the surface occurs, and ultimately bits of metal are knocked from the surface by mechanical cracking and fatigue failure. Ductile metals resist cavitation damage longer than more brittle materials. On the other hand, 11 to 13% chromium martensite stainless steels are often superior to bronze in water service. As might be expected, there is good correlation between fatigue life and cavitation resistance.

Cavitation resistance increases as the grain size of a metal becomes smaller. The nature of the grain boundaries also plays a role. Stainless steels are among alloys more resistant to cavitation damage. This is due to their ductility, toughness, high corrosion fatigue limit, homogeneity, fine grain size, and ability to work harden.

Cavitation can occur in pipes when liquid is moving at high velocity, especially at fittings where the smooth bore of the pipe is interrupted. The effect is usually to produce pitting on the downstream side of the turbulence.

The prevention of cavitation damage requires the use of the most resistant alloys and, where possible, design of the system to avoid turbulence and cavitation. Design changes may aim to raise pressures

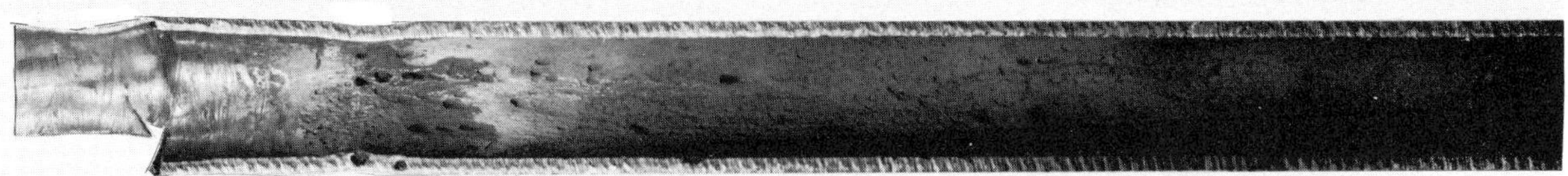

FIGURE 5.16 — Impingement attack near inlet end of condenser tube. [SOURCE: Corrosion Resistance of Metals, ACS Monograph No. 158, Reinhold, New York, N.Y. (1963).]

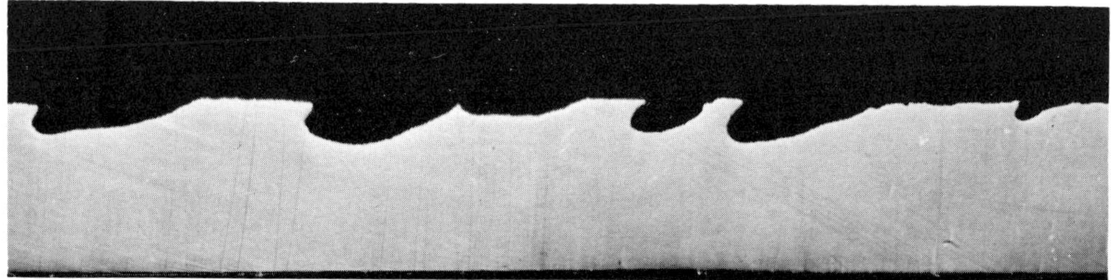

FIGURE 5.17 — Cross section of copper tube attacked by turbulent erosion-corrosion.

or eliminate abrupt changes in cross section (of pipes). The shape of a propeller or impeller also may be altered. (For a more comprehensive review of this subject, refer to relevant works listed in this chapter's bibliography.)

Impingement

Impingement attack is related to cavitation damage, and has been defined in Chapter 1 as "localized erosion-corrosion caused by turbulence or impinging flow." Entrained air bubbles tend to accelerate this action, as do suspended solids. This type of corrosion occurs in pumps, valves, orifices, on heat-exchanger tubes, and at elbows and tees in pipelines.

Impingement corrosion usually produces a pattern of localized attack with directional features. The pits or grooves tend to be undercut on the side away from the source of flow, in the same way that a sandy river bank at a bend in the river is undercut by the oncoming water (Figures 5.16 and 5.17).

When a liquid is flowing over a surface (*e.g.*, in a pipe), there is usually a critical velocity below which impingement does not occur and above which it increases rapidly. Impingement attack first received attention due to the poor behavior of some copper alloys in seawater.

In practice, impingement and cavitation may occur together, and the resulting damage can be the result of both. Impingement may damage a protective oxide film and cause corrosion, or it may mechanically wear away the surface film to produce a deep groove.

The solution to impingement attack is the use of more resistant alloys. When the 70-30 brasses began failing in seawater service (*e.g.*, condenser tubes), it was found that 70-30 cupronickel containing 0.4 to 1.0% Fe was considerably more resistant, while 90-10 cupronickel is even more resistant. A Ti-6Al-4V alloy has a high resistance to impingement attack.

FIGURE 5.18 — Fretting corrosion.

Other Types of Local Attack

Fretting Corrosion

The rapid localized corrosion that occurs on closely fitting surfaces in contact under a load and subject to small amplitude slip (*i.e.*, chafing or vibratory motion) is termed *fretting corrosion* (Figure 5.18). This phenomenon has also been called *false Brinelling* (because of the formation of pits or indentations such as those formed in making a Brinell hardness test) and *friction oxidation* (because the corrosion product is an oxide).

Almost all alloys are subject to fretting corrosion, and its incidence is high in vibrating machinery. It occurs mainly on machine parts such as ball and roller bearings, shafts, gears, etc., although it can also occur on sheets of metal (*e.g.*, aluminum) in packages that are jolted about during shipment.

Fretting corrosion takes the form of local surface discolorations and deep pits. These occur in regions where slight relative movements have occurred between mating, highly loaded surfaces. The pits sometimes provide stress raisers for the initiation of corrosion fatigue.

Two different theories have been advanced to explain the mechanism of fretting corrosion. In the first, it is assumed that pressure causes small areas of the mating surfaces to weld together. Motion then tears a small volume of metal from one of the surfaces (*galling*). Subsequent motion detaches this, and then a small volume of metal from the other surface is removed to produce another discrete fine particle of metal. In the presence of oxygen and moisture, these metal particles oxidize or corrode to produce a metal oxide corrosion product.

The term *fretting* also is used by some persons to describe the result when abrasion due to the relative motion removes protective surface films, leading to oxidation or corrosion of the fresh metal surface to form corrosion products which are detached from the surface by subsequent further abrasion.

Fretting corrosion can be prevented by eliminating any slipping movement between the two surfaces. Thus, it is sometimes possible to overcome fretting by increasing the load on the surfaces, if it prevents relative motion. Alternatively, some authorities recommend *decreasing* the load to minimize the effect of vibratory motion. In other cases, roughening the surfaces can increase the friction between them and stop the movement. Fretting corrosion can be greatly retarded by lubricating the contacting surfaces with an oil or grease with sufficient load-bearing characteristics that it will separate the surface from the environment and thus from local welding even under the loads and vibrations encountered.

In the case of metal sheets in packages, a rigid package and pressure packing of the contents will eliminate relative motion during transit and thus eliminate fretting corrosion. Interleaving with paper and oiling are also beneficial. (For a review on fretting corrosion, refer to the relevant work listed in this chapter's bibliography.)

Intergranular Corrosion

Intergranular corrosion is a form of localized surface attack in which a narrow path is corroded out preferentially along the grain boundaries of a metal. It initiates on the surface and proceeds by local cell action in the immediate vicinity of a grain boundary.

The driving force is a difference in corrosion potential that develops between a thin grain boundary zone and the bulk of the immediately adjacent grains. The difference in potential may be caused by the difference in chemical composition between the two zones. This can develop as a result of migration of impurity or alloying elements in an alloy to the grain boundaries. If the concentration of alloying elements in the grain boundary region becomes sufficient, a second phase or constituent may separate or precipitate. This may have a corrosion potential different from that of the grains (or *matrix*) and cause a local cell to form.

The constituent may be anodic, cathodic, or neutral to the base metal or adjacent denuded zone. Examples of anodic constituents are Mg_5Al_8 and $MgZn_2$ in aluminum alloys and Fe_4N in iron alloys. Examples of cathodic constituents are $FeAl_3$ and $CuAl_2$ in aluminum alloys and Fe_3C in iron alloys. Examples of neutral constituents are Mg_2Si and $MnAl_6$ in aluminum alloys and Mo_6C and W_6C in wrought Ni-Cr-Mo alloys.

If the precipitate is anodic to the denuded zone, it corrodes preferentially, and if it is present in a continuous network throughout the grain boundary, it is removed to produce a thin channel, fissure, or crack between adjacent grains. As the cracks develop inwardly, they follow the grain boundaries in a lateral direction as well, loosening the outer layers of grains (Figure 5.19).

In mild cases, the penetration rate may be extremely slow and even after a considerable time may penetrate to a depth of only a dozen grains or so. In more severe cases, penetration may go quickly completely through sections that are a quarter of an inch thick or more. Shallow intergranular corrosion may escape visual inspection, but is seen easily under a microscope. As attack proceeds, it can be recognized more readily by the naked eye. In an extreme case, an alloy can be reduced to a heap of loose individual grains.

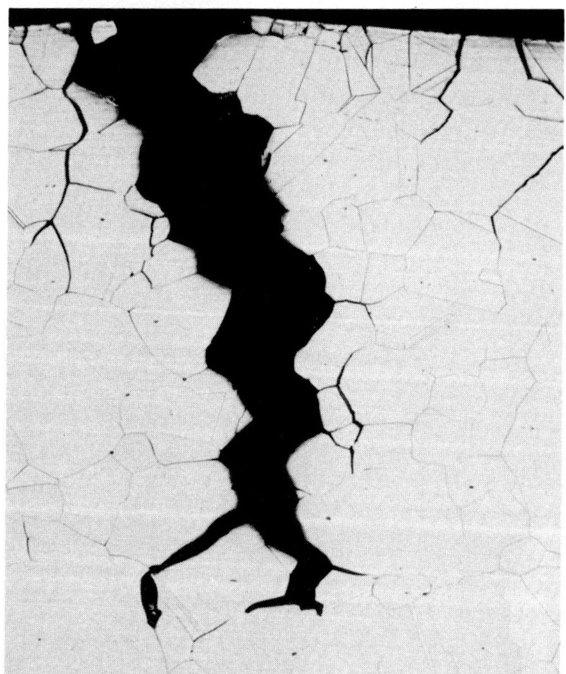

FIGURE 5.19 — Example of exfoliation corrosion: left, top view of surface of sheet; right, edge view.

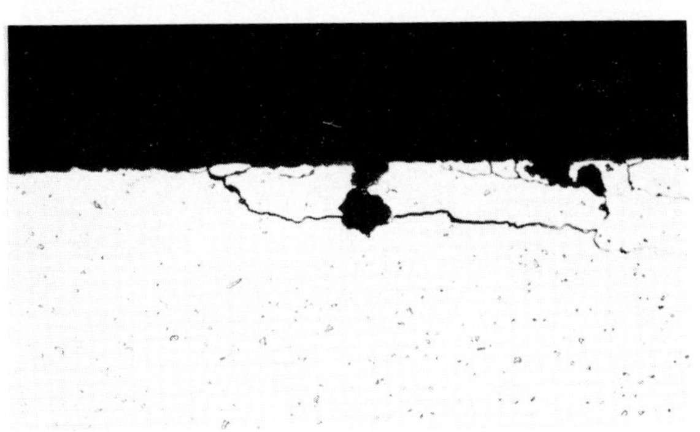

FIGURE 5.20 — Pitting developed after intergranular corrosion in 2024-T4 aluminum alloy.

Intergranular corrosion causes a drop in elongation, which can become appreciable before any significant loss of ultimate tensile stress or yield stress can be detected. In severe cases, there can be a marked loss of tensile properties, even though only a small volume of metal has corroded.

Local action that begins in a grain boundary may develop into a pit, and further intergranular penetration may proceed from the walls of the pit (Figure 5.20).

If the constituent is cathodic to the denuded zone, it remains intact and local cell action develops a crack in the adjacent denuded zone.

Since most intergranular corrosion is the result of small differences in composition at grain boundaries, the metallurgical history of an alloy can be as important as its average chemical composition.

This is because heat treatments and cold working affect not only the size and shape of the grains, but also the composition, location, amount, and size of the constituents. Heating an alloy to a high temperature just below the melting point tends to dissolve the alloying elements. Such a treatment is often carried out in practice and is called a *solution heat treatment.*

If the alloy is then quenched rapidly, the elements are trapped in solid solution. The solution is now supersaturated and there is a tendency for the elements to precipitate out as constituents. This tendency is accelerated by a moderate increase in temperature and is discussed more fully in Chapter 3.

Depending on the composition and location of these precipitated constituents, they may become the cause of intergranular corrosion. The amount and distribution is important, because even anodic and cathodic constituents will not cause intergranular corrosion if they are present in small sizes and are uniformly dispersed throughout the grain structure. This condition can be achieved by *over aging,* which means heating for a time longer than that which develops maximum mechanical properties of the alloy.

A practical example of this is the special T-73 aging treatment for the Al-Zn-Mg-Cu alloy A97075 which almost eliminates susceptibility to intergranular corrosion. If the alloy is quenched slowly from a high temperature, or if there is a delay between removal from a heat-treatment furnace and immersion in the quenching medium, there is a marked tendency toward susceptibility to intergranular corrosion.

Extensive cold work also will tend to cause precipitation in the grain boundaries and susceptibility to intergranular corrosion. An example is Al-5Mg alloy, which should not be used for rivets because the heavily deformed region just under the formed head suffers intergranular precipitation and intergranular corrosion. Stress corrosion also may occur in this case.

A second influence of cold work is to produce an elongated grain structure parallel to the surface. When grain boundary precipitation and susceptibility to intergranular corrosion occur, intergranular penetration tends primarily to be parallel to the surface. The expansive force of insoluble corrosion products tends to force the grains apart and leads to *exfoliation corrosion,* sometimes known as *lamellar* or *layer corrosion* (Figure 5.19).

In extreme cases, the edges of the affected area are leaflike and resemble the separated pages of a wetted book that has become swollen and begun to open up. It may even be possible to insert the tip of a pen knife for some distance into apparently solid metal in a direction parallel to the surface.

Attack on the austenitic stainless steels occurs in certain specific media after sensitization of the alloy has been developed by some heating process, normally welding. In this case, a precipitation of chromium carbide occurs around the grain boundaries of the alloy which reduces the chromium content of the grain immediately adjacent to the boundary. The deficient alloy in this area can then be attacked so severely in certain environments that the entire grain is surrounded and will fall from the structure (Figure 5.21). Welding of the unstabilized stainless steels can be a major source of this problem (Chapters 3 and 15).

Intergranular corrosion occurs most commonly in aluminum, copper, and 18-8 stainless steel alloys (Figure 5.22). A similar subsurface intergranular penetration pattern can be produced by corrosion

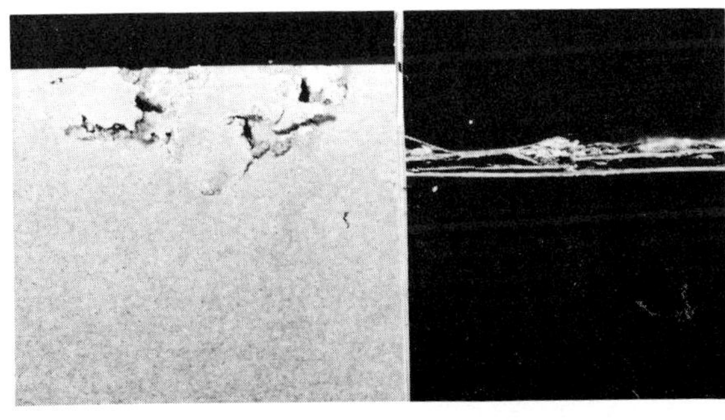

FIGURE 5.21 — Weld decay in the heat-affected zone (HAZ) of an austenitic stainless steel. [SOURCE: Corrosion Resistance of Metals, ACS Monograph No. 158, Reinhold, New York, N.Y. (1963).]

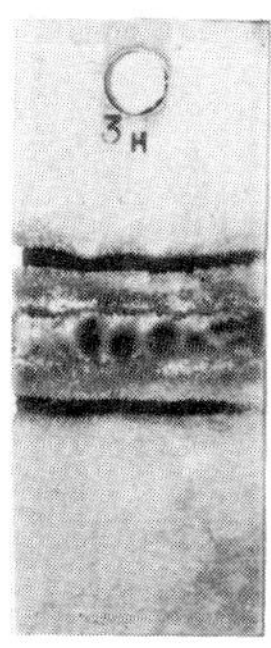
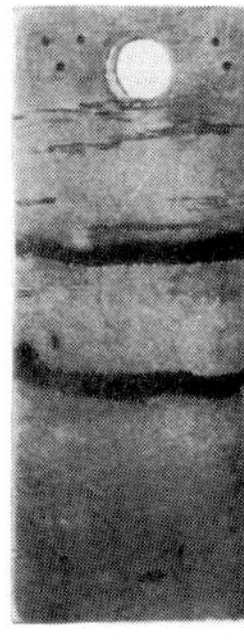

FIGURE 5.22 — An example of intergranular stress corrosion cracking of Type 304 stainless steel by oxygenated high-purity water, 500X.

fatigue and by hydrogen embrittlement (of steel at low temperature).

Intergranular corrosion is prevented by ensuring a metal microstructure that is immune to this type of attack. (For a short review on intergranular corrosion, refer to the relevant work listed in this chapter's bibliography.)

Transgranular Corrosion

Transgranular corrosion is a form of localized subsurface attack in which a narrow path is corroded at random across the grain structure of a metal, disregarding grain boundaries (Figure 5.23). Thus, cracks appear to develop across grains without any apparent effect on the crack direction by the presence of the grain boundary. It initiates on the surfaces and proceeds inward, presumably by local cell action.

Surprisingly, transgranular corrosion is not always discussed as a separate subject in corrosion textbooks, and there remain differences of opinion as to the mechanism and the reason for the course of the corrosion paths. It can occur during the stress corrosion cracking of austenitic stainless steels and less commonly during the stress corrosion cracking of low-alloy steels. It can also occur in the stress corrosion cracking of copper alloys in certain media (*e.g.*, ammonia). It seldom occurs in aluminum alloys.

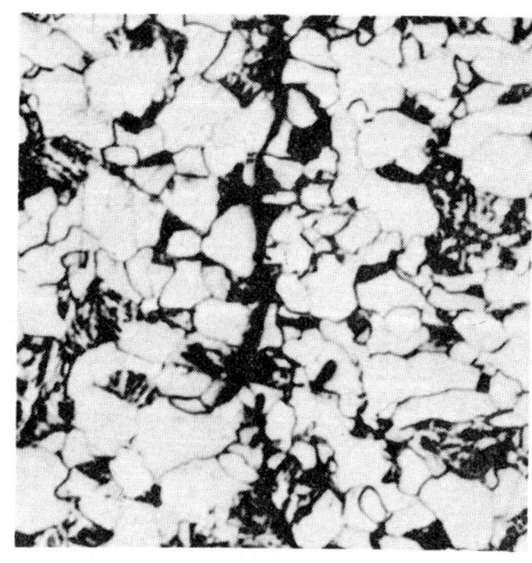

FIGURE 5.23 — An example of transgranular corrosion. The dark vertical line is the path of the corrosion damage.

Corrosion fatigue occurs most commonly by transgranular cracking. However, in the cases of both stress corrosion cracking and corrosion fatigue, penetration sometimes can be intergranular. On other occasions, both this and transgranular attack occur together. The type of penetration that occurs, transgranular or intergranular, varies with the alloy and with the corrosive environment.

Transgranular attack is avoided by the use of suitable nonsusceptible alloys.

Corrosion Fatigue

Fatigue is the failure of a metal by cracking when it is subjected to cyclic stress. The usual case involves rapidly fluctuating stresses that may be well below the tensile strength. As stress is increased, the number of cycles required to cause fracture decreases. There is usually a stress level below which no failure will occur, even with an infinite number of cycles, and this is called the *endurance limit.*

It has become a general practice to give the endurance limit as that stress below which no failures occur in a million cycles. A *fatigue curve,* commonly known as an S-N curve, is obtained by plotting the number of cycles required to cause failure against the maximum applied cyclic stress. A typical S-N curve is shown by the dotted line in Figure 5.24.

When a metal is subjected to cyclic stress in a corrosive environment, the number of cycles required to cause failure at a given stress will be reduced below that shown in Figure 5.24. This acceleration of fatigue by corrosion is called *corrosion fatigue.* A corrosion fatigue S-N curve can also be drawn, as shown by the solid line in Figure 5.24. It is lower than the normal fatigue curve obtained for the same metal in air. In both cases, the frequency of stressing should be reported because this factor influences the endurance limit.

The solid S-N curve obtained in this case is the result of both corrosion and fatigue. The solid curve indicates that metal life under such conditions will

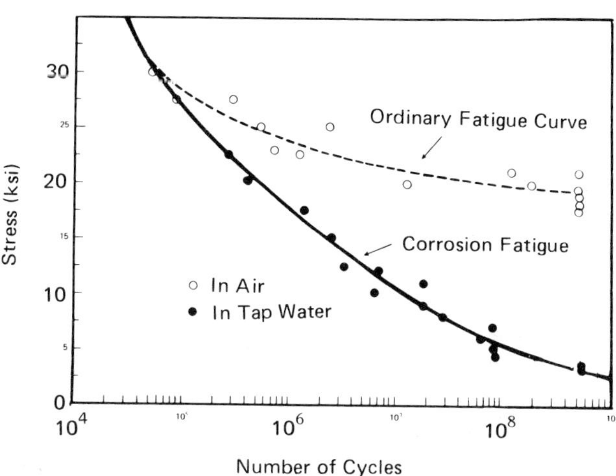

FIGURE 5.24 — Fatigue and corrosion fatigue curves for aluminum alloy.

be lower than (perhaps only 30 to 50% of) the dotted curve established in air. The S-N curve with corrosion tends to keep dropping, even at low stresses, and thus does not level off, as will the ordinary fatigue curve.

A marked drop in or elimination of the endurance limit may occur even in a mildly corrosive environment, especially in the case of a film-protected alloy. For example, deionized (*i.e.,* pure) water, which ordinarily will produce only film growth on an aluminum alloy immersed in it, will appreciably reduce the endurance limit of the same alloy subjected to cyclic stressing. This is because stress reversals cause repeated cracking of the otherwise protective surface film, and this allows access of the water to the unprotected metal below with resultant corrosion.

In aqueous solutions (and also under humid spray conditions when there is a film of water or solution on the metal surface), corrosion fatigue is accelerated by anodic polarization and retarded by cathodic polarization (cathodic protection), unless the metal is susceptible to hydrogen embrittlement, in which case cathodic polarization also will reduce the number of cycles to failure (by a combination of corrosion fatigue and hydrogen embrittlement).

At any given time prior to failure, damage due to corrosion fatigue will be greater than the sum of corrosion damage plus fatigue damage. Localized corrosion, such as pitting or intergranular corrosion, has a greater accelerating action than does uniform surface corrosion. Subsurface cracks develop at the affected sites and are usually transgranular (except for lead and tin) in both ordinary fatigue and corrosion fatigue. In ordinary fatigue, there is rarely more than one major crack that begins at the surface and leads to failure. In corrosion fatigue, there are usually many cracks, which initiate at the sites of local damage (Figure 5.25).

The cracking which develops with corrosion fatigue must be differentiated from the other forms of corrosion-induced cracking which are described in Chapter 6. When cracks are observed, the possibility of corrosion fatigue must be weighed against stress corrosion, hydrogen embrittlement, liquid metal, or *hot short* cracking. (Hot shortness is a metallurgical phenomenon not related to corrosion mechanisms.) Most often, the corrosion fatigue crack can be identified by the "shell" marks remaining on the fractured surface (Figure 5.25).

Failures that occur on vibrating structures (*e.g.,* taut wires or stranded cables) exposed to the weather under stresses below the endurance limit are usually caused by corrosion fatigue. Corrosion fatigue also has been observed in steam boilers, due to alternating stresses caused by thermal cycling.

Since corrosion fatigue limits are dependent on both chemical and physical characteristics of the environment which are difficult to define with accuracy, they are less reliable for use in the design of structures than ordinary fatigue limits.

Shot peening, which induces compressive stresses in the surface, tends to reduce corrosion fatigue, at least until subsurface corrosion penetrates into metal under tensile stress.

Electroplated sacrificial coatings (*e.g.,* zinc or cadmium on steel) protect against corrosion fatigue. Zinc coating on steel by spray metallizing also is effective. In aqueous environments, protection can be achieved by cathodic protection and by inhibitors, although these will not prevent ordinary fatigue that may occur at high cyclic stress levels.

The corrosion fatigue life of high-strength aluminum alloys can be improved by spray metallizing (with an aluminum alloy that is anodic to the base metal). (For a comprehensive review of the literature on corrosion fatigue up to 1955, refer to this chapter's bibliographic listing by Gilbert. A more recent survey of the subject is summarized in the listing by Scharfstein or Staehle.)

Dealloying

Another type of localized corrosion that is completely different from any of those hitherto described, involves the selective removal (*i.e.,* corrosion) of one of the elements of an alloy by either

FIGURE 5.25 — Corrosion fatigue. [SOURCE: Corrosion Resistance of Metals, ACS Monograph No. 158, Reinhold, New York, N.Y. (1963).]

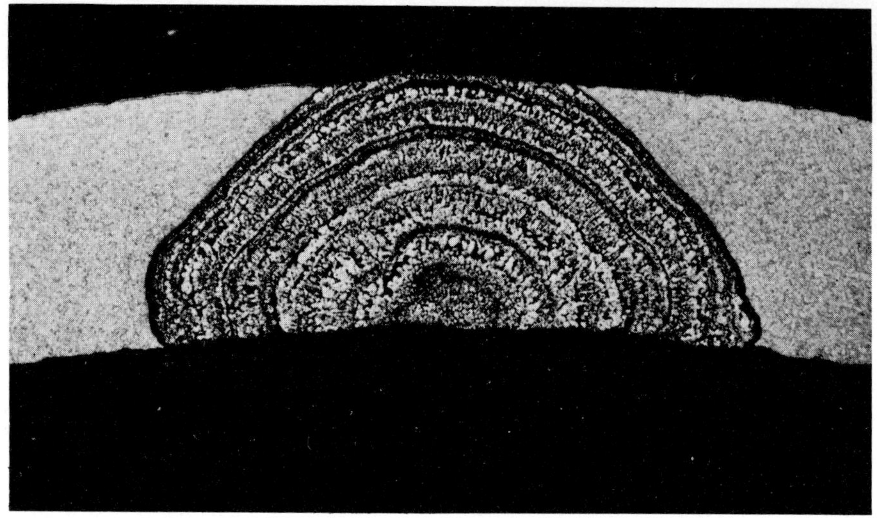

FIGURE 5.26 — Plug-type dezincification of brass. [SOURCE: Corrosion Resistance of Metals, ACS Monograph No. 158, Reinhold, New York, N.Y. (1963).]

preferential attack or by complete dissolution of the matrix followed by redeposition of the cathodic constituent. This has been called *parting* or *dealloying.* The element removed is always anodic to the matrix.

Although the color of the affected area may change, there is no visible evidence of loss of metal in the form of pits, dimension changes, cracks, or grooves. The surface shape and contour remain intact, including fine irregularities or roughness on the original surface. However, the affected metal becomes lighter, porous, and loses its original mechanical properties, *i.e.,* it becomes very brittle and has a very low tensile strength (Figure 5.26).

The various kinds of selective dissolution have been named after the alloy family that has been affected, usually on the basis of the element dissolved (except in the case of graphitic corrosion). Table 5.1 gives several examples, including *dezincification,* which refers to selective removal of zinc from brass.

Since the inference in Table 5.1 that each of several designations indicates a separate mechanism may be misleading, this system of terminology is undesirable and should be replaced by the generally

TABLE 5.1 — Nomenclature of Selective Dissolution Related to Alloy Constituents

Selective Attack Called	Removes	From Alloy System
dealuminization	aluminum	copper-aluminum
decobaltification	cobalt	Stellite (Co-Cr-W-C)
decuprification	copper	copper-silver copper-gold
demanganization	manganese	copper-manganese
denickelification	nickel	copper-nickel
desiliconification	silicon	silicon-copper
(no name yet)	silver	silver-gold
(no name)	tin	lead-tin (solders)
graphitic corrosion	iron	iron-carbon (gray cast iron)

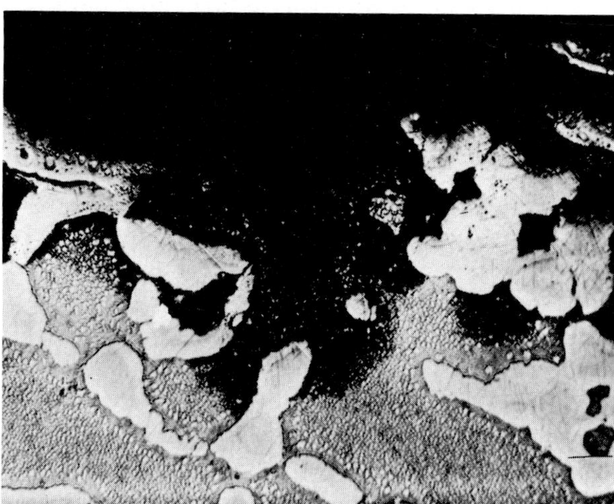

FIGURE 5.27 — Preferential corrosion of a low silicon phase in a silicon alloy. [SOURCE: Corrosion Resistance of Metals, ACS Monograph No. 158, Reinhold, New York, N.Y. (1963).]

applicable terms dealloying, parting, leaching, or selective dissolution (Figure 5.27). (Heidersbach[8] published a literature survey of the phenomenon of selective dissolution.)

Dezincification

The most common example of selective dissolution is the dezincification of brass (the name given to a range of binary copper-zinc alloys which contain 10 to 40% zinc). Dezincification can be recognized by a change from the original color to a distinctly red, coppery appearance. It is more common in brasses which contain more than 20% zinc (the "yellow" brasses). Brasses which contain less than 15% zinc (the "red" brasses) are practically immune to this form of attack, but can dezincify under specific conditions.

The addition of tin to the alloy is beneficial, and a small amount (0.02 to 0.05%) of antimony, arsenic, or phosphorus effects marked improvement. Inhibited admiralty metals, for example, contain 1% tin, 29% zinc, and 70% copper, plus As, Sb, or P, and are much more resistant to dezincification then yellow brass, although not as resistant as red brass.

Dealloying of brasses occurs in soft water service, especially when carbon dioxide content is high. Other factors that increase the susceptibility of brasses to dealloying are high temperature, high chloride content in water, low water velocity, crevices, and deposits such as sand on the metal surface.

The attack may occur in a local area and proceed inward rather than laterally, in which case it is called *plug-type dezincification.* The dezincified zones (plugs) generally remain in place and permit only slight seepage, even though they may completely penetrate the wall of a pipe containing water under pressure. However, if the section is stressed by bending, the plug of porous copper may pop out. The site of plugs may be recognized by the presence above them of a deposit of brownish-white zinc-rich corrosion product. Plug-type dezincification tends to occur with lower zinc alloys. It is also favored by more aggressive conditions such as higher temperatures.

If the selective leaching proceeds over the whole surface, it is termed *uniform* or *layer dezincification.* This type is more common with higher zinc brasses.

Dezincification can be prevented by the proper choice of alloys, as described above. (For further information on dezincification, refer to the relevant work listed in this chapter's bibliography.)

Graphitic Corrosion

Graphitic corrosion, or *graphitization* as it is sometimes called, is a form of dealloying or parting caused by the selective dissolution of iron from some cast irons, usually gray cast iron.

It usually proceeds uniformly inward from the surface, leaving a porous matrix of the remaining alloying element, carbon. Graphitization occurs in salt waters, acidic mine waters, dilute acids, and soils, especially those containing sulfates and sulfate-reducing bacteria.

There is no outward appearance of damage, but the affected metal loses weight, and becomes porous and brittle. Depending on alloy composition, the porous residue may retain appreciable tensile strength and have moderate resistance to erosion. For example, a completely graphitized buried cast iron pipe may continue to hold water under pressure until jarred by a workman's shovel.

The presence of sulfates and sulfate-reducing bacteria in soil stimulates this form of attack. The addition of several percent of nickel to cast iron greatly reduces susceptibility to graphitization. It is possible to produce graphitization in the laboratory by immersing gray cast iron in very dilute sulfuric acid. A completely graphitized cast iron pipe elbow may be used to write on paper as readily as a lead pencil.

Hydrogen Phenomena

There are several categories of hydrogen phenomena which are localized in nature. These are discussed briefly below, while more extended discussion of the cracking phenomena is found in Chapter 6 and of the high-temperature effects (*e.g.,* methanation) in Chapter 13.

The little hydrogen atom (not the molecule) is about one third the size of the crystal cube in a metal lattice and can diffuse easily through the structure. When the crystal lattice is in contact or is saturated with atomic hydrogen, the mechanical properties of many metals and alloys are diminished.

Four separate phenomena are commonly observed:

1. Loss of ductility and strength
2. Hydrogen blistering
3. Hydrogen-induced cracking (HIC)
4. High-temperature effects

The source of the hydrogen may be corrosion, electrochemical treatment (*e.g.,* electroplating or cathodic protection), or a high-pressure, high-temperature gaseous environment containing hydrogen.

In the absence of oxygen and in the presence of a clean metal surface, the hydrogen molecule (H_2) dissociates into two hydrogen atoms (which it is otherwise reluctant to do). When these hydrogen atoms contact the metal structure, many metals suffer a reduction in strength and ductility.

Loss of Ductility and Strength

All metals will lose some portion of their ductility and to a lesser extent, their strength, when exposed to hydrogen. Copper, aluminum, and the austenitic stainless steels are less affected than the iron or nickel-base alloys. The effect can be noted by conducting a slow strain rate tensile test of the materials in a hydrogen environment and comparing the results with the standard data. The changes noted are probably due to the same hydrogen lattice actions as those assumed to be responsible for brittle (delayed) fracture of the alloys. In the latter case, the metal fails under a prolonged static loading without evidence of ductility. The alloy is then said to be hydrogen embrittled (HE) or to have failed by hydrogen-induced cracking (HIC).

Removing the alloy from the source of the hydrogen and baking the metal around 200 C (400 F) to remove any absorbed hydrogen will restore the mechanical properties. Hydrogen in or on a metal has both surface and bulk atom effects. The loss of mechanical properties is probably more closely associated with the surface effect of hydrogen in contact with a metal lattice.

Hydrogen Blistering

In relatively soft steels, hydrogen blistering is observed under mildly corrosive conditions in the presence of specific contaminants (*e.g.*, sulfides, selenides, arsenides, antimony compounds, cyanides, etc.). Such species poison the recombination of atomic (*e.g., nascent*) hydrogen to the molecular gaseous form. Heavier corrosion in the absence of these poisons also can generate sufficient atomic hydrogen to cause the blistering.

When the atomic hydrogen enters the metal structures, nonmetallic inclusions catalyze the formation of molecular hydrogen within the metal lattice, generating tremendous internal pressures and causing splits, fissures, and even blisters on the metal surface (Figure 10.11). Normally an inch or so in diameter, some blisters have been observed as large as four feet in diameter in special cases. The tendency to blister can be combatted to some extent by using steels of the same grain size and cleanliness as are specified for low-temperature service.

Hydrogen-Induced Cracking (HIC)

HIC occurs in hardened or otherwise highly stressed steels, and is similar in many respects to stress corrosion cracking (SCC). However, cathodic protection aggravates the cracking (unlike SCC, which may be prevented by cathodic protection).

A large number of hardened steels, martensitic stainless steels, cold-worked austenitic stainless steels, precipitation hardening stainless alloys, etc. are susceptible to hydrogen-induced cracking. Even copper and nickel alloys and cold-worked nickel-chromium-molybdenum alloys at high strength are susceptible, particularly in a galvanic couple with a less noble material (Chapter 6).

High-Temperature Effects

At elevated temperatures and high partial pressures of hydrogen, the tendency for molecular hydrogen to split into hydrogen atoms is greater and the atoms themselves are more active and mobile. Above about 400 C (750 F), cuprous oxide inclusions in metallic copper can be reduced, generating steam within the metal lattice and causing internal fissuring.

Of more importance from the engineering standpoint is the *methanation* of steel. At temperatures above about 220 C (430 F), atomic hydrogen will react with the iron carbides in steel, forming methane within the structure and causing localized decarburization and fissuring. (The topic is discussed in Chapter 13.)

Other Forms

Other forms of localized attack (stress corrosion cracking, hydrogen embrittlement, and liquid metal embrittlement) are discussed in Chapter 6 as forms of environmental cracking.

References

1. Bothwell, M. R., Pitting Corrosion (of Magnesium), The Corrosion of Light Metals, John Wiley & Sons, New York, N.Y., p. 272 (1967).
2. Godard, H. P., Pitting Corrosion (of Aluminum), John Wiley & Sons, New York, N.Y., p. 49 (1967).
3. Kane, R. L., Localized Corrosion (of Titanium), John Wiley & Sons, New York, N.Y., p. 324 (1967).
4. Uhlig, H. H., Pitting in Stainless Steels and Other Passive Metals, Corrosion Handbook, John Wiley & Sons, New York, N.Y., p. 165 (1948).
5. Romanoff, M., Underground Corrosion, U.S. National Bureau of Standards, Circular 579, pp. 38, 47, 71-73 (1957).
6. Godard, H. P., The Corrosion Behavior of Aluminum in Natural Waters, Can. J. Chem. Eng., Vol. 38, p. 167 (1960).
7. Szklarska-Smialowska, Z., Review of Literature on Pitting Corrosion Published Since 1960, Corrosion, Vol. 27, No. 6, p. 223 (1971).
8. Heidersbach, R., Clarification of the Mechanism of the Dealloying Phenomenon, Corrosion, Vol. 24, p. 38 (1968).

Bibliography

Eisenberg, P., H. S. Preiser, and A. T. Thiruvengadam. How To Protect Materials Against Cavitation Damage. Mater. in Design Eng., p. 86 (March 1967).

Finnegan, J. E., et al. Optical Studies of Dezincification in a Brass. Corrosion, Vol. 37, No. 5, p. 256 (1981).

Gilbert, P. T. Corrosion Fatigue. Metallurg. Rev., Vol. 1, No. 3, p. 379 (1956).

Godard, H. P. Pitting Corrosion. The Encyclopedia of Electrochemistry. Reinhold, New York, N.Y., p. 925 (1964).

Greene, N. D. and M. G. Fontana. A Critical Analysis of Pitting Corrosion. Corrosion, Vol. 15, p. 25 (1959).

Kolotyrkin, Ya. Pitting Corrosion of Metals. Corrosion, Vol. 19, p. 261 (1963).

Lichtman, J. Z. Cavitation Erosion. The Encyclopedia of Electrochemistry. Reinhold, New York, N.Y., p. 159 (1964).

McDowell, J. R. Fretting Corrosion. The Encycolpedia of Electrochemistry. Reinhold, New York, N.Y., p. 159 (1964).

Nole, V. F. Dezincification. The Encyclopedia of Electrochemistry. Reinhold, New York, N.Y., p. 319 (1964).

Scharfstein, L. R. Corrosion Fatigue. The Encyclopedia of Electrochemistry. Reinhold, New York, N.Y., p. 272 (1964).

Staehle, R. W., et al. Corrosion Fatigue. NACE, Houston, Texas (1972).

Staehle, R. W., et al. Localized Corrosion. NACE, Houston, Texas (1979).

Streicher, M. A. Intergranular Corrosion. The Encyclopedia of Electrochemistry. Reinhold, New York, N.Y., p. 720 (1964).

Chapter 6

Environmental Cracking

ENVIRONMENTAL CRACKING

Definition

One of the most important forms of localized corrosion (Chapter 5) is environmental cracking. This is a particularly dangerous type of failure which can occur with a wide variety of metals and alloys in *specific* environments. This specificity (*i.e.,* one or more particular species causing cracking of only some materials) distinguishes the several types of environmental cracking from corrosion fatigue. In the latter type of attack, *any* corrosive environment serves to reduce the endurance limit or fatigue strength which would otherwise be anticipated.

There are three major types of environmental cracking:

1. Stress corrosion cracking (SCC); an anodic process.
2. Hydrogen-induced cracking (HIC); a cathodic process.
3. Liquid metal cracking (LMC); usually a physicochemical process.

All three of these phenomena meet the definition in Chapter 1 for *stress corrosion cracking,* as follows:

> Environmental cracking is a spontaneous *brittle fracture* of a *susceptible material* (usually itself quite ductile) under *tensile stress* in a *specific environment* over a *period of time.*

Mechanical forces (*e.g.,* tensile or compressive forces) will usually have little if any effect on the overall corrosion of metals, as measured for example in mils per year penetration. However, a combination of tensile stresses and a specific corrosive environment is one of the most important causes of sudden cracking-type failures of metal structures. All commercial metals will crack in a brittle manner in some environment. There is no one environment, however, that will crack all metals.

Fracture Mode

Environmental cracks are microscopic in their early stages of development. In many cases, they are not evident on the exposed surface by normal visual examination, and can be detected only by special techniques. Optical or scanning electron microscopy of failed sections may be required to fully identify them. As the cracking penetrates farther into the material, it eventually reduces the effective supporting cross section to the point where the structure fails by overload or, in the case of vessels and piping, escape (seepage) of the contained liquid or gas occurs.

Cracking usually is designated as either intercrystalline (also called intergranular) or transcrystalline (transgranular). Occasionally, but not usually, both types of cracking are observed. Intercrystalline cracks follow the grain boundaries in the metal. Transcrystalline cracks cross the grains without regard for the grain boundaries. The morphology of the cracks may change with the same material in different specific environments. In fact, a change in temperature or pH of the corrodent may change the mode of cracking of an alloy in a single corrodent.

Intercrystalline cracking of an aluminum alloy 2024 (A92024) nut and of a brass is shown in Figures 6.1 and 6.2, respectively; transcrystalline cracking of AISI 304 (S30400) is shown in Figure 6.3.

Environmental cracking should not be confused with *stress-assisted cracking.* This occurs when a material subject to intergranular corrosion (*e.g.,* an austenitic stainless steel suffering from weld decay) breaks more quickly because of a mechanical load superimposed on the area weakened by corrosion.

Sources and Magnitude of Stress

As the term *environmental cracking* implies, failures result from the combined action of tensile stress and a specific corrosive. Let us consider how much stress and how extensive the corrosive action must be to produce cracking.

Compressive stresses do not cause cracking. In fact, shot peening is used in practice to reduce the possibility of fatigue, stress corrosion cracking, etc.

We may say that failures generally do not result from the ordinarily applied engineering loads or stresses, as such. (An exception is the effect of *internal* stresses, evidenced by hardness, on hydrogen-induced cracking of several types.) Engineering loads, however, may be and usually are additive to the residual stresses already present in the structure. These residual stresses result from fabricating processes (*e.g.,* deep drawing, punching, rolling of tubes into tubesheets, mismatch in riveting, spinning, welding, etc.). There may be sufficient residual stresses left in castings to produce stress corrosion cracking. The crack in the Liberty Bell in Philadelphia has the macroscopic characteristics of a stress corrosion crack.

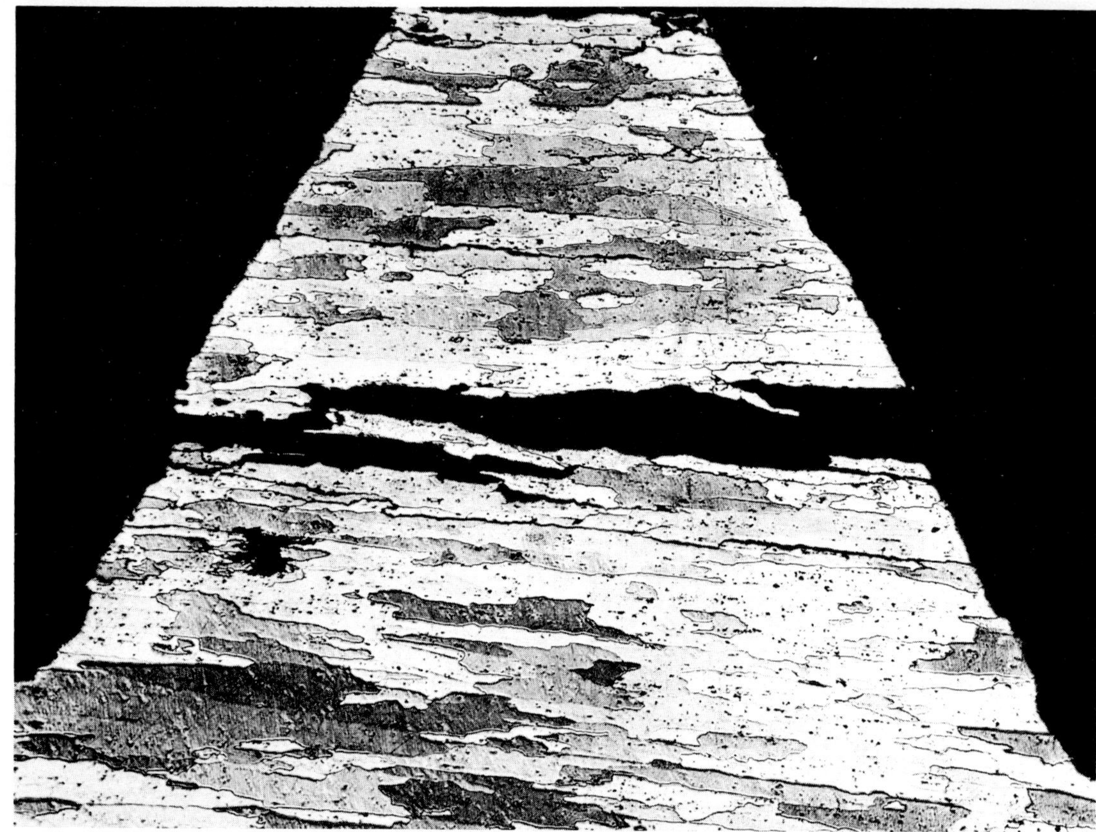

FIGURE 6.1 — Intercrystalline stress corrosion cracking in a nut machined from a 2024 aluminum alloy extrusion. Stresses were in the transverse direction. Note how cracks follow a series of boundaries. Etched 2.5% HNO_3, 1.5% HCl, 1% HF. X 100.

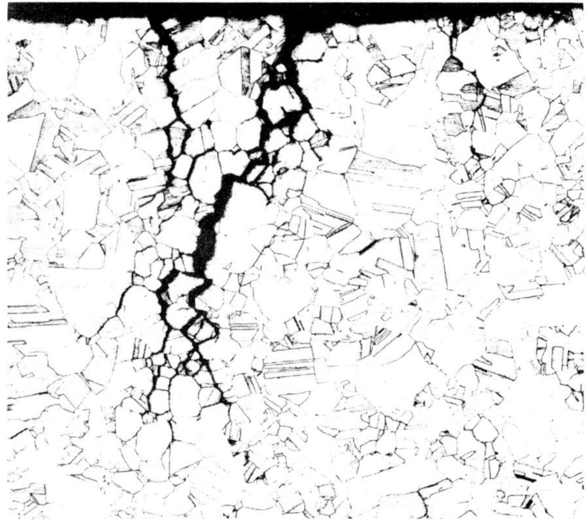

FIGURE 6.2 — Typical intercrystalline stress corrosion cracks in cartridge brass (70 Cu, 30 Zn). Etched 30% H_2O_2, 30% NH_4OH, 40% H_2O. X 75.

Residual stresses will remain in the structure unless it is annealed (or otherwise thermally stress relieved) following fabrication; this, of course, may often be impractical. Cooling from the high temperatures required may also induce internal stresses because of nonuniform cooling (*e.g.*, quenching stresses in the solution annealing and water quenching of austenitic stainless steels). In fact, very slow, controlled cooling is a prerequisite for effective stress relief. Cooling rates after tempering of hardened steel, as a palliative measure for HIC, may be less critical once the proper maximum hardness is attained.

Failures also have resulted from stresses produced by forcing oversized bushings or bearings into aluminum forged or extruded fittings used in aircraft. Corrosion products from general corrosion may build up between mating surfaces and, because they occupy so much more volume than the metal from which they are produced, generate sufficient stresses to cause SCC. Figure 6.4 shows an aluminum alloy nut that failed in this manner. Moisture working down the threads caused enough crevice corrosion to generate high stresses and induce SCC.

Even less is known about the magnitude of these stresses. To be effective, the stresses must be tensile in nature. As mentioned previously, and as will be discussed further, *compressive* stresses actually militate against cracking.

Some alloy/environment combinations seem to require stresses at or above the yield point (*e.g.*, steel in caustic). Others appear to have a definite threshold stress below which cracking will not occur. This is the case for sulfide stress cracking (hydrogen embrittlement) of alloy steels, in which a hardness below about Rockwell C 22 seems to result in practical immunity to this type of attack.

Other systems seem to have no apparent minimum stress. For example, chloride stress corrosion cracking of austenitic stainless steels has been

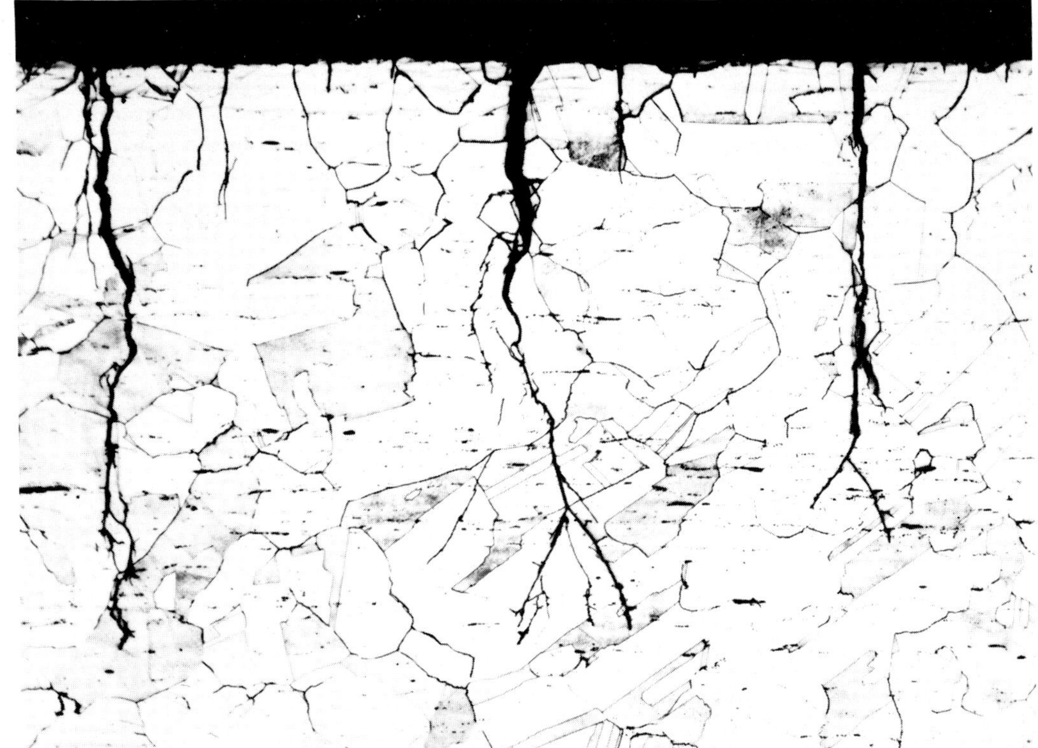

FIGURE 6.3 — Transcrystalline stress corrosion cracking in a Type 304 austenitic stainless steel. Material subjected to tensile stress in boiling 42% $MgCl_2$ solution. Etched electrolytically in a 10% oxalic acid solution. X 100. (Courtesy Ugiansky and Stiefel, National Bureau of Standards.)

FIGURE 6.4 — Stress corrosion crack in an aluminum alloy nut. X 1/4.

reported at stresses as low as one tenth the yield strength. (It should be noted, however, that notches or stress raisers, even of a microscopic nature, can concentrate the apparent stresses locally to a value several times the known macrostress.) Laboratory work suggests that, at least in the region in which the cracks originate, the stresses must be above the yield strength (*i.e.,* the stress must produce some plastic deformation of the material in that region, sometimes very small, at which cracking occurs).

Stress Corrosion Cracking (SCC)

Stress corrosion cracking is an anodic process, a fact which can be verified by the applicability of cathodic protection as an effective remedial measure. SCC may occasionally lead to fatigue, or vice versa. Usually, the true nature of the cracking can be identified by the morphology of the cracks themselves.

Usually, there is an induction period, during which cracking nucleates at a microscopic level. This is followed by actual propagation. Eventually, the cracks may be self-arresting to a large extent, as in the typical multibranched transcrystalline SCC, apparently due to local mechanical relief of stresses.

How severe is the corrosion that is required to produce SCC? Typically, SCC occurs under only very mildly corrosive conditions [although there are exceptions, as in the cracking of Monel Alloy 400 (N04400) in aerated hydrogen fluoride vapors]. The aforementioned Liberty Bell is tarnished, but one would hardly say that it has been severely corroded. The new surfaces resulting from a stress corrosion crack may show evidence of corrosion (*e.g.,* a blue color on brass, rust on steel), but other surfaces of the metal usually do not appear to be corroded.

We can say that SSC occurs in metals exposed in an environment where, if the stresses were nonexistent or even much lower, there would be no damage. On the other hand, if the structure, subject to the same stresses, were in a different environment (*i.e.,* one that did not contain the specific corrodent or corrodents for that material) there would be no failure either.

The term *stress corrosion cracking* implies the formation of cracks, and, as indicated above, there is usually little metal loss or general corrosion associated with it. If there is severe general corrosion, stress corrosion cracking usually will not occur. Thus, the failure of a stress bolt rusted away until it eventually cannot sustain the applied load is not classified as stress corrosion or SCC. However, if products from general corrosion are *trapped* so as to exert stress in a structure, they can cause SCC, as in the case of the aluminum nut previously described.

The idea, once prevalent, that only alloys, not pure metals, are susceptible to SCC is quite possibly correct. The question is, "How pure is *pure*?" Copper containing 0.004% phosphorous or 0.01% antimony is reported to be susceptible to SCC in ammoniacal environments (*i.e.,* those containing ammonia or ammonium ions). Cracking has been produced in a decarburized steel (containing less than 0.01% carbon, but, of course, small amounts of manganese, sulfur, and silicon) in a boiling ammonium nitrate solution.

Stress corrosion cracking has been produced in commercial titanium (containing, among other constituents, 600 ppm of oxygen and 100 ppm of hydrogen). Hence, the idea that a given material cannot fail by SCC because it is *commercially* pure is not correct.

Mechanism of SCC

One may well wonder why SCC occurs as it does under mildly corrosive conditions and under conditions in which engineering stresses are not excessive. Various mechanisms have been proposed, some of which involve a complex dislocation theory. (For an up-to-date review, particularly on chloride cracking of austenitic stainless steels, refer to R. W. Staehle's work, listed in this chapter's bibliography.)

Basically, the mechanism appears to be electrochemical/mechanical. As was pointed out earlier, stress corrosion cracking can be accelerated by the application of an anodic current and stopped by the application of a cathodic current (*i.e.,* cathodic protection). Hence, we must assume that SCC is, in part at least, electrochemical in nature. Six different theories have been advanced to explain the mechanisms.[1]

Electrochemical Aspects

In the electrochemical part of the mechanism, it is assumed that there is at least a thin film of electrolyte on the metal surface (in many cases the surface is completely immersed in the electrolyte) and that both anodic and cathodic areas exist on the surface covered by the liquid film. Almost without exception, very thin oxide films form almost instantaneously on the surface of all metals exposed to moisture. This oxide-covered surface is much less chemically active than a bare or unfilmed surface, and it will be the cathode in an electrolyte.

The anodes may be located at discontinuities in the oxide film covering the metal surface. They may be located at a grain boundary where the mismatch of metal lattices is large, or they may result from inhomogeneous composition of the metal at an area on the surface (Chapter 3).

It is postulated in one of the current theories of the mechanism of stress corrosion cracking (film-rupture mechanism) that stresses applied to susceptible alloys will rupture this oxide film on the metal surface. The film-free surfaces are normal to the applied stress and very narrow compared to the film-covered surfaces separating them. A film-free surface may be 0.15 volt or more anodic to a film-covered surface. Metal ions will be dissolved at those anodes, current will flow to the cathodes, and hydrogen will be discharged there.

Because the anodes are very small compared to the cathodic area, the current density will be large at the anodes and trenches or grooves will be formed. These grooves concentrate the stresses at the groove tips, preventing the film from reforming there. Because effective stresses soon disappear on the sides of the groove, a film will form there and thus produce cathodic areas very close to the anodes to accelerate chemical action. Hydrogen discharged at the cathodes on the sides of the crack may be seen coming out of the stress corrosion crack in Figure 6.5.

In the case of intercrystalline stress corrosion cracking, the initial notch may well form at grain boundaries which may be more chemically active than the grain face because of a discontinuity in the metal lattice there. Once a notch is formed, stress concentrations at its tip will cause the film there to rupture and prevent it from reforming, as was discussed previously.

There may be regions at a grain boundary in an alloy that are depleted in one or more alloying elements, and as a result are either anodic or cathodic to the alloy as a whole. In this case, we may have intercrystalline corrosion in the absence of stress. Stress will simply accelerate the attack by opening up channels for the corrodent to reach the metal. Such an attack is sometimes called *stress-accelerated* intercrystalline corrosion.

It has been suggested also that stresses may produce areas of inhomogeneous metal composition at the metal surfaces which could form anodes. Whether or not such a mechanism can produce as large an electrochemical solution potential difference between such anodes and cathodes as is produced by film rupture is an unanswered question.

Thus, stress corrosion cracking can be initiated and propagated at least in part by an electrochemical mechanism. However, in some work with magnesium alloys, it was calculated that a current density of 14 amp/cm^2 would be required to produce cracking at the measured rate by purely electrochemical processes. Current densities of 1 amp/cm^2 have been reported experimentally, but not 14 amp/cm^2. Thus, it was necessary to assume a mechanical fracture phase in the mechanism of stress corrosion cracking.

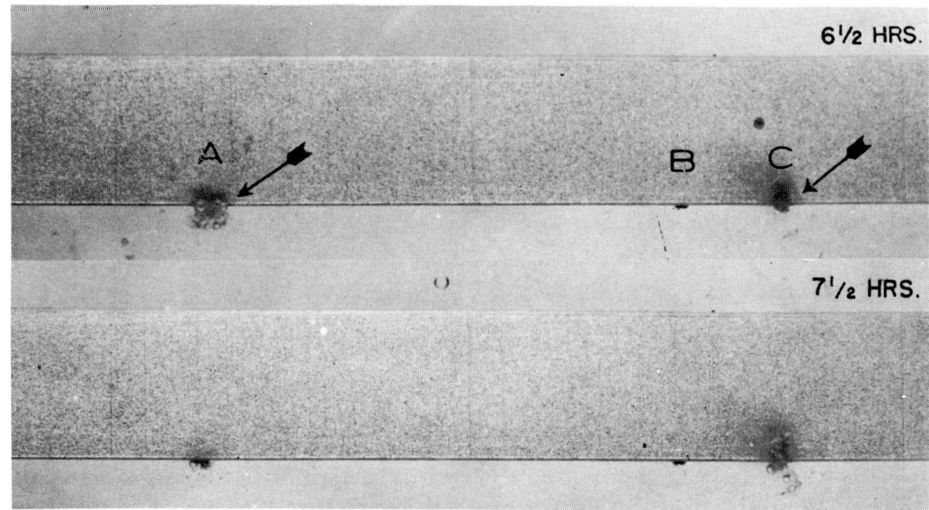

FIGURE 6.5 — Development of stress corrosion cracks in 304 stainless specimen exposed stressed in boiling 0.5 N NaCl, 0.1 N $NaNO_2$ solution. Photographs were taken of specimen edge in the solution without disturbing the experiment. Note hydrogen bubbles escaping from cracks developing at A and C (arrows). Subsequent examination showed a pit developing at B. X 6.

Mechanical Aspects

As the electrochemically propagated stress corrosion crack penetrates into the material, stress concentrations will build up at the crack tip until there is sufficient energy available to initiate mechanical cracking. This energy is required to form new surfaces and to deform material at the tip of the crack. (The energy to form new surfaces is very small compared to that needed to deform the metal at the tip of the crack.) The deformed (work-hardened) material becomes increasingly more difficult to fracture mechanically (the energy available as the result of the stress concentration being used up) and the mechanically propagated crack stops. The electrochemical process again takes over, building up stress concentrations at the tip of the crack, and the process is repeated.

Figure 6.6 shows evidence that crack propagation is in part mechanical. The upper curve is a plot of the elongation of a notched low-carbon steel specimen, exposed in a stressed condition to a boiling 20% NH_4NO_3 (ammonium nitrate) solution. The break in the curve indicates that a small mechanical fracture has occurred. If cracking were purely electrochemical, the curve should fall slowly and uniformly without sudden breaks.

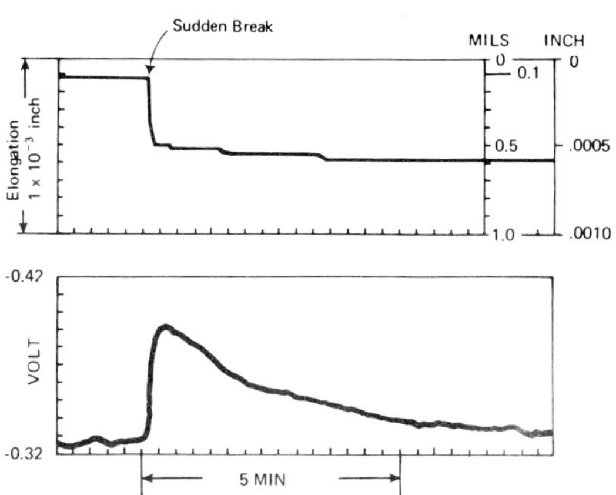

FIGURE 6.6 — Automatically recorded elongation-time and electrochemical solution potential-time curves for notched low-carbon steel specimen exposed in boiling NH_4NO_3 solution. Sudden break in elongation curve indicates mechanical fracture that exposed film-free and anodic metal to corrodent (indicated by change in potential). Fracture was prevented from going to completion by energy required to form new surfaces and plastically deform material at tip of crack.

The lower curve is a plot, recorded simultaneously with the extension curve, of the electrochemical solution potential of the specimen. The change in potential, approximately 0.08 volt, occurring simultaneously with the sudden extension of the specimen, confirms the idea that there had been a small mechanical fracture. This fracture exposed bare metal to the corrodent. This bare metal was much more reactive than the film-covered metal which was the "old surface," hence, the sudden change in potential.

Other Studies

Electron micrographs of the fractured surfaces of brass, magnesium alloys, and stainless steel specimens that had failed by stress corrosion cracking have shown evidence of similar discontinuous propagation of stress corrosion cracks.

Why will only certain corrodents produce cracking in a given material and why are certain materials apparently immune to cracking? There are several possible explanations as to why only certain corrodents produce cracking in a given alloy. There must be a certain driving potential (voltage) between the film-covered and film-free metal if the electrochemical process is to become active. Boiling magnesium chloride ($MgCl_2$) forms an invisible film on stainless steel, while a boiling $MgSO_4$ (same concentration of Mg) does not. Similarly, a boiling LiCl solution did not readily crack the steel until oxygen was bubbled through it and formed a sufficiently good film for cracking to develop.

For film rupture to be effective, stresses must produce inhomogeneous extension of the metal. If the metal is strained uniformly over its surface, surface films will be more or less uniformly ruptured. The film-free and film-covered areas will be more nearly the same size, and the current density at the anodes will be relatively small. Shallow pits rather than sharp notches will form.

Hence, in alloys susceptible to cracking, we will expect inhomogeneous straining of the metal when stresses are applied and narrow regions where stress concentrations are large. The difference in response of alloys to applied stresses will depend upon their submicroscopic characteristics.

Behavior of Engineering Materials

Aluminum Alloys

SCC in aluminum alloys is confined primarily to the higher strength alloys, such as those of the 2000 (A92000) and 7000 (A97000) series, whose strength properties are due to heat treatment rather than cold rolling. However, cracking also may occur in aluminum-magnesium alloys containing more than 3.5% magnesium, and failures have also occurred in certain high-strength casting alloys.

Failures have occurred most commonly in marine atmospheres, but are no longer confined to coastal areas due to the general use of salt for ice and snow removal. Nor, apparently, is a large amount of salt or chloride necessary. Failures of highly stressed castings have occurred in only mildly corrosive industrial atmospheres.

While many alloys' susceptibility to SCC is greater if they are stressed perpendicular to (rather than in the direction of) rolling, this condition is especially pronounced in aluminum alloys. The high-strength alloys generally are quite resistant to cracking if they are stressed in the direction of rolling or extrusion, but most susceptible if stressed in the short transverse direction.

Specifically, the yield strength of the 7075-T6 (A97075) alloy, tested in air, is reported to decrease from 550 MPa (80,000 psi) in the longitudinal direction to 410 MPa (60,000 psi) in the short transverse direction. However, the *threshold stress* (maximum stress that the material can withstand without failure when subjected to cyclic alternate immersion for 10 minutes in a 3.5% NaCl solution, 50 minutes in air) decreased from approximately 410 MPa (60,000 psi) (three-fourths of the yield strength) for material tested in the longitudinal direction to 82 MPa (12,000 psi) (one-fifth of the yield strength) for material tested in the short transverse direction.

Figure 6.7 gives a generalized picture of the grain structure to be expected in a thick aluminum alloy plate or extrusion. The failed aluminum nut shown in Figure 6.1 had been machined from an extruded rod and was stressed by the buildup of corrosion products in the transverse direction.

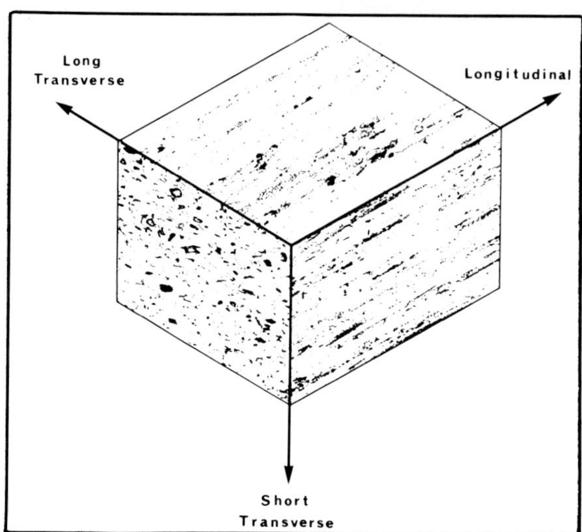

FIGURE 6.7 — Composite micrograph illustrating the grain structure in three orientations at the center of a 7075-T6 (A97075) aluminum alloy extrusion. Note that grain structure is very much elongated in longitudinal direction (direction of extrusion). Etched 2.5% HNO_3, 1.5% HCl, 1% HF in H_2O. X 100.

The grains of an aluminum alloy are very much elongated in the direction of rolling or extrusion (longitudinal direction). Current research indicates that if an alloy is susceptible to stress corrosion cracking, grain boundaries normal (at right angle) to the applied stress are subject to about the same attack in a corrodent whatever their orientation with respect to the direction of extrusion.

The boundaries of grains normal to a stress applied in the longitudinal direction are very short and mostly do not link up with that part of the boundary of the adjacent grain that is attacked. Hence, while there is attack on boundaries of individual grains, the regions attacked do not link up to form a continuous corroded path, and the material, therefore, is relatively immune to stress corrosion cracking if stressed in this direction.

On the other hand, the boundaries are very much elongated normal to a stress applied in the short transverse direction, so a crack formed on the elongated boundary of one grain frequently can join a crack formed at the boundary of the next grain. The continuous crack networks thus formed cause an alloy stressed in the short transverse direction to fail under a much lower stress than is required to cause a failure from stress applied in the longitudinal direction with respect to the rolling or extruded direction. Figure 6.1 shows how cracks can join up to produce a continuous crack in a material stressed in the transverse direction.

If the stress is applied in the long transverse direction, the structural conformation is more favorable to the linking up of corroded grain boundaries than if applied in the longitudinal direction, but less favorable than if applied in the short transverse direction. Hence, the alloy will support higher stresses without failure by stress corrosion cracking when stressed in the long transverse rather than in the short transverse direction, but smaller stresses than when stressed in the longitudinal direction.

Much current research is devoted to developing aluminum alloys that are immune to stress corrosion cracking, or heat treatments for current alloys that will make them completely resistant to stress corrosion cracking in all environments.

Copper-Base Alloys

Ammonia, or amines that may break down to give ammonia, generally are considered to be corrodents active in the cracking of brass and bronze. The resistance of brass to cracking is considered to increase as its copper content increases. Hence, a red brass (copper content, 85%) generally would be more resistant than a yellow brass (copper content, approximately 65%).

However, under certain conditions, even commercial copper tubing can fail by stress corrosion cracking. Two cases of failure have been seen in tubing used to connect homes with gas mains in the street. The copper tubing (deoxidized with and containing 0.033% phosphorus) had been supplied in rolls, and the stresses were believed to result from straightening out the rolled material or from rerolling the tubing after too much had been unrolled for use at a time. The corrodent would come from decayed vegetable matter in the soil. Fortunately, failures in installations like these are rare—only four or five failures have occurred in more than 50,000 installations.

Failures of brass bolts by stress corrosion cracking can occur readily. Two failures occurred after years of exposure to normal conditions, one in a home, the other in the wall of a public building. There was no general corrosion buildup in either case. Cracks started at the roots of threads where stresses tend to be concentrated. The stresses probably were due to tightening of the nuts during assembly.

Susceptibility of cartridge brass (70% copper, 30% zinc; C26000) to stress corrosion cracking was studied extensively during World Wars I and II. Small arms cartridge cases made from this alloy are deep-drawn in a series of steps with intermediate anneals to remove residual stresses. These anneals are expensive, but seem to be necessary to prevent the metal from cracking from excessive deformation and to reduce susceptibility to SCC. As will be discussed further, there are standard laboratory tests to evaluate the susceptibility of cold-worked brass to SCC or liquid metal corrosion.

Steels

The most common types of failure in low-carbon structural steels were those occurring in riveted steam boilers and in welded vats used to contain caustics at high temperatures.

Failures in boilers have been known to occur as the result of fine leaks at riveted joints where deposits from evaporated water could build up and where residual stresses were present in the steel as the result of a mismatch of rivet holes or of extensive caulking to prevent leaking. This stress corrosion cracking (also referred to as caustic embrittlement), in some cases at least, was believed to have triggered boiler explosions.

The replacement of steam locomotives with diesels and of riveted stationary boilers by welded pressure vessels has substantially eliminated this type of failure. However, failures of welded steel tanks containing hot concentrated caustics for de-enameling steel and for treating bauxite ores have been reported.

Stress corrosion cracking of structural steel specimens is easily produced in the laboratory in boiling concentrated nitrate solutions. While some authors have set 0.20% carbon as the upper limit of carbon content of steels susceptible to SCC in nitrates, Logan has produced SCC in a steel containing 0.03% carbon. On the other hand, he has also produced failures in a decarburized steel containing less than 0.01% carbon at the surface.

One special case of failures in low to medium carbon steels should be mentioned; namely, the failures of tanks used for storage and transportation of agricultural ammonia. A number of catastrophic failures triggered by SCC in containers for this service had been reported. After an extensive investigation, the following procedures were recommended to prevent failures of these tanks:

1. Tanks (over 3 feet in diameter) to be stress-relief annealed following fabrication to remove residual stresses.

2. Tanks to be purged to remove air before they were filled.

3. Water content to be maintained above 0.2%.

Elimination of air most probably removes one of the sources of the corrosive process, and the addition of water acts to inhibit the corrosion process in the event of accidental ingress of air to the ammonia.

Steels are now known to suffer environmental cracking in mixtures of carbon monoxide, carbon dioxide, and water (as steam or vapor). The specific mixture is required, the individual components or binary mixtures having no effect.

Steels can also suffer cracking in the presence of carbonates or bicarbonates under some conditions. In some cases, there is obviously reversion to free caustic (as at elevated temperatures). In other cases, as in soils of certain compositions, there may be hydrogen-induced cracking effects predominating.

The problem of combatting SCC in low-carbon steels has been solved in several ways. First, boiler water in locomotives was treated with sodium nitrate, phosphates, tannic acid, and a variety of other materials. Secondly, where welded steel vessels are used to contain hot caustic liquors, many authorities recommend annealing of the welded structure at about 650 C (1200 F) to remove residual stresses. Coatings also have been applied to serve as a barrier between the hot caustic and the stressed steel. Where welded structures are subject to gaseous environments, the welds should be annealed, if possible, or a protective coating should be considered. Another possibility is to use a different material, *i.e.,* one not susceptible to SCC in the particular environment anticipated.

Stainless Steels

Although martensitic and ferritic stainless steels can be stress cracked by hot caustic (and the ferritics by chlorides under some special conditions), the pre-eminent problems as to SCC arise with the austenitic grades.

The 18-8 grades (*e.g.,* AISI 304, 316, and their low-carbon and columbium- or titanium-stabilized variants) are highly susceptible to SCC, particularly by chloride ions. A common problem is SCC of stainless steel heat-exchanger tubes in chloride-bearing cooling waters when the skin temperature is above about 60 C (140 F). Crevices where chlorides can accumulate, *e.g.,* where the tubes are rolled into the tube sheet (Figure 8.7), are particularly susceptible. Calcareous or other films or deposits aggravate the situation by occluding and concentrating chlorides to a level higher than that in the bulk solution.

A particularly obnoxious form of SCC occurs from the *outside* when 18-8 equipment becomes contaminated under insulation with chlorides from water leaks or storm-driven water salts. Chloride cracking is usually transcrystalline (even when the 18-8 is sensitized) in natural waters. There appears to be little difference between conventional 18-8s and their manganese-substituted counterparts (*e.g.,* 204, 216) in this type of environment.

Higher temperatures are not required to produce SCC of the austenitic stainless steels. Highly stressed parts, such as cold-drawn wire cable, have cracked at atmospheric temperatures.

Austenitic stainless steels of the 18-8 variety are also susceptible to SCC by hot caustic solutions (Figure 10.8), or even by traces of caustic entrained in high-pressure steam.

In heavy water plants, it has been observed that the presence of hydrogen sulfide evidently weakens the passive oxide film on 18-8-type stainless, and chloride SCC occurs (or H_2S/Cl^- SCC) at much lower chloride concentrations and temperatures than would otherwise occur in a chloride-bearing natural water.

If the 18-8 is sensitized, it becomes subject to SCC in an intergranular mode in very high temperature water, as in nuclear reactors (Chapter 8), a form of attack known as oxygen stress corrosion cracking.

Likewise, sensitized stainless steels suffer intergranular cracking (IGSCC) in refineries during down time; a mode of attack ascribed to polythionic acid. These polythionic acids are formed by the interaction of hydrogen sulfide or metallic sulfides with sulfur dioxide and water when there is ingress of atmospheric air to the system. Stabilized grades are reportedly immune, but low-carbon grades are unreliable due to the possibility of long-time sensitization at normal service temperatures. This type of attack is combatted by alkaline or ammoniacal washes plus nitrogen inerting of the equipment.

New superferritic stainless steels, special proprietary grades, and "super" stainless steels (Chapter 4) have been developed specifically to resist chloride SCC. In general, those alloys containing more than 29% nickel are immune to chloride cracking. Resistance to caustic and other environments is a more nebulous topic, requiring investigation. (Practical experience with SCC is discussed in a Materials Technology Institute manual listed in this chapter's bibliography.)

Nickel Alloys

Of the chromium-free nickel alloys, nickel (Alloy 200; N02200) and Monel (Alloy 400; N04400) are the most common. These can be stress cracked in very hot, concentrated caustic (*e.g.,* 50% caustic at more than 200 C), but are still an order of magnitude more resistant than the austenitic stainless steels, at least. Nickel is even more resistant than Inconel (Alloy 600; N06600), as will be discussed further. For temperatures above about 450 C (850 F), a low-carbon grade of nickel (less than 0.02% carbon) is recommended. No instances of stress corrosion cracking of Alloy B (N10001) have been reported.

Alloy 400 (N04400) is also susceptible to stress corrosion cracking by the vapors of hydrofluoric acid in the simultaneous presence of oxygen. Apparently, cracking does not occur when the metal is fully immersed. A nitrogen atmosphere may be employed to prevent ingress of atmospheric moisture and oxygen and militate against SCC. As mentioned previously, at least moderate general corrosion may occur simultaneously with the SCC, and copper plating is often observed simultaneously.

Titanium Alloys

Titanium and its alloys are relatively immune to general corrosion in most media. Hence, once the cost had been reduced to the point where these materials could be considered for use where a high strength-to-weight ratio or a chemical inertness, or both, were desirable, they were used or considered for many chemical and aerospace applications.

The first problem developed as the result of contact of one of these alloys with an organic chloride at a temperature above 290 C (550 F). The chloride was believed to have broken down to form HCl. It was found subsequently that stress corrosion cracks developed in highly stressed alloy specimens coated with sea salt, chemically pure sodium chloride, or with a number of other chlorides (barium, cesium, lithium, or strontium) and heated to 400 to 430 C (750 to 800 F) in a laboratory furnace.

To avoid the possibility of failure from high-temperature chloride cracking, titanium alloy parts should be cleaned after fabrication and prior to heat treatment in solvents containing no chlorides. It has even been suggested that perspiration or even fingerprints have caused failures of material subsequently heat treated.

Recently, specimens of a number of titanium alloys containing regions of very high stress concentration have failed in chloride solutions at room temperature. Specimens were notched on one edge and a fatigue crack was started at the root of the notch. A plastic cell containing the corrodent was placed around the notched area and the precracked specimen was loaded as a cantilever beam until failure occurred.

Stress corrosion cracking also has been reported in tanks containing liquid nitrogen-tetroxide (N_2O_4) or methyl alcohol. At present, little has been published to indicate possible causes of the failures in commercial N_2O_4. It cannot, then, be said whether failures result from the action of small amounts of impurities in the N_2O_4, or if cracking has been inhibited in other instances by impurities in the N_2O_4.

Small amounts of water have inhibited cracking of titanium alloys in the methyl alcohol. Cracking has occurred in a much shorter time in specimens of commercially pure titanium and in an alloy exposed to alcohol vapor than in those completely submerged in the liquid. Methyl alcohol is much more potent in producing cracking than the higher alcohols, *e.g.*, ethyl alcohol.

The mechanisms of stress corrosion cracking of titanium alloys in hot salt, N_2O_4, or methyl alcohol have not been explained. There is probably no common mechanism. Cracking of the alloy in chlorides and methanol at room temperature probably can be considered an example of the mechanism of hydrogen-induced cracking, which will be discussed in a later section.

Magnesium Alloys

Certain magnesium alloys are susceptible to stress corrosion cracking in laboratory corrodents. Because cracking can be produced easily and quickly, they have been used in mechanistic studies. There have been no recent reports of service failures of these alloys attributed to stress corrosion cracking, and for that reason stress corrosion of magnesium alloys is not discussed in this chapter.

Table 6.1 is a brief summary of the resistance of metals and alloys, indicating the species which may cause environmental cracking (*i.e.*, stress corrosion cracking, hydrogen-induced cracking, and liquid metal cracking, which will be discussed further).

Plastic Materials

Thermoplastic and thermosetting resins will stress crack in a manner physically comparable to that of the metals. As in the case of the metals, a loss of cohesive bonding in the material allows the matrix to separate without any other evidence of attack. The type of bonding between atoms and chains of atoms may be different in metals and plastics, but the loss of cohesive strength along weak lines in the structure results in the same brittle cracking.

Cracking of stressed plastic structures can occur almost instantaneously when the specific cracking environment is introduced. As an example,

TABLE 6.1 — Materials Susceptibility to Environmental Cracking

Material	SCC	HIC	LMC
Magnesium	$Cl^- + K_2CrO_4$	—	Na, Zn
Aluminum	Cl^- + Oxidants	—	Hg, Ga, Na, In, Sn, Zn
Steels, Soft	OH^-, NO_3^-, CN^-, $CO_3^=$ NH_3, $CO/CO_2/H_2O$	—	Cu, Pb, Sb, Sn, Zn, As, P, Cd
Steels, Hard	OH^-, Cl^-	H^O, H_2S, H_2Sb	Cu, Pb, Sb, Sn, Zn, As, P, Cd
Stainless Steels			
Martensitic	OH^-	H^O, H_2S, Cl^-	?
Ferritic	OH^-, Cl^-	—	?
Austenitic	OH^-, Cl^-, $S_2O_6^{-(1)}$, $O_2/H_2O^{(1)}$	—	Zn, Cd, Al, Pb, Cu, In
Copper Alloys	NH_3, NO_3^-, Steam	H^O	Hg, Li, In
Monel	OH^-, $HF + O_2$	H_2S	Hg, S
Nickel	OH^-	—	S
Inconel	OH^-	—	S
Alloy C276, Alloy 625	OH^-	$H_2S/Cl^{-(2)}$	S
Titanium	HNO_3, Cl^- (300 C)	MeOH	Cd, Zn, Sn, Pb

(1) Sensitized.

(2) Severely cold-worked and in galvanic couple.

simply waving a stressed section of polycarbonate or polysulfone plastic over an open bottle of an oxygenated organic chemical such as a ketone, can produce immediate cracking. Other plastic cracking agent combinations may require a significant period of contact before cracking is obtained (*e.g.*, polyethylene in detergents). It is important to understand that such a mechanism of attack can occur on plastic materials and often must be evaluated before selecting the proper material for a specific service.

Many of the means for alleviating the cracking of metals are applicable for avoiding the cracking of plastics. These include stress relief, coating, alloying, redesign, or a change of material.

Hydrogen-Induced Cracking (HIC) or Hydrogen Embrittlement (HE)

Hydrogen-induced cracking (HIC) is a cathodic phenomenon; anodic protection will ameliorate and cathodic protection aggravate the attack. In fact, in susceptible material, field applications of cathodic protection have been responsible for cracking failures (*e.g.*, in pipelines and ship propellers).

In the sense of ordinary environmental cracking and the specificity to which we have referred, it is usually from the ordinary corrosion process that nascent atomic hydrogen is produced at the local cathodes. If these are in a highly stressed condition, they may be subject to HIC. The situation is aggravated when certain chemical species are present which act as negative catalysts (*i.e.*, poisons) for the dimerization of atomic to molecular hydrogen. The normal formation and evolution of molecular hydrogen,

$$2\,H^{0} \rightarrow H_2 \tag{6.1}$$

is suppressed, the nascent atomic hydrogen diffusing into the interstices of the metal instead of being harmlessly evolved as a gaseous cathodic reaction product.

There are many chemical species which poison this dimerization (*e.g.*, cyanides, arsenic, antimony, or selenium compounds). However, the one most commonly encountered is hydrogen sulfide (H_2S), which is formed in many natural decompositions (*e.g.*, rotten eggs), and in many petrochemical processes.

Processes or conditions involving wet hydrogen sulfide are called *sour services,* and the high incidence of sulfide-induced HIC has resulted in the term *sulfide stress cracking* (SSC). However, whether or not the "poison" on the metal surface is a sulfide, other inhibitors of the H_2 formation, or hydrogen atoms penetrating the steel in the absence of such inhibitors, the presence of hydrogen in the metal creates the conditions for cracking under stress and the same mechanism occurs. The presence of the sulfide or other inhibiting ion simply implies that a greater volume of hydrogen will penetrate the metal during a given time to produce cracking of the steel in a shorter time. The result is the same: HIC.

The SSC of medium strength steels has been a continuing source of trouble in the oil fields, and from these troubles evolved the NACE Standard MR-01-75, "Materials Selection for Sour Service." However, similar problems are encountered wherever wet hydrogen sulfide is encountered (*e.g.*, acid gas scrubbing systems, heavy water plants, waste water treatment). Steels having a hardness of Rockwell C22 or more may fail by SSC in any environment containing moisture and even very small amounts of H_2S. It is not necessary to have a superimposed stress, although this aggravates matters. The hardness of a steel is due to high *internalized* stresses induced by the martensite transformation (Chapter 3).

While structural, bolting, and pressure vessel steels, as well as piping materials, are of primary concern, other materials are also susceptible. MR-01-75 indicates the heat treatments and hardnesses which should be specified for a variety of steels: martensitic and ferritic stainless steels, precipitation hardening steels, etc.

Failures have occurred in the field when storage tank roofs have become saturated with hydrogen by corrosion and then been subjected to a surge in pressure, resulting in a brittle failure of the circumferential welds. Dangerous releases have occurred in chemical plant (*e.g.*, a sour vapor over naphtha) and in a blowdown tank in a heavy water plant.

In rare instances, even copper and Monel 400 (N04400) have been subjected to HIC (usually in sour service or under the influence of a cathodic current in a galvanic couple). More resistant materials, such as Inconels and Hastelloys often employed to combat HIC, can become susceptible under the combined influence of severe cold work, chemical poisons, and a direct current from the galvanic couple due to electrical contact with a more anodic member.

There are some failures which have not yet been distinguished as to whether they involve anodic or cathodic processes or both. Very high strength steels developed for aerospace applications may suffer chloride cracking, presumed to be HIC. Evidence of hydriding in cracked titanium suggests that cracking in methanol, previously discussed, may actually be HIC rather than SCC.

The mechanism of hydrogen-induced cracking has not been definitely established. Various factors are believed to contribute to unlocking the lattice of the metal, such as hydrogen pressure at the crack tip, the competition of hydrogen atoms for the lattice bonding electrons, the easier plastic flow and dislocation formation in the metal at the crack tip in the presence of hydrogen, the formation of certain metal hydrides in the alloy, etc. What is known is that the maximum effect takes place at room temperature, that as little as 5 ppm of hydrogen in the lattice produces the full HIC, that the hydrogen concentrates at stressed areas in the metal matrix, and that carbon and manganese downgrade the resistance of a steel, among other things.

This is the common *hydrogen embrittlement,* a *temporary* loss of ductility when steel becomes saturated with atomic hydrogen (*e.g.,* as in pickling, electroplating). Ductility can be restored, without permanent damage, by heating for a short time at about 200 C (400 F) to drive out the hydrogen.

The outstanding characteristic of this embrittlement is that the loss of ductility can only be detected by a very *slow* strain or a static stress on a notched coupon. It is *not* detected by a Charpy impact test, for example.

Liquid Metal Cracking (LMC)

Certain metals subjected simultaneously to a high tensile stress and specific molten metals suffer a form of environmental cracking known as *liquid metal cracking* (or embrittlement) (LMC). The specificity of the relation between alloy and molten metal is such that it fits our original definition. The mechanism of failure is said by some to be closely related to the hydrogen-induced cracking process.

The particular cracking of highly stressed yellow brass in mercurous nitrate solution (the basis for a test method) is actually a combination of electrochemical attack and LMC. It appears that mercurous ions are cathodically reduced to metallic mercury at local sites. The susceptible material then suffers LMC by the metallic mercury.

Metallic mercury from broken thermometers, blown manometers, or from certain catalysts, can move freely around a process (mercury has a relatively high vapor pressure) and raise havoc with any highly stressed copper alloy or Monel (Figure 6.8).

Mercury can cause LMC of high-strength aluminum alloys (as could sodium, gallium, indium, tin, and zinc). Molten copper (as from welding of copper-clad steel) will cause LMC of steels or stainless steels. Cadmium, aluminum, and lead can cause LMC of austenitic stainless steels,

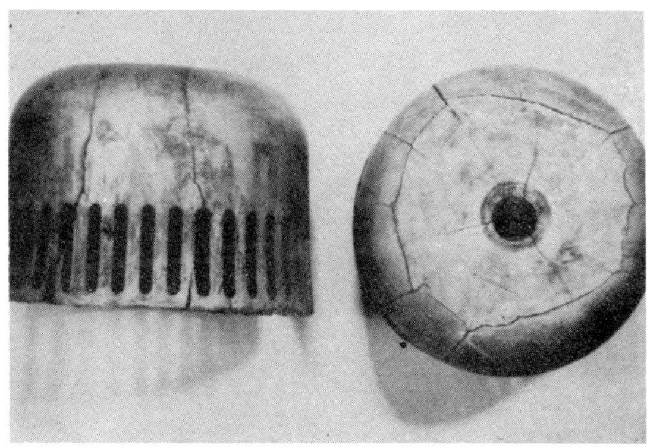

FIGURE 6.8 — Stressed Monel part from a distillation column which cracked from exposure to mercury.

and cadmium can attack titanium. Steel can be cracked by exposure to tin, cadmium, zinc, lead, bismuth, copper, and antimony.

The biggest problem, however, in engineering practice is zinc, because it is widely employed in galvanizing and in the zinc-pigmented paint system. Fortunately, zinc as a molten metal is only encountered during welding or brazing, or in the unfortunate event of a fire or explosion.

Nevertheless, overspray of zinc-rich paint from adjacent structures will render stainless steel or high-nickel alloys unweldable. Although most people should know better, problems still arise from those trying to weld galvanized structural members to alloy equipment. (The subject of LMC is addressed in R. W. Staehle's work and the Materials Technology Institute manual listed in this chapter's bibliography.)

A related subject is the cracking induced in high-nickel alloys by sulfur contamination. A low melting sulfur eutectoid is formed, which will penetrate the stressed metal in an intergranular mode of cracking akin to LMC. Any sulfur contamination results in immediate cracking during the welding of nickel and its alloy, so that scrupulous cleanliness must be observed. Even traces of sulfate ions have been known to cause cracking during inert gas tungsten arc welding of Alloy C (N10276).

Corrosion Testing for Environmental Cracking

Corrosion testing and test methods are the subject of a later chapter. However, a few remarks on testing as it pertains to environmental cracking are pertinent here.

Tests to determine the susceptibility of a material to environmental cracking are performed with several objectives:

1. In alloy development work, to compare the cracking resistance of new alloys with those currently in use.

2. To determine whether stock materials in hand are sufficiently resistant to justify their being fabricated into finished equipment or structures.

3. To determine whether new materials, as they are developed, are more suitable for use in a given operation than those currently employed.

In alloy development work, numerous alloy compositions must be screened. This suggests the need for testing many specimens in a laboratory environment. The more promising materials will then be further exposed (*e.g.,* to industrial or marine environments) in the field.

If the purpose of the test is to determine whether or not the material in the warehouse is sufficiently resistant to justify its use in the final product, the test must be of relatively short duration and, at the same time, correlate well with results that would be obtained in several years of ordinary exposure. Designing such tests and interpreting their results is a major problem today. If the corrosion engineer is to take full advantage of new materials as they are developed, the more promising materials must be kept continually, for at least the more critical areas of service, in test under real or simulated plant conditions.

One difficulty is that a negative proposition cannot be proven. If a test results in a cracking failure, one knows that failures may occur in service. If no cracking results, it may simply be that the test was not run long enough, or at a suitable stress, or did not really reproduce the true service conditions. The development of the Slow Strain Rate Test has done much to ameliorate this problem. When a tensile coupon is pulled *very* slowly in a corrosive environment, if it does *not* develop cracks it is unlikely to do so in service. If it *does* crack, it may not necessarily do so under less severe stress or strain in service in that environment.

There are a number of standardized tests available to study different modes of environmental cracking. Mixtures of sodium chloride and oxidants are used to study stress corrosion cracking of aluminum alloys; a boiling 42% magnesium chloride test is used to study SCC of stainless steel and related alloys; SSC is studied in a brine solution saturated with hydrogen sulfide; liquid metal cracking is studied in an internationally standardized mercurous nitrate test for cold-worked brasses; and ammonia cracking of the brasses may be evaluated by a standard test.[2]

Standardized tests are used mostly for alloy development and quality assurance. Of course, tests for environmental cracking may be run with any alloy system in specific environments for purposes of materials selection.

Prevention of Environmental Cracking

It is something of a joke among practicing corrosion engineers to say, "It's easy to prevent environmental cracking. Simply eliminate the stress, isolate the metal from the environment, or change the environment!" This is, of course, much easier said than done. Nevertheless, the five basic methods of corrosion control, as previously discussed, apply.

Change of Material

A total or even partial change of material (*e.g.*, "safe-ending" of heat-exchanger tubes) is a common approach. For complete reliability, use of a totally crack-resistant material is often the most economical approach.

Change of Environment

The removal of chlorides, caustic, or other major cracking-type species is an effective solution where possible. However, much less drastic changes are often effective as well; the removal of oxygen or oxidizing agents, a change in pH, or additions of inhibitors may be effective (nitrates have been used as an inhibitor against caustic embrittlement of steel; and chromates have controlled cracking of die steels and maraging steels in 3% sodium chloride solutions). Unfortunately, except in the laboratory, no inhibitors have been found to prevent chloride stress cracking of stainless steels in water services. Water inhibits cracking of titanium by methanol and of steel by ammonia.

Barrier Coatings

Even temporary rust preventative greases have been found helpful against atmospheric cracking of high-strength steels, although good commercial organic coatings are much more reliable. Certain silicone-base paints are routinely used to prevent *external* chloride stress cracking of insulated stainless steel vessels and piping.

Electrochemical Techniques

Cathodic protection has been found effective against anodic stress corrosion cracking. Lead-tin solders and nickel plating have been used to protect stainless steel tube ends against SCC by water. This technique must be carefully controlled to be effective. Austenitic stainless steel has been known to suffer chloride SCC while simultaneously causing galvanic corrosion of steel components.

Design

This is probably the most important consideration when use of suitable alternative materials of construction is not possible.

Residual stresses may be minimized by careful thermal stress relief appropriate to the alloy. Compressive stresses may be introduced by controlled shot peening. Stress raisers may be eliminated or minimized. Crevices should be eliminated, as well as areas where deposits can accumulate.

Tube wall temperatures can be minimized by proper sizing of condensers. Waters that will cause SCC of an 18-8 stainless steel tube bundle in less than two years at high heat flux can be rendered innocuous by designing to a maximum tube wall temperature of about 50 C (120 F).

Conditions conducive to evaporation and concentration of corrosive species can be minimized. Vertical condensers with water on the tube side should be pressed full, the tube sheets being adequately vented. Tube-to-tube sheet rolls can be seal-welded or strength-welded. All welds should be full penetration to prevent accumulation of species in the crevice which otherwise exists. Alternate steam and water service for 18-8 stainless steel exchangers should be avoided like the plague. Drive pins should be of constant diameter (*i.e.*, rolled) rather than tapered. Dead areas should be avoided around exit nozzles.

In short, every effort should be made to minimize stress as well as to reduce the concentration of specific species in the equipment to a level compatible with engineering performance.

Conclusion

Stress corrosion and other types of environmental cracking are among the most insidious of the localized forms of corrosion. Usually, equipment leaks before it ruptures, but this is not always the case. Since there is usually little or no loss of thickness before penetration, astute materials selection, proper design, and assiduous inspection of critical equipment (Chapter 14) should be the order of the day.

References

1. Brown, B. F., Stress Corrosion Cracking Control Measures, NBS Monograph 156 (1977).
2. ASTM, Use of Mattson's Solution of pH 7.2 to Evaluate the Stress Corrosion Cracking Susceptibility of Copper-Zinc Alloys, ASTM Standard G37-73 (1979).

Bibliography

ASTM. Review of Recent Studies on the Mechanics of Stress Corrosion Cracking of Austenitic Stainless Steels. STP610 (1976).

ASTM. Stress Corrosion Cracking of Metals—A State of the Art. STP-518 (1972).

Bowman, R. W., et al. CO/CO_2 Cracking in Inert Gas—Miscible Flooding. Materials Performance, Vol. 16, No. 4, p. 28 (1977).

Cotterill, P. The Hydrogen Embrittlement of Metals. Pergamon Press, New York, NY, 1961.

Jewett, R. P., et al. Hydrogen Environmental Embrittlement of Metals. NASA Technology Report, B73-10168, June, 1973.

Materials Technology Institute. Guidelines for Control of Stress-Corrosion Cracking of Nickel-Bearing Stainless Steels and Nickel-Base Alloys. Manual No. 1. Columbus, OH, 1978.

NACE. Fundamental Aspects of Stress Corrosion Cracking. Houston, TX, 1969.

Smialowski, M. Hydrogen in Steel. Pergamon Press, New York, NY, 1962.

Staehle, R. W., Ed. Stress Corrosion Cracking and Hydrogen Embrittlement of Iron Base Alloys. NACE, Houston, TX, 1977.

NOTE: References of Chapter 5 contain considerable data on SCC, HIC, LME, caustic cracking, CO-CO_2 cracking of steel, and excellent bibliographies for these subjects.

NOTES

Chapter 7

Inhibitors

INHIBITORS

Introduction

Definition of Corrosion Inhibitor

An inhibitor is a substance which retards or slows down a chemical reaction. Thus, a *corrosion inhibitor* is a substance which, when added to an environment, decreases the rate of attack by the environment on a metal. Corrosion inhibitors are commonly added in small amounts to acids, cooling waters, steam, and other environments, either continuously or intermittently to prevent serious corrosion.

It would be awkward to include mechanisms of inhibition in the definition of a corrosion inhibitor because inhibition is accomplished by one or more of several mechanisms. Some inhibitors retard corrosion by *adsorption* to form a thin, invisible film only a few molecules thick; others form visible bulky precipitates which coat the metal and protect it from attack. Another common mechanism consists of causing the metal to corrode in such a way that a combination of adsorption and corrosion product forms a *passive* layer.

We also include in the definition those substances which, when added to an environment, retard corrosion but do not interact directly with the metal surface. This type of inhibitor causes conditions in the environment to be more favorable for the formation of protective precipitates or it removes an aggressive constituent from the environment.

Presentation

The use of corrosion inhibitors has grown to be one of the foremost methods of combating corrosion. To use them effectively, the corrosion engineer must, first of all, be able to identify those problems which can be solved by the use of corrosion inhibitors. Second, the economics involved must be considered, *i.e.,* whether or not the loss due to corrosion exceeds the cost of the inhibitor and the maintenance and operation of the attendant injection system. Third, the compatibility of inhibitors with the process being used must be considered to avoid adverse effects such as foaming, decreases in catalytic activity, degradation of another material, loss of heat transfer, etc. Finally, the inhibitor must be applied under conditions which produce maximum effect.

This chapter on inhibitor fundamentals was written with the above tasks in mind. Corrosion inhibitors are discussed from four points of view:

1. Their effects on the corrosion process.
2. Their interactions with various aggressive environments.
3. Properties of the inhibitors themselves.
4. Possible effects of inhibitors on unit operations.

The chapter emphasizes the protection of steel. Other materials, such as aluminum, copper, brass, and zinc can be protected by suitable inhibitors. The references and bibliography at the end of the chapter give sources of specific information.

Basic Types of Inhibitors and How They Work

Polarization Diagrams

The electrochemical nature of corrosion and the concept of mixed potentials were explained in preceding chapters. In this chapter, the relationship of corrosion inhibitors to anodic and cathodic polarization will be explained.

Of the four components of a corrosion cell (anode, cathode, electrolyte, and electronic conductor), three may be affected by a corrosion inhibitor to retard corrosion. The inhibitor may cause:

1. Increased polarization of the anode (anodic inhibition),
2. Increased polarization of the cathode (cathodic inhibition), or it may
3. Increase the electrical resistance of the circuit by forming a thick deposit on the surface of the metal.

Of course, the bulk film-formers also restrict diffusion of depolarizers (such as dissolved oxygen) to the surface of the metal; hence, they may play a dual role. The resistance of the electronic conductor connecting the anodes and cathodes (*i.e.,* usually the resistance of the metal itself) is very low and cannot be changed by corrosion inhibitors.

The effects of a corrosion inhibitor on a corrosion cell are conveniently determined by polarizing the corroding metal in a suitable electrolyte with varying amounts of current from an external source such as a battery. A simple laboratory apparatus for polarization measurements is shown in Figure 7.1. The force which must be applied to stimulate the anodic or cathodic reactions is measured by the potential difference between the working electrode and a reference electrode. No current is applied to the reference electrode; hence, it is used as a standard against which the potential of the working electrode is measured.

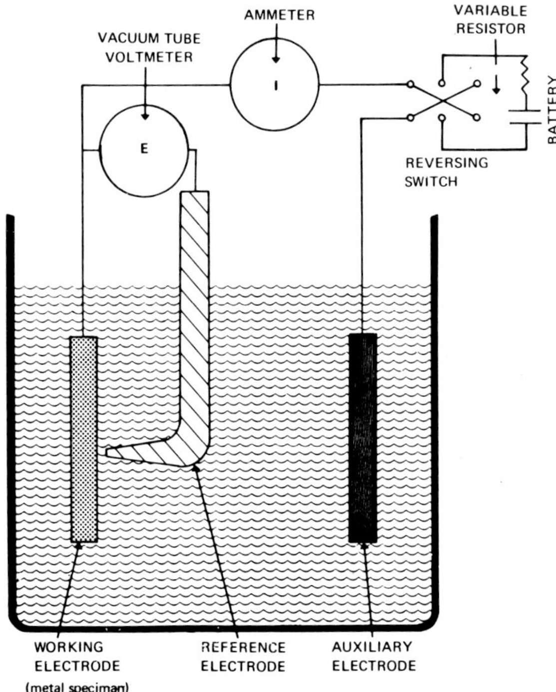

FIGURE 7.1 — Laboratory apparatus for polarization measurements.

In an experiment to compare the effects of several inhibitors on the polarization of steel in a corrodent, such as dilute acid, the apparatus would be assembled using steel as the working electrode, dilute acid as the electrolyte, and an inert material such as carbon or platinum as an auxiliary electrode. Then the current would be increased in steps by means of the variable resistor, and the potential of the working electrode read at each step. A plot of the current versus potential in the absence and presence of an inhibitor shows the effects of the inhibitor on the polarization characteristics of the steel. Typical curves are shown in Figure 7.2.

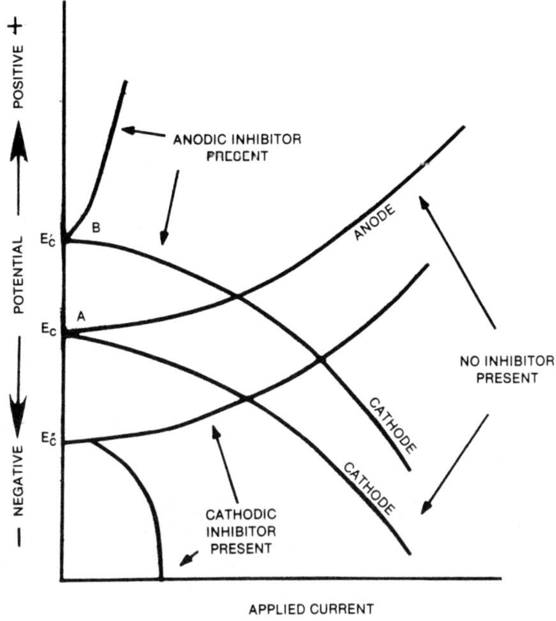

FIGURE 7.2 — Effect of inhibitors on polarization curves.

When no current is applied and the working electrode has achieved a steady state, then the potential is the *corrosion potential,* E_C, of the electrode material. The corrosion potential is the mixed potential to which the anodes and cathodes are polarized by the corrosion reaction. It is also an indication of the effects of corrosion inhibitors.

Types of Inhibitors

Six classes of inhibitors will be discussed: (1) passivating (anodic), (2) cathodic, (3) ohmic, (4) organic, (5) precipitation inducing, and (6) vapor phase inhibitors. While some authors may use slightly different classes, these will illustrate the complexity of the inhibitor picture.

Anodic Passivating Inhibitors

Anodic inhibition is illustrated in Figure 7.2, which shows an increase in the polarization (a large potential change results from a small current flow) of the anode in the presence of an anodic inhibitor. Addition of the inhibitor causes the corrosion potential to shift in a cathodic direction from E_C to E_C'.

Anodic inhibitors which cause a large shift in the corrosion potential are called *passivating inhibitors.* They are also called *dangerous inhibitors* because, if used in insufficient concentrations, they cause pitting and sometimes an increase in corrosion rate. There are two types of passivating inhibitors: oxidizing anions such as chromate, nitrite, and nitrate which can passivate steel in the absence of oxygen; and the nonoxidizing ions such as phosphate, tungstate, and molybdate which require the presence of oxygen to passivate steel. With careful control, however, passivating inhibitors are frequently used because they are very effective in sufficient quantities.

Passivating inhibitors such as sodium chromate (Na_2CrO_4) and sodium nitrite ($NaNO_2$) do not require oxygen to be effective. They increase the rate of anodic passivation to the extent that the anodes are polarized to a passive potential (or Flade Potential), indicated by E_f in Figure 7.3. The cathode will be depolarized, as indicated by the cathodic lines in Figure 7.3. The depolarized cathodic curve then intersects the anodic curve in the passive region at E_C. Adsorption of the inhibitor on anodic areas also plays a part in the process because it decreases the current (or corrosion rate) required for the anode to reach the critical passive potential (E_f).

Note that when an insufficient amount of inhibitor is used, the cathodic curve will intersect the anodic curve in an active region or in both active and passive regions, indicated by A and B in Figure 7.3. In the former case, corrosion will proceed at a high rate; in the latter case, passivity is unstable, and the corrosion potential will oscillate between A and B, usually resulting in pitting of the metal. Measurement of the corrosion potential when using

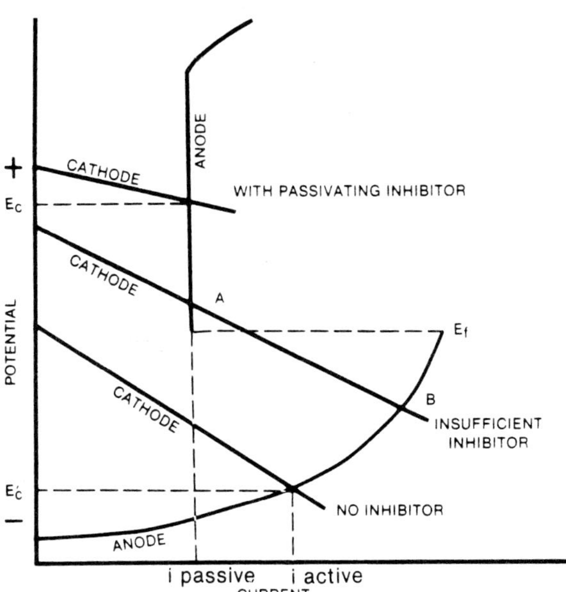

FIGURE 7.3 — Effect of passivating inhibitors on the corrosion of iron.

passivators is a good way to determine if the inhibitor is doing its job because a large positive shift in potential should occur if the metal passivates. Passivation by inhibitors is more difficult at higher temperatures, higher salt concentrations, lower pH, and lower dissolved oxygen concentrations.

Mechanism. The mechanism by which chromate passivates steel has been studied extensively, and it appears likely that protection is afforded by a combination of adsorption and oxide formation on the steel surface. Adsorption helps to polarize the anode to sufficient potentials to form very thin hydrated ferric oxides which protect the steel. Because the oxide film is invisible on steel, articles protected by chromate remain bright in otherwise aggressive environments. The oxide film is a mixture of ferric and chromic oxides and is kept in good repair by adsorption and oxidation with very little loss of metal as long as sufficient chromate remains in solution.

Chromates are accelerators of corrosion at low concentration because they are good cathodic depolarizers. The passive oxide film is conductive and cathodic to steel; therefore passive steel consists almost entirely of cathodic areas. When the passive film is penetrated by scratching or by dissolution, and when insufficient chromate is present to repair the film, the exposed steel becomes a small anodic area on which accelerated localized corrosion can occur, resulting in pitting of the metal (Chapter 2, Area Effects). Mechanisms similar to that proposed for chromate are believed to apply to nitrites and nitrates.

It is practical to passivate steel in any aqueous solution, except those which contain easily oxidized substances in solution or high concentrations of chloride ions. For example, chromate should not be used in hydrogen sulfide (H_2S)-containing (or *sour*) environments because it is lost by oxidation of the sulfide to free sulfur. High concentrations of chloride ions prevent passivation because they compete with chromate for adsorption, thus preventing polarization of the anodes. They also prevent deposition of protective oxides by forming a soluble complex with ferric ions.

For a given concentration of passivating inhibitor, there is a concentration of chloride and sulfate ions which will cause depassivation. Table 7.1 lists the "critical concentrations" of sodium chloride (NaCl) and sodium sulfate (Na_2SO_4) required to cause pitting of steel in the presence of various concentrations of sodium chromate and sodium nitrite. The critical concentrations will vary depending on other factors; for example, more chloride or sulfate will be required for depassivation as the temperature is lowered, the oxygen concentration is increased, or the pH is increased.

Chromate concentration should be maintained at at least twice the level required to prevent pitting. In calculating the amount of chromate needed for passivation, allowance should be made for the inhibitor which is consumed initially in establishing the passive film. The amount of sodium chromate consumed in passivating steel in sodium chloride solutions is given in Table 7.2. For example, if it is

TABLE 7.1 — Critical Concentration of Sodium Chloride or Sodium Sulfate Above Which Pitting of Armco Iron Occurs in Chromate or Nitrite Solutions

5-Day Tests, 25 C, Stagnant Solutions

Inhibitor	Concentration ppm	Critical Concentration, ppm NaCl	Critical Concentration, ppm Na_2SO_4
Na_2CrO_4	200	12	55
	500	30	120
$NaNO_2$	50	210	20
	100	460	55
	500	2000	450

[SOURCE: H. H. Uhlig, Corrosion and Corrosion Control, John Wiley & Sons, Inc., New York, N.Y., p. 232 (1963).]

TABLE 7.2 — Amount of Sodium Chromate Consumed by Steel in 50 Days in Establishing Passivity in Sodium Chloride Solutions

Sodium Chromate ppm	kg/1000 m² (lb/1000 sq ft) Sodium Chromate Consumed in Water Containing 10 ppm NaCl	1000 ppm NaCl
25	4.88 (1.0)	—
50	6.35 (1.3)	—
100	2.44 (0.5)	5.37 (1.1)
250	1.46 (0.3)	6.35 (1.3)
500	0.49 (0.1)	3.42 (0.7)
1000	0.49 (0.1)	3.42 (0.7)

[SOURCE: M. Darrin, Ind. Eng. Chem., Vol. 38, p. 368 (1946).]

desired to maintain 250 ppm of sodium chromate in water containing 1000 ppm of sodium chloride, then referring to Table 7.2, 6.35 kg of sodium chromate per thousand square meters of exposed steel should be added in addition to the quantity required to achieve a concentration of 250 ppm.

Nonoxidizing passivators such as sodium benzoate, polyphosphate, and sodium cinnamate require the presence of oxygen to cause passivation. They do not inhibit corrosion in the absence of oxygen. They apparently function by promoting the adsorption of oxygen on the anodes, thereby causing polarization into the passive region. Nonoxidizing passivators are also dangerous when used in insufficient amounts because the oxygen which is required for passivation is a good cathodic depolarizer.

Cathodic Inhibitors

Cathodic inhibitors either slow the cathodic reaction itself, or they selectively precipitate on cathodic areas to increase circuit resistance and restrict diffusion of reducible species to the cathodes. The effects of cathodic inhibitors on cathodic polarization are shown in Figure 7.2. In this case, where the anodic polarization is unaffected, the corrosion potential is shifted to more negative values from E_C to E_C''.

The cathodic reaction is often the reduction of hydrogen ions to form hydrogen gas. Some cathodic inhibitors make the discharge of hydrogen gas more difficult, and they are said to increase the hydrogen overvoltage. Compounds of arsenic and antimony are examples of this type of inhibitor which are often used in acids or in systems where oxygen is excluded. Another possible cathodic reaction is the reduction of oxygen. The inhibitors for this cathodic reaction are different from those mentioned for the more acidic systems.

Other cathodic inhibitors utilize the increase in alkalinity at cathodic sites to precipitate insoluble compounds on the metal surface. The cathodic reaction, hydrogen ion and/or oxygen reduction, causes the environment immediately adjacent to the cathodes to become alkaline; therefore, ions such as calcium, zinc, or magnesium may be precipitated as oxides to form a protective layer on the metal. Many natural waters are self-inhibiting due to the deposition of a scale on metals by precipitation of naturally occurring ions.

Inhibition by polarization of the cathodic reaction can be achieved in several ways, and several examples already have been given. The three main categories of inhibitors which affect cathodic reaction are cathodic poisons, cathodic precipitates, and oxygen scavengers.

Cathodic Poisons. Cathodic poisons are substances which interfere with the cathodic reduction reactions, *i.e.*, hydrogen atom formation and hydrogen gas evolution. The rate of the cathodic reaction is slowed, and because anodic and cathodic reactions must proceed at the same rate, the whole corrosion process is slowed.

While some cathodic poisons such as sulfides and selenides are adsorbed on the metal surface, compounds of arsenic, bismuth, and antimony are reduced at the cathode to deposit a layer of the respective metals. Sulfides and selenides generally are not useful inhibitors because they are not very soluble in acidic solutions, they precipitate many metal ions, and they are toxic. Arsenates are used to inhibit corrosion in strong acids, but in recent years the trend has been to rely more on organic inhibitors (discussed later in this chapter; also Table 7.4) because of the toxicity of arsenic.

A serious drawback of the use of cathodic poisons is that they sometimes cause hydrogen blistering of steel and increase its susceptibility to hydrogen embrittlement. Since the recombination of hydrogen atoms is inhibited, surface concentration of hydrogen atoms is increased, and a greater fraction of the hydrogen produced by the corrosion reaction is absorbed into the steel. Note that hydrogen is *ad*sorbed on the surface, but some of it is *ab*sorbed into the steel.

Hydrogen atoms which penetrate steel may pass through and diffuse out the other side if it also is not corroding to produce hydrogen. Blisters are formed when hydrogen atoms combine to form hydrogen molecules (H_2) inside the steel. Molecular hydrogen does not diffuse through steel; therefore, it collects at defects or voids to create pressures which may reach a million psi or more.

Figure 7.4 shows the increase in the fraction of produced hydrogen which penetrates the steel as sulfide or arsenic concentrations are increased. Only small amounts of sulfide or arsenic are required to increase the amount of hydrogen penetrating the steel, which accounts for the frequent occurrence of blistering and hydrogen embrittlement in the presence of these poisons.

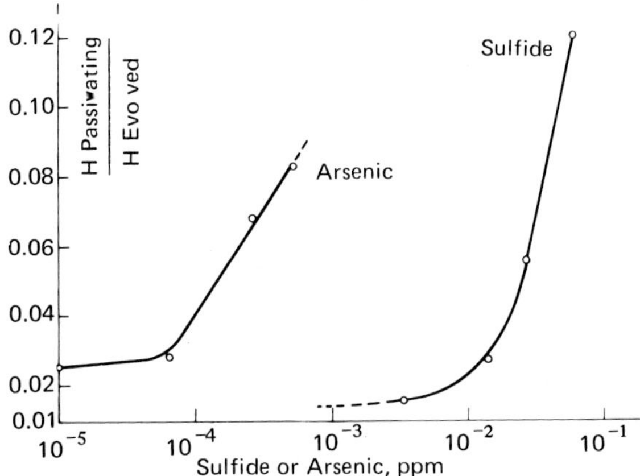

FIGURE 7.4 — Effect of sulfide and arsenic on the fraction of corrosion-produced hydrogen which enters steel. [SOURCE: Hudson, Snavely, Payne, Fiel, and Hackerman, Absorption of Hydrogen by Cathodically Protected Steel, Corrosion, Vol. 24, No. 7, pp. 189-196 (1968).]

It has been suggested that the absorption of hydrogen by steel could be used to advantage in sealed absorption-type refrigeration machines. Corrosion occurs very slowly in these systems, but sufficient hydrogen is produced to lower the efficiency of the thermal cycle, thus requiring an occasional pump-down of the units. The use of an inhibitor, such as an antimony compound, would cause the hydrogen to pass through the steel piping and vessels so that pumping would not be required. In this application, it would be important to use a steel which is not susceptible to blistering or embrittlement.

Cathodic Precipitates. The most widely used cathodic precipitation-type inhibitors are the carbonates of calcium and magnesium because they occur in natural waters and their use as an inhibitor usually requires only an adjustment of pH. Zinc sulfate ($ZnSO_4$) precipitates as zinc hydroxide $Zn(OH)_2$ on cathodic areas and is considered an inhibitor of this type. Phosphates and silicates are not distinctively cathodic or anodic inhibitors, but appear to be a combination of both types, so they will be considered later in the section on precipitation inhibitors.

Many natural waters and municipal water supplies contain calcium carbonate (limestone, $CaCO_3$) in solution. Limestone is dissolved in water to form soluble calcium bicarbonate [$Ca(HCO_3)_2$]. Limestone can be caused to precipitate again, forming a milky-white suspension by making the calcium bicarbonate solution more alkaline or by adding more calcium. Usually, lime is added to accomplish both objectives.

The objective of corrosion-inhibiting water treatment is to increase the alkalinity of the water to a pH at which precipitation of $CaCO_3$ is just about to occur. If the appropriate pH is exceeded, $CaCO_3$ will precipitate to form a slimy, porous deposit which does not provide corrosion protection and may increase corrosion by creating concentration cells involving oxygen. At the correct pH, the deposit will be fairly hard and smooth and similar to an eggshell. Once a protective deposit is formed, the pH of the water must be maintained at the equilibrium level, because if it is allowed to become acidic, it will redissolve the protective deposit.

A convenient way to express the condition of a water with respect to its tendency to deposit $CaCO_3$ is the Langelier Index, which is the difference between the pH of the water and the pH required to precipitate $CaCO_3$ (Table 8.2).[1]

Although many metal ions form insoluble hydroxides, few are useful cathodic corrosion inhibitors. Zinc sulfate, which is a good example, when added to neutral water, causes polarization of the cathodic reaction by precipitating zinc hydroxide.

Oxygen Scavengers. Corrosion of steel in waters above about pH 6.0 is due to the presence of dissolved oxygen which depolarizes the cathodic reaction and increases corrosion. Neutral water of low salt content in equilibrium with air at 21 C (70 F) will contain about 8 ppm of dissolved oxygen. The concentration of oxygen decreases with increasing salt concentration and increasing temperature. Only 0.1 ppm of oxygen is required to increase corrosion rates seriously in a dynamic system. In static systems, a higher concentration of oxygen is required to increase the corrosion rate seriously because the corrosion reaction soon depletes the oxygen supply in the immediate vicinity of the metal.

Oxygen scavengers help inhibit corrosion by preventing the cathodic depolarization caused by oxygen. Oxygen scavengers are added to water, either alone or with a corrosion inhibitor, to retard corrosion. Organic corrosion inhibitors alone in aerated brine water will slow general corrosion, but will not always prevent pitting attack. The most common oxygen scavengers used in water at ambient temperatures are sodium sulfite (Na_2SO_3) and sulfur dioxide (SO_2). At elevated temperatures, hydrazine is used to remove oxygen.

The reaction rate of sulfites with oxygen at low temperature is slow, so a catalyst is usually added. Cobalt, manganese, and copper salts are the best catalysts. Cobalt gives the greatest increase in reaction rate. Copper should not be added to water which contacts steel or aluminum because it lowers the hydrogen overvoltage and increases corrosion rate. Consequently, cobalt is preferred and manganese is a close second choice.

Hydrazine reacts very slowly with oxygen in water at low temperatures in the absence of a catalyst, and thus is not often used at low temperatures.[2] The hazards of hydrazine are significant, especially in the hands of untrained personnel. In high-pressure boilers, hydrazine is the preferred oxygen scavenger.

Ohmic Inhibitors

Inhibitors which increase the ohmic resistance of the electrolyte circuit already have been considered to some extent in discussions of anodic and cathodic film-forming inhibitors. Because it usually is impractical to increase resistance of the *bulk* electrolyte, increased resistance is practically achieved by the formation of a film, a microinch thick or more, on the metal surface. If the film is deposited selectively on anodic areas, the corrosion potential shifts to more positive values; if it is deposited on cathodic areas, the shift is to more negative values; and if the film covers both anodic and cathodic areas, there may be only a slight shift in either direction.

Organic Inhibitors

Introduction. Organic compounds constitute a broad class of corrosion inhibitors which cannot be designated specifically as anodic, cathodic, or

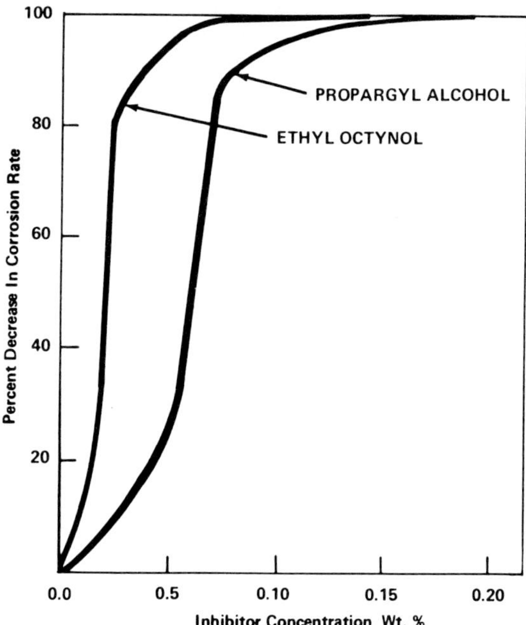

FIGURE 7.5 — Effect of concentration of organic inhibitor on corrosion rate [SOURCE: J. G. Funkhouser, Corrosion, Vol. 17, p. 283 (1961).]

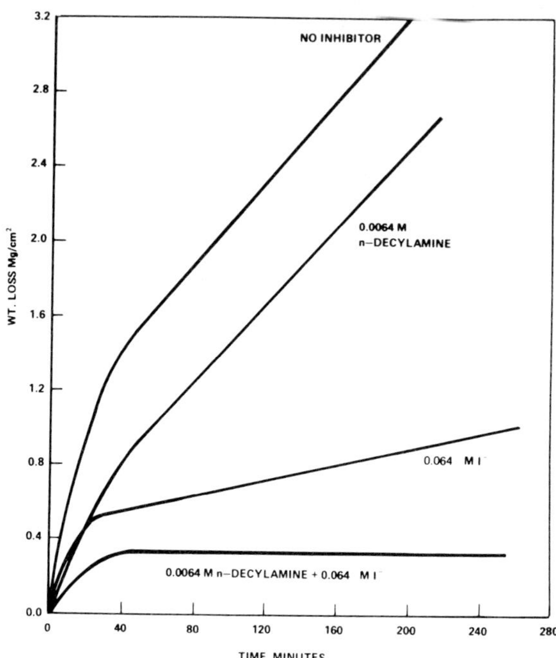

FIGURE 7.6 — Corrosion of mild steel in 3 M phosphoric acid (30%) showing synergistic effect of n-decylamine and iodide ion. [SOURCE: Hackerman, Snavely, and Payne, J. Electrochem. Soc., Vol. 113, p. 677 (1966).]

ohmic. Anodic or cathodic effects alone are sometimes observed in the presence of organic inhibitors, but, as a general rule, organic inhibitors affect the entire surface of a corroding metal when present in sufficient concentration. Both anodic and cathodic areas probably are inhibited, but to varying degrees, depending on the potential of the metal, chemical structure of the inhibitor molecule, and size of the molecule.

The typical increase in corrosion inhibition with inhibitor concentration, as shown in Figure 7.5, suggests that inhibition is the result of adsorption of inhibitor on the metal surface. The film formed by adsorption of soluble organic inhibitors is only a few molecules thick and is invisible.

Organic inhibitors will be adsorbed according to the ionic charge of the inhibitor and the charge on the metal surface. Cationic inhibitors (positively charged, +), such as amines, or anionic inhibitors (negatively charged, −), such as sulfonates, will be adsorbed preferentially, depending on whether the metal is charged negatively or positively (opposite sign charges attract). The in-between potential at which neither cationic nor anionic molecules are preferred is known as the zero point of charge or ZPC. Thus, a combination of cathodic protection and an inhibitor which is adsorbed more strongly at negative potentials gives greater inhibition than either cathodic protection or an inhibitor when used alone.

Synergism with Halogen Ions. The efficiency of organic amines as corrosion inhibitors is improved when certain halogen ions are present. Halogen ions alone inhibit corrosion to some extent in acid solutions. The iodide (I^-) ion is the most effective, followed by bromide (Br^-) and chloride (Cl^-). Fluoride (F^-) does not have significant inhibitive properties. Chloride ions, for example, lower the rate of attack on steel by sulfuric acid. A combination of amine and iodide may be more inhibitive than either additive alone, *i.e.*, the two additives are *synergistic*.

Figure 7.6 shows a comparison of corrosion rates of mild steel in 3.0 M phosphoric acid (H_3PO_4) in the presence of n-decylamine, iodide ion, and a combination of n-decylamine and iodide ion. One explanation for synergism is that steel adsorbs iodide ions whose charge shifts the surface potential in a negative direction, thereby increasing adsorption of the cationic amine.

Effects of Molecular Structure. How the size of organic molecules influences their effectiveness as corrosion inhibitors has been investigated many times. However, the results are not consistent enough to permit formulation of a general rule regarding the effect of increasing molecular weight.

Structures of some organic inhibitors are given in Figure 7.7. Primary amines such as n-decylamine become more efficient inhibitors as the chain length is increased, but, in contrast, primary aliphatic mercaptans such as n-butyl mercaptan and some aldehydes decrease in efficiency as the chain length is increased. These results probably are due to the interaction of various factors which influence the strength of the adsorption bond, compactness of the adsorbed layer, and tendency of the adsorbed molecules to cross-link or otherwise interact with neighboring molecules.

There is little doubt that the bonding of amines to a metal surface is through the nitrogen atom. For example, in the series of saturated cyclic imines, in-

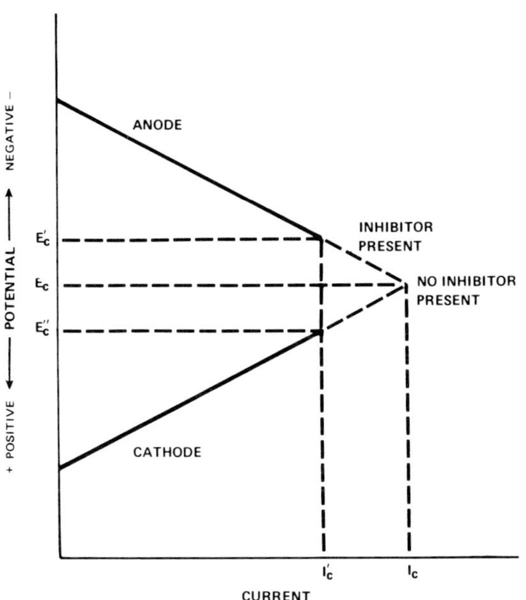

FIGURE 7.7 — Structures of some organic inhibitors.

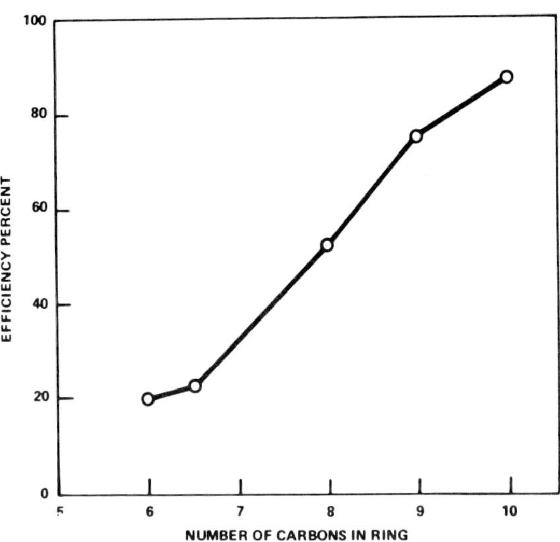

FIGURE 7.8 — Inhibitor efficiencies of saturated cyclic imines for corrosion of steel in 6 M hydrochloric acid (22%). [SOURCE: Hackerman, Hurd, and Annand, Corrosion, Vol. 18, p. 37 (1962).]

hibitor efficiency increases, as shown in Figure 7.8, as the number of carbon atoms is increased in the ring up to at least 10.

Adsorption. The observations given previously indicate that for soluble organic inhibitors, the strength of the adsorption bond is the dominant factor. Adsorption of an inhibitor from solution establishes the following *equilibrium:*

$$I_{solution} \rightleftharpoons I_{surface} \qquad (7.1)$$

where I is the concentration of a soluble organic inhibitor.

It is characteristic of an equilibrium that if the concentration of one species is changed, then the concentration of species with which it is in balance will change in the same direction to preserve equilibrium. From Equation (7.1), it is seen that the amount of inhibitor on the surface is increased by increasing $I_{solution}$, *i.e.,* the concentration of inhibitor used in the environment to be inhibited. Inhibitor efficiency increases with concentration until the surface is saturated, *i.e.,* it has adsorbed inhibitor molecules on all available sites. Therefore, the stronger the adsorption bond, the lower the concentration of inhibitor in solution required to achieve a given coverage of the surface.

Soluble organic inhibitors form a protective layer only a few molecules thick, but if an insoluble organic inhibitor is added by dispersion as fine droplets, the film may continue to build to a thickness of several thousandths of an inch. Such films show good *persistence, i.e.,* they continue to inhibit corrosion for a time when an inhibitor is no longer injected into the environment. Persistency is an important property when inhibitors can be injected into a system only in slugs.

Precipitation Inhibitors

Precipitate-inducing inhibitors are film-forming compounds which have a general action over the metal surface and which, therefore, interfere with both anodes and cathodes indirectly. The most common inhibitors of this class are the silicates and phosphates.

In waters with a pH near 7.0, a low concentration of chlorides, silicates, and phosphates cause passivation of steel when oxygen is present; hence, they behave as anodic inhibitors. Another anodic characteristic is that corrosion is localized in the form of pitting when insufficient amounts of phosphate or silicate are added to saline water. However, both silicates and phosphates form deposits on steel which increase cathodic polarization. Thus, their action appears to be *mixed, i.e.,* a combination of both anodic and cathodic effects.

Silicate is used most often in low salinity waters which contain oxygen. It has the rare property of inhibiting corrosion of steel which is already scaled with rust. While the concentration of silicate required for protection depends on the salinity of the water, for most city water supplies, 5 to 10 ppm are required initially, followed by a gradual reduction to 2 to 3 ppm after a protective deposit is established. High concentrations of calcium and magnesium interfere with inhibition by silicates, but this problem is often overcome by adding 2 to 3 ppm of polyphosphates in addition to the silicates.

Sodium silicate is used in many private water softeners to prevent occurrence of *red water* or *rust water* which is caused by suspended ferric hydroxide. The silicates remove iron by precipitation, so they should not be used where formation of scales cannot be tolerated.

In aerated hot water systems, sodium silicate protects steel, copper, and brass. However, protection is not always reliable and depends on pH and composition of the water. The best procedure is to adjust the saturation index as already described to facilitate formation of a protective silicate-containing film.

Phosphates, like silicates, require oxygen for effective inhibition. A concentration of sodium hexametaphosphate, a typical polyphosphate, of about 10 ppm provides corrosion inhibition in aerated water if the water is in motion. In stagnant areas, corrosion might be increased due to the establishment of oxygen concentration cells.

In addition to motion, the presence of calcium ions is essential to inhibition. Phosphates inhibit the deposition of $CaCO_3$; therefore, the concentration of phosphate and calcium must be in proper balance to obtain effective inhibition. A rule of thumb is that the phosphate must not exceed twice the concentration of calcium carbonate. Therefore, if a water contains 10 ppm of calcium carbonate, then up to 20 ppm of phosphate can be used for inhibition. If the water is exceedingly soft, calcium can be increased by the addition of lime.

If chromate cannot be used because windage losses or disposal causes problems due to the toxicity of chromate, industrial cooling towers can be inhibited with about 50 ppm of sodium hexametaphosphate. Addition of a soluble zinc salt often improves inhibition by polyphosphates (Figure 8.19). Pitting and excessive scale formation are prevented by maintaining the pH in a range of 6 to 7.

The silicates and phosphates do not afford the degree of protection that can be obtained with chromates and nitrites; however, they are very useful in situations where nontoxic additives are required. Their main drawbacks are their dependence on water composition and the careful control required to achieve maximum inhibition.

Vapor Phase Inhibitors

Vapor phase inhibitors (VPI), also called volatile corrosion inhibitors (VCI), are compounds which are transported in a closed system to the site of corrosion by volatilization from a source. In boilers, volatile basic compounds such as morpholine or octadecylamine are transported with steam to prevent corrosion in condenser tubes by neutralizing acidic carbon dioxide. Compounds of this type inhibit corrosion by making the environment alkaline. In closed vapor spaces, such as shipping containers, volatile solids such as the nitrite, carbonate, and benzoate salts of dicyclohexylamine, cyclohexylamine, and hexamethylene-imine are used. The mechanism of inhibition by these compounds is not entirely clear, but it appears certain that the organic portion of the molecules merely provides volatility.

On contact with a metal surface, the inhibitor vapor condenses and is hydrolyzed by any moisture present to liberate nitrite, benzoate, or bicarbonate ions. Since ample oxygen is present, nitrite, and benzoate ions are capable of passivating steel as they do in aqueous solution. The mechanism for carbonate may not be the same, and here the organic amine portion of the VPI may serve to aid inhibition by adsorption and by providing alkalinity.

It is desirable for a VPI to provide inhibition rapidly and to have a lasting effect. Therefore, the compound should have a high volatility to saturate all of the accessible vapor space as quickly as possible, but at the same time it should not be too volatile, because it would be lost rapidly through any leaks in the package or container in which it is used. The optimum vapor pressure of VPI then would be just sufficient to maintain an inhibiting concentration on all exposed metal surfaces.

Vapor pressures and other properties of some VPIs are given in Table 7.3. Note that the vapor pressure of cyclohexylamine carbonate is 2000 times higher than dicyclohexylamine nitrite, thus making it a better choice for containers which are opened occasionally because it will resaturate the vapor space rapidly.

Dicyclohexylamine nitrite is advantageous for once-opened containers which may be stored for extended periods. The amount of VPI required depends on conditions, but 2.2 kg per 100 m^2 (1 lb per 500 ft^2) of surface has been suggested for dicyclohexylamine nitrite and 2.2 kg per 30 m^3 (1 lb per 500 ft^3) space for cyclohexylamine carbonate. Vapor phase inhibitors attack nonferrous metals to varying degrees, so it is suggested that a potential user test several of the commercially available VPIs for his particular application. Compatibility of the amines and nitrites with the copper alloys should especially be considered.

Environmental Factors

Aqueous Systems

Introduction

Aqueous systems are by far the most common corrosive environments to which corrosion inhibitors are applied. Water is a powerful solvent capable of carrying many different ions at the same time, so requirements for corrosion inhibition may vary greatly, depending on the type and amount of dissolved species present. Because there is no

TABLE 7.3 — Properties of Vapor Phase Inhibitors

Compound	Vapor Pressure mm Hg at 25 C	Remarks
Dicyclohexylamine nitrite	0.0002	Protects steel, aluminum, and tinplate. Increases corrosion of zinc, magnesium, cadmium, lead, and copper. Discolors some plastics.
Cyclohexylamine carbonate	0.4	Protects steel, aluminum, solder, tin, and zinc. No effect on cadmium. Increases corrosion of copper, brass, and magnesium.

universal inhibitor for water systems, an inhibitor which may be satisfactory for one system may be ineffective or even harmful in another. The main factors which must be considered in the application of corrosion inhibitors to aqueous systems are salt concentration, pH, dissolved oxygen concentration, and the concentration of interfering species.

This survey of the use and properties of corrosion inhibitors in aqueous solutions illustrates some common inhibitor-environment interactions. A process should be analyzed carefully and some tests made before a large-scale program of corrosion inhibition is initiated. When working with natural water, special attention must be given to its composition, particularly in regard to possibilities of natural inhibition and the presence of interfering ions.

Effects of Various Dissolved Species

Demineralized water is relatively noncorrosive toward steel because of its high electrical resistance (ohmic control) and low hydrogen ion concentration. However, when demineralized water is in contact with the atmosphere, it will absorb carbon dioxide and form carbonic acid which will decrease its resistance so that significant corrosion of steel will occur, the cathodic reaction being primarily reduction of dissolved oxygen rather than reduction of hydrogen ions. In this case, minimal concentrations of inhibitors such as sodium chromate, sodium nitrite, polyphosphates, sodium benzoate, or borax are effective. Steel is easily passivated in demineralized or distilled water because the pH is neutral and there are no dissolved ions to interfere with formation of the passive layer.

Industrial and domestic waters contain dissolved substances which affect their aggressiveness and corrosion inhibitor requirements in various ways, depending on the nature of the substances. The most common dissolved substances and their effects on corrosion inhibition are as follows.

Oxygen (O_2). In neutral water, oxygen causes corrosion; therefore, if it is reduced to less than 0.1 ppm by scavenging compounds or by stripping, sufficient control is thereby provided for some systems; *e.g.,* in boilers and hot water supplies. Oxygen can be utilized in passivating steel by adding a passivating inhibitor. Organic inhibitors are seldom effective against oxygen-caused attack unless they contain passivating groups such as benzoate or nitrite.

Chloride (Cl^-). Steel, as well as many other metals, is more difficult to passivate in the presence of the chloride ion; therefore, a higher concentration of passivating inhibitor is required if chlorides are present. Nonpassivating inhibitors must be used in higher concentrations also because chloride ions are strongly adsorbed by steel.

Sulfate ($SO_4^=$). The effects of sulfate on passivity are similar to those of chloride, but to a lesser degree. Sulfates or chlorides must not be allowed to build up in a system by evaporation because depassivation may occur.

Bicarbonate (HCO_3^-). Bicarbonate in hard waters can be utilized for natural inhibition by forming precipitates. In soft water, corrosion inhibitors must be used if excess carbon dioxide is present because of the acidic condition it produces.

Sulfides ($S^=$). Sulfides precipitate many metal ions; *e.g.,* inhibitors which contain zinc cannot be used. Oxidizing inhibitors are reduced by sulfide to form free sulfur. They are effective only if an excess above the amount required to react with sulfide is used and the colloidal precipitate of free sulfur can be tolerated.

Metal Cations. Sodium (Na^+) and potassium (K^+) ions have no particular effects on inhibitors; calcium (Ca^{++}) and magnesium (Mg^{++}) may be used to form protective precipitates, but at high concentrations they interfere with inhibitors by precipitating nonprotective deposits and also by precipitating inhibitors such as phosphate ($PO_4^=$) and silicate ($SiO_3^=$). Very small concentrations of heavy metal ions, such as copper and mercury, can cause severe interference with inhibitors.

Acid (H^+). Hydrogen ions increase corrosion rates and increase the difficulty of passivating steel. Passivation is used in sulfuric (H_2SO_4) and phosphoric acids (H_3PO_4), but not in hydrochloric acid (HCl). Nonpassivating organic or cathodic inhibitors (*e.g.,* guanidine or sodium arsenate) are preferred in pickling acids to avoid the disastrous consequences of depassivation.

Alkali (OH^-). In alkaline solutions, corrosion of steel is controlled by the rate of oxygen diffusion through the precipitated corrosion product [usually ferrous hydroxide, $Fe(OH)_2$], so corrosion rates are low. Steel is easily passivated in alkaline solutions. Amphoteric metals such as aluminum, zinc, and lead corrode slowly at low alkali concentrations, but above pH 9.0 their rates are very high and inhibitors are required.

Waters of Low-to-Moderate Salt Concentrations

Waters of low-to-moderate salt concentrations are encountered in municipal water systems, cooling waters, marine and offshore activities, and oilfield water injection systems. Because metals adsorb ions of dissolved salts in water, an inhibitor has more difficulty in reaching the metal surface and displacing adsorbed ions than it has in demineralized water; hence, a higher concentration of inhibitor is required. Furthermore, chloride ions have a *depassivating* effect, *i.e.,* they make it more difficult to control corrosion by passivating inhibitors. Thus, it is important to maintain inhibitor concentrations at a safe level in waters containing dissolved salts, particularly the chlorides.

Municipal drinking water cannot be treated with most inhibitors because of their toxicity. Fortunately, treatment with lime to raise the pH usually affords sufficient protection to steel or cast iron water pipes. If the water is high in chlorides or sulfates, then polyphosphates are used for added inhibition. Silicates also may be used in municipal waters, but they have the disadvantage of forming precipitates with iron and calcium which scale pipes and heat transfer surfaces.

Cooling water systems may be either recirculating or once-through types. In closed recirculating systems, oxygen can be excluded, and corrosion often can be controlled by adjusting the pH to an alkaline value. Recirculating systems are more easily controlled by inhibitors since higher concentrations can be applied because the water is reused. Sodium chromate or sodium nitrite are both effective in all-steel, closed recirculating systems. Sodium nitrite may form ammonia by reduction at cathodic sites; therefore, it should not be used in systems which include brass or copper, since these materials are subject to stress corrosion cracking by ammonia.

Glycol-water mixtures, such as those used to cool engines and to transfer solar heat, cannot be inhibited with oxidizing inhibitors such as chromate or nitrite because the glycol is oxidized. This not only consumes the inhibitor, but also forms organic acids which attack the cooling system. Such cooling systems usually are inhibited with a mixture of borax (for maintaining an alkaline pH) and mercaptobenzothiazole, which inhibits the corrosion of brass and copper. Borax alone is satisfactory for steel in contact with glycol-water mixtures, but borax and glycol attack zinc galvanizing rapidly and attack the zinc in brass due to the formation of complex zinc compounds at low temperatures.

So, the addition of mercaptobenzothiazole is necessary in mixed metal cooling systems. A soluble oil also is often added to increase protection and to lubricate moving parts in the cooling system. In some mixed metal systems, silicates and nitrates are now used. Amine phosphates have also long been used in such systems.

Once-through cooling systems require inexpensive corrosion inhibitors. In open systems, corrosion is more severe and good inhibition is imperative. The situation is similar to that in municipal water supplies, so comparable remedial measures, namely addition of lime or polyphosphates, are used. In waters which are very corrosive due to high chloride concentrations, chromates or nitrites may be required in addition to polyphosphates.

Waters which may contain appreciable quantities of organic matter, such as seawater and oil field injection brines, usually are not inhibited with oxidizing inhibitors such as chromate and nitrite because of the high consumption of inhibitor through oxidation of the organics. Nonoxidizing inorganic inhibitors such as sodium silicate also must be used in high concentrations due to the high chloride content of brines. Generally, organic inhibitors offer the best means for protection in organic-contaminated brines. Concentrations of only 10 to 20 ppm of inhibitors such as the fatty amines often effectively control corrosion in oil field brines.

High Salt Concentrations

Extremely high salt concentrations are used in aqueous solutions for heat transfer in refrigeration systems. The temperatures encountered are always low, and since the brines are recirculated, a high concentration of inhibitor can be maintained economically. Sodium chromate is effective in refrigeration brines, provided there is no limitation due to its toxicity. If physiological effects are a factor, then disodium phosphate can be used, although it is not as effective as sodium chromate in controlling corrosion.

Effects of pH

The pH of aqueous solutions is extremely important in determining the type of corrosion inhibitor which is most effective and most economical. Natural hard waters retain compounds of calcium, including calcium carbonate ($CaCO_3$) and calcium bicarbonate [$Ca(HCO_3)_2$], along with carbon dioxide, in solution. There is an equilibrium which occurs among these species as shown by Equation (7.2).

$$\underset{\text{insoluble}}{CaCO_3} + CO_2 + H_2O \rightleftharpoons \underset{\text{soluble}}{Ca(HCO_3)_2} \qquad (7.2)$$

At high temperatures the reverse reaction occurs, and heated surfaces become coated with $CaCO_3$. A protective scale is produced also when $Ca(HCO_3)_2$ becomes alkaline in the region of a cathodic area. The scale thus deposited inhibits corrosion by reducing the cathodic area, restricting diffusion of cathodic depolarizers, and increasing ohmic resistance. This scale is often developed on cathodically protected steel surfaces in seawater, and is often called a calcareous deposit. For this reason, some natural hard waters are less corrosive than softened waters. The addition of zinc sulfate ($ZnSO_4$) in alkaline solutions also inhibits corrosion by precipitating insoluble zinc hydroxide [$Zn(OH)_2$] on the cathodic area.

Hydrogen sulfide is a particularly troublesome problem. The dissolved gas attacks steel only slowly when first exposed due to the formation of a protective layer of iron sulfide. The iron sulfide film affords only temporary protection, however, because it becomes permeable to hydrogen sulfide, and the corrosion rate increases with time, producing blistering, high metal loss, and possibly hydrogen embrittlement. Organic corrosion inhibitors prolong the interval preceding higher corrosion rates, but the iron sulfide film eventually prevents access of the inhibitor to the steel surface and, as a result, the corrosion process can proceed uninhibited.

The most effective chemical control measures against hydrogen sulfide (sour) corrosion are removal of the hydrogen sulfide from the water by countercurrent gas stripping or by cleaning the steel periodically with acid to allow access of the inhibitor to the metal surface. Steel sometimes can be cleaned sufficiently for inhibition to be effective by the use of a powerful wetting agent.

Strong Acids

High acid concentrations are encountered in pickling processes, oil well acidizing, and during the transportation of acids for use in chemical processes. Hydrochloric acid of all concentrations requires an inhibitor if steel is to be used. The use of an inhibitor in pickling processes also allows the acid to dissolve scale from steel without appreciable attack on the metal. Pickling acids are effectively inhibited by adding about 0.2% of an organic compound such as anilines, pyridines, thiourea, or sulfonated castor oil. Cathodic inhibitors such as arsenates (As_2O_3), *e.g.*, sodium arsenate (Na_3AsO_4), are also good inhibitors for pickling acids, but are less popular than organics because they cause blistering and hydrogen embrittlement of some steels. Arsenic compounds should never be used in fluids which are to be catalytically processed because arsenic is a poison to most catalysts.

Sulfuric and phosphoric acid concentrations up to 70% can be inhibited by methods similar to those used for hydrochloric acid. Concentrations of sulfuric acid higher than 70% are strongly oxidizing, attack steel slowly, and do not require inhibition. Fertilizer grade phosphoric acid (73% black acid) attacks steel readily and usually is inhibited with potassium iodide. Organic inhibitors are not effective in concentrated phosphoric acid when used alone, but it has been reported that a lower concentration of potassium iodide is required for inhibition if a fatty amine is also added.

Nonaqueous Systems

Corrosion in nonaqueous liquids such as fuels, lubricants, and edible oils is usually caused by the small amounts of water often present. Water is slightly soluble in petroleum products, and its solubility increases with temperature. If a nonaqueous solvent is saturated with water and the temperature is lowered, then some of the water will separate to attack steel that it contacts. Oils that have been subjected to high temperatures in air will contain organic acid that will be extracted by any water present to increase the rate of attack on steel.

Corrosion in steel systems handling wet oils can be inhibited with both organic and inorganic compounds. Effective organic compounds include various amines, lecithin, and mercaptobenzothiazole. The inorganic inhibitors include sodium nitrite and sodium nitrate. Chromates are not used because of their instability in the presence of organics.

Small amounts of water inhibit corrosion in some nonaqueous solvents. Halogenated (containing chlorides, fluorides, bromides, or iodides), nonaqueous solvents can be particularly troublesome. Organic amines are effective inhibitors for steel degreasing vessels that contain hot chlorinated solvents.

Inhibition of corrosion of aluminum when in contact with halogenated solvents is more difficult. Aluminum in contact with many of the one and two carbon chlorocarbons can explode after a varying incubation period if the solvent is dry. A few parts per million of water will inhibit the reaction. On the other hand, if water is added exceeding the limited solubility in the solvent, the water layer will become highly acidic from hydrolysis of the organic compound. Thus, inhibition in systems containing these materials must be approached with caution.

Water also inhibits the stress corrosion cracking of steel in ammonia and titanium in methanol, as well as the attack on titanium by "dry" chlorine.

A trace of water (0.001%) in liquid hydrogen fluoride (HF) behaves as a passivating inhibitor for nickel. This is an extreme example of the importance of solvent-inhibitor interactions. The exact mechanism of inhibition by water in HF is unknown, but the passivating effect is similar to that observed on steel in the presence of chromates in aqueous solutions.

Solubility is an important factor to be considered in evaluating corrosion inhibitors for nonaqueous fluids, because they do not have the tremendous solvent power range of water. Because an inhibitor must be transported through the environment to sites where corrosion occurs, it must be either soluble in the environment or sufficiently dispersed in fine droplets that settling does not occur. Also, the inhibitor must not form filter-plugging insoluble products by reaction with metals or components of the nonaqueous fluid. Some corrosion inhibitors formerly used in gasoline were found to react with zinc in galvanized fuel tanks to form a precipitate that clogged fuel filters.

Testing inhibitors in nonaqueous media is more difficult than in aqueous solutions, especially if corrosion is due to water separating to form a two-phase system. This condition is difficult to duplicate in the laboratory, and polarization curves cannot be used effectively because most nonaqueous solvents are nonconductors. Furthermore, corrosion coupons placed in pipes or tanks carrying fuel or similar products may give misleading results because they may not be contacted or wetted by the water phase.

A widely accepted laboratory test for two-phase systems (called the *wheel* test) consists of alternately wetting a corrosion specimen by the organic and water phases by rotating a bottle containing the two phases. This test is very useful, but not always reliable because of the restricted volume of the solvents and the difficulty in duplicating velocity and stagnation effects in real systems.

Gaseous Environments

Introduction

Gaseous environments include the open atmosphere, the vapor phase in tanks, natural gas in wells, and the empty space in packaging containers. Here again, water and oxygen are the principal corrosive agents, but the main problem in providing inhibition is to transport the inhibitor from a source to the sites where corrosion may occur.

The Open Atmosphere

Inhibitors for corrosion in the open atmosphere are applied directly to the metal surfaces to be protected. The most common method is the use of chromates in paints. Zinc chromate and red lead are used in primer coats. Rivet heads are coated with a slurry of microencapsulated zinc chromate. When the rivet is driven against another surface, the capsules rupture to provide lasting inhibition to the crevice under the rivet head. Volatile inhibitors are never used in the open atmosphere because they are impractical and cannot saturate the vapor space.

Closed Vapor Spaces

The walls of tanks above a water line are subject to extensive corrosion because the relative humidity is always high and oxygen is plentiful if the tank is vented to the atmosphere. Where water contamination is not a factor, a layer of oil on the surface helps to maintain a low humidity and, as the level is raised and lowered, the walls are coated with a layer of oil. The oil may contain an organic inhibitor and an agent (usually an amine) to cause the oil to spread on the metal surface. An oil layer containing about 15% lanolin has been used in ship ballast tanks to control corrosion.

Gas wells corrode mostly in the *reflux zone,* which is an area of the well somewhere between the bottom and the wellhead where condensation occurs. As the gas flows up the well, its temperature drops due to expansion, and this causes condensation when the temperature reaches the dew point of the gas. Volatile inhibitors such as formaldehyde and ammonia, injected into gas wells have been used successfully to inhibit corrosion. Many gas wells today are protected by injecting amine inhibitors continuously, in slugs, or by *squeeze* which means they coat the well when injected and also enter the gas stream partially by vaporization and partially by entrainment.

Packaged materials may be protected from corrosion in several ways. Packages which can be sealed and which contain parts that cannot be coated with an inhibitor or exposed to volatile inhibitors (such as electronic parts) are protected by placing a desiccant, such as silica gel, in the package to maintain the humidity at a low level.

Vapor phase inhibitors (VPI) can be placed in a package in bulk or by wrapping an article in paper impregnated with a VPI. These compounds are volatile organics, so the package in which they are used must be fairly well sealed. The most common VPIs in use are dicyclohexylamine nitrite (DHN) and cyclohexylamine carbonate (CHC). These inhibitors are very effective for steel, but they should be tested if metals other than steel are present because they attack some nonferrous metals.

Inhibited coatings provide a cheap, effective method for controlling corrosion of packaged materials. Easily strippable coatings which do not harden are called *soft coatings* or *slushing compounds.* Oils and greases containing amines may be used. Steel and zinc articles can be protected with a thickened aqueous solution of sodium benzoate or sodium nitrite.

Metals which are very sensitive to hydrogen sulfide, such as copper and silver, are protected by enclosing them in paper impregnated with copper or zinc compounds. These materials are not corrosion inhibitors in a strict sense since they adsorb gaseous sulfur compounds to prevent reaction with the silver or copper.

Effect of Elevated Temperatures

Most effects of elevated temperatures are adverse to corrosion inhibition. High temperatures increase corrosion rates (about double for a 15 C rise at room temperature), and they decrease the tendency of inhibitors to adsorb on metal surfaces. Precipitate-forming inhibitors are less effective at elevated temperatures because of the greater solubility of the protective deposit. Thermal stability of corrosion inhibitors is an important consideration at high temperatures. Polyphosphates, for example, are hydrolyzed by hot water to form orthophosphates which have little inhibitive value. Most organic compounds are unstable above about 200 C (400 F); hence, they may provide only temporary inhibition at best.

In neutral or slightly alkaline, oxygen-free aqueous systems, corrosion of fairly clean steel occurs at a very low rate at elevated temperatures. This principle is the basis for most boiler water treatment to prevent corrosion, *i.e.,* treatment is designed to provide alkalinity, to remove oxygen, and to prevent scale deposition. Other additives are also used to prevent foaming, but these will not be considered here.

In oxygen-free hot water, steel is protected by formation of a natural coating of magnetite (Fe_3O_4 or black rust) formed by the reaction:

$$\underset{\text{Iron}}{3Fe} + \underset{\text{Water}}{4H_2O} \rightarrow \underset{\text{Iron Oxide}}{Fe_3O_4} + \underset{\text{Hydrogen}}{4H_2} \qquad (7.3)$$

If oxygen is present, then nonprotective Fe_2O_3 (red rust) is formed. At elevated temperatures, oxygen is removed readily from boiler waters by reaction with sodium sulfite or hydrazine. Hydrazine is preferred because it does not increase the salt content of the boiler water; moreover, it reacts faster than sodium sulfite at elevated temperatures, the dosage required to react with a given amount of oxygen is lower, and it is easier to apply because it is a liquid.

Boiler waters are maintained at an alkaline pH to facilitate formation and maintenance of the Fe_3O_4 protective film (Chapter 8). It is desirable to use an additive which will be carried into steam condensate lines to maintain an alkaline condition in these areas also. Volatile amines such as ammonia, morpholine, cyclohexylamine, and octadecylamine are used.

Deposition of scales in boilers reduces heat transfer and produces pitting-type corrosion. Water softeners often are used to treat boiler feed to remove objectionable ions such as calcium, magnesium, and iron, but usually a scale inhibitor such as sodium phosphate also is added. The phosphates prevent scaling by increasing the supersaturation of $CaCO_3$ and $CaSO_4$ in water.

In high-pressure, high-temperature boilers, demineralized water often is used. In these cases, only oxygen removal with hydrazine is required.

High temperatures always increase the rate of attack of metals in acids because the driving force for the anodic and cathodic reactions is greater and hydrogen overvoltage is decreased. Other factors such as the greater solubility of corrosion products and the higher rate of solution of metal oxides also increase the rate of attack.

While hot acids are handled best by resistant alloys or coated steel, an exception is the acidizing of oil wells. It is impractical to coat or line oil well casing because the coating is soon destroyed by the insertion of tools, etc. Bottomhole temperatures of oil wells are often 93 to 150 C (200 to 300 F) and sometimes as high as 230 C (450 F). Inhibitors for acids used to increase the permeability of oil-producing formations (usually HCl) are effective, but have temperature limitations. Amine-type pickling acid inhibitors are often used for oil well acidizing.

Utilization

Application Techniques

Continuous Injection

Continuous injection of corrosion inhibitors is practiced in once-through systems where slugs or batch treatment cannot be distributed evenly through the fluid. This method is used for water supplies, oil field injection water, once-through cooling water, open annulus oil or gas wells, and gas lift wells. Liquid inhibitors are injected with a chemical injection pump. These pumps are extremely reliable and require little maintenance. Most chemical injection pumps can be adjusted to deliver at the desired injection rate.

Another form of continuous application is by the use of slightly soluble forms of solid inhibitors. The inhibitor (such as a glassy phosphate or silicate in the form of a cartridge) is installed in a flow line where the inhibitor is continuously leached out by passage of fluid through the cartridge. Inhibitors in the form of sticks or pellets are used in oil and gas wells to supply inhibitor continuously by their natural slow dissolution.

Boilers, closed cooling water systems, and other closed circulating fluid systems can be treated with inhibitors with continuous injection. When such systems are started up after construction or major maintenance, the inhibitor is often injected at higher-than-normal concentration to permit rapid development of protective films.

Batch Treatment

The most familiar example of batch treatment is the automobile cooling system. A quantity of inhibitor is added at one time to provide protection for an extended period. Additional inhibitor may be added periodically, or the fluid may be drained and replaced with a new supply. In most aerated, closed-loop cooling systems, it is important that the inhibitor concentration be measured occasionally to ensure that a safe level is maintained.

Batch treatment is also used in treating oil and gas wells. An inhibitor is diluted with an appropriate solvent and injected into the annulus of open hole wells or into the tubing of gas wells that have a packer. In this application, it is important that the inhibitor contact all surfaces and that it has good persistence. Most wells require treatment about every two weeks.

Squeeze Treatment

The squeeze treatment is a method of continuously feeding an inhibitor into an oil well. A quantity of inhibitor is pumped into a well and is followed by sufficient solvent to force the inhibitor into the formation. The inhibitor is absorbed by the formation from which it slowly escapes to inhibit the produced fluids. Protection applied in this manner has been known to last for a year.

Volatilization

Volatilization has already been discussed under vapor phase inhibitors in connection with boilers and closed containers. Another application is the inhibition of gas condensate corrosion. However, the treatment here is essentially the same as used in batch or squeeze treatments.

Coatings

Inhibitors are used in coatings exposed to the open atmosphere. When moisture contacts the paint, some inhibitor is leached out to protect the metal. Thus, the inhibitor must be soluble enough to be leached out in sufficient amounts to protect the metal, but not so soluble that it will be lost rapidly. The most common coating inhibitors are zinc chromate and plumbous orthoplumbate (red lead), which passivate steel by providing chromate and plumbate ions, respectively, as well as the zinc and lead cathodic inhibitors. These inhibitors are not effective against attack by seawater or brines because the high chloride concentration prevents passivation of steel.

Recently, heavy coatings which act as sealants for crevices have been developed for the aircraft and aerospace industries. These coatings contain proprietary inhibitor formulations which are especially effective in minimizing corrosion associated with dissimilar metal fasteners.

System Condition

A system must be carefully examined before a program of corrosion inhibition can be planned effectively. The examination must include a survey of any adverse effects an inhibitor may have on the process in which it is to be used and an analysis to detect the presence of interfering substances.

Another possible adverse effect of inhibition is an increased rate of corrosion of a metal in the system other than the one for which the inhibitor was selected to protect. For example, some amines protect steel admirably, but will severely attack copper and brass. Nitrites may attack lead and lead alloys such as solder. In some cases, the inhibitor may react in the system to produce a harmful product. An illustration of this is the reduction of nitrite inhibitors to form ammonia which causes stress corrosion cracking of copper and brass. The only way to avoid these problems is to know the metallic components of a system and be thoroughly familiar with the properties of the inhibitor to be used.

The examination must include preparation of a complete list of materials, both metallic and nonmetallic, which will be in contact with the fluid that will be inhibited. Such small items as gaskets, instrument probes, and control devices may be made from materials that will not be compatible with some inhibitors. The results from this examination may suggest that certain parts of a system should be changed to permit the use of a particular inhibitor.

The examination must also include a determination of the cleanliness of the surfaces of the system that will be in contact with the inhibited fluid. A system can be plugged as the result of an inhibitor loosening scale and suspending it in the fluid. This problem is best avoided by planning ahead. The best preventive measure is to clean the system thoroughly, if possible, before an inhibitor is applied.

Cleaning may be accomplished with chemical cleaners, mechanical cleaners, ultrasonic energy, or thermal shock. Inhibitors can reach cleaned metal surfaces much more easily than they can reach heavily fouled or scaled surfaces.

Selection

There are several approaches to follow in selecting a proper inhibitor for a given system. An outside consultant can be asked for advice. Various suppliers can be asked for advice. Tabulations such as Tables 7.4 and 7.5 can be consulted. Table 7.4 is arranged by environments, while Table 7.5 is arranged by material. The extent of the list of references with Table 7.5 indicates the broad range of material-environment-inhibitor combinations that are possible.

Often, the particular combination at hand cannot be found in the literature. In such cases, testing must be conducted to determine which inhibitor and what concentration to use. Standard tests can be found in American Society for Testing and Materials (ASTM) publications and in NACE Standard Test Methods.

A frequent cause of ineffective inhibition is loss of the inhibitor before it has a chance to contact the metal surfaces or effect the desired changes in the environment. An inhibitor might be lost by precipitation, adsorption, reaction with a component of the system, or by being insufficiently soluble or too slow to dissolve.

Typical examples of losses of an inhibitor due to these factors are precipitation of phosphates by the calcium ion, reaction of chromates with sulfides or organics, adsorption of inhibitors on suspended solids, and injection of a poorly soluble inhibitor without an adequate dispersing agent. To avoid these problems, inhibitors should be tested in the actual fluids to be treated rather than in simulated environments. If possible, testing should be done in the process stream or in a small side stream.

Concentration and Performance Technology

Corrosion inhibitors are sold in solid or liquid form. Most solids are relatively pure, but sometimes a solid inhibitor is fused with another ingredient or encapsulated where a controlled rate of solubility is required. Liquids are usually preferred because of the ease with which they can be transported, measured, and dispersed.

Liquid inhibitors are rarely pure for several reasons. Organic inhibitors seldom have optimum characteristics of viscosity, or freezing or boiling points; therefore, they are dissolved in an appropriate solvent to achieve the properties desired. Furthermore, it is often desirable to blend the inhibitor with a demulsifier, dispersant, surfactant, antifoaming agent, or synergistic agent.

Liquid inhibitors are sold by the gallon, part of which is solvent. The amount of inhibitor present is expressed as *percent active, i.e.,* a gallon of inhibitor

TABLE 7.4 — Some Corrosive Systems and the Inhibitors That Have Been Used to Protect Them

System	Inhibitor	Metals Protected	Concentration
Water, Potable	$Ca(HCO_3)_2$	Steel, Cast Iron + Others	10 ppm
	Polyphosphate	Fe,Zn,Cu Al	5-10 ppm
	$Ca(OH)_2$	Fe,Zn,Cu	Sufficient for pH 8.0
	Na_2SiO_3	Fe,Zn,Cu	10-20 ppm
Water, Cooling	$Ca(HCO_3)_2$	Steel, Cast Iron + Others	10 ppm
	Na_2CrO_4	Fe,Zn,Cu	0.1%
	$NaNO_2$	Fe	0.05%
	NaH_2PO_4	Fe	1%
	Morpholine	Fe	0.2%
Boilers	NaH_2PO_4	Fe,Zn,Cu	10 ppm
	Polyphosphate	Fe,Zn,Cu	10 ppm
	Morpholine	Fe	Variable
	Hydrazine	Fe	O_2 Scavenger
	Ammonia	Fe	Neutralizer
	Octadecylamine	Fe	Variable
Brines	$Ca(HCO_3)_2$	Fe,Cu,Zn	10 ppm
	Na_2CrO_4	Fe,Cu,Zn	0.1%
	Sodium Benzoate	Fe	0.5%
	$NaNO_2$	Fe	(NaCl, 5%)
Oil Field Brines	Na_2SiO_3	Fe	0.01%
	Na_2SO_3 (or SO_2)	Fe	O_2 Scavenger (O_2 × 9) ppm
	Quaternaries	Fe	10-25 ppm
	Imidazoline	Fe	10-25 ppm
	Rosin Amine Acetate	Fe	5-25 ppm
	Coco Amine Acetate	Fe	5-15 ppm
	Formaldehyde	Fe	50-100 ppm
Seawater	Na_2SiO_3	Zn	10 ppm
	$NaNO_2$	Fe	0.5%
	$Ca(HCO_3)_2$	All	pH Dependent
	NaH_2PO_4 + $NaNO_2$	Fe	10 ppm + 0.5%
Engine Coolants	Na_2CrO_4	Fe,Pb,Cu,Zn	0.1-1%
	$NaNO_2$	Fe	0.1-1%
	Borax	Fe	1%
Glycol/Water	Borax + Mercaptobenzothiazole	All	1% + 0.1%
Acids, HCl	Ethylaniline	Fe	0.5%
	Mercaptobenzothiazole	Fe	1%
	Pyridine + Phenylhydrazine	Fe	0.5% + 0.5%
	Rosin Amine + Ethylene Oxide	Fe	0.2%
H_2SO_4	Phenylacridine	Fe	0.5%
Con. H_3PO_4	NaI	Fe	200 ppm
Most Acids	Thiourea	Fe	1%
	Sulfonated Castor Oil	Fe	0.5-1%
	As_2O_3	Fe	0.5%
	Na_3AsO_4	Fe	0.5%
Vapor Condensate	Morpholine	Fe	Variable
	Ammonia	Fe	Variable
	Ethylenediamine	Fe	Variable
	Cyclohexylamine	Fe	Variable
Enclosed Atmosphere	Cyclohexylamine Carbonate	Fe	1 lb per 500 cu ft
	Dicyclohexylamine Nitrite	Fe	1 lb per 500 sq ft
	Amylamine Benzoate	Fe	Variable
	Diisopropylamine Nitrite	Fe	Variable
	Methylcyclohexylamine Carbonate	Fe	Variable
Coating Inhibitors	$ZnCrO_4$ (yellow)	Fe,Zn,Cu	Variable
	$CaCrO_4$ (white)	Fe,Zn,Cu	Variable
	Red Lead	Fe	Variable

TABLE 7.5 — Corrosion Inhibitor Reference List

Metal	Environment	Inhibitor	Reference
Admiralty	Ammonia, 5%	0.5% hydrofluoric acid	54
Admiralty	Sodium hydroxide, 4° Be	0.6 moles H_2S per mole NaOH	71
Aluminum	Acid hydrochloric, 1 N	0.003 M α-phenylacridine, β-naphthoquinone, acridine, thiourea, or 2-phenylquinoline	39
Aluminum	Acid nitric, 2-5%	0.05% hexamethylene tetramine	22
Aluminum	Acid nitric, 10%	0.1% hexamethylene tetramine	22
Aluminum	Acid nitric, 10%	0.1% alkali chromate	16
Aluminum	Acid nitric, 20%	0.5 hexamethylene tetramine	22
Aluminum	Acid phosphoric	Alkali chromates	52
Aluminum	Acid phosphoric, 20%	0.5% sodium chromate	16, 60
Aluminum	Acid phosphoric, 20-80%	1.0% sodium chromate	16, 60
Aluminum	Acid sulfuric, conc.	5.0% sodium chromate	45
Aluminum	Alcohol anti-freeze	Sodium nitrite and sodium molybdate	6
Aluminum	Bromine water	Sodium silicate	10
Aluminum	Bromoform	Amines	44
Aluminum	Carbon tetrachloride	0.05% formamide	55
Aluminum	Chlorinated aromatics	0.1-2.0% nitrochlorobenzene	21
Aluminum	Chlorine water	Sodium silicate	10
Aluminum	Calcium chloride, sat.	Alkali silicates	59
Aluminum	Ethanol, hot	Potassium dichromate	52
Aluminum	Ethanol, commercial	0.03% alkali carbonates, lactates, acetates, or borates	50
Aluminum	Ethylene glycol	Sodium tungstate or sodium molybdate	41
Aluminum	Ethylene glycol	Alkali borates and phosphates	52
Aluminum	Ethylene glycol	0.01-1.0% sodium nitrate	7
Aluminum	Hydrogen peroxide, alkaline	Sodium silicate	75
Aluminum	Hydrogen peroxide	Alkali metal nitrates	20
Aluminum	Hydrogen peroxide	Sodium metasilicate	59
Aluminum	Methyl alcohol	Sodium chlorate plus sodium nitrite	42
Aluminum	Methyl chloride	Water	72
Aluminum	Polyoxyalkene glycol fluids	2% Emery's dimer acid (dilinoleic acid), 1.25% $N(CHMe_2)_3$, 0.05-0.2% mercaptobenzothiazole	43

(continued)

TABLE 7.5 (continued)

Metal	Environment	Inhibitor	Reference
Aluminum	Seawater	0.75% sec. amyl stearate	5
Aluminum	Sodium carbonate, dilute	Sodium fluosilicate	67
Aluminum	Sodium hydroxide, 1%	Alkali silicates	59
Aluminum	Sodium hydroxide, 1%	3-4% potassium permanganate	17
Aluminum	Sodium hydroxide, 4%	18% glucose	57
Aluminum	Sodium hypochlorite contained in bleaches	Sodium silicate	58
Aluminum	Sodium acetate	Alkali silicates	59
Aluminum	Sodium chloride, 3.5%	1% sodium chromate	22
Aluminum	Sodium carbonate, 1%	0.2% sodium silicate	28
Aluminum	Sodium carbonate, 10%	0.05% sodium silicate	28
Aluminum	Sodium sulfide	Sulfur	46
Aluminum	Sodium sulfide	1% sodium metasilicate	59
Aluminum	50% sodium trichloracetate soln.	0.5% sodium dichromate	1
Aluminum	Tetrahydrofurfuryl alcohol	1% sodium nitrate or 0.3% sodium chromate	15
Aluminum	Triethanolamine	1% sodium metasilicate	22
Brass	Carbon tetrachloride, wet	0.001-0.1 aniline	53
Brass	Furfural	0.1% mercaptobenzothiazole	36
Brass	Polyoxyalkene glycol fluids	2.0% Emery's acid (dilinoleic acid), 1.25% $N(CHMe_2)_3$, 0.05-0.2% mercaptobenzothiazole	43
Brass	50% sodium trichloracetate soln.	0.5% sodium dichromate	1
Cadmium plated steel	55/45 ethylene glycol—water	1% sodium fluorophosphate	12
Copper	Fatty acids as acetic	H_2SO_4, $(COOH)_2$, or H_2SiF_6	63
Copper	Hydrocarbons containing sulfur	p-hydroxybenzophenone	61
Copper	Polyoxyalkene glycol fluids	2% Emery's acid (dilinoleic acid), 1.25% $N(CHMe_2)_3$, 0.05-0.2% mercaptobenzothiazole	43
Copper & brass	Acid sulfuric, dil.	Benzyl thiocyanate	68
Copper & brass	Ethylene glycol	Alkali borates & phosphates	22
Copper & brass	Polyhydric alcohol anti-freeze	0.4-1.6% Na_3PO_4 plus 0.3-0.6 sodium silicate plus 0.2-0.6% sodium mercaptobenzothiazole	62
Copper & brass	Rapeseed soil	Succinic acid	9
Copper & brass	Sulfur in benzene solution	0.2%, 9, 10-anthraquinone	29
Copper & brass	Tetrahydrofurfuryl alcohol	1% sodium nitrate or 0.3% sodium chromate	15
Copper & brass	Water-alcohol	0.25% benzoic acid, or 0.25% sodium benzoate at a pH of 7.5-10	23
Galvanized iron	Distilled water	15 ppm. mixture calcium and zinc metaphosphate	77
Galvanized iron	55/45 ethylene glycol—water	0.025% trisodium phosphate	12
Iron	Nitroarylamines	Dibenzylaniline	19
Lead	Carbon tetrachloride, wet	0.001-0.1% aniline	53
Magnesium	Alcohol	Alkaline metal sulfides	16
Magnesium	Alcohol, methyl	1% oleic or stearic acid neutralized with ammonia	13
Magnesium	Alcohols, polyhydric	Soluble fluorides at pH 8-10	26
Magnesium	Glycerine	Alkaline metal sulfides	16
Magnesium	Glycol	Alkaline metal sulfides	16
Magnesium	Trichlorethylene	0.05% formamide	55
Magnesium	Water	1% potassium dichromate	8
Monel	Carbon tetrachloride, wet	0.001-0.1% aniline	53
Monel	Sodium chloride, 0.1%	0.1% sodium nitrite	72
Monel	Tap water	0.1% sodium nitrite	72
Nickel & silver	Sodium hypochlorite contained in bleaches	Sodium silicate	58
Stainless steel	Acid, sulfuric, 2.5%	5-20 ppm, $CaSO_45H_2O$	35
Stainless steel	Cyanamide	50-500 ppm, ammonium phosphate	65
Stainless steel, 18-8	Potassium permanganate contained in bleaches	Sodium silicate	58
Stainless steel, 18-8	Sodium chloride, 4%	0.8% sodium hydroxide	49
Steel	Acid citric	Cadmium salts	37
Steel	Acid sulfuric, dil	Aromatic amines	51
Steel	Acid sulfuric, 60-70%	Arsenic	74
Steel	Acid sulfuric, 80%	2% boron trifluoride	4
Steel	Aluminum chloride—hydrocarbon complexes formed during isomerization	0.2-2.0% iodine, hydriodic acid, or hydrocarbon iodide	30
Steel	Ammoniacal ammonium nitrate	0.2% thiourea	40
Steel	Ammonium nitrate—urea solns.	0.05-0.10% ammonia 0.1% ammonium thiocyanate	18
Steel	Brine containing oxygen	0.001-3.0 methyl, ethyl, or propyl substituted dithiocarbamates	48
Steel	Carbon tetrachloride, wet	0.001-0.1% aniline	53
Steel	Caustic—cresylate solution as in regeneration of refinery caustic wash solutions, 240-260 F	0.1-1.0% trisodium phosphate	3
Steel	Ethyl alcohol, aqueous or pure	0.03% ethylamine or diethylamine	25
Steel	55/45 ethylene glycol—water	0.025% trisodium phosphate	12
Steel	Ethylene glycol	Alkali borates & phosphates	22
Steel	Ethylene glycol	Guanidine or guanidine carbonate	
Steel	Ethyl alcohol, 70%	0.15% ammonium carbonate plus 1% ammonium hydroxide	56
Steel	Furfural	0.1% mercaptobenzothiazole	36
Steel	Halogenated dielectric fluids	0.05-4% $(\gamma C_4H_3S)_4$ Sn, $\gamma(C_4H_3)_2$Sn, or γC_4H_3S $SnPh_3$	24
Steel	Halogenated organic insulating materials as chlorinated diphenyl	0.1% 2, $4(NH_2)_2C_6H_3NHPh$, o—MeH_4NH_2, or p—$NO_2C_6H_4NH_2$	31

(continued)

TABLE 7.5 (continued)

Metal	Environment	Inhibitor	Reference
Steel	Herbicides as 2, 4-dinitro-6-alkyl phenols in aromatic oils	1.0-1.5% furfural	32
Steel	Isopropanol, 30%	0.03% sodium nitrite plus 0.015% oleic acid	72
Steel	1:4 methanol—water	To 4l. water and 1l. methanol add 1 g. pyridine and 0.05 g. pyragallol	66
Steel	Nitrogen fertilizer solutions	0.1% ammonium thiocyanate	2
Steel	Phosphoric acid, conc.	0.01-0.5% dodecylamine or 2 amino bicyclohexyl and 0.001% potassium iodide, potassium iodate, or iodacetic acid	47
Steel	Polyoxyalklene glycol fluids	2% Emery's acid (dilinoleic acid) 1.25% $N(CHMe_2)_3$ 0.05-0.2% mercaptobenzothiazole	43
Steel	Sodium chloride, 0.05%	0.2% sodium nitrite	72
Steel	50% sodium trichloracetate soln.	0.5% sodium dichromate	1
Steel	Sulfide containing brine	Formaldehyde	14
Steel	Tetrahydrofurfuryl alcohol	1% sodium nitrate or 0.3% sodium chromate	15
Steel	Water	Benzoic acid	70
Steel	Water for flooding operations	Rosin amine	38
Steel	Water saturated hydrocarbons	Sodium nitrite	73
Steel	Water, distilled	Aerosol (an ionic wetting agent)	27
Tin	Carbon tetrachloride, wet	0.001-0.1% aniline	53
Tin	Chlorinated aromatics	0.1-2.0% nitrochlorobenzene	21
Tinned copper	Sodium hypochlorite contained bleaches	Sodium silicate	58
Tin plate	Alkali cleaning agents as tri-sodium phosphate, sodium carbonate, etc.	Diethylene diaminocobaltic nitrate	34
Tin plate	Alkaline soap	0.1% sodium nitrite	64
Tin plate	Carbon tetrachloride	2% mesityl oxide, 0.001% diphenylamine	76
Tin plate	Sodium chloride, 0.05%	0.2% sodium nitrite	72
Titanium	Hydrochloric acid	Oxidizing agents as chromic acid or copper sulfate	33
Titanium	Sulfuric acid	Oxidizing agents or inorganic sulfates	33
Zinc	Distilled water	15 ppm. mixture calcium and zinc metaphosphates	77

[SOURCE: Maxey Brooke, Corrosion Inhibitor Checklist, Chem. Eng., pp. 230-234 (December, 1954).]

References

1. Alquist, F. N. and Wasco, J. L., Corrosion, 8, 410-12 (1952).
2. J. T. Baker Chemical Co. Product Bull. 101, p. 6.
3. Baer, M. S., U.S. Pat. 2,513,131 (June 27, 1950).
4. Baker, R. A., U.S. Pat. 2,572,301 (Oct. 23, 1951).
5. Baldeschwigler, E. L. and Zimmer, J. C., U.S. Pat. 2,396,236 (March 12, 1946).
6. Bayes, A. L., U.S. Pat. 2,147,395 (Feb. 14, 1939).
7. Bayes, A. L., U.S. Pat. 2,153,952 (Apr. 11, 1939).
8. Beck, A. and Dibelka, H., U.S. Pat. 1,876,131 (Sept. 6, 1933).
9. Bhatnagar, S. S. and Krishnamurth, K. G., J. Sci. Ind. Res. (India) 4,238-40 (1945), C.A. 40, 2968.
10. Bohner, Hauszeit v., A. W. v. Erftwerk, A. G., Aluminum, 3, 347-8 (1931).
11. Brooke, M., Chem. Eng., 59, 286-7 (Sept. 1952).
12. Brophy, J. E. et al, Ind. Eng. Chem., 43, 884-96 (1951).
13. Bushrod, C. J., U.S. Pat. 2,325,304 (July 27, 1944).
14. Clay, J. A., Jr., Pet. Engr., 18, No. 2, 111-14 (1946).
15. Clendenning, K. A., Can. J. Research, 26 F, 209-20 (1948).
16. Colegate, G. T., Metallurgia, 39, 316-18 (1949), C.A. 43, 4622 c.
17. Collari, N. and Fongi, N., Alluminio, 16, 13-21 (1947), C.A. 41, 4759 f.
18. Crittenden, E. D., U.S. Pat. 2,549,430 (Apr. 17, 1951).
19. E. I. du Pont de Nemours and Co., Brit. Pat. 439,055 (Nov. 28, 1935).
20. E. I. du Pont de Nemours and Co., Fr. Pat. 792,106 (Dec. 23, 1935).
21. Egerton, L., U.S. Pat. 2,391,685 (Dec. 25, 1945).
22. Elder, J. A. Jr., Pats. 2,487-755-6 (Aug. 9, 1949).
23. Eldredge, G. G. and Mears, R. B., Ind. Eng. Chem., 37, 736-41 (1945).
24. Ellenburg, A. M., U.S. Pat. 2,573,894 (Nov. 6, 1951).
25. L'Etat Francais Represente par le Ministre de la Marine, Fr. Pat. 785,177 (Aug. 2, 1935).
26. Farbenind, I. G., Ger. Pat. 636,912 (Oct. 17, 1936).
27. Fontana, M. G., Ind. Eng. Chem., 42, 7, 65A-66A (July, 1950).
28. Geller, Vereningte Aluminum-Werke, A. G. Lautawerk, Ger. Office of Technical Services, PB-70016, Frames 6609-6610 (July 1937).
29. Gindin & Sil's Doklady Akad. Nuak SSSR, 63, No. 6, 685-8 (1948), C.A. 43, 4207 c.
30. Glassmire, W. F. and Smith, W. R., U.S. Pat. 2,586,323 (Feb. 19, 1952).
31. Hanson, W. J. and Nex, R. W., U.S. Pat. 2,623,818 (Dec. 30, 1952).
32. Hardy, E. E. and Jackson, E. F., U.S. Pat. 2,617,770 (Nov. 11, 1952).
33. Harple, W. W. and Kiefer, G., Chem. Eng. Prog., 49,566 (1953).
34. Henkle et Cie, G.M.B.H., Brit. Pat. 415,672 (Aug. 30, 1934).
35. Hetherington, H. C., U.S. Pat. 2,462,638 (Feb. 22, 1949).
36. Hillyer, J. C. and Nicewander, D. C., U.S. Pat. 2,473,750 (June 21, 1949).
37. Hoar, T. P. and Havenhand, D. J., Iron Steel Inst. (London) 133, 252 (1936).
38. Howell, W. E., Barton, J. K. and Heck, E. T., Producers Monthly, 13, No. 7, 27-34 (1949).
39. Jenckel & Woltmann, Z. anorg. allgem. Chem., 217, 298 (1934).
40. Keener, F. G., U.S. Pat. 2,238,651 (April 15, 1941).
41. Lamprey, H., U.S. Pat. 2,147,409 (Feb. 14, 1939).
42. Lamprey, H., U.S. Pat. 2,153,961 (Apr. 11, 1939).
43. Langer, T. W. and Mago, B. F., U.S. Pat. 2,624,708 (Jan. 6, 1953).
44. Lichtenberg, H., Aluminum, 19, 504-9 (1937).
45. Lichtenberg, H., Aluminum, 20, 264-5 (1938).

(continued)

TABLE 7.5 (continued)

46. Lichtenberg, H. and Geier, Aluminum, 20, 784 (1938).
47. Malowan, J. E., U.S. Pat. 2,567,156 (Sept. 4, 1951).
48. Mann, C. A., Lauer, B. E., and Hultin, C. T., Ind. Eng. Chem., 28, 1048-51 (1936).
49. Marsh, G. A., U.S. Pat. 2,574,576 (March 13, 1951).
50. Mathews, J. W., and Uhlig, H. H., Corrosion, 7, 419-22 (1951).
51. McDermatt, F. A., U.S. Pat. 1,927,842 (Sept. 26, 1933).
52. Mears, R. G. and Eldredge, G. G., Trans. Electrochem. Soc., 83, 403 (1943).
53. Ohlmann, E. O., U.S. Pat. 2,387,284 (Oct. 23, 1945).
54. Phillips, C. Jr., U.S. Pat. 2,550,425 (Apr. 24, 1951).
55. Plueddemann, E. P. and Rathmann, R., U.S. Pat. 2,423,343 (July 1, 1947).
56. Rae, Mfg. Chem., 16, 394 (1945).
57. Rhodes, F. H. and Berner, F. W., Ind. Eng. Chem., 25, 1336-7 (1933).
58. Robinson, E. A., Soap and Sanitary Chem., 28, 34-6 (Jan. 1952).
59. Rohrig, H., Aluminum, 17, 559-62 (1935).
60. Rohrig, H. and Gaier, K., Aluminum, 19, 448-50 (1937).
61. Schultz, W. A., U.S. Pat. 2,089, 579 (Aug. 10, 1937).
62. Smith, W. R., U.S. Pat. 2,524,484 (Oct. 3, 1950).
63. Soc. Continentale Parker, Fr. Pat. 732,230 (Apr. 27, 1931).
64. Stone, I., U.S. Pat. 1,168,722 (July 31, 1934).
65. Swain, R. C. and Paden, J. H., U.S. Pat. 2,512,590 (June 20, 1950).
66. Swiss Pat. 247,835 (Dec. 16, 1947).
67. Thomas, J. F. J., Can. J. Research, 21B, 43-53 (1943).
68. Uhlig, H. H., Corrosion Handbook, 1st Ed., John Wiley & Sons.
69. Unknown, Light Metals, 12, No. 134, 165-8 (1949), see Corrosion 6, 42 (1950).
70. Vernon, W. H. J., Corrosion, 4, 141-8 (1948).
71. Viles, P. S. and Walsh, D. C., U.S. Pat. 2,550,434 (Apr. 24, 1951).
72. Wachter, A., Ind. Eng. Chem., 37, 749-51 (1945).
73. Wachter, A. and Smith, S. S., Ind. Eng. Chem., 35, 358-67 (1943).
74. Wachter, Treseder, and Weber, Corrosion, 3, 406 (1947).
75. Weiler, W., Dent. Farber-Ztg., 74, 27 (1938).
76. Williams, U.S. Signal Laboratory, Fort Monmouth, N.J., Technical Memorandum No. M-1156 (1948).
77. Worspop, F. E. and Kingsburg, A., Chem. Eng. Mining Rev. (Australia), 173-176 (1950).

which is 20% active contains 20% by weight of inhibitor. In cold climates where inhibitors are likely to be stored or used in subfreezing temperatures, it may be impossible to use as concentrated a solution as in warmer climates without resorting to more expensive solvents.

Corrosion inhibitors are usually compared on the basis of their *inhibitor efficiency,* which is the percentage that corrosion is lowered in their presence as compared with the corrosion rate which occurs in their absence. The inhibitor efficiency is calculated from the formula:

$$E = \frac{R_o - R_i}{R_o} \times 100 \tag{7.4}$$

where:
E = inhibitor efficiency
R_o = corrosion rate in absence of inhibitor
R_i = corrosion rate in presence of inhibitor

Example: Mild steel corroded in a cooling water at a rate of 1650 μm/y (65 mpy). When 10 ppm of an inhibitor was added, the corrosion rate dropped to 380 μm/y (15 mpy). What is the inhibitor efficiency?

$$R_o = 1650 \text{ and } R_i = 380$$

substituting in Equation (7.4),

$$E = \frac{1650 - 380}{1650} \times 100 = 77\%$$

Inhibitor concentrations are expressed as parts per million (ppm); for solids the units are on a weight basis, *e.g.,* kilograms (or pounds) of inhibitor per million kilograms (or pounds) of fluid; and for liquid inhibitors, volumes are used, *e.g.,* liters of inhibitor per million liters of fluid. To obtain the amount of inhibitor required for a given system, simply divide the amount of fluid to be inhibited by 1,000,000 and multiply by the ppm desired:

$$Q = \frac{V}{1{,}000{,}000} \times \text{ppm} \tag{7.5}$$

where Q is quantity of inhibitor required, V is the amount of fluid to be inhibited, and ppm is the inhibitor concentration in parts per million. Note that the quantity of inhibitor must be in the same units as those for the amount of fluid.

Example: What is the dosage of sodium chromate (a solid) required to add 10 ppm to 620,000 liters (165,000 gallons) of water? The volume of water is first converted to weight.

Kilograms of water = 620,000 liters × 1
(Pounds) = (165,000 gallons × 8.316/gal)

then:

V = 620,000 kg (1,369,500 lbs.)
ppm = 10

substituting in Equation (7.5):

$$Q = \frac{620{,}000}{1{,}000{,}000} \times 10 = 6.2 \text{ kg}$$

$$\left(\text{or } Q = \frac{1{,}369{,}500}{1{,}000{,}000} \times 10 = 13.7 \text{ lbs.}\right)$$

Safety Precautions

Safety precautions cover personnel safety, environmental protection, and equipment operational protection.

Handling

Toxic effects of inhibitors must be considered in processes where the compounds may be inhaled or contacted. Extreme care must be taken to avoid the use of toxic compounds in or near food processing equipment, ice plants, and potable water supplies.

When inhibitor solutions are prepared for injection, care should be taken to follow label instructions regarding skin contact, eye contact, ingestion, and inhalation. Read all safety information labelling before opening containers of inhibitors.

Disposal

Because a number of inhibitors contain ions and compounds which may be toxic, it is often difficult to dispose of inhibited fluids which are drained, dumped, or leaked from systems. Chromates are a prime example of this problem. In fact, the use of chromate-containing inhibitors has been severely reduced in recent years by this disposal problem. Disposal must be considered during inhibitor selection.

Heat Transfer

Heat transfer is an important consideration in applying corrosion inhibitors. Scaling of heat transfer surfaces, which can reduce heat transfer, should be avoided or held to a minimum. Excessive deposits of phosphates, silicates, or sulfates should be avoided; the latter in particular because they are difficult to remove by chemical means.

Foaming

The most appropriate action to take in avoiding difficulty from foaming is to determine where foam-forming conditions exist in the system. These will consist of places where the inhibitor-containing fluid is agitated with a gas such as in a gas separator, in a counter-current stripper, or in an aerator. The next step is to obtain a sample of the fluid and gas from the process, add the inhibitor in question, adjust the temperature to that corresponding to the process step, and shake vigorously. If this test produces a stable foam, a potential problem exists.

There are three alternative remedies: (1) an antifoaming agent may be added (this must be tested also), (2) tests can be made to select an inhibitor which does not cause foaming, and (3) the system can be shut down periodically and treated with a slug of persistent inhibitor. The last two remedies are the least palatable because the need for an inhibitor is at hand, and there are few processes which can be shut down with sufficient frequency to maintain effective inhibition by slug treatment.

Emulsions

Emulsions are similar to foams in that they consist of one phase dispersed in another. Emulsions consist of two liquid phases that will not mix, while foams are composed of a gas and a liquid phase. Conditions favorable for emulsion formation are the presence of two liquid phases, agitation, and an emulsion stabilizer. In this case, the corrosion inhibitor may behave as an emulsion stabilizer, so the procedure is to shake the two liquid phases with inhibitor and measure the time required for them to separate. If the time required for the separation is longer in the presence of the inhibitor than in its absence, the inhibitor is an emulsion stabilizer. The remedy is the same as for foams: add a demulsifier, use another inhibitor, or inhibit with slugs during shutdown.

Economics of Inhibition

Prevention of corrosion by inhibition may be desirable for several reasons:

1. To extend the life of equipment
2. To prevent shutdowns
3. To prevent accidents resulting from brittle (or catastrophic) failures
4. To avoid product contamination
5. To prevent loss of heat transfer
6. To preserve an attractive appearance

Potential savings for each of these goals must be evaluated to determine if a program of corrosion inhibition will be economical. Because costs are sometimes difficult to estimate, the best method is to obtain data on maintenance, replacements, etc., from past history of the system which is to be protected or from a similar system. Literature on the economics of inhibition is a tremendous aid in estimating costs. Bregman,[3] for instance, discusses the economics of inhibitors in boilers, cooling towers, closed cooling cycles, petroleum production, and petroleum refining.

There are several costs associated with the use of inhibitors. In fact, the cost of one or more of the following must be factored into any economic evaluation of corrosion inhibition.

1. Installation of injection equipment
2. Maintenance of injection equipment
3. Purchase of inhibitor chemical(s)
4. Monitoring inhibitor concentration(s)
5. System changes to accommodate the inhibitor
6. Operational changes to accommodate the inhibitor
7. System cleaning
8. Waste disposal
9. Personnel safety equipment

In cases where major shutdowns can be avoided through the use of inhibitors, the economic advantages of inhibition undoubtedly will be clear. Other cases will require detailed economic evaluations (Chapter 15).

Summary

This chapter has presented technical background on the fundamentals of corrosion control by inhibition. Basic approaches to the selection and utilization of inhibitors have been outlined. The importance of the Safety Precautions and Economics sections for successful use of inhibitors cannot be overemphasized.

Because of the complexity of corrosion inhibition, it is advisable to seek assistance from individuals with specific knowledge when considering the use of inhibitors.

References

1. Langelier, W. F., J. Am. Water Works Assoc., Vol. 28, p. 1500 (1963).
2. Noack, M. G., Evaluation of Catalyzed Hydrazine as an Oxygen Scavenger, Materials Performance, Vol. 21, No. 3, p. 26 (1982).
3. Bregman, J. I., Corrosion Inhibitors, Macmillan Company, New York, N.Y. (1963).

Bibliography

Barron, Bruce M. Oil & Gas J., Vol. 63, No. 20, p. 128 (1965).

Dean, S. W., Jr., R. Derby, and G. T. von dem Bussche. Inhibitor Types. Materials Performance, Vol. 70, No. 12, p. 47 (1981).

Evans, U. R. The Corrosion and Oxidation of Metals. Edward Arnold (Publishers) Ltd., London, England, 1960.

Greene, N. D. Mechanism and Application of Oxidizing Inhibitors. Materials Performance, Vol. 21, No. 3, p. 20 (1982).

Hausler, R. H. Inhibition of Corrosion Reaction by Corrosion Product Layers of Type Metal Ion Chelating Agent. Corrosion/81, Paper No. 253, NACE, Houston, Texas, April, 1981.

NGAA Condensate Well Corrosion Committee. Condensate Well Corrosion. Natural Gasoline Association of America, Tulsa, Oklahoma, 1953.

Oakes, B. D. Historical Review of Inhibitor Mechanisms. Corrosion/81, Paper No. 248, NACE, Houston, Texas, April, 1981.

Ostroff, A. G. Introduction to Oilfield Water Technology. Prentice-Hall, Inc., Englewood Cliffs, N.J., 1965.

Ranny, M. W. Corrosion Inhibitors—Manufacture and Technology. Noyes Data Corp., Park Ridge, N.J., 1976.

Roller, D. and W. R. Scott. Corrosion Technology, Vol. 8, No. 3, pp. 71-76 (1961).

Shreir, L. L. Corrosion. Vol. 2. John Wiley & Sons, Inc., New York, N.Y., 1963.

Speller, Frank N. Corrosion: Causes and Prevention. McGraw-Hill Book Company, Inc., New York, N.Y., 1951.

Tomashov, N. D. Theory of Corrosion and Protection of Metals. Macmillan Company, New York, N.Y., 1966.

Trabanelli, G. and F. Zucchi. Fundamentals of Inhibition with VCI's. Corrosion/76, Paper No. 82, NACE, Houston, Texas, March, 1976.

Uhlig, H. H. Corrosion and Corrosion Control. John Wiley & Sons, Inc., New York, N.Y., p. 232, 1963.

EDITOR'S NOTE: In addition to the references given, useful information on inhibitors has been published by the National Association of Corrosion Engineers in committee reports and in hundreds of technical articles. This information is readily retrieved by using the indexes to journals. Copies of the referenced articles and reports can be secured at modest costs. For a reasonably complete list of books about inhibition, refer to the NACE Bibliography of Books on Corrosion available from NACE, Box 218340, Houston, Texas 77218.

NOTES

Chapter 8

Corrosion by Water and Steam

CORROSION BY WATER AND STEAM

Introduction

A major use of water in industry is the transfer of heat and the production of steam. There is extensive use of cooling water in almost every manufacturing process, in commercial air conditioning, and even a substantial percentage in domestic air conditioning. Water is used in the passive sense for potable and for fire control purposes. Fossil and nuclear fuel steam plants are encountered in the heating and power generating fields. Saline waters of the world corrode ships and many other man-made structures.

Water can be corrosive to most metals. Pure water, without dissolved gases (*e.g.*, oxygen, carbon dioxide, and sulfur dioxide) does not cause undue corrosion attack on most metals and alloys at temperatures up to at least the boiling point of water. Even at temperatures of about 450 C (850 F), almost all of the common structural metals, except magnesium and aluminum, possess adequate corrosion resistance to high-purity water and steam. Naturally occurring or man-made contaminants make water corrosive, as will be discussed later in the chapter.

In summary, the factors influencing the corrosion of materials in water systems are:

1. The physical configuration of the system
2. The chemistry of the water (hardness salts, chlorides, and dissolved gases being the most important)
3. The presence of solids in the water
4. The flow rate
5. The temperature of the water
6. The presence of bacteria

This chapter will discuss the corrosion problems associated with natural and treated waters at temperatures ranging from ambient up to about 540 to 600 C (1000 to 1100 F), the means of preventing or slowing down corrosion, and the proper choice of materials for each environment.

The Role of Contaminants

The effect of contaminants is controlling in regards to both corrosion and scaling phenomena. The specific water chemistry must be known before an accurate appraisal can be made of its probable corrosivity. At a minimum, the analytical data required for such an appraisal should include the information presented in Table 8.1.

TABLE 8.1 — Data Needed for Analyzing Specific Water Chemistry

Total Dissolved Solids	TDS[1]	as ppm
Calcium	Ca[1]	as ppm $CaCO_3$
Magnesium	Mg	as ppm $CaCO_3$
Total Hardness	TH	as ppm $CaCO_3$
Phenolphthalein Alkalinity	P Alk	as ppm $CaCO_3$
Methyl Orange Alkalinity	MO Alk[1]	as ppm $CaCO_3$
Chlorides	Cl	as ppm Cl^-
Sulfates	SO_4	as ppm $SO_4^=$
Silica	SiO_2	as ppm SiO_2
Iron	Fe	as ppm Fe
Dissolved Oxygen	D.O.	as ppm O_2
H^-/OH^- Concentration	pH[1]	

[1]Used in scaling calculations, as explained in the text.

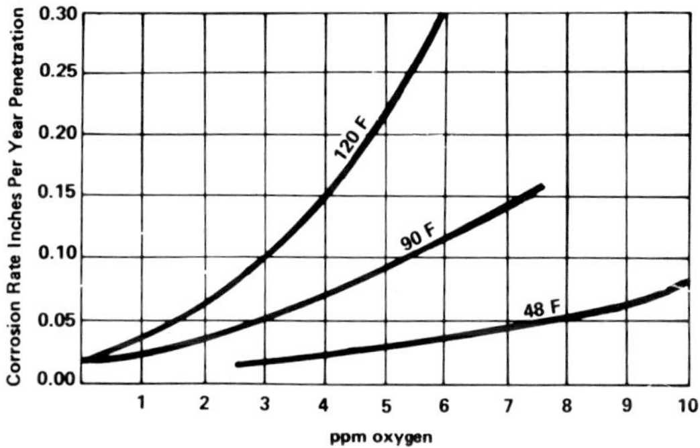

FIGURE 8.1 — Effect of oxygen concentration on the corrosion of low-carbon steel in tap water at different temperatures. [SOURCE: Betz Laboratories, Inc. advertisement, Materials Protection, Vol. 4, No. 10, p. 21 (1965).]

From a corrosion standpoint, the most significant contaminant is oxygen from the air dissolved in water. This is illustrated in Figure 8.1. As described in Chapter 2, oxygen is a cathodic depolarizer that reacts with and removes hydrogen from the cathode during electrochemical corrosion, thereby permitting corrosion attack to continue.

Other contaminants that contribute to corrosion are chlorides (usually sodium chloride, commonly referred to as "table salt") and carbon dioxide. Calcium hardness has an indirect effect, as will be explained later. Other contaminants include sulfides or ammonia from industrial or natural sources. Of course, many other man-made contaminants can be found in local areas if industries are permitted to discharge their waste products into water resources. As with other chemical reactions, corrosion increases with elevated temperature, unless stifled by insoluble scales or the loss of corrosive gases.

Scales precipitated onto metal surfaces can provide excellent protection of the substrate or may accentuate pitting at pores, cracks, or other voids in the film. If the film attains any significant thickness, the loss of heat transfer through the metal and deposited scale can be a problem in certain applications. Thus, the development of scales on metal surfaces is an important consideration when using metals in waters.

The factors that affect scaling (*i.e.,* the deposition of primarily calcareous deposits) are the total dissolved solids, calcium ion concentration, methyl orange alkalinity (bicarbonates and carbonates), and temperature. Calcium and magnesium carbonates have an inverse solubility with temperature, therefore they tend to deposit scale with rising temperature. In extreme cases, scaling by insoluble calcium sulfate also can be a problem, although it is not a prevalent one under most conditions.

The tendency of a water to deposit such scale, or of the temperature required to initiate such deposition to exist, can be estimated from the Langelier-Ryznar Index, as shown in Table 8.2. From the relationship between the controlling factors, one can estimate the pH_S (*i.e.,* the pH which the water in question would acquire if saturated with calcium carbonate). This can then be used to calculate either the Corrosion Index (which indicates the corrosivity to ordinary cement or concrete), the Saturation Index, *or* (by substituting the measured pH for the pH_S) the temperature at which deposition will be initiated (Table 8.3).

The effect of oxygen on corrosion with increasing temperature is also shown in Figure 8.2, indicating the results in a closed versus an open container (the latter permitting natural deaeration by ebullition). In a closed vessel, corrosion continues to increase with temperature, whence the requirement for removing dissolved oxygen from hot water systems and boilers.

TABLE 8.2 — Scaling by Water

Corrosion Index = $pH - pH_S$ Where 0 indicates balanced conditions,
+ indicates scale formation
− indicates scale dissolution or corrosion

Stability Index = $2\,pH_S - pH$ Where values greater than 7.5 indicate increasing corrosivity

$pH_S = (9.3 + A + B) - (C + D)$

Total Solids	A	Ca as $CaCO_3$	C	M.O.Alk.	D
50-330	0.1	10-11	0.6	10-11	1.0
400-1000	0.2	12-13	0.7	12-13	1.1
		14-17	0.8	14-17	1.2
Temp. °F	**B**	18-22	0.9	18-22	1.3
		23-27	1.0	23-27	1.4
32-34	2.6	28-34	1.1	28-35	1.5
36-42	2.5	35-43	1.2	36-44	1.6
44-48	2.4	44-45	1.3	45-55	1.7
50-56	2.3	56-69	1.4	56-69	1.8
58-62	2.2	70-87	1.5	70-88	1.9
64-70	2.1	88-110	1.6	89-110	2.0
72-80	2.0	111-138	1.7	111-139	2.1
82-88	1.9	139-174	1.8	140-176	2.2
90-98	1.8	175-220	1.9	177-220	2.3
100-110	1.7	230-270	2.0	230-270	2.4
112-122	1.6	280-340	2.1	280-340	2.5
124-132	1.5	350-430	2.2	360-440	2.6
134-146	1.4	440-450	2.3	450-550	2.7
148-160	1.3	560-690	2.4	560-590	2.8
162-178	1.2	700-870	2.5	700-880	2.9
178-194	1.1	880-1000	2.6	890-1000	3.0
194-210					

NOTE: Use actual pH of water when calculating "B," the temperature at which the water will cause scale.

TABLE 8.3 — Calculation of Scaling Temperature

Conditions:

pH = 7.5	
TDS − 300 ppm	A = 0.1
Ca = 45 ppm	C = 1.3
M.O.Alk. = 120 ppm	D = 2.1

B = pH − 9.3 − A + C + D
= 7.5 − 9.3 − 0.1 + 1.3 + 2.1
= 10.9 − 9.4
= 1.5
= 1.5 132 F or 55 C

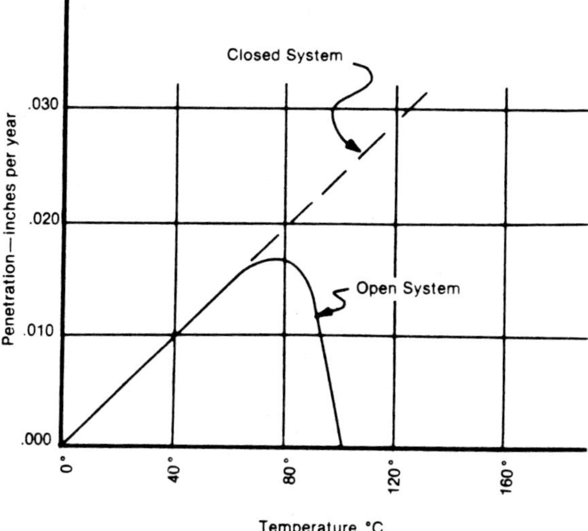

FIGURE 8.2 — Effect of oxygen on corrosion of steel.

The effect of oxygen and pH is shown in Figure 8.3. In a broad range of about pH 5 to 9, the corrosion rate can be expressed simply in terms of the amount of dissolved oxygen present (*e.g.,* μm/y

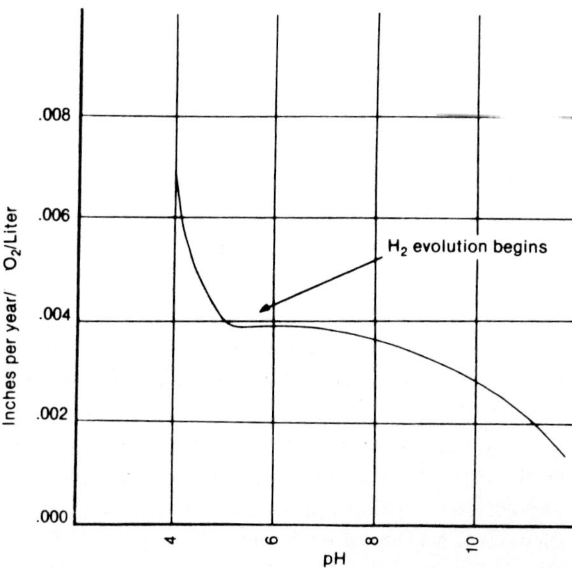

FIGURE 8.3 — Effect of pH and O_2 on the corrosion of steel.

per ml. D.O. per liter of water). At about pH 4.5, acid corrosion is initiated, overwhelming the oxygen control. At about pH 9.5 and above, deposition of insoluble ferric hydroxide tends to stifle attack.

Types of Water

In common terms, water is frequently described either in terms of its nature, usage, or origin. Ground waters originate in subterranean locations (*e.g.*, wells), while surface waters comprise the lakes, rivers, and seas.

Fresh Water

Fresh water may come from either a surface or ground source, and typically contains less than 1% sodium chloride. It may be either "hard" or "soft," *i.e.*, either rich in calcium and magnesium salts and thus forming insoluble curds with ordinary soap, or not. Actually, there are gradations of hardness, which can be estimated from the Langelier-Ryznar Index or accurately determined by titration with standardized chelating agent solutions (*e.g.*, "versenates").

Brackish Water

Brackish water contains between 1 and 2.5% sodium chloride, either from natural sources (*e.g.*, salt deposits) around otherwise fresh water or by dilution of seawater (*e.g.*, tidal rivers).

Seawater

Seawater typically contains about 3.5% sodium chloride, although the salinity may be weakened in some areas by dilution with fresh water or concentrated by solar evaporation in others. Seawater is normally more corrosive than fresh water because of the higher conductivity and the penetrating power of the chloride ion through surface films on a metal. The rate of corrosion is controlled by the chloride content, oxygen availability, and the temperature.

The 3.5% salt content of seawater produces the most corrosive chloride salt solution that can be obtained (Figure 8.4). The combination of high conductivity and oxygen solubility is at a maximum at this point (oxygen solubility is reduced in more concentrated salt solutions). The corrosion of numerous metals in a wide range of saline waters is reported in Table 8.4.

It might be assumed that the corrosion of steel in seawater would increase with an increase in temperature; however, oxygen solubility in the water increases *inversely* with the temperatures. As a result, some deep, cold currents of seawater are more corrosive than warmer bodies of salt water. Large areas of ocean have been surveyed at various depths regarding temperature, salt content, oxygen content, and other relevant factors.

The varying corrosion rates to be expected at differing levels of a coastline have been generalized by Edwards (Figure 8.5). Note the two maxima in the curve. Essentially all of the corrosion mechanisms discussed in this book are accelerated in the seawater electrolyte. Greater pitting, more rapid dealloying, shorter fatigue life, and more severe galvanic problems are typical problems in seawater as compared with fresh water exposures.[1] Sulfur compounds or other contaminating ions in the water can complicate the corrosion of most metals.

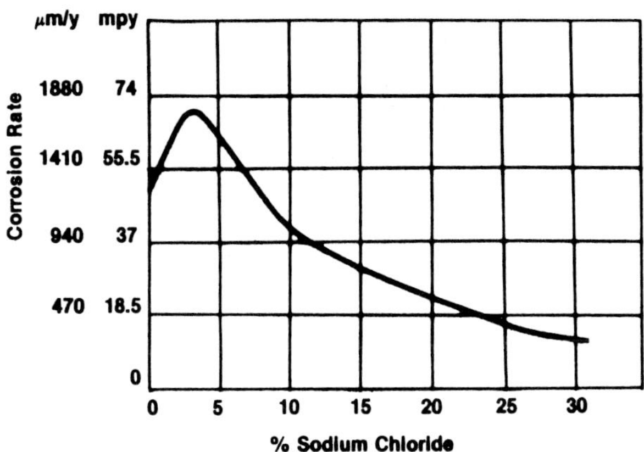

FIGURE 8.4 — Corrosion of steel in various sodium chloride solutions. Note the peak in corrosion rate at approximately 3% sale. [SOURCE: Guidelines for Selection of Marine Materials, INCO, New York, N Y (1971).]

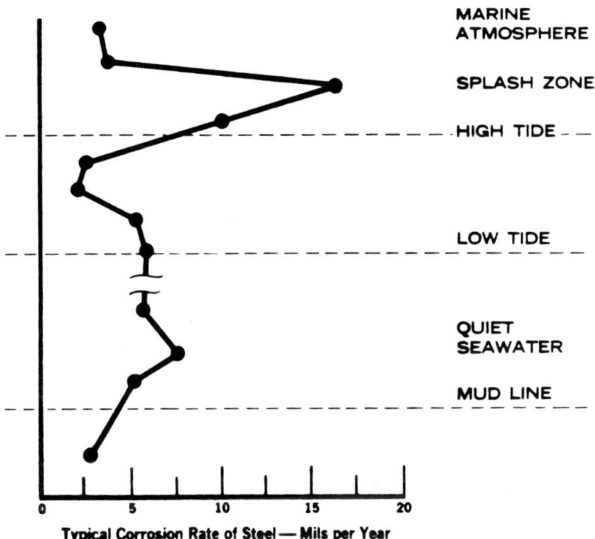

FIGURE 8.5 — Typical corrosion rates for steel at various depths in seawater. [SOURCE: Guidelines for Selection of Marine Materials, INCO, New York, N Y (1971).]

As mentioned earlier, water may also be described in terms of its end use. The implications in these descriptions range from being highly specific to so general as to be nondefinitive.

Distilled or Demineralized Water

The total mineral content of water can be removed by either distillation or mixed-bed ion exchange. In the first case, purity is described qualitatively in some cases (*e.g.*, triple-distilled water), but is best expressed, for both distilled and demineralized water, in terms of specific conductivity. Water also can be demineralized by reverse osmosis or electrodialysis.

TABLE 8.4 — Corrosion Rates of Metals in Seawater, mpy (Multiply mpy by 25.4 to obtain μm/y)

Alloy and Composition	Laboratory Nat.	Laboratory Water	Laboratory Syn.	Great Salt Lake	New York Harbor	New York Harbor	Hudson River (near New York City)	Hudson River (near New York City)	Miami, FL	Miami, FL	Haifa, Israel	Avg. in Seawater
Exposure, yrs	0.67	0.66	0.33	0.26	0.30	0.68	0.31	0.69	0.19	0.39	0.25	—
pH	7.4	7.3	—	—	~7.4	~7.4	~7.3	~7.3	—	—	—	—
Solid Cont. %	2.51	1.26	3.45	varies	~2.51	~2.51	~1.26	~1.26	~3.5		~3.5	—

ALUMINUM ALLOYS

Al	Mn	Mg	Cr	Si	Cu	Description	Lab. Nat.	Lab. Water	Lab. Syn.	Great Salt Lake	New York Harbor (0.30)	New York Harbor (0.68)	Hudson River (0.31)	Hudson River (0.69)	Miami, FL (0.19)	Miami, FL (0.39)	Haifa, Israel	Avg. in Seawater
98.98	1.02					—	0.119	0.110	0.153	4.86	0.258	0.319	0.378	0.126	1.77	1.22	1.16	0.747
97.87		1.03	0.25	0.59	0.26	—	0.237	0.087	0.080	5.89	2.44	1.46	1.93	1.03	3.19	3.50	0.910	2.07
97.37		2.42	0.21			—	0.054	0.038	—	4.36	0.630	0.634	0.521	0.165	1.38	1.22	0.850	0.743

COPPER ALLOYS

Cu	Al	Sn	Pb	Fe	Ni	Mn	Si	Zn	P	Be	Description	Lab. Nat.	Lab. Water	Lab. Syn.	Great Salt Lake	New York Harbor (0.30)	New York Harbor (0.68)	Hudson River (0.31)	Hudson River (0.69)	Miami, FL (0.19)	Miami, FL (0.39)	Haifa, Israel	Avg. in Seawater
99.96									0.2		—	0.520	0.472	0.968	9.25	2.74	1.02	1.77	1.00	1.81	1.65	2.65	1.81
99.94											—	0.641	0.291	0.661	—	2.43	0.954	1.85	1.03	1.97	1.65	2.13	1.72
97.65					0.35					2.0	Be copper	0.641	0.397	0.853	3.93	2.08	0.638	0.630	0.449	1.06	0.866	1.29	0.736
97.65					0.35					2.0	Be copper[1]	0.208	0.279	0.925	3.93	1.48	0.619	0.028	0.826	1.04	0.827	1.97	0.828
95.50		4.25							0.25		Phos Bronze	0.622	0.551	0.907	7.10	1.99	0.783	0.985	0.708	1.58	1.14	1.39	1.23
89.90		10.0							0.10		Phos Bronze	0.760	0.512	0.689	1.17	1.51	0.571	0.029	1.30	0.866	0.905	0.774	0.684
95.79				0.07		0.93	3.11	0.10			—	0.484	0.708	1.57	8.13	2.61	1.06	1.85	0.865	1.65	1.26	2.24	1.65
94.94	4.85			0.06	0.05			0.10			—	0.141	0.385	0.327	0.330	0.752	0.343	0.457	0.189	0.590	0.385	0.974	0.527
94.13	5.8		tr	0.02	tr	tr	tr	0.05			—	0.176	0.401	0.250	0.290	1.26	0.512	0.693	0.488	0.865	0.524	1.17	0.787
93.68	5.83		tr	0.05		0.05	0.04	0.10			—	0.155	0.248	0.767	—	1.11	0.252	0.346	0.181	—	—	1.15	0.608
85.23			tr	0.01				14.76			Red brass	0.361	0.220	0.669	9.01	2.71	1.07	1.89	1.14	2.48	1.58	1.85	1.82
84.67			tr	tr	tr			15.33			Red brass	0.212	0.108	0.541	1.001	2.74	1.03	1.10	1.03	2.32	1.58	2.16	1.71
77.17	2.05		0.02	0.02				20.69			Alum brass	0.031	0.059	0.082	0.815	0.984	0.102	0.355	0.154	0.748	0.362	0.488	0.456
76.46	2.00		tr	0.02	tr	0.02	0.03	21.45			Alum brass	0.022	0.033	0.178	—	1.05	0.248	0.629	0.260	—	—	1.15	0.608
76.14	1.98	0.02	0.02	0.02				21.78			Alum brass	0.023	0.087	0.084	0.330	2.35	0.551	0.276	0.228	0.708	0.512	0.659	1.11
71.46		1.0	0.05	tr				27.45			Admir brass	0.271	0.076	0.977	0.887	2.73	0.985	1.73	1.10	1.81	1.10	1.55	1.57
71.27		1.05	0.02	0.02				27.59			Admir brass	0.282	0.193	0.259	8.00	2.73	1.07	1.65	0.945	1.69	1.02	1.41	1.50
70.74		1.09	0.02	0.01				28.11			Admir brass	—	—	0.886	9.01	—	—	—	—	—	—	1.24	—
70.00		1.0						28.97	0.03		Phosphorized Admir brass	0.306	0.098	—	7.39	1.67	1.19	1.85	0.985	1.46	1.10	1.65	1.42
70.00	2.00							27.97			Alum brass	0.121	0.067	—	0.524	1.29	0.449	0.669	0.268	0.905	0.787	0.569	0.705
66.52			0.15	0.04	0.03			33.26			Yellow brass	0.118	0.034	0.385	9.22	2.90	1.00	1.26	1.08	2.05	1.02	0.722	1.43
61.93		0.7						38.00			Naval brass	0.240	0.250	0.904	0.890	—	—	—	—	—	—	1.01	—
61.93		0.72						37.28	0.07		Phosphorized Admir brass	0.280	0.310	0.946	8.70	—	—	—	—	—	—	2.06	—
60.00			tr	tr		tr		40.00			Muntz Metal	0.373	0.385	0.861	—	5.74	2.84	4.05	2.25	3.90	4.13	5.17	4.01

(continued)

TABLE 8.4 (continued)

IRONS AND STEELS

Fe	Mn	Si	Cr	Ni	P	S	Cb	C	Cu	Mo	Description												
100.00											Iron	1.30	1.30	6.84	2.22	—	42.5	—	2.52	4.82	5.90	7.68	5.03
99.54	0.30				0.01	0.03		0.08	0.04		Low carb steel	1.34	1.40	11.60	3.23	2.60	2.37	2.00	0.988	5.12	6.30	6.78	4.31
99.36	0.27	tr		0.04	0.01	0.02		0.08	0.22		Cu-bearing steel	1.26	1.40	10.90	3.35	—	2.79	—	0.748	5.90	6.69	8.45	4.92
70.33	1.25	0.25	18.4	9.6	0.02	0.02		0.05	0.08		304 stainless	sl	sl	0.004	0.153	—	sl	nil	sl	—	—	0.009	sl
70.33	1.25	0.25	18.4	9.6	0.02	0.02		0.05	0.08		304 stainless[2]	0.088	0.098	0.009	0.287	0.382	0.351	0.551	0.347	3.07	3.46	0.032	1.17
68.80	2.0	1	18	10	0.04	0.04	0.8	0.10			347 stainless	0.230	0.181	sl	0.282	0.654	0.703	0.614	0.468	4.45	4.53	0.064	1.64
67.70	1.36	0.37	17.0	11.3	0.02	0.01		0.06		2.30	316 stainless	0.074	0.017	0.004	0.043	0.011	0.100	0.351	0.028	2.01	1.65	0.014	0.605
64.87	1.14	0.37	18.0	13.3	0.02	0.02		0.02		2.26	316 stainless	0.036	nil	0.008	0.058	nil	0.002	nil	nil	1.89	1.06	0.027	0.426
62.97	1.50	0.42	18.6	14.0	0.01	0.01	0.32	0.02		2.15	316 stainless	0.070	nil	0.010	0.055	nil	0.038	0.015	0.005	2.64	1.69	0.022	0.630
65.20	2.0	0.75	20	11				0.08			308 stainless	0.128	0.039	—	—	0.882	0.326	0.748	0.307	—	—	sl	0.453
61.78	1.56	0.17	19.3	13.3	0.02	0.02		0.07	0.18	3.60	317 stainless	sl	sl	nil	—	—	0.162	nil	sl	—	—	sl	sl
61.78	1.56	0.17	19.3	13.3	0.02	0.02		0.07	0.18	3.60	317 stainless	0.001	nil	0.006	0.013	nil	nil	nil	nil	0.017	nil	0.022	0.006

NICKEL AND CUPRONICKEL ALLOYS

Ni	Cu	Mn	Mg	Zn	Al	Fe	Pb	Cr	Si	C	Description												
99.40	0.1	0.2				0.15			0.05	0.1	—	0.022	0.009	0.022	0.003	0.468	0.335	0.061	0.039	3.35	2.25	0.126	0.946
78.50	0.2	0.25				6.5		14.0	0.25	0.08	Inconel	0.013	nil	0.006	0.375	sl	0.003	nil	0.010	2.44	0.905	0.036	0.485
67.00	30.0	1.0				1.4			0.1	0.15	Monel	0.053	0.057	0.204	0.360	0.826	0.606	0.826	0.213	2.48	1.81	0.151	0.99
30.53	67.86	1.06	tr	0.06	tr	0.49	tr				Cupro-	—	—	0.104	0.120	—	—	—	—	—	—	0.333	—
30.32	68.62	0.58				0.48	tr				nickel	—	—	0.061	0.643	—	—	—	—	—	—	0.327	—
30.07	68.69	0.48		0.32		0.44	tr				30%	0.090	0.181	0.303	1.04	0.870	0.299	0.421	0.339	0.550	0.343	0.390	0.460
19.78	79.51	0.6		0.05		0.06	tr				Cupro-	0.368	0.295	0.404	5.13	1.76	0.669	0.905	0.535	1.46	1.06	1.66	1.15
19.51	78.76	0.98	tr	0.10		0.55	tr				nickel 20%	—	—	1.45	1.21	—	—	—	—	—	—	0.711	—
18.00	65.0			17.0							—	0.066	0.087	0.191	4.72	2.33	0.842	0.945	0.555	1.38	0.827	1.05	1.13
18.00	55.0			27.0							—	0.056	0.055	0.155	4.67	2.80	0.860	0.905	0.547	0.945	0.590	1.09	1.11
10.20	89.02	0.24		0.10		0.44	tr				Cupro-	0.340	0.292	0.344	2.13	1.66	0.512	0.748	0.528	2.17	1.22	0.840	1.10
9.86	89.25					0.89	tr				nickel	—	—	0.523	0.955	—	—	—	—	—	—	0.597	—
9.84	89.06	0.19		0.20	0.71						10%	—	—	0.221	1.00	—	—	—	—	—	—	0.874	—

[1]Maximum hardness.

[2]Similar materials from different suppliers, or duplicate specimen.

[SOURCE: Hugh B. Gordon, How Metals Resist Sea Water, Materials Engineering, Vol. 65, No. 5, pp. 82-83 (1967).]

Steam Condensate

Water condensed from industrial steam is called steam condensate. It approaches distilled water in purity, except for contamination (as by dissolved oxygen or carbon dioxide) and the effect of deliberate additives (*e.g.,* neutralizing or filming amines).

Boiler Feedwater Make-up

The feedwater make-up for boilers is always softened and subsequently deaerated, as described in more detail later. It may vary in quality from fairly high dissolved solids (*e.g.,* Zeolite-Treated), to very pure demineralized feed for high-pressure boilers. It tends to be *highly* corrosive, because of its softness, until thoroughly deaerated. This term is more precise than *Boiler Feedwater,* which may include recirculated steam condensate in various ratios to fresh make-up water.

Potable Water

Potable water is fresh water that is sanitized with oxidizing biocides like chlorine or ozone to kill coliform and other bacteria and make it safe for drinking purposes. By definition, certain mineral constituents are also restricted. For example, the chloridity will be not more than 250 ppm chloride ion in the United States or 400 ppm on an international basis. There is at least an implied control of water chemistry in the definition itself.

Process or Hydrotest Water/Firewater

These terms are essentially nondefinitive, since the water employed may be of almost any chemistry, ranging from demineralized water to quite saline fresh water or even seawater in some cases. *Produced Water* is that which originates in oil and gas production, emanating from geological sources with the hydrocarbons.

Cooling Water

Cooling water is another undefined term, although it implies that any necessary treatment against excessive scaling or corrosion has been applied, or corrosion-resistant material selected. This may include anything from fresh water to seawater, and may comprise either an open or closed system, or a once-through system, as will be described later.

Waste Water

By definition, waste water is any water that is discarded after use. Sanitary waste from private or industrial applications is contaminated with fecal matter, soaps, detergents, etc., but is quite readily handled from a corrosion standpoint. Industrial wastes from chemical or petrochemical sources can contain strange and specific contaminants which greatly complicate materials selection, especially in the uses of plastics and elastomers.

Corrosion of Materials

Surprising as it may seem, while many contaminants in water promote corrosion, they often are not found in the corrosion products. The corrosion products are usually oxides, hydroxides, or hydrated oxides of the metal components of the alloy in use.

Zinc

Zinc is employed primarily in the form of the hot dip coating (*i.e.,* galvanizing) or an electroplate. Useful in cold water applications, its potential application for hot water systems should be carefully evaluated. In waters of a particular chemical makeup and at temperatures of approximately 80 C (170 F), zinc may become *cathodic* to steel and cause accelerated corrosion. Dissolved oxygen and carbon dioxide aggravate the corrosion of zinc, as do extremes of pH. Zinc is amphoteric, *i.e.,* attacked by both free acid and free alkali.

Aluminum

Unalloyed aluminum and those alloys containing combinations of magnesium, zinc, and silicon are most resistant to water. Aluminum alloys containing appreciable percentages of copper may be severely corroded. Aluminum and its alloys often pit in seawater and brackish waters, especially because of oxygen concentration cells under deposits. Recall that the metal is amphoteric and can be attacked in moderately acid or alkaline media.

In both fresh and salt water, extreme care must be exercised to avoid galvanic couples of aluminum to more noble metals (*e.g.,* steel or copper). The copper-aluminum couple is particularly bad. Copper or iron ions in the system can cause attack even when not directly connected; the heavy metal ions plate out on the aluminum by reduction effects and cause pitting or other types of localized corrosion at holes in the metallic, cathodic plate (deposition corrosion).

Aluminum corrosion products in water are white to translucent. At low temperatures, an amorphous (*i.e.,* noncrystalline) alumina film (Al_2O_3) forms at first. This may revert to boehmite ($Al_2O_3 \cdot H_2O$ or $AlO \cdot OH$) and finally bayerite ($Al_2O_3 \cdot 3H_2O$). At higher temperatures (above the boiling point of water), the oxide film consists of an inner layer of amorphous alumina and randomly oriented boehmite, and an outer layer of highly oriented boehmite.

Aluminum is not used in water above 200 C (390 F).

Iron and Steel

Iron and steel corrode in water to form various compounds, depending on the temperature and other environmental conditions. In its simplest concept, water ionizes to produce a hydrogen ion and a hydroxyl ion. A small fraction of these ions are always present in the free state where they can react with a metal.

$$\underset{\text{Water Molecule}}{H_2O} \rightarrow \underset{\text{Hydrogen Ion}}{H^+} + \underset{\text{Hydroxyl Ion}}{OH^-} \quad (8.1)$$

Theoretically, iron should react with water in the absence of air to form ferrous hydroxide, at least under extreme conditions.

$$\underset{\text{Iron}}{Fe} + \underset{\text{Water}}{2\,H_2O} \rightarrow \underset{\text{Ferrous Hydroxide}}{Fe(OH)_2} + \underset{\text{Hydrogen}}{H_2} \quad (8.2)$$

Pure ferrous hydroxide is white. However, even extremely small quantities of dissolved oxygen in the water will produce ferric ions, which color the ferrous hydroxide green, brown, or black, depending on the quantity of the chromophoric ferric ions.

The oxides normally found include *lepidocrocite* (gamma FeO • OH), a yellow to orange rust, such as that found on well-used train rails after a rain, and ordinary rust ($Fe_2O_3 \cdot 3H_2O$). At higher temperatures, one may encounter *magnetite* (Fe_3O_4), a black magnetic film strongly cathodic to steel and therefore conducive to localized attack in aerated water. Other high-temperature forms include gamma Fe_2O_3 (a brown, protective, slightly magnetic product) and alpha Fe_2O_3 (a brick-red oxide). Its presence in steam generating equipment indicates oxygen contamination.

Cast Iron

In most cases, ordinary gray cast iron is about twice as resistant to natural waters as steel because of the bonding of rust products by the graphite flakes. Soft or low pH waters may cause graphitic corrosion (Chapter 5). In modern engineering practice, most cast iron piping receives an internal cementatious coating for improved water side corrosion resistance.

Wrought Iron

Wrought iron is no longer commercially available in the United States. In general, it will have improved resistance, as compared with carbon steel, only in very soft waters (*e.g.*, steam condensate).

Carbon Steel

Carbon steel is severely corroded by raw waters containing oxygen. It is completely resistant in soft, anaerobic waters. With the exception of specifically treated boiler feedwater make-up, duly deaerated, or water specifically inhibited for cooling purposes, one should assume that ordinary surface or ground waters will be corrosive to steel. Rates may be as low as 130 μm/y (5 mpy) at room temperature, but escalate rapidly with increasing temperature. Pitting rates will be about ten times the general rate of corrosion. Deposition corrosion can occur if copper ions are present.

Low-alloy steels offer little improvement, although the "weathering steels" of ASTM A-242 (Chapter 11) resemble wrought iron in their improved resistance to steam condensate. Alloy steels of the chromium-molybdenum family are employed in parts of boilers because of their improved high-temperature properties, the steam and water having already been rendered substantially noncorrosive.

Stainless Steels

Martensitic stainless steels of the 11 to 13% chromium variety (*e.g.*, AISI 410; S41000) are resistant to water, but very prone to severe localized chloride pitting (Chapter 5). In deaerated boiler feedwater make-up, the cast varieties (*e.g.*, CA6NM) are finding successful application.

The ferritic grades (*e.g.*, AISI 430; S43000) are more reliable than the lower chromium grades (*e.g.*, AISI 410, 405, and 409), but still are prone to pitting in chloride-bearing waters. The low interstitial superferritics (*e.g.*, S44625), however, are extremely resistant to both pitting and stress corrosion cracking in otherwise aggressive waters, while the 29-4 varieties will even cope with hot seawater.

The garden variety 18-8 grades are useful in water, depending upon chloridity and aeration, as well as such extraneous factors as heat flux in coolers and condensers, but are subject to pitting and stress corrosion cracking (Figure 8.6).

The 18-8 stainless steels containing molybdenum (*e.g.*, AISI 316L and CF3M) may be successfully used in seawater, but their performance is contingent upon high-velocity (4 mps or more), aeration (or total deaeration), and freedom from deposits of scale or marine organisms. The regular carbon grades, unless stabilized, may be subject to intergranular corrosion in natural waters under some conditions (*e.g.*, high chloridity, bacterial action, oxygen cells, etc.).

Perhaps the major loss of austenitic stainless steels has been by exposure to heating or cooling waters when used as tubing in tube-and-shell heat exchangers. Figure 8.7 shows the actual problem area. If the tube sheet is sufficiently hot, salts can accumulate on the tube in the crevice or immediately behind the sheet as water is evaporated. Chlorides contained in these salts will crack the stressed metal in this area.

When using the austenitic stainless steels in tube-and-shell heat exchangers, the design factors become most important. Preventing the accumulation of silt or evaporated salt deposits on the metal surface may, for example, require running the water on the tube side (Figure 8.8) and the process stream on the shell side of condensers. With water on the shell side, void spaces above the cooling water may be eliminated by placing the exchanger in a horizontal rather than vertical position. Other approaches have been used to circumvent this problem.

FIGURE 8.6 — Split austenitic stainless steel pipe which failed by stress corrosion cracking. [SOURCE: B. B. Peters, J. A. H. Carson, and R. D. Barer, Bird Droppings, Sea Water and Even Hot Water Cause Stress Corrosion Cracking in Marine Service, Materials Protection, Vol. 4, No. 5, pp. 24-37 (1965).]

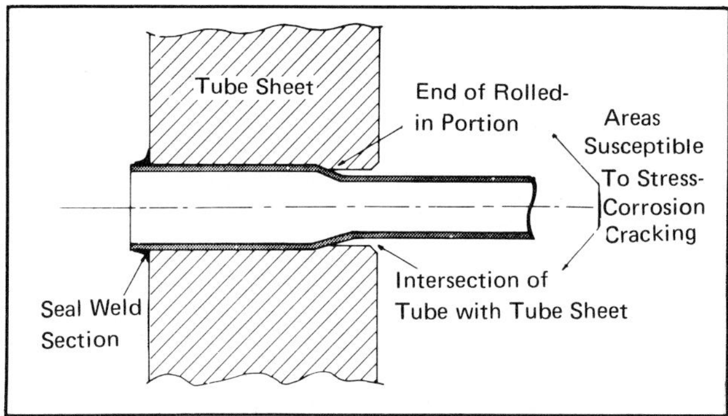

FIGURE 8.7 — Stainless steel tube-to-tube sheet joint (exaggerated).

The superstainless steels exemplified by the 20 Cr-25Ni-4Mo-2Cu proprietary alloys (*e.g.*, UNS 90403) are very resistant to aggressive waters, as are the 20 alloys and similar alloys intermediate between high stainless and low nickel specialty materials. Alloy 825 (N80825) will withstand high heat flux in aerated waters of quite high chloridity (*e.g.*, at least 1500 ppm Cl^-).

Copper Alloys

The corrosion product formed on copper alloys in low-temperature water is reddish cuprous oxide (Cu_2O). It may be present as a very thin film in the

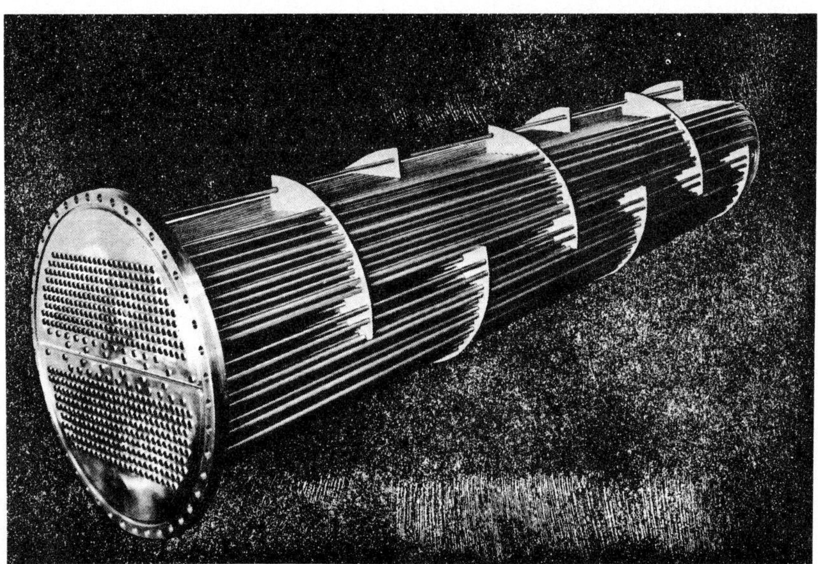

FIGURE 8.8 — Photograph of a heat exchanger (titanium) with shell and header removed to show tubes, tube sheet, and baffles. [SOURCE: Wolverine Tube Div., Calumet & Hecla advertisement, Materials Protection, Vol. 4, No. 11, p. 33 (1965).]

form of an iridescent tarnish. Continued exposure to oxygenated water at low temperatures, or to high-temperature water, can produce black cupric oxide (CuO). Usually, CuO forms on the surface of Cu_2O to produce a grayish-black powdery appearance. The corrosion resistance of copper and its alloys depends on the retention of such protective films, which may be removed by acidic dissolution or by velocity effects, or may suffer chemical conversion due to contamination (*e.g.*, by traces of hydrogen sulfide).

As the oxygen content of the water is increased and the chloridity increases, increasingly resistant alloys are demanded. Although copper and red brass are used extensively in fresh water (*e.g.*, potable water), the more resistant inhibited yellow brasses (*e.g.*, admiralty; C44300) are used to withstand aerated moving cooling waters. In seawater, copper may be used up to about 8 mps (25 fps) under wholly anaerobic conditions. In conventional heat exchangers, however, at least arsenical aluminum brass (C68700) is employed. This is largely being replaced in current seawater applications requiring copper alloys for the process side with the iron-modified 90-10 cupronickel (C70600) or the related 70-30 alloy (C71500). The tin-modified aluminum bronze (C61300) is also used in severe seawater applications.

Waters in which copper alloys perform poorly are those contaminated with carbon dioxide, sulfide, or ammonia.

Soft, aerated waters containing free carbon dioxide often result from decomposition of organic products such as in reservoirs and certain rivers. The settling treatment used in treating reservoir waters may release copious quantities of carbon dioxide. Such potable waters may cause green staining of copper plumbing fixtures.

Sulfides may be found in certain well, surface, and mine waters and may be produced by sulfate-reducing bacteria in many waters. Sulfides interfere with the otherwise protective film on copper alloys. Even 90-10 cupronickel may suffer pitting if seawater becomes contaminated with as little as 10 to 15 ppm hydrogen sulfide. Fresh waters contaminated with sulfides are better resisted by brasses than by copper.

Polluted seawater and brackish water containing sulfides and nitrogenous material (*e.g.*, ammonia) are particularly corrosive. Improved resistance to polluted water has been reported for brass alloys containing more than about 4% tin.

The status of copper alloy use in seawater has been well summarized in other works.[2]

Nickel Alloys

In the chromium-free alloys, the oxide formed on nickel is NiO. In low-temperature waters, a thin, barely visible yellow tint develops on nickel. At boiler water temperatures, the oxide film on nickel is dark gray and slightly powdery.

Nickel itself (N02200) is rarely used in water services. The 67Ni-30Cu alloy (N04400 or Monel) has been widely used in both fresh water and seawater. It is less resistant to fouling by marine organisms than the copper-base alloys. As with many other materials, crevices must be avoided to prevent pitting in water containing dissolved oxygen. However, Monel has been used for saltwater evaporators, with the cupronickels used for piping and heat-exchanger tubing.

Even the high-molybdenum grade, N10001 (Hastelloy B), can pit in fresh water.

In the nickel alloys containing chromium, the spinel oxides (M_3O_4), where M is a metal atom such as Cr, Ni, or Fe, are formed in high-temperature water. These alloys without molybdenum, such as N06600 (Inconel) and N08800 (Incoloy) are prone to pitting in chloride-bearing waters. The molybdenum-bearing grades (*e.g.*, N08825 or Incoloy 825) are a substantial improvement, and the high-molybdenum grades like Alloy C-276 (N10276) and Alloy 625 (N06625) are resistant even to hot seawater.

Reactive Metals

Titanium is highly resistant to aggressive waters, even more so than zirconium. It develops a tarnish film of TiO_2 only after long-time exposure to steam at temperatures of approximately 400 C (750 F). Titanium is an excellent choice for heat-exchanger tubing in seawater-cooled exchangers, provided that the process side is also compatible. Despite its position in the galvanic series, little or no galvanic effects by titanium (*e.g.*, on bronze tube sheets) have been observed.

Zirconium is occasionally specified for seawater-cooled exchangers where the process side is incompatible with titanium or Alloy C276. The greatest use of zirconium is as the tin-bearing alloy, Zircaloy (R60704), in nuclear reactors modified with water. A bright luster is maintained at temperatures well above boiling. At higher temperatures of about 260 C (500 F), a black, shiny film of zirconium oxide develops. The oxide-to-zirconium ratio is slightly less than 2:1, and the film has some metallic properties (*e.g.*, luster and reasonably good heat transfer properties). After long-time exposure of several years at such temperatures [or several days at 425 C (800 F)], the black oxide changes to white, powdery ZrO_2. This is more of an insulator, but less protective because of its powdery nature.

Concrete

Concrete is readily attacked by soft, aggressive waters, which tend to leach calcium from the structure. Acidic waters corrode concrete by direct attack on calcium carbonate constituents.

A special problem is the direct attack on tricalcium aluminate constituents by sulfate ions (Cement Bacillus), requiring increasingly resistant grades of concrete as the sulfate ion content increases. The Type V portland cement will tolerate up to 1500 ppm sulfate ion. Above that concentration in *fresh* waters, special types of cement are required (*e.g.*, Pozzolans and Brown-millerite). Oddly, the sulfate attack is not encountered in seawater, probably because of the chloridity/alkalinity/hardness ratios.

The most common problem is internal corrosion occasioned by rusting of the steel *rebar* used to strengthen the concrete, which requires zinc coating or epoxy coating for good resistance to water penetrating the structure.

Plastic and Elastomers

Plastics and elastomers are inherently resistant to natural waters up to a minimum of about 80 C (175 F), or within their inherent temperature/pressure limitations. The selection of materials may start with solid plastic (*e.g.*, polypropylene, polyethylene, and PVC), then move to the reinforced plastics (*e.g.*, glass-fiber reinforced plastic or GFRP) and thence to the plastic-lined steel (*e.g.*, polypropylene or Saran-lined steel) pipe.

The intrinsic problem with plastics and elastomers is their capability of selectively absorbing parts per million of organic species from industrial waste water to their ultimate detriment. Traces of chlorinated solvents or aromatic hydrocarbons will ultimately attack many elastomers, while polypropylene has been destroyed by accumulations of sorbic acid. In industrial wastes, the plastic or elastomer selected should be resistant to a *100%* concentration of any contained organic solvents at ambient temperature.

Wood

Although modern plastics are finding a place in cooling tower construction, wood has been the traditional material of construction, and large amounts are in service today. Chemically, the cellulose constituents in wood are susceptible to acid attack and the lignin by alkali, especially in the presence of oxidizing agents.

Deterioration in cooling towers is usually in the form of delignification, occasioned by the concentrated alkalinity in the water, and aggravated by the requisite chlorination. A rot occasioned by fungus attack is shown in Figure 8.9, while delignification is shown in Figure 8.10. The latter form of attack occurs at the surface of the wood, which then may suffer erosion by the cascading water.

Biological attack may be in the form of internal decay or surface (soft) rot. In soft rot, the organisms consume cellulose and leave lignin behind. The wood turns dark; becomes brittle, soft, punky, and cross-checked or *fillibrated;* and loses its strength. Soft rot is usually found in the flooded section of the tower. Internal decay is usually found in plenum areas, such as cell partitions, doors, drift eliminators, decks, and fan housing. This type of decay is difficult to detect because the external surface of the wood is unattacked.

FIGURE 8.9 — Showing advanced biological surface attack on wood. Surface has brash, dark colored appearance in a checkered pattern. Attack caused by living organisms. [SOURCE: Wood Maintenance for Cooling Water Towers, Materials Protection, Vol. 1, No. 10, pp. 32-39 (1962).]

FIGURE 8.10 — Typical chemical surface attack on wood, showing early stages of delignification. Alkaline agents such as sodium carbonate and bicarbonate remove lignin (the cement for wood's cellulose fibers) and expose fibers that eventually are washed away by water. [SOURCE: Wood Maintenance for Cooling Water Towers, Materials Protection, Vol. 1, No. 10, pp. 32-39 (1962).]

Chemical attack on the wood can be minimized by keeping the water below pH 8, preferably between pH 6 and 7, if this is compatible with the other requirements. Biological attack shown in Figure 8.9 can be controlled by pressure impregnation of the wood with chemicals (such as chromate and zinc), chlorophenols, or by using a nonoxidizing biocide in the water. As with slime control, chlorine residuals should be kept below 1 ppm.

Localized Forms of Corrosion

Most of the localized forms of corrosion described in Chapters 2, 5, 6, and 13 can be found in water services. Corrosion in seawater (Tables 8.4 through 8.7) is particularly severe.

Pitting Corrosion

Pitting corrosion is quite common. It can occur on a metal surface because of localized differences in composition or surface preparation, deposition or incorporation of metallic and nonmetallic particles, or contact with a dissimilar metal or nonmetallic

TABLE 8.5 — Galvanic Series of Various Alloys in Seawater Flowing at 13 ft/sec (74 to 80 F)

Material	Days in Test	Potential, -volt vs SCE[1]
Zinc	5.7	1.03
2% Ni Cast Iron	16	0.68
Cast Iron	16	0.61
Carbon Steel	16	0.61
Type 430 Stainless Steel (Active)	15	0.57
Ni-Resist Type 2	16	0.54
Type 304 Stainless Steel (Active)	15	0.53
Type 410 Stainless Steel (Active)	15	0.52
Type 3 Ni-Resist Iron	16	0.49
Type 4 Ni-Resist Iron	16	0.48
Type 1 Ni-Resist Iron	16	0.46
Tobin Bronze	14.5	0.40
Copper	31	0.36
Red Brass	14.5	0.33
Aluminum Brass	14.5	0.32
"G" Bronze	14.5	0.31
Admiralty Brass	19.9	0.29
90-10 Cu-Ni + 0.8 Fe	15	0.28
70-30 Cu-Ni + 0.45 Fe	15	0.25
Type 430 Stainless Steel (Active)	15	0.22
Type 316 Stainless Steel (Active)	15	0.18
Inconel nickel-chromium alloy	15	0.17
Type 410 Stainless Steel (Passive)	15	0.15
Titanium	41	0.15
Type 304 Stainless Steel (Passive)	15	0.084
Hastelloy "C"	15	0.079
Monel nickel-copper alloy	6	0.075
Type 316 Stainless Steel (Passive)	15	0.05

[1]SCE = Saturated calomel electrode (Chapter 2).

TABLE 8.6 — Galvanic Corrosion of Cast Iron Coupled to 70-30 CuNi in Quiet Seawater

Alloy	Corrosion Rate, (mpy) Uncoupled	Coupled[1]
Gray Iron	11.5	16.3
Ni-Mo cast iron	9.1	15.6
Cu-Mo cast iron	12.6	16.5
Type 1 Ni-Resist iron	3.2	3.7

[1]Area ratio-Cu/Ni:Iron = 1.32.

conductor. The formation of *tubercules,* a localized blister-like oxide formation, is a form of pitting corrosion.

Crevice Corrosion

Crevice corrosion also often occurs in water service, usually as a particular form of the oxygen concentration cell, which also occurs under deposits

TABLE 8.7 — Av. 60-Day Corrosion Rates With Controlled Seawater Temperature[1]

Material	Mils Per Year 75 F	100 F
Ni-Al Bronze	15	17
90/10 Cu-Ni	17	38
Ductile Iron	47	113
Gray Iron	34	103
Mild Steel	72	259
Ni-Resist, Type 1	13	23

[1]Specimen tip velocity 27 fps.

(*e.g.,* sand, dirt, sludge, barnacles, debris, etc.). A small area of metal (*i.e.,* in the crevice or under the deposit) becomes depleted of oxygen, while the rest of the surface has ready access to well-aerated water and functions as the cathode of the cell. Rapid attack occurs in the oxygen-depleted area because of the large cathode:anode area ratio.

Metal Ion

Metal ion concentration effects can also occur in crevices, and in a direction opposing the oxygen cell. Under these conditions, metal ions go into solution and build up to a high concentration in the crevice. The corrosion pitting then occurs just *outside* the crevice opening, where the metal ion concentration is low. This type of cell has a much lower incidence in practical engineering applications than does the oxygen concentration cell.

Galvanic Corrosion

Galvanic corrosion occurs on an anodic metal in electrical contact with either a more noble metal or an electrically conductive nonmetal (*e.g.,* mill scale or carbon). This can occur in domestic hot water pipes when a portion of galvanized pipe is in contact with bronze valves, or has been replaced with a section of copper or brass pipe. Even if electrically isolated, as they should be, an aggressive water may take cupric ions in solution, which can then plate out on the galvanizing (deposition) and cause a small localized galvanic cell, causing eventual pitting.

Impingement Corrosion or Velocity Attack

Impingement corrosion or velocity effects (Figures 8.11 to 8.13) is typified by the behavior of certain copper alloys, which corrode excessively when the water velocity exceeds about 1 mps (3 fps). Dissolved oxygen or chlorides in the water and low pH increase the sensitivity to velocity attack. Under certain conditions cavitation damage may occur (Figure 15.13).

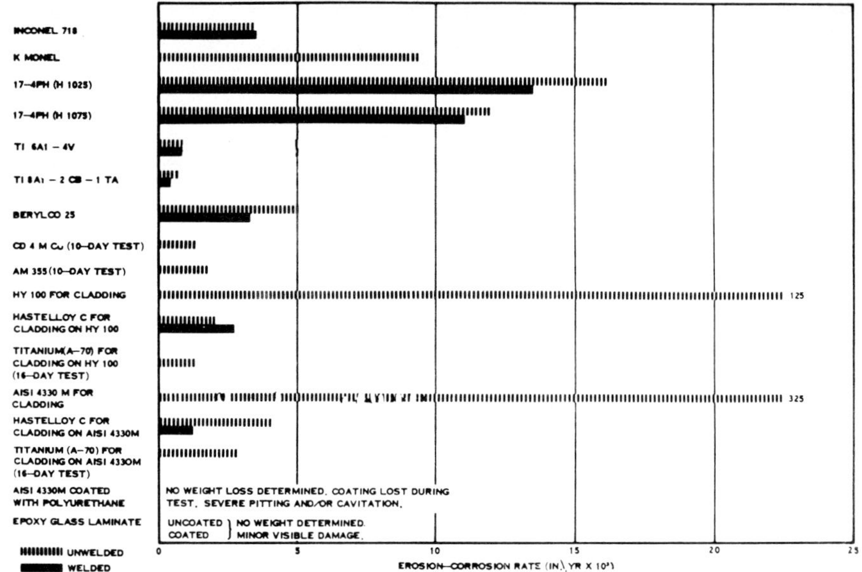

FIGURE 8.11 — Rates for jet erosion-corrosion in seawater, exposure 30 days at 90 knots. [SOURCE: A. E. Hohman and W. L. Kennedy, Corrosion and Materials Selection. Problems on Hydrofoil Craft, Materials Protection, Vol. 2, No. 9, pp. 56-68 (1963).]

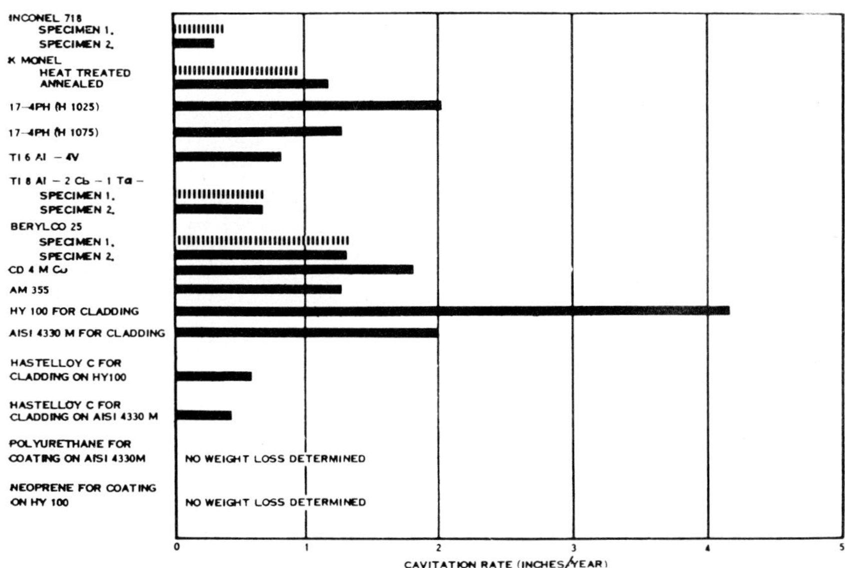

FIGURE 8.12 — Cavitation rates in seawater: double amplitude 0.001-inch, frequency 22,000 cycles per second. [SOURCE: A. E. Hohman and W. L. Kennedy, Corrosion and Materials Selection Problems on Hydrofoil Craft, Materials Protection, Vol. 2, No. 9, pp. 56-68 (1963).]

Dealloying

Dealloying can occur in all types of aqueous exposures. Dezincification of uninhibited brasses is prevalent, and even denickelification of a 70-30 cupronickel can occur in a brackish water.[3]

Intergranular Corrosion

Intergranular corrosion may also occur in waters. It normally is found only at high temperatures, except when low-temperature water is highly contaminated. All but a few of the aluminum alloys may fail catastrophically by intergranular corrosion in water at about 315 C (600 F). The 18-8 stainless steels (*e.g.*, S30400) may suffer IGC or weld decay if exposed to static hydrotest water of relatively high chloridity for more than a couple of days. (They are even more prone to pitting under such circumstances.) Very high-purity, high-temperature water will also attack the austenitic steel intergranularly.

Stress Corrosion Cracking

Stress corrosion cracking (Chapter 6) can occur in waters at moderately elevated temperatures, depending on chloridity and other factors. The situation is aggravated where boiling or flashing of steam can concentrate the impurities present in the water. Chlorides or caustic (from sodium carbonate or sodium hydroxide) concentrated in this fashion can cause stress corrosion cracking of stainless steels, steel, and even nickel alloys under some circumstances. Ammonia in water can cause cracking of high-strength copper alloys, especially the yellow

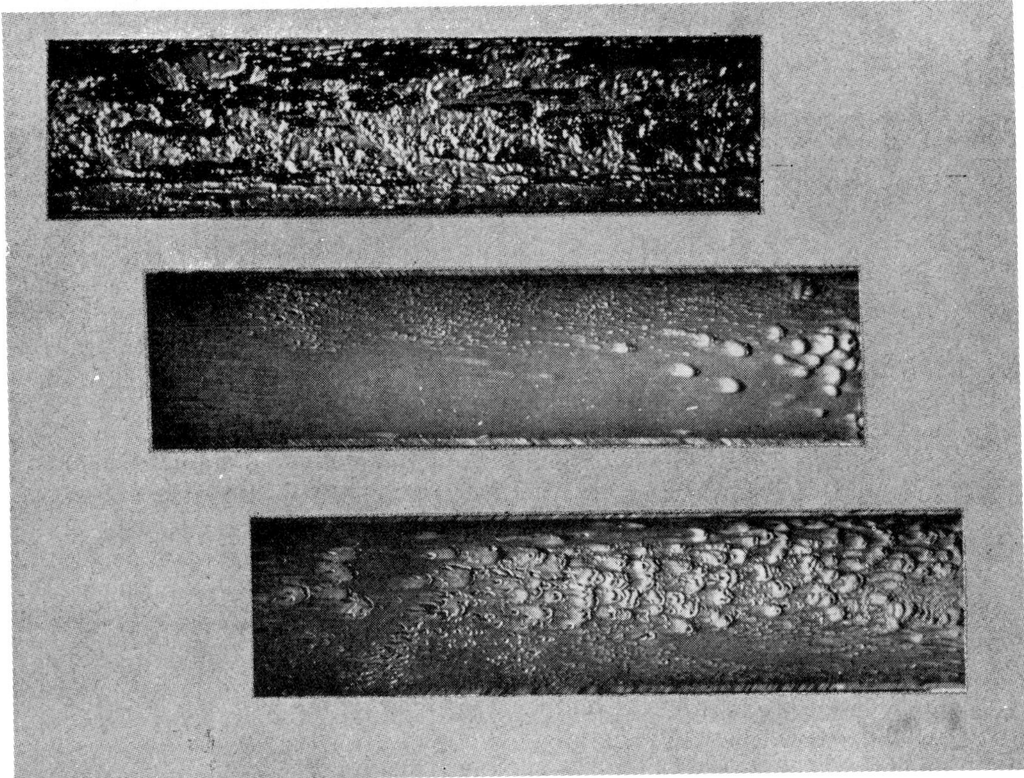

FIGURE 8.13 — Typical examples of erosion-corrosion in copper tube heaters. [SOURCE: D. B. Gardner, Equipment Room Corrosion Problems Solved by Mechanical Design, Materials Protection, Vol. 2, No. 4, pp. 54-61 (1962).]

brasses. Seawater can cause SCC of high-strength steels [*e.g.*, with yield strengths in excess of 1034 MPa (150,000 psi)] and of certain high strength aluminum alloys.

Corrosion Fatigue

Corrosion fatigue also lowers the fatigue life of many alloys in water service, as compared with their fatigue strength in dry air.

Corrosion Control

All of the standard measures for corrosion control apply in water services.

Change of Materials

Corrosion-resistant materials must be chosen for water systems that do not lend themselves to other protection techniques (*e.g.*, raw, aerated waters).

Environmental Control

Control of the chemical make-up of the water system is frequently chosen as the most economical approach (*e.g.*, inhibition of recirculated cooling water; deaeration of boiler feedwater; and miscellaneous applications of pH control, chelation, and buffers).

Barrier Coatings

Effective barrier paints or coatings are often used to effect resistance of iron or steel for immersion service (*e.g.*, coal tar epoxy coatings for sea-going barges). The hotter the water, the more difficult it is to find an effective coating. Also, and perhaps more importantly, the *purer* the water, the more difficult it is to find an effective coating. For example, it is difficult to find a coating that will resist clean steam condensate at temperatures close to the atmospheric boiling point.

Cathodic Protection

Cathodic protection is widely used in the form of galvanized steel in *cold* water services, as are the inorganic zinc-pigmented paint systems. Cathodic protection by sacrificial anodes is routine (*e.g.*, magnesium in low-chloridity waters, such as domestic hot water tanks; zinc in seawater applications; and speciality aluminum alloys in hot waters or seawater). Impressed current systems are also employed (*e.g.*, in fire control or potable water elevated storage tanks and in long-term seawater systems) (Chapter 9).

Design

Good design in water services should include minimizing velocity effects, eliminating crevice conditions wherever possible, and avoiding galvanic couples insofar as practical.

Cooling Systems

Water treatment is done mainly with fresh water. Corrosion tendencies relate to the degree of aeration, hardness phenomena, and specific salts. Many municipal water plants adjust the relationships between pH and calcium salts so that the water is sufficiently scaling to minimize corrosion, but not to plug up the piping. This type of control is obviously not suitable in cooling systems. If the supply is nonscaling, deposits will form in the hottest sections; if the hot water is adjusted, the cooler water is corrosive.

Once-Through Systems

Traditionally, industry has tended to develop in areas with an adequate supply of cooling water. Originally, it was sufficient to pipe water through the plant and discharge it back to its natural source. Only nominal attention was paid to control of water chemistry, and it is in fact economically ridiculous to attempt the chemical treatment of large volumes of once-through water.

Even such minor additions as the ''threshold'' treatment with 1 to 2 ppm sodium hexametaphosphate, for example, would be unacceptable in today's ecology. Many modern countries, states, or provinces forbid the return to source even of higher concentrations of *natural* constituents than were in the intake water. In some areas, thermal pollution is forbidden since the discharge of the *same* water at a higher temperature than the inlet temperature may be harmful to certain marine species (*e.g.*, oyster beds).

The bottom line today for once-through cooling systems is that, where permitted, the materials of construction must be chosen to be resistant to the water, be it fresh water or seawater. *All* natural waters must be presumed to be corrosive to iron and steel by virtue of aeration. Historically, acceptable performance of bare steel in natural waters has been associated with biological or chemical oxygen demands (*i.e.*, BOD or COD) that are today ecologically unacceptable. River waters which were once noncorrosive have become corrosive as they are sufficiently cleaned to provide a suitable environment for game fish.

Biocidal control is likewise inimical. Treatments against bacteria (discussed later) are incompatible with ecological considerations, unless the effluent is treated for their removal.

Recirculated Systems

Recirculated systems are divided essentially into two types, closed and open.

Closed

Closed recirculated systems are characterized by an essentially permanent charge of water, greatly facilitating the selection of chemical control and permitting the use of relatively large amounts of chemical additives. Typical closed recirculated systems are automobile radiator cooling systems, which are air cooled, and engine jacket cooling systems, which may either be air cooled or have the sensible heat removed by exchange with another type of cooling water system. In some locations, plants may be cooled with a closed loop of treated fresh water, and the fresh water in turn cooled in large exchangers carrying a once-through seawater coolant.

Open

Open recirculated cooling water systems remove the heat picked up in plant by evaporative cooling. This may be done by a spray pond, for example, combining air conditioning needs with aesthetic considerations in industrial parks. The most common type of evaporative cooling, however, is effected in cooling towers of one type or another.

Cooling towers may operate on natural draft, as in the case of wind-cooled towers for small home air conditioning systems or the large concrete hyperbolic towers used in power generating stations. In process plants, the towers are more often aided by fans, either forced or induced draft operations, to improve the cooling capacity (Figure 8.14).

There are certain fundamental considerations which should be understood in relation to open recirculated systems. First is the concept of *Cycle of Concentration.* If three cups of boiling water in a tea kettle were allowed to boil away to one cup, the residual cup would contain a three-fold concentration of soluble water salts (*e.g.*, sodium chloride), assuming that only steam (*i.e.*, pure H_2O) was driven off. The water would be said to be at three cycles of concentration.

To prevent this accumulation from becoming unacceptable from the standpoint of scale and corrosion, a small amount of blow-down (bleeding of the system) is maintained to control the number of cycles of concentration from evaporation. This means that make-up water must be added to equal the evaporation and blow-down losses, but this is a minor amount compared to the volume of the total system.

For example, if we needed 5000 gpm of cooling water in a system, the cost for treatment in a once-through design would be excessive. However, in a recirculating system, the make-up may only be 100 gpm, of which only 25 gpm may need to be treated with inhibitors. This brings chemical treatment into the range of economic feasibility, as compared with a once-through system.

Not only are there tangible limits, imposed by water chloridity and hardness, as to how far one can concentrate the water, but the savings effected by a recirculating system versus a once-through system are maximized at about 4 to 6 cycles of concentration. Below this range, treatment costs become prohibitive. At high cycles (*e.g.*, 8 to 10), the additional water savings are not commensurate with the increased difficulty of effective treatment. If the blow-down is shut off entirely, there is still an effective upper limit of concentration dictated by water losses from drift or windage. The normal upper limits might be about 20 to 22 cycles of concentration for a mechanical-draft tower.

FIGURE 8.14 — Typical cooling water tower. This two cell, induced draft tower is of redwood construction with redwood fill, distributors, and louvers; aluminum fan blade; and concrete basin. Rated to circulate 3000 gallons per minute, it was designed for 104 F (40 C) in, 85 F (30 C) out but operates at 120 F (49 C) in, 100 F (38 C) out with a maximum process side temperature in the heat exchanger of 225 F (107 C). [SOURCE: O. W. Siebert and W. C. Engman, Case History on Economics of Chemical Treatment of a Recirculating Water Cooling Tower, Materials Protection, Vol. 3, No. 10, pp. 20-25 (1964).]

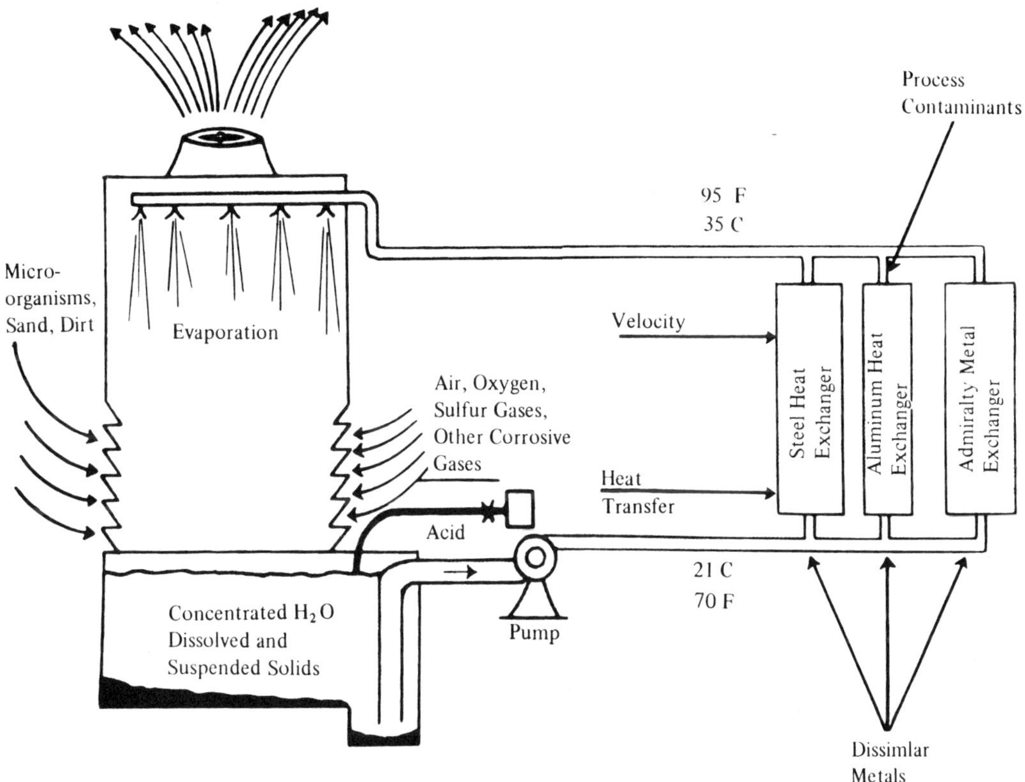

FIGURE 8.15 — Schematic flow diagram of an industrial cooling tower system. [SOURCE: A. J. Freedman, W. J. Ryzner, and J. D. Knapp, Design of Equipment and Operation of Laboratory Tests for Heat Transfer Surface Inhibition in Cooling Water, Materials Protection, Vol. 1, No. 10, pp. 22-30 (1962).]

The advantages of water savings effected by the cooling tower impose certain inherent disadvantages as well. The water becomes air saturated, ensuring its full corrosion potential, its natural alkalinity tends to increase and aggravate scaling tendencies; and the air scrubbing action contaminates the water with airborne materials, notably dust fines (which form silt in the tower basin) and spores of slime, algae, and fungi which can reproduce in the warm nutrient water of the system (Figure 8.15).

Other Water Systems

Miscellaneous water systems, which bridge the cooling water regimens and the high-temperature operations, range from cooling brines through tempered water to waste heat boilers, which last approximate the problems inherent in direct-fired boilers.

Brines

Low-temperature brines are most often formulated at about 25% calcium chloride to maintain below freezing temperatures for the control of exothermic reactions. These solutions are corrosive to common materials of construction unless pH is adjusted to about 8.5 and inhibited with about 2000 ppm sodium chromate. Care must be taken to exclude entry of air, from which the absorption of carbon dioxide tends to lower the pH and encourage the deposition of calcium carbonate. The salt brines may also be rendered innocuous by completely removing all oxygen or oxidants from the system.

Methanol or glycol brines are sometimes employed. Chromates are basically incompatible with the alcohol or glycol brines, accelerating the formation of organic acids and being reduced in turn by reaction with the hydroxyl groups, although the reaction is slow at low temperature. Consequently, the organic water heat-exchanger agents are inhibited with organic, inorganic, and metallorganic compounds. Mercaptobenzothiazole is normally included as an inhibitor to protect any copper alloys in the system.

Automotive Radiators (and Other Closed Glycol-Water Systems)

''All-weather'' ethylene glycol-type solutions require complex mixtures of additivies, including corrosion inhibitors, stabilizers, and buffering agents. Without proper commercial inhibition, a 40% glycol solution at 70 C (160 F) would corrode iron and steel at 250 to 500 μm/y (10 to 20 mpy), while also attacking the copper, brass, solder, and/or aluminum components at 25 to 50 μm/y (1 to 2 mpy). Details concerning the various inhibitors employed are given in Chapter 7.

Hot Water Systems

Tempered water systems are often employed to regulate chemical reactions at some moderately elevated temperature, typically 60 C (140 F) to 80 C (175 F). As previously explained, water is highly corrosive to ferrous metals at such temperatures, so the systems must be either strongly inhibited or deaerated.

In domestic and industrial hot water systems, corrosion tends to be self-arresting because the dissolved oxygen initially present is consumed by corrosion. However, if there should happen to be regular ingress of oxygen for any reason, corrosion can be disastrous. Corrosion failures in a large research center were traced to the ''stealing'' of hot water from the heating circuit for the preparation of distilled water for laboratory use. The make-up water was not deaerated.

Waste Heat Boilers

In many chemical or petrochemical processes, economy dictates that superfluous exotherms be utilized to generate steam as an energy conservation measure. Examples are the cooling of a butane oxidation reaction or the condensation of hot sulfur vapors, steam pressure being generated on the shell side of specially designed heat exchangers.

Corrosion problems in waste heat boilers usually arise either from unusual materials of construction or from inattention to the required details of water treatment. Austenitic stainless steels may be required from the process side corrosion aspects, yet be highly susceptible to stress corrosion cracking from boiler feedwater. The most frequent problem, however, is that the operating department personnel, whose primary concern is the manufacture of marketable products, simply do not give adequate technical attention to the details of boiler feedwater chemistry. Many a chemical converter, oxycat exchanger, or other waste heat boiler have failed from steam-generating side corrosion due to such inattention.

The intimations of this discussion lead naturally into the major topic of water treatment for steam generation.

Water Treatment

Inhibition

Corrosion inhibition in recirculated cooling water systems historically depended on the oxidizing inhibitors such as chromates and nitrites. Nitrites cannot be effectively used in open recirculated systems because they are highly conducive to organic growths, which just love nitrogenous compounds as fertilizer.

The higher the chloridity of the water, the more it seems to require a good dosage of sodium chromate. In fact, even refrigerating brines (*e.g.*, 25% calcium chloride) can be effectively inhibited with about 2000 ppm $CrO_4^{=}$. However, in an effort to make them economically more practical, combinations of inhibitors were developed which permitted lower concentrations of chromate to be effective (*e.g.*, chromates combined with polyphosphates and traces of zinc). Today, chromates are usually forbidden for ecological reasons, unless the tower blow-down can be chemically treated to destroy them. This may sometimes be the only feasible approach to a corrosive, high-chloride cycle water.

The synergistic effect of zinc additions is observed in both chromate and polyphosphate systems in Figures 8.16 through 8.20.

Traditional glassy metaphosphates lose much of their effectiveness because they revert in time to orthophosphates. There is also an attendant possible danger of precipitation of tricalcium phosphate. This is a function of pH, calcium, orthophosphate, and solids concentration of the water.

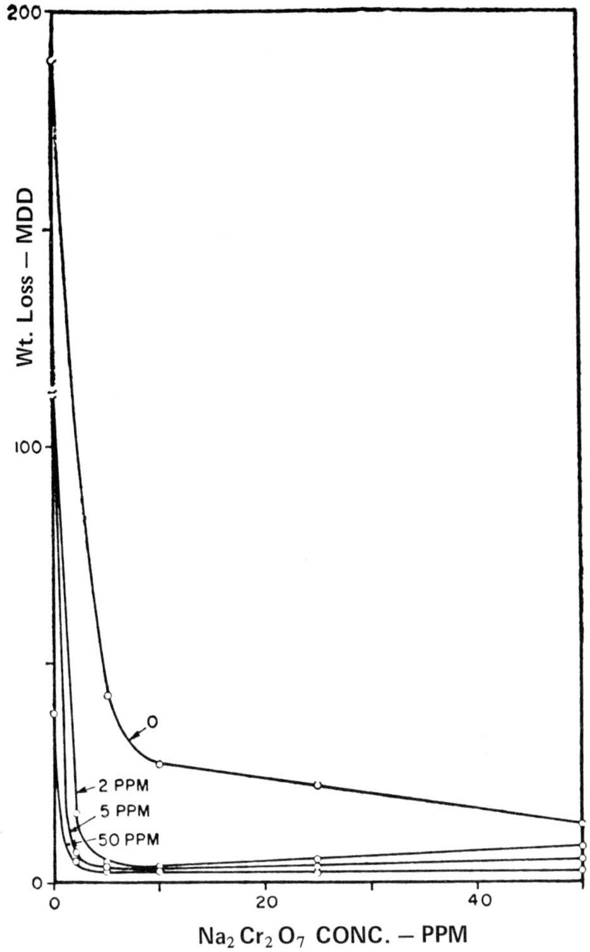

FIGURE 8.16 — Inhibition of attack on steel in 35 C tap water with dichromate at several zinc levels (pH = 6; zinc added as the sulfate). Curves suggest the dichromate-zinc ratio is determinant factor rather than total inhibitor concentration alone. [SOURCE: G. B. Hatch, Low-Level Dichromate-Zinc Inhibition in Recirculation Cooling Water Systems, Materials Protection, Vol. 4, No. 7, pp. 52-56 (1965).]

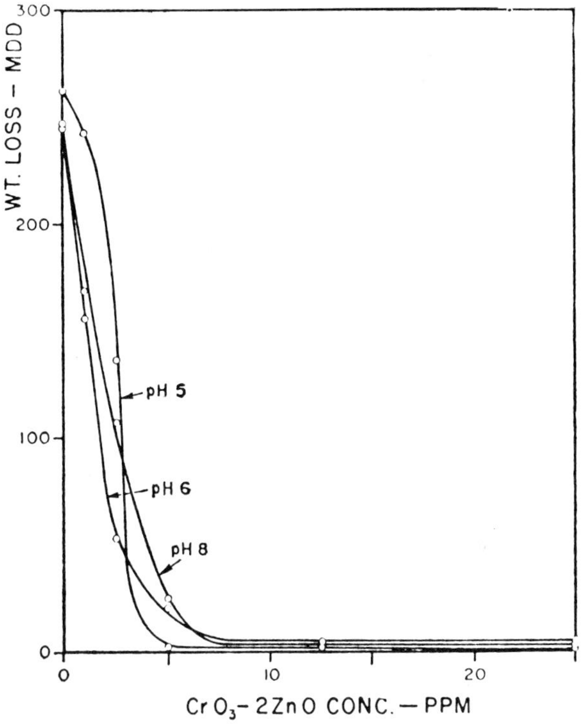

FIGURE 8.18 — Inhibition of attack on steel in 35 C tap water with 1:4 dichromate-zinc at several pH levels. Shape of curve at pH 5 differs slightly from those at two higher levels but not sufficiently to warrant particular consideration. [SOURCE: G. B. Hatch, Low-Level Dichromate-Zinc Inhibition in Recirculating Cooling Water Systems, Materials Protection, Vol. 4, No. 7, pp. 52-56 (1965).]

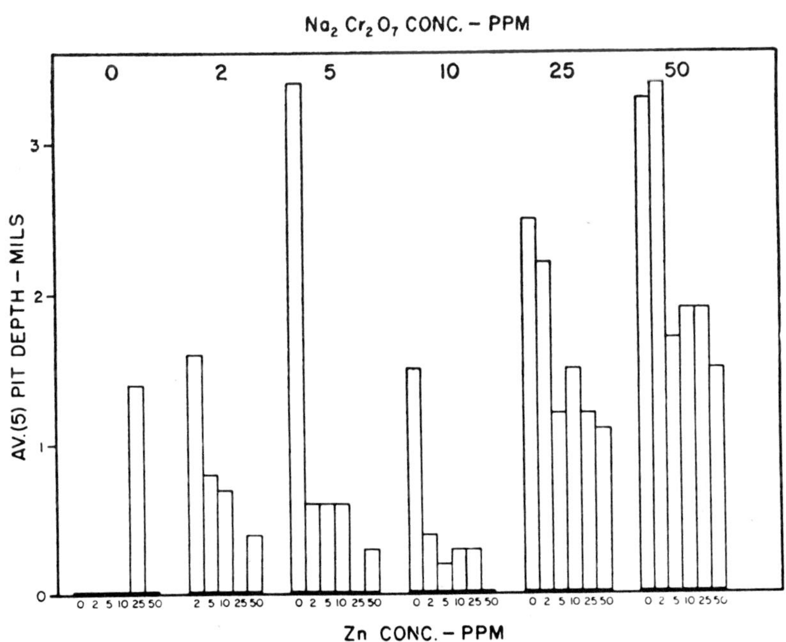

FIGURE 8.17 — Influence of zinc on pitting of steel at several dichromate levels in 35 C tap water (pH = 6). Tests last five days. Values represent average of five deepest pits. [SOURCE: G. B. Hatch, Low-Level Dichromate-Zinc Inhibition in Recirculating Cooling Water Systems, Materials Protection, Vol. 4, No. 7, pp. 52-56 (1965).]

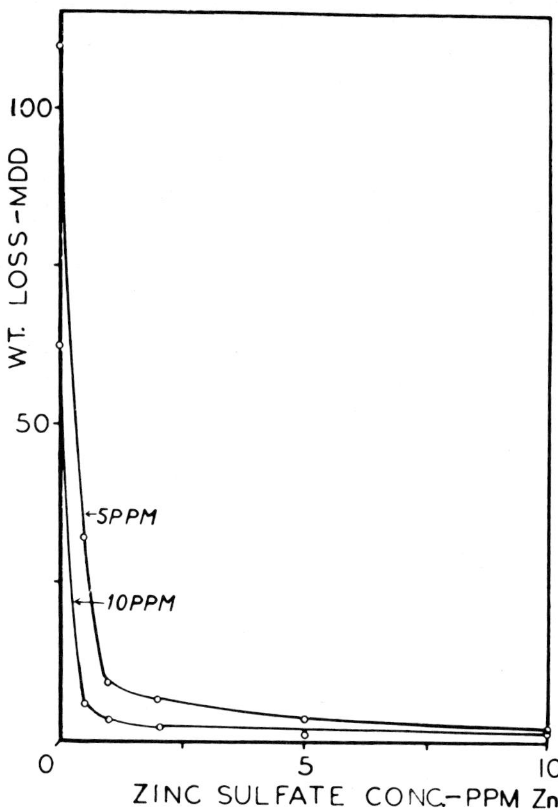

FIGURE 8.19 — Influence of zinc sulfate on inhibiting corrosion of steel at two glassy phosphate levels. Test 5 days at 35 C, solution (Pittsburgh tap water) had pH of 6.9. [SOURCE: G. B. Hatch and P. H. Ralston, Oxygen Corrosion Control in Flood Waters, Materials Protection, Vol. 3, No. 8, pp. 35-41 (1964).]

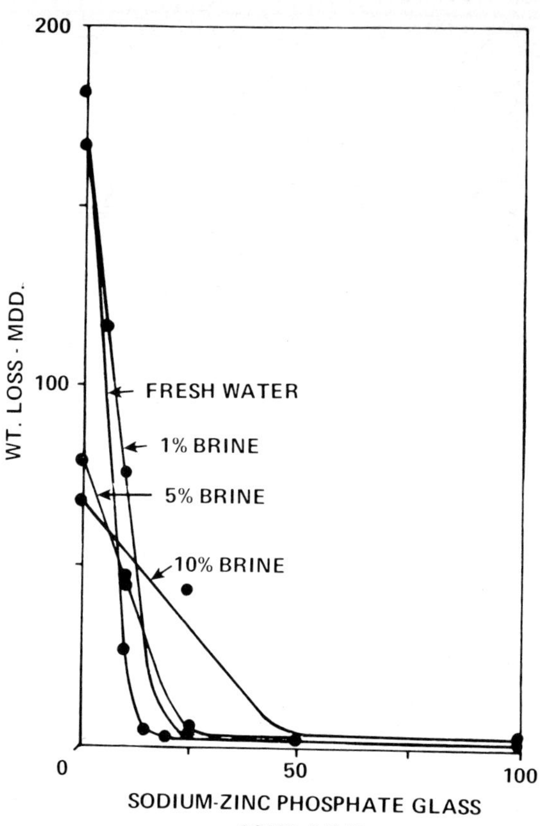

FIGURE 8.20 — Inhibition of corrosion of steel with a sodium zinc phosphate glass at several brine concentrations. Test conducted five days at 35 C, pH 6.5. [SOURCE: G. B. Hatch and P. H. Ralston, Oxygen Corrosion Control in Flood Waters, Materials Protection, Vol. 3, No. 8, pp. 35-41 (1964).]

Silicates have also been used in water systems, but tend to have an adverse effect on heat transfer because of the deposition of relatively heavy films.

Because of the ongoing search for effective nonoxidizing inhibitors over the past several decades, many new formulations have been developed; notably, polyphosphonate and organophosphate combinations. A careful study should be made of the behavior of specific cooling water systems, as it is extremely difficult to predict from water chemistry the suitability of complex commercial formulations. Laboratory, pilot plant, and field tests are usually required to select the proper system. Even then, its effectiveness will depend upon the care taken in control of operating conditions.

There is no magic method of water treatment available. Beware of magnetic treatments that claim some beneficial change in the properties of the water.[4,5]

Scale Control

Scale control in open recirculated systems is effected primarily by limiting the concentration of the scale-forming species and related parameters, especially by controlled acid additions to effect a suitable pH range. Many of the commercial inhibitor formulations include scale-controlling additives (*e.g.*, polyphosphates and chelating agents).

Slime and Algae

Algae and slimes are found in cooling water towers and spray ponds where sunlight and airborne contamination are to be expected. Slimes normally contain fungi, yeasts, bacteria, and entrapped quantities of inorganic and/or organic material. Some of these are listed in Table 8.8.

Slime deposits can cause oxygen concentration cell-type corrosion and pitting, as shown in Figure 8.21. If sloughed off, they can contribute towards fouling of filters and other equipment.

TABLE 8.8 — Some Troublesome Microorganisms

Bacteria (Slimeforming)	**Algae**
Flavobacterium	Chroococcus
Mucoids	Oscillatoria
Aerobacter	Chlorococcus
Pseudomonas	Ulothrix
B. Subtilis	Scenedesmus
B. Cereus	Navicula
Bacteria (Corrosive)	**Fungi**
Desulfovibrio	Aspergillus
Clostridia	Alternaria
	Penicillium
Bacteria (Iron Depositing)	Trichoderma
Gallionella	Torula
Crenothrix	Monilia

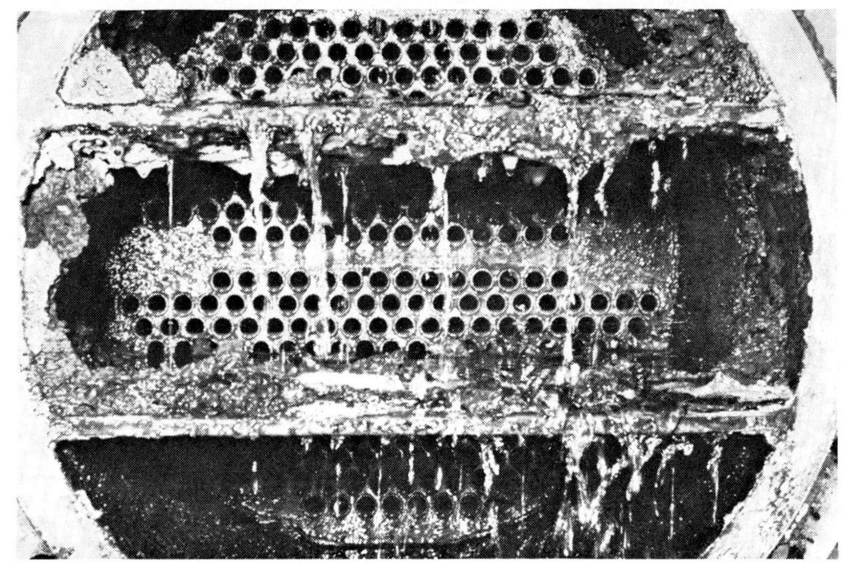

FIGURE 8.21 — Example of biological growths in a heat exchanger. Slime growths such as this retard heat transfer and cause a serious loss of efficiency. [SOURCE: J. M. Donohue, A. J. Piluso, and J. R. Schrieber, Materials Protection, Vol. 5, No. 7, pp. 22-24 (1966).]

Chlorination to a residual chlorine content of a few tenths of a part per million is the usual treatment for control of biological growths in open recirculated systems. In order to discourage the development of chlorine-resistant strains, it is advisable to surprise the microbes at irregular intervals with other biocides (*e.g.*, quaternary amines). Other oxidizing biocides (*e.g.*, chlorine dioxide and ozone) used in municipal waters are normally too expensive for cooling water systems.

Historically, copper sulfate had been used to control algae in some specific applications. It is effective at concentrations as low as 1 ppm, but must be used at a sufficiently low pH to prevent precipitation of insoluble copper hydroxide. As suggested earlier, the use of copper sulfate as a biocide is incompatible with aluminum equipment because of deposition and subsequent galvanic corrosion. Similar copper plating would occur on zinc (as in galvanized steel piping) and even steel.

Bacterial action (*e.g.*, sporovibrio desulfuricans, the sulfate-reducing bacteria which generates hydrogen sulfide by metabolizing sulfates; and crenothrix, the "iron-eating" bacteria which metabolizes soluble iron and manganese, excreting the oxides) poses a further threat to such systems, but usually are effectively controlled by the chlorination procedures mentioned earlier.[6]

Silt Control

The accumulation of silt in the bottom of the tower sump is due to the scrubbing of large volumes of air, especially in a mechanical draft tower. This can be controlled to a certain extent by the addition of polyelectrolytes (*e.g.*, polyacrylate structures) which inhibit its aggregation from a dispersed form. In areas subject to large amounts of dust in the air, side stream filtration of the recirculated water may be required.

Steam Generation

The greatest use of high-temperature water and steam is in electrical power generation. Historically, fossil fuels (*i.e.*, wood, coal, gas, and oil) were used almost exclusively to heat water and make steam. However, in recent decades nuclear power steam generators have moved into the picture. The two types of power plants have much in common, but are sufficiently different to be discussed separately. Both, however, presuppose technically advanced water treatment and control for successful operation.

Treatment of Boiler Feedwater Make-Up

To an extent determined by specific requirements of boilers operating in different temperature and pressure ranges, the boiler feedwater make-up and boiler feedwater must be softened and deaerated. Low hardness salts to prevent scaling in the system and deaeration to prevent corrosion of steel are the goals.

A number of lime softening treatments were used in the past, but these have given way to more sophisticated treatments. Probably the most common for boilers up to 2750 to 4000 kPa (400 to 600 psig) is *Zeolite* softening. In this treatment, a sodium salt of a long-chain polymeric organic molecule comprises the ion exchange bed. As the feedwater passes through the bed, sodium ions are exchanged for the calcium and magnesium ions which comprise the water hardness. Other cations are also exchanged. The water then exits from the Zeolite bed softened and at a higher pH than it entered. The beds are regenerated intermittently with sodium chloride solution to backwash the hardness salts and reform the sodium salt.

For purer water quality, the water may be totally demineralized by mixed beds of polymeric resins which exchange in turn hydrogen ions for all cations and hydroxyl ions for all anions, effectively producing pure H_2O from a raw water stream. Such highly purified waters are required for boilers operating from about 6000 kPa (900 psig) and for all nuclear boilers (to avoid radioactive half-life of water-borne salts).

The softened water is now at its most corrosive, being still saturated with dissolved oxygen and having no hardness which could even hope to slow down attack. Many plant problems arise from people stealing boiler feedwater, under the impression that it is "the best water in the plant." So it will be, but only after further treatment.

For reasons of economy, the removal of dissolved oxygen is usually first effected in part by thermomechanical deaeration. The boiler feedwater is preheated, then flashed in a deaerator to remove any free carbon dioxide and most of the dissolved oxygen. Then the last traces of dissolved oxygen are chemically scavenged with sodium sulfite:

$$2\,Na_2SO_3 + O_2 \rightarrow 2\,Na_2SO_4$$

$$\text{Sodium Sulfite} + \text{Oxygen} \rightarrow \text{Sodium Sulfate} \quad (8.3)$$

or, if nil solids are desired, with hydrazine

$$2\,N_2H_2 + O_2 \rightarrow 2\,H_2O + 2\,N_2$$

$$\text{Hydrazine} + \text{Oxygen} \rightarrow \text{Water} + \text{Nitrogen} \quad (8.4)$$

Both sulfite and hydrazine are available in catalyzed form to promote more rapid reaction rates. A boiler designed to be operated with a catalyzed oxygen scavenger must *never* be operated on the uncatalyzed grades, or severe corrosion will be encountered in the economizers or even the steam drum.

A final step in boiler feedwater treatment consists of pH adjustment as a further aid to corrosion control. Usually the pH is adjusted to a range of 10 to 11 with trisodium phosphate (or combinations of caustic with sufficient mono- or di-sodium phosphate to form trisodium phosphate upon inadvertent evaporation of the water). The *Coordinated Phosphate* treatment is intended to preclude environmental cracking of steel by free sodium hydroxide (caustic embrittlement), as discussed further in Chapter 6. Caustic carry-over with the steam can present severe corrosion problems (Figure 8.22).

Nuclear requirements are such that zero solids treatment is required, precluding the addition of sodium salts and necessitating the use of ammonium hydroxide for pH adjustment.

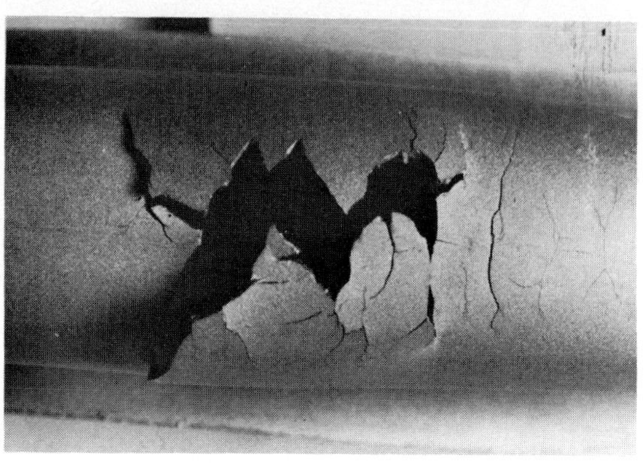

FIGURE 8.22 — Type 321 stainless steel expansion joint in 2750 kPa (400 psi) steam service cracked from caustic carry-over in the steam.

Fossil Fuel Steam Plants

A conventional steam generation plant (Figure 8.23) consists of feedwater heaters with water inside and steam outside the tubes; a boiler (water inside the tubes, hot combustion product gases outside), where water is heated to high temperatures under pressure and is sometimes flashed to steam; a steam drum, wherein steam is formed from water and water is separated from the steam (Note: The steam drum is omitted in certain once-through systems); a superheater, where the steam is further heated to even higher temperatures; a turbine, where the steam expands against the vanes of a wheel to drive the turbine which generates electricity; and a condenser, where the low-pressure steam is condensed to water and returned to the feedwater heaters.

There are special corrosion problems in each of these sections. There are also special problems

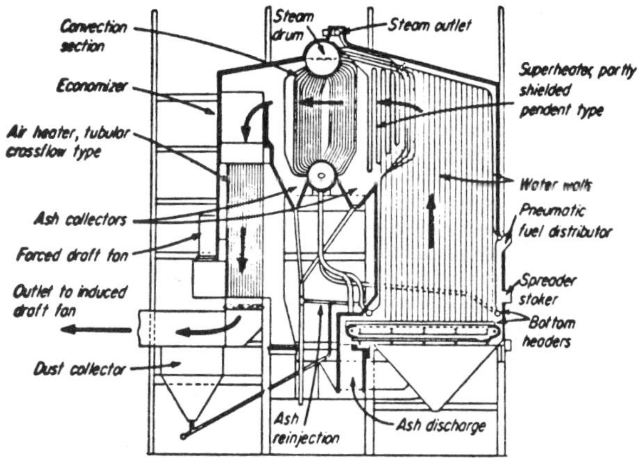

FIGURE 8.23 — Schematic of modern steam generator using specialized heating surfaces to achieve high efficiency. Waterwalls depend mainly on furnace radiant heat; superheater may use both radiant and convection transfer. Convection surface works in somewhat cooler zone; economizer and air heater use still cooler convection heat because they transfer it to low temperature level fluids. [SOURCE: R. H. Marks, Steam Generation—A Special Report, Power, No. 6 (1964).]

associated with exposure to hot combustion gases. The latter is discussed in Chapter 13.

The materials used in high-temperature steam and water include steels, stainless steels, and nickel-base alloys. Copper-base alloys are employed in the intermediate and low temperature ranges. Aluminum alloys generally are not used because of their poor resistance above about 200 C (390 F).

Materials of Construction

Boilers. The boiler proper is most often fabricated from ordinary carbon steel, which is useful under controlled conditions up to about 450 C (840 F). Alloy steels (*e.g.,* 1 1/4 Cr-1/2 Mo; 2 1/4 Cr-1 Mo) are used at the 565 C (1050 F) maximum operating temperatures of most steam plants.

In a plant operated under properly degassed conditions, the steels develop a thin black protective film of magnetic iron oxide (magnetite, Fe_3O_4). If appreciable dissolved oxygen effects entry, brick-red Fe_2O_3 will also be found.

Steel may experience an occasional failure from on-load pitting during steady operations. This failure has been attributed to a variety of causes including copper deposits, thermogalvanic effects caused by temperature differences, incipient attack started during shutdown, and localized caustic concentration.

There is no question that severe pitting can occur during shutdown if the boiler tubes are not completely dried out or completely filled with sulfite-treated water (Figures 8.24 through 8.29). Failure to do this during shutdown can result in the formation of air bubbles in the tubes, and attendant *air bubble pitting.*

Considerable quantities of copper may sometimes be found interspersed in the oxide scale in boilers. The source of the copper is usually the condenser and feedwater heater tubes, and indicates inadequate dissolved oxygen control, often in conjunction with amine-type inhibitors used for condensate corrosion control.

Copper-free corrosion products are often removed from the boiler by treatment with inhibited hydrochloric acid or other specialized acid formulations. If considerable copper is present, this will lead to deposition attack; the copper plating out on the boiler tubes. Copper may be removed from the boiler by first oxidizing it to cupric oxide by means of oxidants (*e.g.,* chlorates, bromates, or persulfates), and then dissolving it in an ammoniacal solution, forming the typical royal blue copper-ammonium complex.

Steam Drums. Corrosion problems in steam drums are similar to those in the boiler tubes. Localized concentration of boiler water chemicals (due to entrainment) and increased corrosion of the steel can occur at the water/steam interface or at steam separation plates.

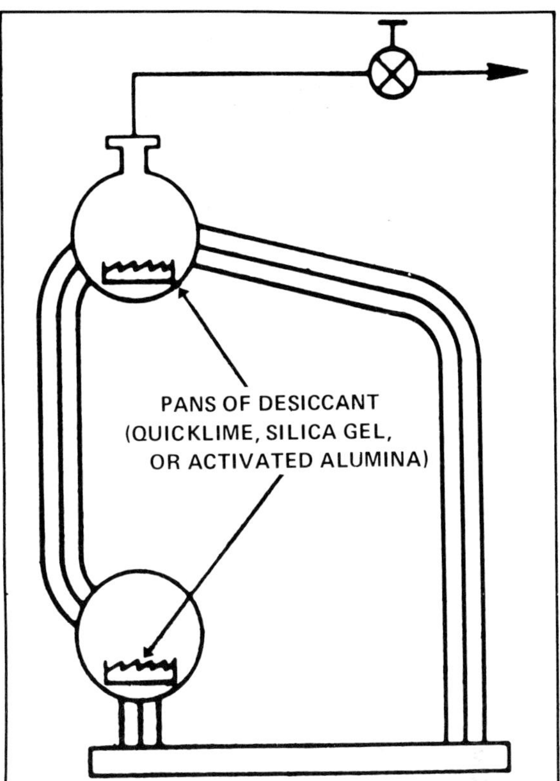

FIGURE 8-24 — Diagram of boiler furnace in dry standby. Pans of desiccant, as indicated by arrow, are placed in unit which is sealed airtight. [SOURCE: William T. Reid, Protecting Standby Equipment, Materials Protection, Vol. 6, No. 7, pp. 42-45 (1967).]

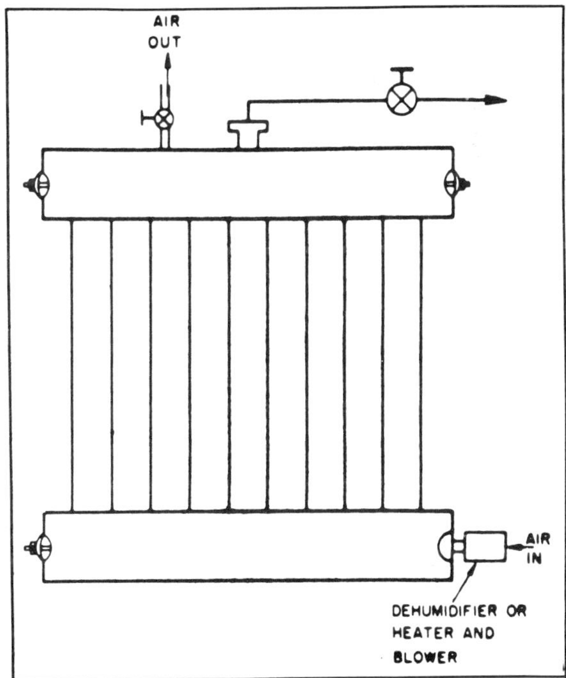

FIGURE 8.25 — Diagram of boiler drum in dry standby with pans of desiccant as in Figure 8.24. Dry, heated air is circulated by blower. [SOURCE: William T. Reid, Protecting Standby Equipment, Materials Protection, Vol. 6, No. 7, pp. 42-45 (1967).]

Superheaters. As indicated in Table 8.9, the 2 1/4% chromium steel is resistant to general corrosion by clean deaerated steam up to about 650 C

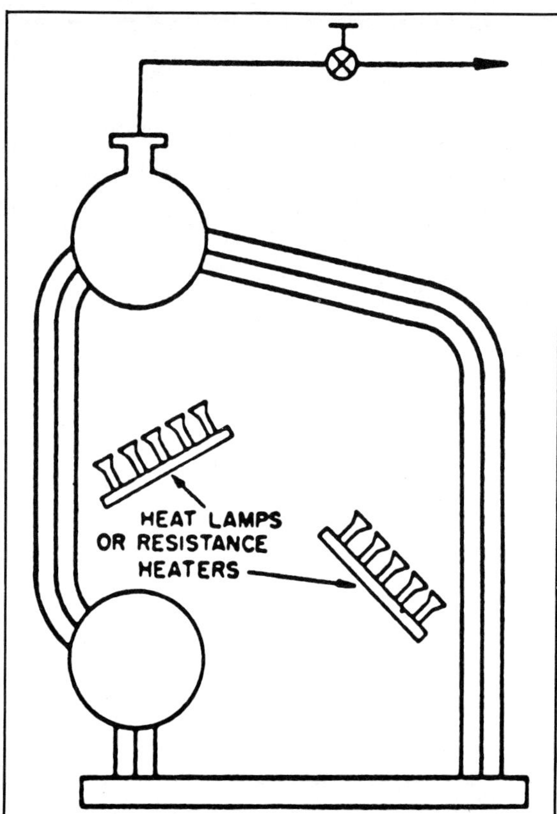

FIGURE 8.26 — Diagram of boiler furnace in dry standby. Heating devices prevent condensation. [SOURCE: William T. Reid, Protecting Standby Equipment, Materials Protection, Vol. 6, No. 7, pp. 42-45 (1967).]

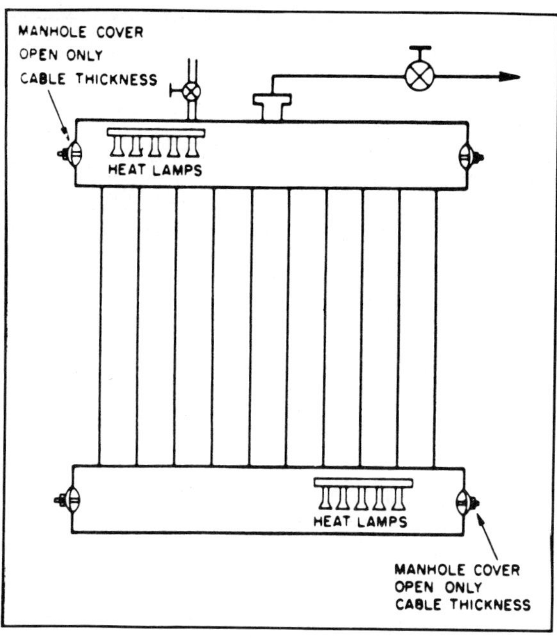

FIGURE 8.27 — Diagram of boiler drum in dry standby. With some outside air circulating, heat lamps prevent condensation. [SOURCE: William T. Reid, Protecting Standby Equipment, Materials Protection, Vol. 6, No. 7, pp. 42-45 (1967).]

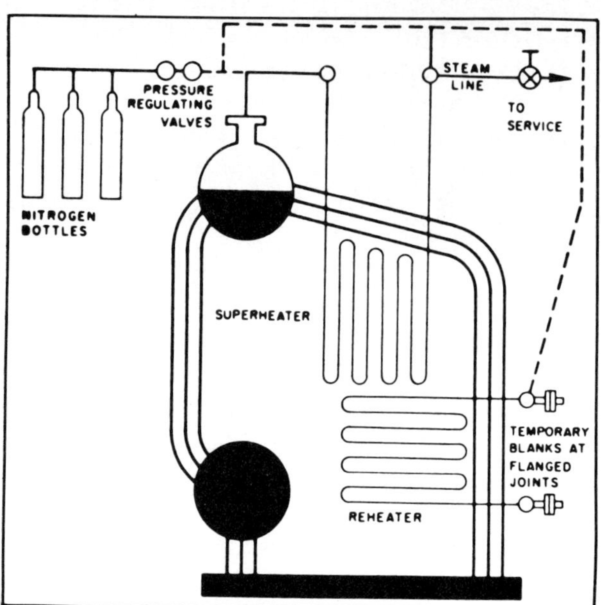

FIGURE 8.28 — Diagram of boiler system in wet standby, using nitrogen blanketing technique. Water is maintained at or near operating level and nitrogen introduced at a positive pressure of 5 to 10 psig. [SOURCE: William T. Reid, Protecting Standby Equipment, Materials Protection, Vol. 6, No. 7, pp. 42-45 (1967).]

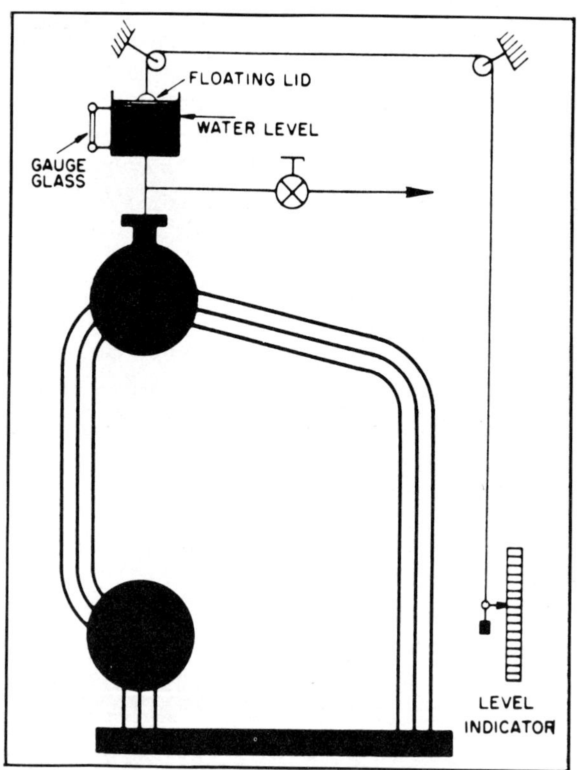

FIGURE 8.29 — Diagram of boiler in wet standby using complete flooding technique. This method is used when system includes a drainable superheater. Nonvolatile materials such as caustic soda and catalyzed sodium sulfite are present for pH adjustment and oxygen scavenging. [SOURCE: William T. Reid, Protecting Standby Equipment, Materials Protection, Vol. 6, No. 7, pp. 42-45 (1967).]

(1200 F). Accelerated attack can occur when the water from the boiler is not completely separated from the steam, and the chemicals in the entrained water become concentrated in localized areas. Concentration of caustic by this mechanism can produce pitting. The corrosion resistance of steel is optimum at pH 10 to 11, decreasing both above and below this range.

Turbines. In conventional steam stations, the areas of the turbine that are most susceptible to

TABLE 8.9 — Maximum Recommended Operating Temperatures for Steels in Steam Service

Steel Composition	Maximum Operating Temperature F	C
Carbon Steel	850	454
1-1/4Cr-1/2Mo	1100	593
2-1/4Cr-1Mo	1200	649
3Cr-1Mo	1200	649
9Cr-1Mo	1200	649
18Cr-8Ni (stainless steel)	1500	816

deterioration are the turbine blades because of erosion-corrosion. Figure 8.30 and 8.31 show erosion-corrosion of associated piping. To resist this type of attack, the turbine blades or blade tips are fabricated from hardened martensitic stainless steel (*e.g.*, S41000).

Deposits on the turbine blades are also a problem if there is carry-over of entrained moisture, such as may occur when saturated steam directly from the steam drum is used to drive the turbine. Silica (SiO_2) carry-over and deposition may also be a problem unless the water chemistry is carefully controlled. At pressures above 4000 kPa (600 psi), silica will carry and escape with the steam. It then deposits on low-pressure areas of the turbine. Silica may be controlled by eliminating it from the boiler feedwater, blowing down a portion of the steam, or by removing it from the steam by water washing.

Condensers. Condensers are shell-and-tube heat exchangers with steam condensing in the tubes and cooling water on the shell (Figures 8.8 and 8.32). Condenser tubes are frequently made of copper-base alloys, such as C443300 (arsenical admiralty) or C70600 (90-10 cupronickel). Selection of condenser tube material is often dictated by the

FIGURE 8.30 — Corrosion of elbow in steam line: (1) indicates impingement on backing ring; and (2) erosion-corrosion along bottom. [SOURCE: Henry Suss, Steam, Water System Corrosion, Materials Protection, Vol. 6, No. 6, pp. 46-47 (1967).]

cooling water which may be more corrosive than the condensing steam.

Corrosion by the condensed steam, as shown in Figure 8.33, is usually the result of dissolved carbon dioxide. The CO_2 is released from carbonates in the boiler (especially in Zeolite-treated BFW of high methyl orange alkalinity) and, being volatile, passes through the turbine to contaminate the condensate and cause low pH (acidic) conditions. Grooving and thinning of the tubes will occur if protective measures are not taken. The usual remedy is to make the condensate alkaline with neutralizing amines (*e.g.*, morpholine) to a pH of about 8.5 to 8.8, or to add filming amines such as octadecylamine. Higher pH values are not recommended because of the intrinsic danger to copper alloys in the event of ingress of oxygen to the condensate.

FIGURE 8.31 — Wet steam impingement attack in section of carbon steel pipe: (1) indicates zone of secondary impingement attack; and (2) primary impingement attack zone. [SOURCE: Henry Suss, Steam, Water System Corrosion, Materials Protection, Vol. 6, No. 6, pp. 46-47 (1967).]

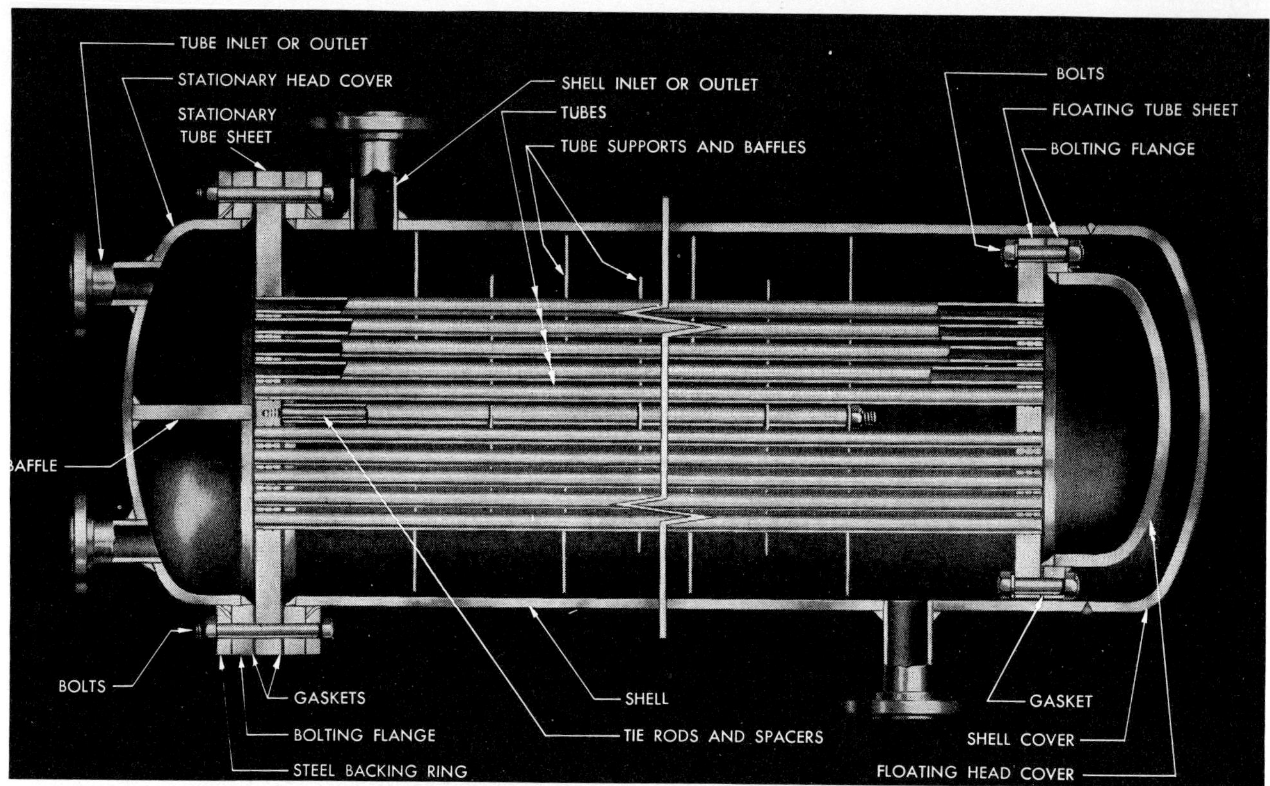

FIGURE 8.32 — Schematic cross section view of heat exchanger. [SOURCE: Aluminum Company of America advertisement, Corrosion, Vol. 16, No. 6, p. 66 (1960).]

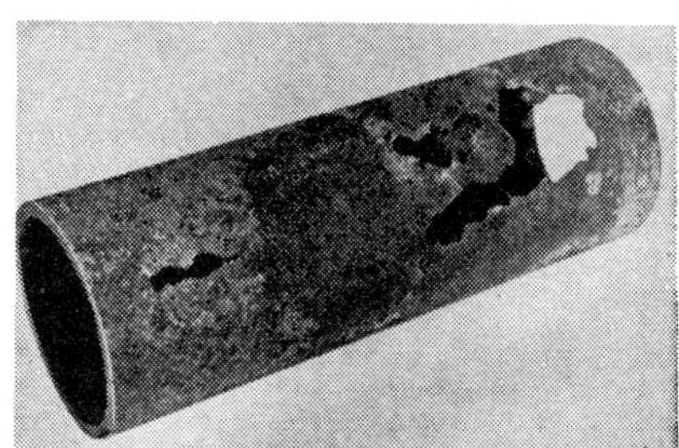

FIGURE 8.33 — Condensate line attacked by carbon dioxide. [SOURCE: Betz Laboratories, Inc. advertisement, Materials Protection, Vol. 6, No. 6, pp. 46-47 (1965).]

Feedwater Heaters. Feedwater heaters have also been tubed with copper-base alloys, usually cupronickels, although modern practice tends towards stainless steel or even titanium. Water at about 260 C (500 F) is inside the tubes, with steam outside the tubes. Corrosion problems are encountered on either or both sides.

Corrosion inside the tubes can be attributed mainly to dissolved oxygen introduced with make-up water or by air leaks to the condenser which operates under a partial vacuum. Sodium sulfite additions may be required to remove the dissolved oxygen and prevent corrosion.

Corrosion outside the tubes occurs as layers of oxides (exfoliation) in installations which are operated in *peak* service (*i.e.,* shut down overnight and weekends). This type of attack is severe on some cupronickels, but not so much with C70600 or Alloy 400 (N04400). Both oxygen, which enters the shell during shutdown, and the presence of contaminated condensate appear to be required for exfoliation (Figure 8.34) to occur. The use of cupronickels containing 4 to 5% iron and manganese will largely eliminate the exfoliation problems in such instances.

Supercritical Steam Plants

Supercritical steam plants operating above the critical temperatures 375 C (705 F) and critical pressure 22000 kPa (3200 psi) of water have the same mechanical components as conventional steam plants. Major differences are greater wall thicknesses to withstand the pressure, more corrosion-resistant materials, and lower dissolved solids in the water and steam.

Almost all of the components in a supercritical steam plant are made of austenitic stainless steels of the 18-8 variety. These materials are employed to minimize corrosion products and their transport through the system. The steam temperature is no higher than in an ordinary superheater, but the pressures are such that many chemicals and corrosion products may show appreciable solubility in the steam.

Many chemicals exhibit an inverse solubility when the temperature of the steam is above the critical temperature and will therefore deposit out at these higher temperatures. In one supercritical plant, about 140 ppm caustic was accidentally introduced into the plant. Within 30 minutes, caustic stress corrosion cracking occurred in that part of the

FIGURE 8.34 — Exfoliation attack on 70-30 cupronickel tubes of a high pressure feed water stage heater. [SOURCE: J. F. Wilkes, Corrosion and Embrittlement Protection for Boiler Systems, Materials Protection, Vol. 1, No. 5, pp. 18-26 (1962).]

plant where the temperature was about 425 C (800 F). This temperature corresponds to the minimum in the caustic solubility-temperature curve. The supercritical steam also undergoes a marked density increase above this temperature range which could cause the deposition of chemicals.

For such reasons, the dissolved solids content of the water must be kept at or near zero. A large fraction of the water is continuously cleaned up in a by-pass circuit containing ion exchange (demineralizer) beds. At start-up, 100% of the water may be passed through the clean-up beds. Chlorides and caustic are the most undesirable salts because they are known to cause stress corrosion cracking of the austenitic stainless steels (Chapter 6). Dissolved oxygen may be reduced by the methods described to only a few parts per billion.

Copper deposition on turbine blades was an early problem with supercritical units. It was found that trace amounts of copper were dissolved from the condenser tubes and recirculated into the boiler section. Because of its excellent solvent properties, the superheated steam carried the dissolved copper into the turbine where, at the lower pressure, the copper was deposited upon the turbine blades. This not only affected the efficiency, but threatened to destroy the turbine by mechanical imbalance due to uneven deposition. The problem has been largely eliminated by using stainless steel or titanium in the condenser.

Nuclear Fuel Steam Plants

There are several types of nuclear fuel steam plants. In some designs, the heat from the nuclear reaction is removed by nonaqueous media which then exchange heat with a water system to generate steam. Molten sodium, organic liquids, and helium gases have been used in this way. In other designs, an aqueous media (*e.g.,* heavy water or pressurized light water) removes the heat of reaction, but the steam is generated in a heat exchanger. There also are designs (*e.g.,* boiling water reactors) in which the reactor cooling water is used to generate steam directly.

Nuclear and conventional plants have many components in common: the boiler, steam drum, turbine, condenser, and feedwater heaters. In the nuclear plant, however, most of these components, with the possible exception of the condenser, are made of austenitic stainless steel or some other nickel-chromium-iron alloy to minimize corrosion products in the system. Low corrosion is desired, because if corrosion products become radioactive by transport through the reactor core, they constitute a radiation hazard to personnel when deposited again in areas remote from the core.

When a water system surrounds the core, whether to generate steam directly or indirectly, specific problems arise which are peculiar to irradiated water.

Boiling Water Reactors

In boiling water reactors, the steam is generated directly in the pressure vessel and, after separation, goes directly to the turbine. This results in boiling occurring on the fuel elements (corrosion characteristics are discussed later), the possibility of solids' deposition on the fuel elements, the stripping

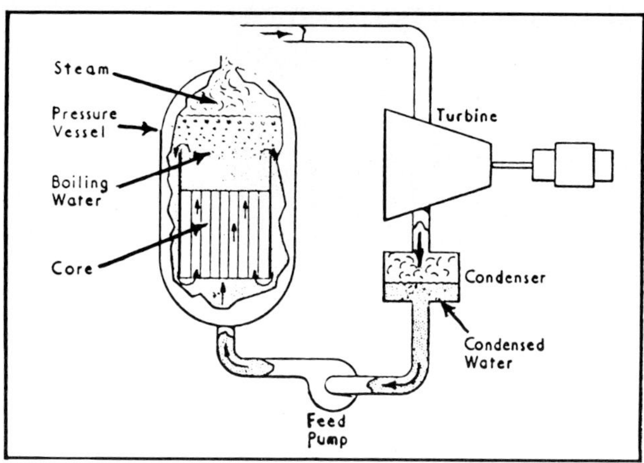

FIGURE 8.35 — Diagram of a boiling water reactor. [SOURCE: Nuclear Power—Today and Tomorrow, K. Jay, ed., Methuen & Co., Ltd., London, pp. 127, 129, 136 (1962).]

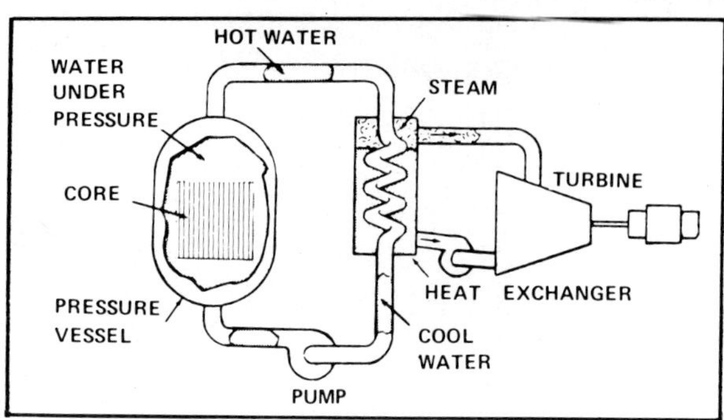

FIGURE 8.36 — Diagram of a pressurized water reactor. [SOURCE: Nuclear Power—Today and Tomorrow, K. Jay, ed., Methuen & Co., Ltd., London, pp. 127, 129, 136 (1962).]

of volatiles from the water, and the production of 0.2 to 0.3 ppm oxygen in the water and up to 20 ppm in the steam by radiolytic decomposition of water (*i.e.,* radiolysis).

High-purity neutral water is used because of the possible deposition of solid chemicals and the stripping of volatile components from any chemicals that would otherwise be used in water treatment.

Although the water and steam contain appreciable amounts of oxygen, carbon steels and stainless steels exhibit equivalent or superior corrosion resistance to that observed in pressurized water at the same temperatures. However, special problems do exist. Stainless steel fuel cladding has failed by stress corrosion cracking in boiler water reactors.

Stress corrosion cracking has also occurred in sensitized stainless steels (*i.e.,* those with chromium carbide precipitation at the grain boundaries) as a result of heating the steel in the 425 to 825 C (800 to 1520 F) range by welding or stress relieving. For this reason, Zircaloys are the preferred cladding materials. However, care must be exercised that alkaline components do not enter the water and become concentrated by boiling on the cladding surface because Zircaloys have poor corrosion resistance in high-temperature caustic solutions.

In a boiling water reactor, the nuclear fuel boils the water and the steam goes directly to the turbine (Figure 8.35). Temperatures are approximately 230 to 290 C (450 to 550 F), and with saturated steam from 2750 to 7200 kPa (400 to 1050 psi). As described with conventional boilers, the silica and copper content of the water must be kept low to prevent their transport by steam and deposition in the turbine.

The boiling water reactor design can be modified to accomodate a superheater to raise the steam temperature for improved efficiency, if desired. Experimental nuclear superheaters have been built to superheat the steam from the boiling water reactor. Corrosion rates for stainless steels and nickel alloys under heat transfer in the 600 C (1100 F) temperature range are about 25 μm/y (1 mpy). However, experience has shown that even minute amounts of entrained moisture in the entering steam can carry sufficient chlorides that cracking of the stainless steel cladding of the nuclear fuel elements may ultimately ensue. Alloys with higher nickel contents (*e.g.,* Alloy 800 or Alloy 600) have resisted stress corrosion cracking under these conditions.

Pressurized Water Reactors

In the pressurized water reactor (Figure 8.36), high-purity water is pumped past nuclear fuel elements where it becomes heated. The heated water then goes to the tube side of a heat exchanger, giving up its heat to boil water on the shell side. Steam is generated from the boiler water.

The water adjacent to the core is called *primary* water. It is usually treated with hydrazine to remove oxygen, with lithium hydroxide or ammonium hydroxide to pH 10 to 11 to minimize corrosion product transport, and with 25 to 50 cc hydrogen per kilogram of water to suppress radiolytic decomposition of water (thereby minimizing corrosion product transport).

Fuel and Fuel Cladding. The nuclear fuel must be clad with a corrosion-resistant material to prevent the release of radioactive gases and fission products to the primary water. The fuel itself is usually uranium oxide, which is quite resistant to the primary water. The fuel is clad with austenitic stainless steel or Zircaloy-2 (Zr-1.5 Sn-0.12 Fe-0.05 Ni) or the extra-low nickel Zircaloy-4 (Zr-1.5 Sn-0.24 Fe).

The Zircaloys are preferred because they do not "poison" the nuclear reaction as much as do the stainless steels. The Zircaloys perform well in the 285 to 315 C (545 to 600 F) temperature of the primary water for the 2 to 3 year expected life of a nuclear core. They develop a shiny, adherent black oxide film, which is protective and has excellent heat transfer properties. However, after long exposure times (2 to 4 years), or shorter times at higher temperatures [*e.g.,* 40 to 120 days at 360 to 400 C (680 to 750 F)], the corrosion rate of the Zircaloys increases and a white, relatively nonadherent insulating film develops.

The stainless steels, on the other hand, develop a relatively thick tarnish film with a thin, powdery surface film. Both films have a nominal composition of M_3O_4, where M represents iron, nickel, or chromium. The corrosion rate of stainless steel is not greatly affected by temperature in the range of 260 to 400 C (500 to 750 F), and it does not exhibit the marked effect of temperature on corrosion rate that is characteristic of Zircaloys.

Control Rod Mechanisms. The drives and bearing surfaces for inserting and withdrawing the control rods in the core are made of hardenable alloys. AISI Type 410 (S41000) and 17-4 PH (S17400) stainless steels have exhibited stress corrosion cracking when heat treated to hardness in excess of Rockwell C 35-40. This is probably the cathodic, hydrogen-induced cracking (Chapter 6). Lowering the strength level (lower hardness) has eliminated the problem. Chromium plating has been ineffective. Hard cobalt-base alloys have been used in bearing surfaces, but tend to contribute the long-lived radioactive isotope, cobalt 60, as a contaminant in the water.

The control material itself (neutron absorbers such as boron or cadmium) is clad with stainless steel for corrosion protection. Hafnium is sometimes used bare or unclad as a control material because its corrosion resistance is superior to the Zircaloys.'

Piping and Steam Generator. The piping, steam generator, and pressure vessel comprise much of the total exposed area in the primary section of a pressurized water reactor. The internal surface of the pressure vessel is clad with austenitic stainless steel. The piping and steam generator components are primarily austenitic stainless steel or Alloy 600 (N06600). Carbon steel has been used in a few systems.

The corrosion rates of austenitic stainless steels and Alloy 600 are about the same, approximately 1.5 μm/y (0.06 mpy). The corrosion rate of carbon steels is 5 to 10 times higher, to a maximum of about 13 μm/y (0.5 mpy). These rates are acceptable from a structural standpoint. However, as much as one-half of this oxide may not remain on the surface, but becomes a radioactive "crud" in its passage through the reactor core. Its subsequent deposition constitutes a personnel hazard.

The isotopes in the corrosion products which contribute most to radioactivity are Co-60, Co-58, Fe-59, Mn-54, and Cr-51. These isotopes are formed from the elements in stainless and nickel-base alloys. Their *half-lives* (*i.e.*, the period over which one-half of the initial activity decays) range from 27 days to more than 5 years. The Co-60 has a long half-life, and this is the reason for minimizing cobalt content in alloys and for using cobalt-base alloys sparingly.

It is generally agreed upon that pH in the range of 6 to 10, oxygen up to 5 ppm, or irradiation have little or no effect on the corrosion rate of stainless steels. However, these factors do affect the amount of crud released into the water, the amount increasing with decreasing pH, increasing oxygen content, and irradiation.

Since the primary water is pressurized, it does not boil. There is, therefore, no opportunity for traces of chloride contamination or caustic from the water treatment to concentrate and cause stress corrosion cracking. The dissolved chloride content is usually restricted to less than 0.1 ppm. Flow rates are of the order of 7 to 10 mps (20 to 30 fps), and the temperature is about 260 to 290 C (500 to 550 F), with water treatment as described for boiling water reactors.

Most structural components are also fabricated from stainless steel with corrosion characteristics as described for the piping.

Secondary (Boiler) Water. There is no radioactivity on the secondary (shell) side of the heat exchanger. Crud transfer is therefore no problem. This secondary water is treated like any conventional boiler water.

The usual conditions are 230 to 260 C (450 to 500 F), pH 10 to 11, 100 to 300 ppm phosphate, less than 0.1 ppm chloride, and sodium sulfite treatment to an oxygen content of less than 0.1 ppm.

Under these conditions, the oxide film that forms on the stainless steel of the steam generator and on the carbon steel of the steam drum is the protective black magnetic M_3O_4. Any evidence of the brick-red Fe_2O_3 indicates that there have been traces of dissolved oxygen in the water.

Both caustic and chloride stress corrosion cracking (Figure 8.6) have been observed in the stainless steel tubed heat exchanger (steam generators) in pressurized water reactors. The cracks have initiated from the secondary water side (*i.e.*, outside the tubes). The cracks have been found in the crevices where the tubes pass through the baffle plates or where the tubes are rolled into the tube sheet. Cracking has also been found where steam blanketing and localized boiling has occurred on the tubes.

Just a few parts per million of chlorides or free caustic in the boiler water can become sufficiently concentrated at these areas to cause cracking. Dissolved oxygen in the water contributes to chloride cracking. Analyses of corrosion products in the vicinity of chloride stress corrosion cracks have revealed the presence of several thousand ppm chloride. Caustic, on the other hand, although it may become concentrated in such areas, does not remain in the corrosion product.

Double tube sheets have been suggested as a remedy for this problem, but are not necessarily effective. Experience has shown that some leakage occurs past the expanded tubes in the inner tube sheet and into the drain area between the two tube sheets. This water flashes to steam, and the resulting concentration of corrosive species can still cause stress corrosion cracking.

Supercritical Reactors

Supercritical reactors are in the development stage. Many of the same problems exist with nuclear supercritical power plants as with the fossil fueled variety: (1) the supercritical steam is an excellent solvent, so solids in the water must be kept to a minimum to avoid deposition in the turbine; and (2) changes in density can result in localized build-up of any chlorides or caustic, with the attendant possibility of eventual stress corrosion cracking by these ions.

Corrosion Tests

The determination of the corrosion behavior of candidate metals and alloys can be determined by laboratory, pilot plant, or in-plant tests. Variables can be changed or controlled as desired in the laboratory and pilot plant. In-plant tests, however, are conducted under actual operating conditions and more accurately evaluate materials performance for the specific application.

Corrosion testing is covered in detail in Chapter 14.

References

1. Janssen, W. S., Corrosion Control in a Refinery Salt Water Cooling System, Materials Protection, Vol. 1, No. 10, pp. 42-53 (1962).
2. Gilbert, R. T., A Review of Recent Work on Corrosion Behavior of Copper Alloys in Sea Water, Materials Performance, Vol. 21, No. 2, p. 47 (1982).
3. Bird, Donald B. and Moore, Kenneth L., Brackish Cooling Water Versus Refinery Heat Exchangers, Materials Protection, Vol. 1, No. 10, pp. 70-75 (1962).
4. Godard, H. P., Watch Out for Wondrous Water Treatment Witchcraft, Materials Performance, Vol. 13, No. 4, p. 9 (1974).
5. Performance Analysis of Permanent Magnet-Type Water Treatment Devices, Research Report, South Dakota School of Mines and Technology, July, 1981.
6. Brooke, M., Cooling Water Treatment, Process Industries Corrosion, NACE, Houston, Texas (1975).

Bibliography

Bender, Rene J. Steam Generation, a Power special report. June, 1964.

Betz Laboratories, Inc. Betz Handbook of Industrial Water Conditioning. 6th Ed. Philadelphia, PA, 1962.

Bregman, J. I., Ed. Corrosion Inhibitors. MacMillan, New York, N.Y., 1963.

Butler, G. and C. K. Ison. Corrosion and Its Prevention in Waters. Reinhold Publishing, New York, N.Y., 1966.

DePaul, D. J., Ed. Corrosion and Wear Handbook for Water Cooled Reactors. USAEC Report TID 7006, March, 1967.

Glasstone, Samuel and Alexander Sesonske, Eds. Nuclear Reactor Engineering. D. van Nostrand, Princeton, New Jersey, 1963.

Horst, R. and S. McLain, Eds. Progress in Nuclear Energy. Series IV, Technology and Eng. McGraw-Hill, New York, N.Y., 1956.

James, G. V. Water Treatment. 3rd Ed. The Technical Press, Ltd., London, England, 1965.

NACE. Primer on Cooling Water Treating. A Report of NACE Technical Unit Committee T-5C. Houston, Texas, 1966.

Nordell, Eskel, Ed. Water Treatment for Industrial and Other Uses. 2nd Ed. Reinhold Publishing Corporation, New York, N.Y., 1961.

Proc. Inst. Mech. Eng. London, England, Vol. 160, pp. 341-350 (1949).

Tatnall, R. E. Fundamentals of Bacteria Induced Corrosion. Materials Performance. Vol. 20, No. 9, p. 32 (1981).

Yoel, William. Determination and Prevention of Fungal Deterioration of Cooling Towers. Materials Protection. Vol. 2, No. 4, pp. 65-68 (1968).

NOTES

Chapter 9

Cathodic Protection

CATHODIC PROTECTION

The purpose of this chapter is to introduce the concept of cathodic protection, one of the widely used "tools" for the control of electrochemical corrosion throughout industry. After a brief description of the meaning of cathodic protection and how it works, the balance of the chapter will consist of a discussion of the practical aspects of cathodic protection systems, as well as various factors that may influence cathodic protection designs. Remember that, although the great majority of cathodic protection is used on underground structures, the process has potential use in any electrolyte.

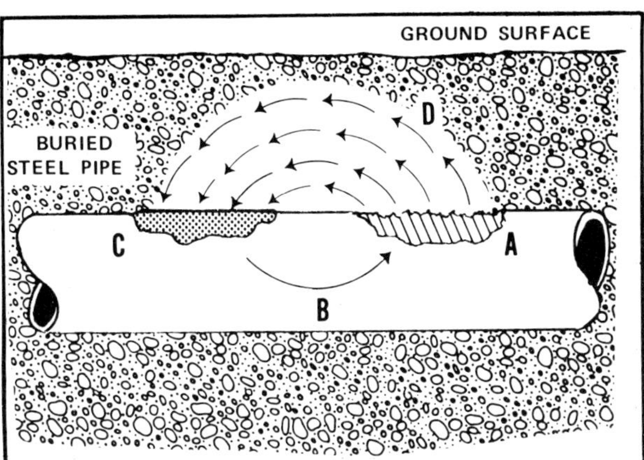

FIGURE 9.1 — Direct current flow on a typical corroding structure: (C) cathodic area, no corrosion at pipe surface; (B) current flows through the pipe steel from cathodic area back to anodic area to complete the circuit; (A) anodic area, where current leaves the steel to enter the surrounding earth, steel is being corroded here; (D) current flow through earth *from* anodic area *to* cathodic area.

Definition

First, it is necessary to agree on a definition for *cathodic protection* in order to establish a basis on which to build a better knowledge of the manner in which it functions and of its practical use.

In the Glossary of Corrosion-Related Terms in Chapter 1, cathodic protection is defined as: *Reduction or elimination of corrosion by making the metal a cathode by means of an impressed direct current or attachment to a sacrificial anode (usually magnesium, aluminum, or zinc).*

Referring again to the glossary, a *cathode* is the electrode where reduction (and practically no corrosion) occurs. Prior to applying cathodic protection, most corroding structures will have both cathodic areas and anodic areas (those areas where corrosion is occurring). It follows, then, that if all anodic areas can be converted to cathodic areas, the entire structure will become a cathode and corrosion will be eliminated.

How Cathodic Protection Works

The second step is to show how application of direct current electricity to a corroding metallic structure can cause it to become a cathode throughout its area. To begin with, direct current electricity is associated with the corrosion process on a buried or submerged metallic structure. This is illustrated by Figure 9.1 which shows the flow of direct current between anodic and cathodic areas on a section of buried pipe.

As shown, there is direct current flowing from the anodic areas into the soil, through the soil and onto the cathodic areas, and back through the pipe itself to complete the circuit. For a given driving voltage (the galvanic potential between anode and cathode), the amount of current flowing is limited by such factors as the *resistivity of the environment* (expressed normally as ohm-centimeters) and the *degree of polarization* at anodic and cathodic areas.

Corrosion occurs where the current discharges from metal into the soil at anodic areas. Where current flows from the environment onto the pipe (cathodic areas), there is no corrosion. In applying cathodic protection to a structure, then, the objective is to force the entire structure surface exposed to the environment to collect current from the environment. When this condition has been attained, the structure's entire exposed surface becomes a cathode, and corrosion is then successfully mitigated.

The basic manner in which cathodic protection is accomplished is illustrated in Figure 9.2, which shows how the originally corroding section of pipe used in Figure 9.1 becomes a cathode with cancellation of all current discharging areas on the pipe surface.

From inspection of Figure 9.2, it can be seen that the cathodic protection current must flow into the environment from a special ground connection (usually called a *ground bed*) established for the purpose. By definition, the materials used in the ground beds are anodes, and material consumption (corrosion) must occur. It follows, then, that corrosion has not been eliminated by the application of cathodic protection, but has been transferred from the structure to be protected to known locations (the ground beds) which can be designed to discharge the cathodic protection current for a reasonably long time, and which, when consumed, may be

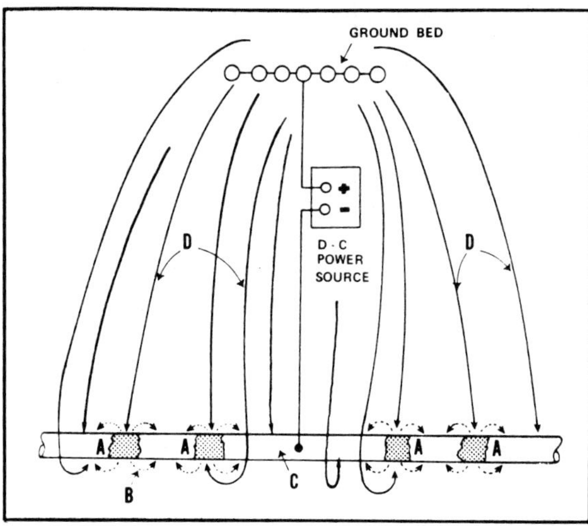

FIGURE 9.2 — Use of applied direct current to stop corrosion: (A) originally anodic areas discharging current and corroding; (B) dotted lines represent lines of current flow which existed prior to applying protection and which have been neutralized by the cathodic protection current; (C) protected structure; (D) current flowing from ground bed to surface of protected structure.

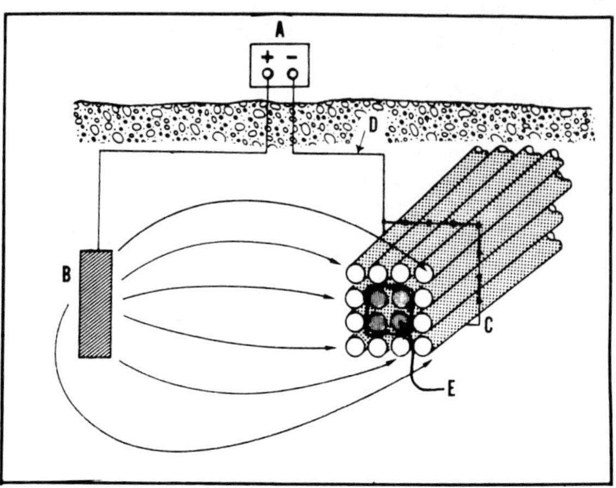

FIGURE 9.3 — Limitation of cathodic protection on congested structures: (A) DC power source; (B) ground bed anode; (C) group of cables to be protected; (D) negative cable connected to all sixteen cables in group (all connections not shown); (E) inner cables shielded by outer cables do not receive effective cathodic protection.

replaced without interruption of the normal function of the structure under protection.

From what we have said up to this point, it becomes obvious that the basic principle of applying cathodic protection is quite simple. In attempting the practical application of this corrosion control method, however, some questions naturally arise. These include, for instance: How much minimum current is needed to protect a given structure; what source of direct current should be used; how should the installation actually be designed; and how can it be ascertained, on a completely buried structure, whether or not the entire structure surface has, in fact, been made a cathode and all corrosion stopped? Information intended to answer these questions is developed in the following sections of this chapter.

There are limitations to what cathodic protection can do. The cathodic protection current must be able to flow through a conducting environment onto the metallic surface being protected. Taking a pipeline as a convenient example, a basic cathodic protection installation, as shown by Figure 9.2, can protect only those external surfaces of the pipe which are in contact with the conducting environment. The internal surfaces of the pipe will receive no protection. This will be true even though the pipe may contain a conductive material, since the current is intercepted by the pipe and carried back to its source.

Any portions of the pipeline system in air (such as aboveground portions, valves, aboveground header manifolds, etc.) cannot receive any of the protection current since air will not carry current to the unburied surfaces. If a closely spaced bank of structures such as a number of lead-sheathed power or communication cables is to be protected, current from the simple cathodic protection system shown in Figure 9.2 will flow mostly to the outermost surfaces of the cables in the bank. Relatively little current will flow to the innermost structures in the bank. This is illustrated by Figure 9.3.

Since the current flow to the innermost structures in the bank is limited, the amount reaching these structures may not be sufficient to adequately protect them, even though the outermost surfaces in the bank are receiving full cathodic protection. This is referred to as *electrical shielding* since the metallic structure surfaces on the outermost portions of the bank intercept the flow of protective current and shield the innermost structures.

Where special problems are encountered (such as the electrical shielding mentioned above and protecting the inner surfaces of pipes carrying corrosive materials), special types of cathodic protection systems may be used. These special systems will be described later in the chapter.

Cathodic Protection Current Sources

This section outlines the various sources of electrical energy needed to overcome corrosion currents flowing on a structure and thus cause it to become a cathode.

Galvanic Anodes

In the glossary in Chapter 1, the term *galvanic* often refers to a dissimilar metal contact which results in electrolytic potential. An *anode* is the corroding member in a dissimilar metal combination. A *galvanic* (also known as *sacrificial*) *anode* can be described as a metal which will have a voltage difference with respect to the corroding structure and will discharge current that will flow through the environment to the structure. To do this the galvanic anode must be electrically connected to the structure to be cathodically protected and also must be in con-

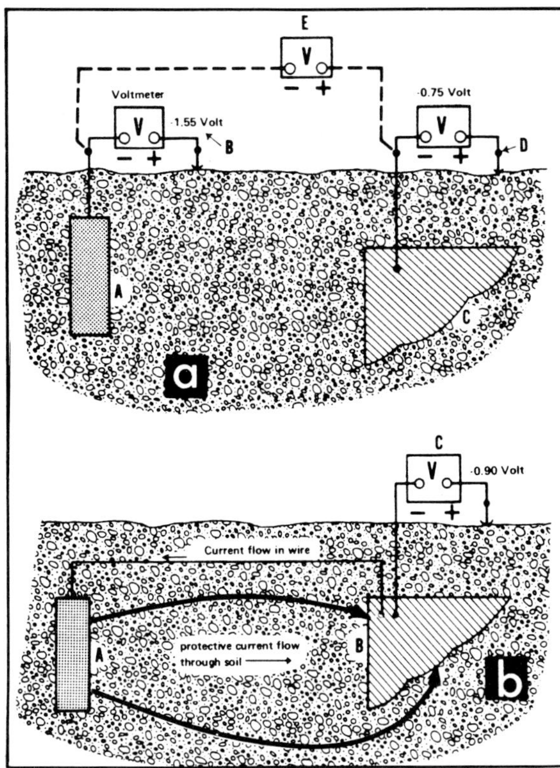

FIGURE 9.4 — Principles involved in using galvanic anodes to protect underground structures.

(a) Before connecting anode to structure to be protected: (A) magnesium galvanic anode *not* connected to structure (except through voltmeter, which does not comprise a load on the anode), no cathodic protection results; (B) this reading shows that the anode is more negative than the structure to be protected and would have to discharge current if the two are connected with wire; (C) structure to be protected; (D) copper sulfate electrode (Figure 9.9); (E) −0.80 volt between structure and anode, structure positive.

(b) After connecting structure to anode: (A) magnesium anode electrically connected to structure to provide cathodic protection; (B) cathodically protected structure; (C) voltmeter showing protective potential on structure.

tact with the conducting environment containing that structure.

This is not as complicated as it sounds. Figure 9.4 contains the essential elements. As shown in the figure, a voltage can be measured between the corroding structure and the material to be used as an anode. The structure must be positive (+) with respect to the anode before the anode can discharge current.

There are certain requirements for a metal to be a practical galvanic anode material.

1. The potential between the anode and corroding structure must be large enough to overcome the anode-cathode cells on the corroding structure.

2. The anode material must have sufficient electrical energy current to permit reasonably long life with a practical amount of anode material.

3. Anodes must have good efficiency, meaning that a high percentage of the electrical energy content of the anode should be available for useful cathodic protection current output. The balance of the energy that is consumed in self-corrosion of the anode itself should be very small.

Materials suitable for use as galvanic anodes include aluminum, magnesium, and zinc.

The amount of electrical energy that can be obtained from a galvanic anode depends on the electrochemical equivalent of the metal used (in terms of ampere-hours per pound) and the efficiency of the working anode. An *ampere-hour* is 1 ampere flowing for 1 hour, or any combination that will be equal to this, such as 0.5 ampere flowing for 2 hours, 2 amperes flowing for 0.5 hour, etc.

The efficiency of a galvanic anode is the ampere-hours per pound actually obtained for cathodic protection purposes divided by the total theoretical ampere-hours per pound for the material used. Galvanic anode materials are subject to self-corrosion which uses energy. This is why the efficiency is less than 100%.

For example, pure zinc has a theoretical maximum electrical energy content of 372 ampere-hours per pound. This means that if a zinc anode were to discharge one ampere continuously, one pound would be consumed in 372 hours—or, if it were discharging one tenth of an ampere, it would take 3720 hours (about 22 weeks) to consume a pound. Actually, zinc anodes operate, typically, at about 95% efficiency. This means that the energy content available for useful current output would be 372 × 0.95, or 353 ampere-hours per pound.

Another way of expressing this is in terms of pounds per ampere-year. At 353 ampere-hours per pound useful output for zinc, the conversion would be:

hr per yr	÷	amp-hr/lb	=	lb/amp-yr	
8760		353		248	(9.1)

meaning that this weight of zinc would be eaten away from an anode discharging one ampere for one year. The "pounds per ampere-year" value is useful in calculating the expected life of a galvanic anode installation.

Zinc anodes are made of high-purity zinc (99.99% purity or better) for soil use. For marine use, small amounts of aluminum and cadmium may be added to ensure maximum efficiency. If zinc of lesser purity is used, efficiency may suffer and the anodes may tend to "go passive," *i.e.*, cease to discharge useful amounts of protective current. Zinc anode working potential with respect to a copper sulfate electrode is in the order of −1.10 volts.

Aluminum anodes have a theoretical energy content of 1345 ampere-hours per pound. In recent years, aluminum anodes for seawater use have been developed that can operate at 95% efficiency with a useful output of approximately 1278 ampere-hours per pound. Working potential is about the same as that of zinc, although there are variations depending on the alloy used.

Aluminum anodes may have small amounts of alloying elements, and in some cases also are heat treated in order to obtain desired characteristics.

Magnesium anodes have a theoretical energy content of 1000 ampere-hours per pound. Efficiency varies with current output density in terms of milliamperes per square foot of anode surface, but typically can be about 50% (useful output of 500 ampere-hours per pound) at a current output density of 30 milliamperes per square foot. Anode working potentials to copper sulfate electrode can range from approximately −1.45 volts for standard alloy magnesium to approximately −1.70 volts for so-called high-potential anodes.

Standard magnesium anodes have appreciable amounts of alloying materials. Typically, these may be 6% aluminum, 3% zinc, and 0.2% manganese. The high-potential magnesium mentioned is a proprietary alloy.

Anode materials are cast in numerous weights and shapes to meet cathodic protection design requirements. Data on available anodes can be obtained from suppliers of cathodic protection materials.

The appearance of typical magnesium anodes is shown in Figure 9.5.

Impressed Current Anodes

The illustration used in Figure 9.2 to show how cathodic protection is applied also is typical of an *impressed current* cathodic protection system. With such a system, the ground bed anodes are not depended upon as a source of electrical energy. Instead, some external source of direct current power is connected (or *impressed*) between the structure to be protected and the ground bed anodes.

The positive terminal of the power source must always be connected to the ground bed, which is then forced to discharge as much cathodic protection current as is desirable. This is important. If a mistake is made and the positive terminal is connected erroneously to the structure to be protected, the structure will become an anode instead of a cathode and will corrode actively, which is the opposite of the desired results.

FIGURE 9.5 — Typical magnesium anodes: (1) 5 × 5 × 20 1/2 in, 32 lb; (2) 2 × 2 × 60 in, 16 lb; (3) 4 × 4 × 17 in, 17 lb; (4) 3 × 3 × 8 in, 5 lb; (5) 3 × 3 × 14 in, 9 lb; (6) 4 × 4 × 17 in, packaged; (7) 7 × 7 × 16 in, 50 lb. (SOURCE: Photograph Federated Metals Division, American Smelting & Refining Co.)

Ground bed anodes forced to discharge current will corrode. It is important to provide anode materials which are consumed at relatively low rates and thus permit designing ground beds which can discharge large amounts of current and still have a long life expectancy.

While scrap steel pipe, rail, rod, or other similar iron or steel material may be used, such materials are consumed at the rate of about 20 pounds per ampere-year (one ampere flowing for one year). This means a relatively large amount of material is needed to achieve the desired operating life for an installation, and may be justified only when it is available as scrap at low cost.

There is a special type of anode material made of cast iron having about 14.5% silicon and other alloying elements that is used up at a low rate (typically, a few tenths of a pound per ampere-year). This material is available commercially in various anode sizes to meet design requirements.

Another widely used commercially available anode material is carbon or graphite (a form of carbon). Anodes of these materials also are consumed at low rates (a pound or less per ampere-year for practical installations). Various shapes and sizes are available.

A sintered iron oxide (magnetite) electrode also is available.[1]

For marine work, considerable success has been achieved with lead-silver alloys. Although lead can be consumed at around 70 pounds per ampere-year, suitable lead-silver alloys will, when operated in accord with manufacturers' recommendations, form a lead peroxide film on this outer surface which discharges current at a low consumption rate for the base anode material.

Platinum works very well as an impressed current anode, being consumed at an extremely low rate. Because of its cost, it is usually plated onto a less expensive material. Titanium and niobium are used successfully for this. Platinized titanium and niobium are used in marine work on ships, hulls, and for applications involving the interior parts of structures such as condenser water boxes, internal parts of pumps, certain pipeline interiors, etc.

Rectifier Current Sources

When impressed current cathodic protection systems are specified, the most common power source used is a *rectifier*. This is a device which is provided with power from electric utility system lines and which converts the alternating current to a lower voltage direct current by means of a stepdown transformer and a rectifying device utilizing, commonly, selenium or silicon elements. These elements have a low resistance to current flow in one

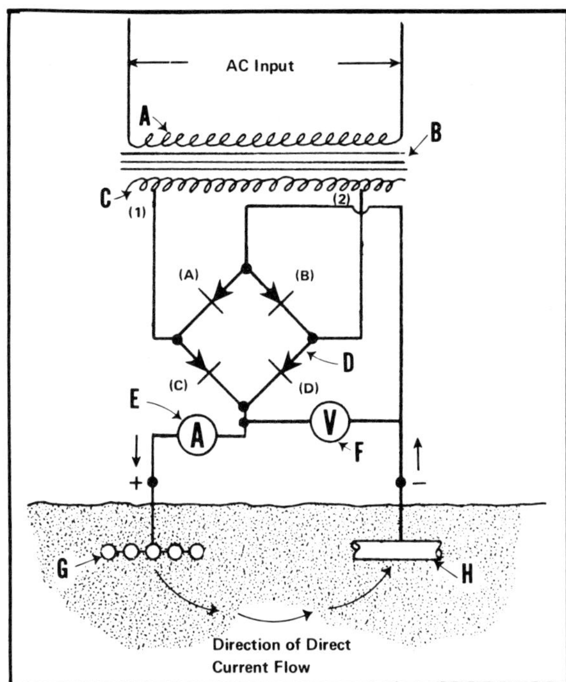

FIGURE 9.6 — Basic circuitry of cathodic protection rectifier: (A) primary winding; (B) step-down transformer; (C) tapped (1,2) secondary winding; (D) bridge-connected rectifying elements (A,B,C,D); (E) DC ammeter; (F) DC voltmeter; (G) ground bed; (H) cathodically protected structure.

direction and a high resistance in the opposite direction. Figure 9.6 shows a simplified circuit of a rectifier utilizing a "bridge-"connected rectifying element which is used in most cathodic protection rectifier systems.

The arrowhead symbols on the rectifying elements show the direction that current can flow readily. For the usual 60-cycle alternating current power source, the direction of current flow reverses 120 times per second. Referring to Figure 9.6, at some instant in time, current may be originating at connection (1) on the transformer secondary. The only path that this current can take will be through leg (C) of the bridge-connected rectifier, through the external circuit (ground bed to protected structure), and through leg (B) of the rectifier to return to the secondary winding at (2).

A one hundred and twentieth of a second later, the direction of the alternating current flow will have reversed and it will originate at connection (2). Under this condition, the only path that it can take to reach connection (1) will be through rectifier leg (D), the external circuit, and rectifier (A). For either direction of alternating current flow, the current flow through the external circuit is in one direction only—direct current.

Rectifiers operate at less than 100% efficiency, meaning that the direct current output power is less than the alternating current input power from the supply line. Rectifiers operate at maximum efficiency when operated at full-rated load. Where large rectifiers are used (as for large, bare structures), the rectifier rating should be selected so that it will

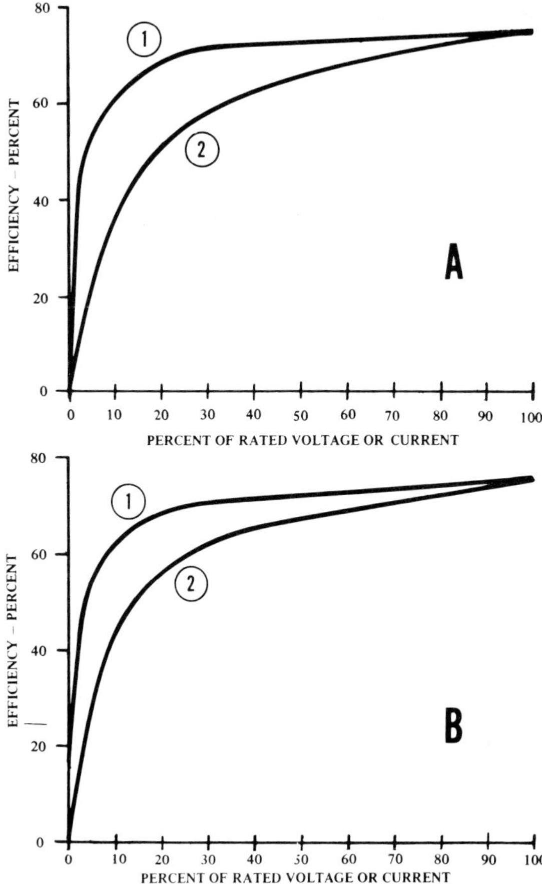

FIGURE 9.7 — Rectifier efficiencies: (A) Selenium—typical overall efficiency curves for custom line air-cooled selenium bridge-connected rectifier unit rated at 60 V, 60 amp full load DC output and 230 V, single phase AC input; (B) Silicon—typical overall efficiency curves for custom line air-cooled silicon bridge-connected rectifier unit rated at 40 V, 34 amp full load DC output and 230 V, single phase AC input (Information made available by Good-All Corrosion Control Co.); (A and B) (1) Efficiency vs percent rated current at full rated voltage, (2) Efficiency vs percent rated voltage at full rated current.

operate at close to the full-load rating. The effect of reducing either current output or voltage is illustrated by Figure 9.7.

The efficiency of a cathodic protection rectifier may be determined by measuring the AC power input with a watt meter, measuring the DC volts and amperes output, and calculating the efficiency as follows:

$$\%\ \textbf{Efficiency} = \frac{\textbf{D-C volts} \times \textbf{D-C amps} \times 100}{\textbf{A-C watts}} \qquad (9.2)$$

For example, a rectifier operating at 20 amperes and 32 volts DC output is found to have an a-c input of 940 watts. The efficiency will then be $(20 \times 32 \times 100) \div 940 = 68\%$.

The difference between AC input power and DC output power represents the power loss in the rectification process. This power loss becomes heat. Cooling is sometimes necessary to prevent the rectifying components from becoming too hot and failing. One cooling method utilizes an air-cooled assembly with the rectifier cabinet arranged so that there will be a natural draft of air through the rectifier components. In another method, the rectifier components are immersed in electrical insulating oil in a large steel case. The rectifying elements give up their excess heat to the oil. The oil is in turn cooled by radiation from finned outer surfaces of the case.

One precaution bears repeating: Never connect the positive (+) terminal of an impressed current power source to the structure to be protected. If this is done, it will rapidly corrode instead of being protected.

Other Current Sources

In some instances where an impressed current cathodic protection system is desired, a commercial alternating current power line may not be available. In such instances, there are other direct current power sources that may be used instead of rectifiers. These include:

1. *Batteries* (limited applications where the current drain is very low).
2. *Engine generator sets* (for large blocks of power—an engine, for which fuel must be provided, driving a direct current generator, or an alternating current generator which is used with a rectifier).
3. *Thermoelectric generators* (using heat, for which a fuel supply must be provided, to activate thermocouples which generate direct current).
4. *Fuel cells* (a device that permits the generation of direct current through the reaction of two gases, such as hydrogen and oxygen or natural gas and air).
5. *Solar cells* (solar panels used with and without battery storage facilities have been most successful and must be evaluated for use in any location where electric power is not available).

Criteria for Cathodic Protection

Any criterion used in connection with cathodic protection is a means that can be used to determine whether or not a structure supposedly under cathodic protection is actually fully protected against corrosion.

Potential to Environment

Potential measurements are used most commonly as a criterion of protection. The basis for this is that if current is flowing onto a protected structure, there must be a change in the potential of the structure with respect to the environment. This is because the current flow causes a potential change, which is a combination of the voltage drop across the resistance between the protected structure and the environment, and the polarization potential developed at the structure surface. (If necessary, review Chapter 2 for a discussion of polarization.)

The resistance between the protected structure and the environment includes the resistance of any electrically insulating paint or coating on the structure. The net result is that the structure being protected will become more negative with respect to its environment. This is illustrated by Figure 9.8.

To expand a bit on the matter of polarization, if just enough current were collected by cathodic areas to polarize them exactly to the open circuit potential of the anodic areas, corrosion would stop because there would no longer be a driving potential to cause corrosion current to flow.

In actual cathodic protection applications, however, such a delicate balance is seldom feasible, and the usual case is to have a net current flow onto the originally anodic areas. This does, however, indicate that measurement of polarization potential should be a good way to tell when the minimum amount of current giving full protection is reaching the metal surface. On this basis, the desirable place to measure potential would be across the interface between the pipe and the environment, as is represented by the terminals marked "polarization potential" on the equivalent circuit of Figure 9.8.

Actually, in practice, this is not feasible in a great many cases (particularly on buried structures), and it becomes necessary to resort to potentials measured between the structure and the environment surface directly above (or otherwise nearest to) the structure. The measured potential now includes the polarization potential plus a portion of the voltage drop across the structure-to-remote-earth resistance, as shown by the potential measured between the structure and point "G" on Figure 9.8. Under some conditions, measurement of the potential between structure and remote earth, point "H," is desirable.

The expression *remote earth* used above and in Figure 9.8 means the infinite conductor (resistanceless, for practical purposes) which is the earth's mass.

With some idea as to where to measure the potential, the next questions are how to do it and how much of a voltage reading is needed to indicate protection. Although in measuring a potential between a buried structure and earth it is relatively easy to visualize a wire (or other metallic conductor) connecting a voltmeter to the structure, actually connecting a wire from the voltmeter to the earth is another matter. This is done by contacting the earth (or other conducting environment) with a *reference electrode.* This is a stable device, suitable for field use, which will permit reproducible results. Figure 9.9 shows how the measurement may be taken.

Reference Electrodes

A common reference electrode (sometimes referred to as a *half-cell*), shown in the detail of Figure 9.9, is a copper-copper sulfate reference electrode widely used in cathodic protection work. Any

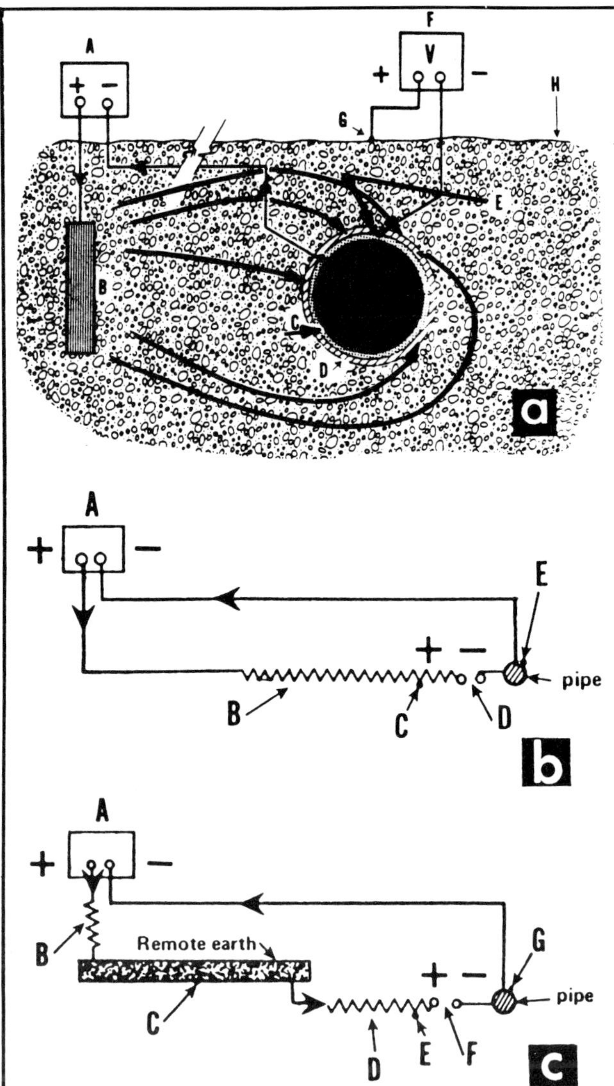

FIGURE 9.8 — Structure-to-environment potential change with flow of cathodic protection current.

(a) Overall schematic: (A) cathodic protection power source; (B) ground bed anode; (C) pipe; (D) coating on pipe; (E) contact with pipeline; (F) voltmeter, when connected as shown, needle will move upscale when protection system is energized and forces current to flow to the pipe surface; (G) contact with "close" earth; (H) contact with "remote" earth (+) terminal of voltmeter.

(b) Equivalent Circuit 1. Pipeline within "area of influence" of ground bed: (A) power source; (B) combination of anode-to-earth resistance and pipe-to-earth resistance which is *less* than the sum of pipe and anode resistance to remote earth; (C) point "G" on earth above pipe; (D) polarized potential; (E) point "E" contact with pipeline.

(c) Equivalent Circuit 2. Ground bed electrically remote from pipeline: (A) power source; (B) resistance of ground bed anodes to remote earth; (C) point "H" in (a); (D) resistance between pipeline and remote earth; (E) point "G" in (a); (F) polarization potential; (G) point "E" in (a).

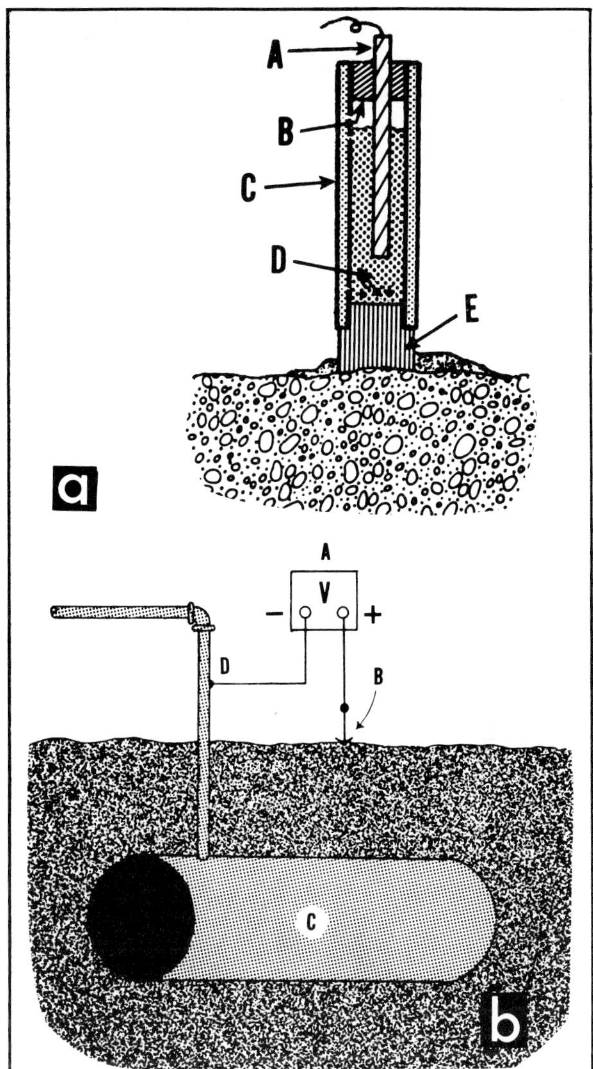

FIGURE 9.9 — Measuring potential of structure to environment.

(a) Detail of copper sulfate reference electrode: (A) copper rod connected to voltmeter; (B) sealing plug or cap; (C) container made of insulating material; (D) saturated solution of copper sulfate with excess crystals to insure complete saturation; (E) porous plug allowing copper sulfate solution to seep slowly through and make electrical contact with soil or water.

(b) (A) high-resistance voltmeter (refer to section on instrument requirements.); (B) stable reference electrode such as the copper sulfate electrode; (C) buried or submerged metal tank to be protected; (D) good electrical contact with structure.

voltage reading taken between the structure and the copper sulfate electrode will consist of two parts (half-cell potentials). One of these will be the half-cell potential between the electrode itself and earth through the porous plug contact; this is a constant value, for all practical purposes, under most field conditions. The other will be the half-cell potential between structure and earth; this is the variable that we are interested in.

In practice, it is not necessary to separate the two half-cell potentials since we are interested primarily in changes in the structure-to-earth portion, and this change shows up well enough in the total value registered by the voltmeter.

If we are talking about a steel structure, a value of −0.85 volt or more read on the voltmeter of Figure 9.8 would indicate full cathodic protection. This is considered to mean that the structure potential is −0.85 volt with respect to a copper sulfate electrode and is based on the value −0.80 volt assigned to the most highly anodic steel found in practical situations. Because in many cases the copper sulfate electrode cannot be placed close to the structure being checked (such as when it has to be placed on the earth surface above a buried structure), and to allow for some latitude in the potential of the most highly anodic areas, the practical value of −0.85 volt has been adopted.

When metals other than steel are to be cathodically protected, different potentials to a copper sulfate reference may be used as a criterion of protection. For lead-sheathed cables, a customary value is −0.70. For aluminum, the potential should be held between the limits of −1.00 and −1.20.[2]

Notice the upper limit given for aluminum. This is because the alkalinity built up at the aluminum surface by the cathodic protection process is corrosive to aluminum. This metal, if subjected to excessive cathodic protection, may actually corrode more rapidly than if not protected at all. Steel is not subject to this effect. Lead may be to some extent, but usually is not a problem unless the cathodic protection current source is interrupted for an extended period. In this case, the alkalinity developed at the lead surface will have an opportunity to corrode the lead.

The degree of cathodic protection attained also may be evaluated using other stable reference electrodes for which the potentials indicative of protection will be different when their half-cell potentials differ from that of the copper-copper sulfate electrode. The calomel electrode is used by some workers for field investigations, although it is much more commonly used as a reference in laboratory studies. The silver-silver chloride electrode is frequently used in work in seawater where the electrode must be submerged, since it is not subject to contamination by seawater as a copper-copper sulfate electrode can be (pressure at moderate depth forces seawater into the cell, causing the contamination).

Pure zinc in packaged backfill material makes a good permanently installed reference electrode for periodic checking of protective potentials at key points. Pure zinc without backfill may be used in seawater, but should be checked against another type of reference prior to each use. Contacting the earth with a steel pin is not suitable; the steel-to-earth half-cell potential is not stable.

A practical guide to the potential differences among the reference electrodes mentioned is included in Table 9.1.

Test Coupons

Better proof that cathodic protection is working can be obtained by using coupons of the same metal as that in the protected structure. These are weighed carefully beforehand and are electrically connected to the protected structure. The coupons should be placed where they can receive the same exposure to cathodic protection current as does the structure; then, after a known exposure time, they may be removed and weighed. Any loss of weight will indicate incomplete cathodic protection.

Coupons may be placed where protective potentials are lowest (as indicated by measurements to a suitable reference electrode) or at locations where it is suspected that the degree of protection may not provide complete freedom from corrosion.

Potential Change

Some workers, when dealing with large *bare* structures, use a potential change of 0.2 to 0.3 volt when protective current is applied as an indication of a reasonable degree of cathodic protection. This

TABLE 9.1 — Comparison of Other Reference Electrode Potentials with That of the Copper-Copper Sulfate Reference Electrode at 25 C

Type of Comparative Reference Electrode	Structure-to-Comparative Reference Electrode Reading Equivalent to −0.85 Volt with Respect to Copper Sulfate Reference Electrode	To Correct Readings Between Structure and Comparative Reference Electrode to Equivalent Readings with Respect to Copper Sulfate Reference Electrode
Calomel (saturated)	−0.776 volt	Add −0.074 volt
Silver-Silver Chloride (0.1 N KCl Solution)	−0.822	Add −0.028
Silver-Silver Chloride (Silver screen with deposited silver chloride)	−0.78	Add −0.07
Pure Zinc (special high grade)	+0.25[(1)]	Add −1.10

[(1)]Based on zinc having an open circuit potential of −1.10 volts with respect to copper sulfate reference electrode.

involves measuring the structure-to-reference electrode potential with the current off, and again with the current on. A potential change in the negative direction of 0.2 to 0.3 volt indicates that the structure is collecting current.

This may not mean that the minimum protective value of −0.85 volt with respect to a copper sulfate electrode (for steel) has been reached, and as a result it cannot be said that corrosion is completely stopped. The corrosion rate may be reduced to a great extent, however, in many situations. The −0.85 volt criterion (for steel) is preferred by most corrosion engineers.

Other criteria are used for evaluating the protection of aluminum and copper.[3]

Other relationships that can be used to establish the protection potential required are the E-log I curve (discussed later) and the potential-pH diagram. To achieve complete protection of steel, what potential to a $CuSO_4$ half-cell is required, according to the potential-pH diagram of Chapter 2 (Figure 2.24)?

Design of a Cathodic Protection System

In many instances, prior to designing a cathodic protection installation, it is necessary to conduct a field survey. Such a survey is made to assemble the data necessary to permit a workable design.

Current Requirement Tests

The current needed to protect a given buried metallic structure may vary over wide limits, depending on the nature of the environment and whether or not it has a protective coating, and if coated, on the quality and effectiveness of the coating as applied.

For example, if it is assumed that a steel structure to be protected is buried in corrosive soil and has an exposed surface area of 1000 square feet, the current required (assuming reasonably uniform distribution of that current) could range from about 3 amperes if the structure is bare, to as low as 30 microamperes or less if the structure has a superior coating. This means that a bare structure may require 100,000 times as much current as the same structure would if it were well coated.

It cannot be assumed, however, that just because a structure is coated, it will take a small amount of current to cathodically protect it. A poor coating material (or even an excellent material poorly applied) can take much more current than the low figure given above.

For the same 1000-square-foot structure, a relatively poor coating could result in a current requirement for cathodic protection in the order of 15 milliamperes or more. A current of 15 milliamperes may not sound like much, but it is 500 times as great as the 30 microampere figure given for a superior coating. The difference is considerable when working with large structures (such as large diameter cross-country pipelines).

Tests for a Coated Line

Obviously, with the wide range of current requirements, some kind of test is needed to determine just how much current is really needed to give adequate protection to a given structure. This may be done by actually applying current using a temporary test setup and adjusting the current from the power source until suitable protective potentials are obtained. This might be accomplished as shown in Figure 9.10 for a 500-foot long section of coated 36-inch diameter pipe crossing under a river.

With the test set up as shown for a coated pipeline, batteries may be used as the power supply. The batteries may be automobile storage batteries or heavy-duty dry batteries (such as ignition batteries of the "hotshot" type). Current flow may be regulated by heavy-duty adjustable resistors. With the switch closed, the current may be increased gradually until the voltmeter at position B reaches −0.85 volt with respect to a copper sulfate electrode placed directly over the pipe under test at that point. Having an observer at position B to check the potential change is most helpful, particularly if voice, radio, or other means of signaling is possible between positions A and B.

The 500 feet of 36-inch diameter pipe used in the illustration would have an external surface area of 4712 square feet. This means that the average current requirement per square foot for cathodic protection would be 25 milliamperes divided by 4712 square feet, or only 0.0053 milliamperes per square foot. This is equivalent to 5.3 millionths of an amp (microamps) per square foot.

Tests for a Bare Structure

The preceding discussion dealt with a coated structure. When current requirement tests are made on an all-bare structure, results may be quite different. Assume, as shown by the illustration in Figure 9.11, that a large buried bare steel tank (which has been in the ground for several years and has developed a corrosion leak) is to be cathodically protected, that it is electrically isolated from all other metallic structures, and that it has an exposed surface area of 5000 square feet.

A current requirement test can be made using a temporary ground bed and power source as described for the preceding coated pipeline example with two exceptions. The half-cell must be moved to remote earth from the tank, and a reading of the polarized potential (current off) of the tank should be made. The latter requires a quick reading (within one second) after the current source switch is opened because depolarization begins immediately.

A well-coated structure usually will polarize very rapidly when a cathodic protection test current

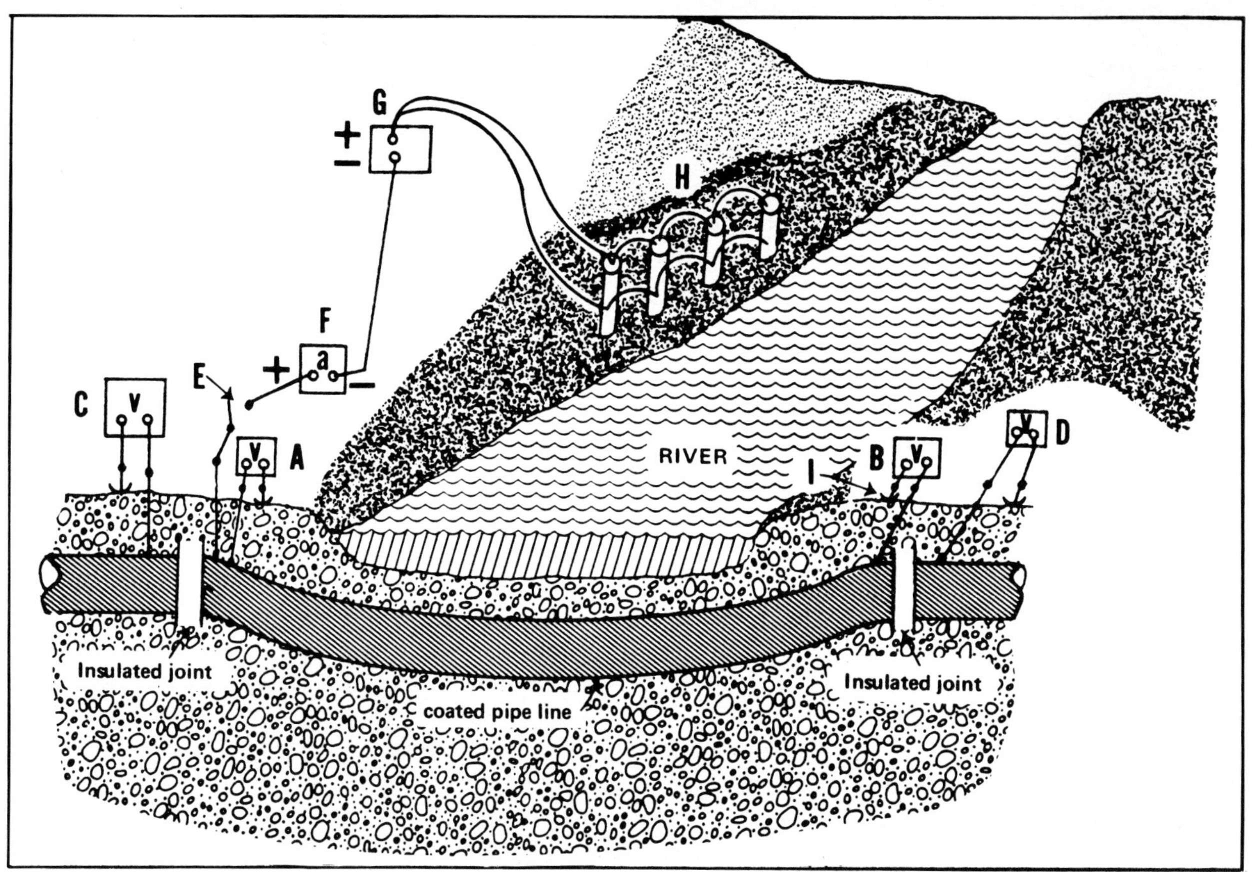

FIGURE 9.10 — Measuring current required to protect a coated pipeline; (A-D) voltmeters; (E) switch; (F) DC ammeter; (G) adjustable DC power source; (H) temporary ground bed of driven rods in low-resistance soil 100 feet or more from pipeline; (I) close copper sulfate reference electrode (directly over pipe).

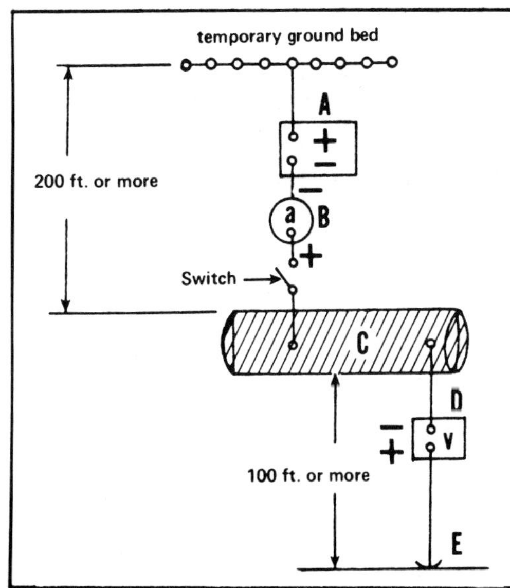

FIGURE 9.11 — Cathodic protection requirement test on bare (uncoated) structure: (A) DC power source; (B) ammeter; (C) buried steel tank, 5000-sq-ft surface area; (D) high-resistance voltmeter; (E) remote copper sulfate reference electrode.

If 12.5 amperes were required to produce adequate polarization of the structure as measured at (E), how many milliamps per square foot are required to protect the steel?

If the power source showed 18 volts required to produce the 12.5 amperes, what is the total resistance of the polarized circuit?

is applied. This means that, with a given value of applied current, the potentials observed between structure and earth right after the current is applied will, for all practical purposes, represent the final polarized protective potentials obtained with a permanent cathodic protection installation. However, when making a test on a large bare structure, it will be found that, in most cases, the observed potential between structure and earth with a given value of test current will increase (become more negative) with time. This is caused by the development of a polarization film.

Under some conditions, full polarization may not be obtained for days, weeks, or even months. This suggests that making a current requirement test on a large bare structure (as was shown in Figure 9.11) could be a long process. The procedure can be shortened substantially by making a polarization curve which, although not complete, can be extrapolated (extended) to give the probable final results for a given value of test current.

The test current used should be a reasonable approximation of the probable current requirement. This can be estimated by allowing one milliampere per square foot for a new, bare steel, buried structure, or three milliamperes per square foot if it has been in the ground for a number of years.

In the example of the old bare tank discussed (Figure 9.11), a reasonable test current could be

5000 square feet (the exposed outer surface area of the tank) multiplied by 3 milliamperes per square foot. This requires 15,000 milliamperes, which is 15 amperes. The power source for the test must have enough capacity to produce this current for a substantial time (possibly 24 hours or more) if a reasonable polarization curve is to be obtained. An engine generator set, a portable rectifier (if AC power is available), or a DC welder may be used as power sources.

Characteristics of Curve

A polarization curve taken on the tank might look something like that shown in Figure 9.12.

The projected curve indicates that the maximum protective potential will be in the order of −0.82 volt at 15 amperes, rather than −0.85 volt to a copper sulfate electrode as desired. Note that the total voltage change from the original structure potential (before any protection was applied) is 0.26 volt. Since 15 amperes produced this result, the average potential change per ampere is 0.26 ÷ 15 = 0.0173 volt.

An additional potential change of 0.03 volt is needed to raise the −0.82 potential (indicated by the curve) to the −0.85 volt minimum. This, then, would indicate that the additional current needed would be 0.03 volt ÷ 0.0173 volt per ampere = 1.73 amperes. On this basis, the total current needed for protection of the tank would be 16.73 amperes, rather than 15.

In practice, it is not necessary to adjust the power source so exactly; a setting at approximately 17 amperes would be reasonable. This method of modifying the test data to obtain the correct value of current for field protection may not be strictly accurate, but is considered sufficient for practical purposes, as long as the value selected for the current requirement test is reasonably close to the final value.

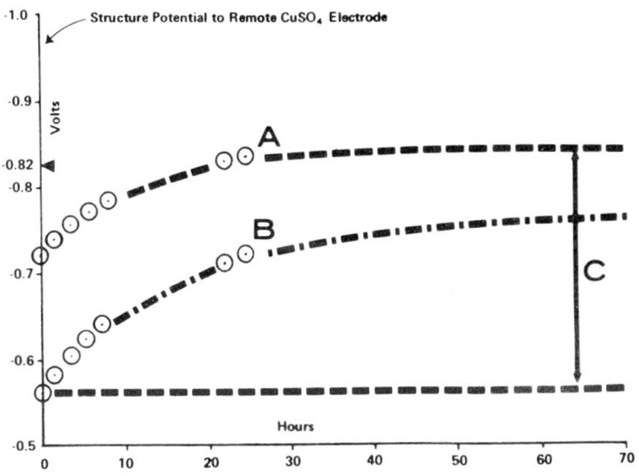

FIGURE 9.12 — Polarization curve for a bare steel structure current requirement test; structure—5000-sq-ft bare steel tank; applied test current—15 amperes: (A) potential with test current on; (B) potential with test current interrupted momentarily (polarization potential); (C) total current change at 15 amperes = 0.82 − 0.56 = 0.26 volt.

At this point, it would be well to check back and compare the results of the two current requirement tests described: the coated pipeline example illustrated by Figure 9.10, and the bare tank example illustrated by Figure 9.11. The surface areas of the two structures are approximately the same, but the bare structure requires 16.73 amperes for protection, while the coated one needs only 25 milliamps (0.025 amp). This means that the bare structure, in this particular case, needs 16.73 ÷ 0.025, or approximately 670 times as much current as the coated one! This point is made to illustrate the great current limiting capability of a good coating.

The curves shown in Figure 9.12 include a polarization potential curve (the lower one) which shows that the maximum polarized potential (taken immediately after opening the power source switch) is considerably less than −0.85 volt. Even though we may get −0.85 volt to remote earth with current on, there still may be small local corrosion cells on the pipe surface which have not been stifled completely. The structure's entire surface would need to be polarized to −0.85 volt or better to do a perfect job of stopping corrosion. It can be seen, however, from a study of Figure 9.12, that it would take 50% more current to raise the polarized potential (with current off) to −0.85 volt to a copper sulfate electrode.

In actual practice, it has been found that the −0.85 volt to remote electrode on bare structures with cathodic protection current on provides good control of the corrosion rate, and for that reason is widely used because it is less expensive. It costs substantially more to protect a large, uncoated structure than a coated one.

Coating Conductance Tests

In the preceding section we talked about current requirement tests on coated structures, and indicated that the coating accounted for most of the resistance between structure and earth. Coatings are subject to gradual deterioration with time, with the deterioration rate depending on quality and environmental conditions. Coating conductance tests are helpful on newly coated structures and at intervals during their operating life to evaluate coating performance. This information is particularly useful when selecting coatings for subsequent projects.

These tests are performed by determining the total current passing from the soil to the pipe surface in a specially designated pipe section that has been prepared for this test. The net average change in pipe-to-soil potential is measured and coating conductance is calculated. Periodic tests will indicate the long-term ability of a coating to maintain low conductance (desirable from a cathodic protection current demand viewpoint) over long periods of time.

Environmental Resistivity

Measurements of the electrical resistivity of the soil or water environment must be made as part of the corrosion survey conducted prior to designing a cathodic protection system. As a minimum, resistivity data are necessary at locations where ground beds (or single anodes) are to be installed. Such data permit the designer to determine the type, size, and number of sacrificial galvanic anodes, if used, or to select the most favorable combination of ground bed, size, and power source DC output rating if an impressed current system is to be used.

The unit of measurement which is probably the most widely used in field practice is the ohm-centimeter. Visualize an isolated cube of the material being evaluated, with the length of each side being one centimeter. The resistance between opposite faces of this one-centimeter cube would be the unit resistivity of the material in ohm-centimeters. In practice, no attempt is made to isolate a single centimeter cube and measure its resistance. Instead, the resistance of larger masses of material is measured and the unit resistivity calculated from the results of this larger mass measurement. This results in an average value of resistivity for the material mass measured.

On very large structures, such as long bare pipelines, soil resistivity measurements made at intervals along the structure may be used to help locate areas where corrosion can be most severe. In general, a structure that is subject to wide variations of environmental resistivity from point to point along its surface will be particularly subject to corrosion damage at the low resistivity areas, which tend to be the anodic portions of the structure.

Soils are particularly apt to have widely varying electrical resistivities within very short distances. For this reason, taking a single soil sample from the surface at a proposed ground bed location and measuring its electrical resistivity can be extremely misleading because it is rarely the case that such a sample will represent a reasonable average of resistivity conditions at the location.

To obtain meaningful data, it is necessary to know the average soil resistivity at various depths below the surface. For example, at the usual ground bed installation involving anodes (impressed current or galvanic) buried in the first ten feet below the surface, it is helpful to have figures on average resistivities by increments of 2.5 feet, down to depths of 15 to 25 feet.

If wide changes in resistivity are suspected, it may be desirable to determine the average resistivity to depths of 50 feet, 100 feet, or even more, particularly for large ground beds.

Fortunately, it is possible to measure soil resistivity to various depths without actually digging down to the depths in question. This is done, basically, as shown in Figure 9.13.

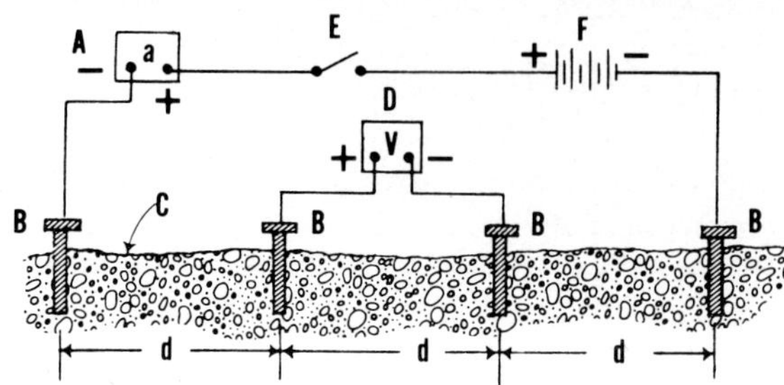

FIGURE 9.13 — Determining soil resistivity: (A) ammeter; (B) pins spaced at equal distances (d) along straight line; (C) surface of earth; (D) voltmeter; (E) switch; (F) battery, voltage sufficient to provide meaningful amount of current for instruments. (SOURCE: ASTM, Field Measurement of Soil Resistivity Using Wenner Four-Electrode Method, ASTM Standard G57-78.)

Four pins are driven into the earth (to a depth of a foot or so—depth usually is not critical) along a straight line, with distance between pins equal to the depth below the surface to which the average resistivity is desired. Current is passed between the outside pins. This current will cause change in voltage ($\triangle V$) between the center pair of pins, with the amount of change dependent on the resistivity of the material being measured.

For any pin spacing, an effective resistance in ohms between the center pair of pins is determined by dividing the change in voltage ($\triangle V$) by the test current causing that change ($R = \triangle V/I$). If $\triangle V$ is expressed in volts, I must be in amperes. If $\triangle V$ is expressed in millivolts, I must be in milliamperes. To convert the measured resistance value to soil resistivity, it must be multiplied by a factor. For any pin spacing used, this factor is the pin spacing in feet multiplied by 191.2, a constant.

As an example, assume that at a particular location, pins are spaced 5 feet apart for a soil resistivity test. Then 500 milliamperes of direct current is passed between the outside pins and causes a voltage change from 50 millivolts to 460 millivolts (which is a $\triangle V$ of 460 to 50 or 410 millivolts) as the current is switched on and off. The resistance value in ohms is then 410 millivolts ÷ 500 milliamperes or 0.82 ohms. The factor for a 5-foot spacing is 191.2 × 5 or 596. The soil resistivity is 0.82 × 956 or 784 ohm-centimeters. Although the resistance measurement may be made with separate voltage and current measurements as described, instruments are commercially available which read directly in ohms.

Some sort of guide to the meaning of soil resistivity determinations is needed. Designations in Table 9.2 can be used as a rough indication. Values encountered in practice can range from less than 20 ohm-centimeters for warm seawater to millions of ohm-centimeters for some soils and rocks.

Those working in areas subject to freezing should know that when soil freezes, its resistivity increases. (Notes regarding temperature effects are in-

TABLE 9.2 — Rough Indications of Soil Corrosivity vs Resistivity

Ohm-cm	Description
Below 500	Very corrosive
500 to 1000	Corrosive
1000 to 2000	Moderately corrosive
2000 to 10,000	Mildly corrosive
Above 10,000	Progressively less corrosive

cluded in a later section of this chapter, "Factors Affecting Cathodic Protection Design.")

Finally, some general guidance is needed in interpreting a series of soil resistivity determinations made by the 4-pin method. As our first example, assume that the data in Table 9.3 (Example 1) have been taken. These data show progressively higher resistivities with depth. Actual values at the lower depths are even *greater* than the values shown. This is because, in the 4-pin test method, the measurement is electrically "looking at," in effect, a mass of earth shaped like a half-cylinder, with the flat side of this half-cylinder at the earth's surface and the radius of this half-cylinder equal to the pin spacing used.

As shown by Figure 9.14, when we go from a 2.5-ft depth to a 5.0-ft depth, the reading at the 5.0-ft depth is looking at a larger volume of earth from the surface down to the 2.5-ft line than from the 2.5-ft line down to the 5.0-ft line. For this reason, in the sample set of data, the soil resistivity from 2.5 ft to 5.0 ft must be higher than 1500 ohm-

TABLE 9.3 — Four-Pin Test Data (Example 1)

Pin Spacing, Feet	Resistivity, ohm-cm
2.5	500
5.0	1500
7.5	3000
10.0	6000

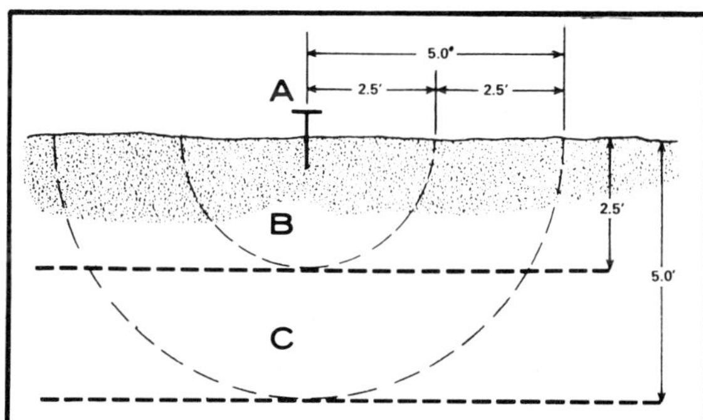

FIGURE 9.14 — Effect of depth increments on 4-pin soil resistivity measurements: (A) temporary electrode; (B) earth mass contained in this line is that affected by soil resistivity readings to 2.5-ft depth; (C) earth mass contained in this line is that affected by soil resistivity reading to 5.0-ft depth.

TABLE 9.4 — Four-Pin Test Data (Example 2)

Pin Spacing, Feet	Reistivity, ohm-cm
2.5	2500
5.0	2000
7.5	1500
10.0	1000

centimeters in order to give a 1500-ohm-centimeter average to the 5.0-ft depth. This same reasoning applies to the remaining readings in the set of data given.

Assume that at another location a set of soil resistivity data such as that in Table 9.4 (Example 2) is obtained. In this case, the readings show progressively lower resistivities with depth. Here, for the reasons given for Example 1 data, actual resistivities at each lower depth are *less* than the values shown. In general, readings showing progressively lower resistivity with depth are most favorable to the best performance of anodes in galvanic anode or impressed current cathodic protection systems.

Local Conditions Affecting Design

During cathodic protection surveys, observations must be made regarding local conditions which may have a bearing on the cathodic protection design finally selected. These include the following:

1. Availability of electric power for impressed current cathodic protection systems.
2. Locations suitable for cathodic protection installations (subject to satisfactory soil resistivity) that will be accessible for routine inspection and maintenance and which are not subject to construction or other activities which may make the site untenable within the expected life of the installation.
3. Presence of metallic structures, other than the one to be protected, which might be affected by the planned system.
4. Presence of cathodic protection systems on other structures which may have an effect on the structure to be protected.
5. Presence of sources of stray direct current such as DC railway systems.
6. Presence of AC systems which can have an effect on the readings taken and on the criteria for full protection of the metal.[4]
7. Unusual environmental conditions such as acid waste discharge from local manufacturing operations.

A later section ("Factors Affecting Cathodic Protection Design") will cover some of the above points in greater detail.

Instrumentation Requirements

In order that necessary corrosion survey work can be done before application of cathodic protection and so that performance tests can be made on completed protection systems, specialized equipment and knowledge of its use is needed. A complete discussion of this subject requires more space than is available in this chapter. (A relatively complete discussion is available, however, in the publication "Control of Pipeline Corrosion," which may be obtained from the National Association of Corrosion Engineers. This publication also covers much more complete information on cathodic protection design details than is possible in this chapter and is suggested as a useful supporting reference.) Basic, necessary equipment includes the following.

1. A sensitive DC voltmeter and ammeter for measuring protective potentials, making soil and water resistivity tests, measuring current flow on structures, and other tests as required. Although separate instruments may be used, combination instruments have been developed and are commercially available for this purpose. These instruments incorporate voltmeter and ammeter circuits, have characteristics designed specifically for cathodic protection test work, and are so arranged as to permit rapid selection of the desired test circuit. Such an instrument is the basic requirement of a test kit. A typical example is shown in Figure 9.15.

2. Suitable reference electrodes such as copper-copper sulfate reference electrodes available through equipment suppliers.

3. A pipe and cable locator is particularly useful when working on such structures and for maintenance work on cathodic protection installations when it becomes necessary to trace buried conductors.

4. An AC-DC volt-ohm-milliammeter is useful for maintenance testing on impressed current cathodic protection installations and for general test purposes.

5. One or more recording DC voltmeters if variable structure-to-environment potentials may be expected as a result of stray direct current flowing in the earth. If recorders are to be used where 120-volt electric power is not readily available (such as in pipeline work), self-powered recorders may be used.

Multirange instruments are desirable for general use. Full-scale ranges from 1 to 2 millivolts up to 200 millivolts are useful for measuring current flow on structures (by recording the voltage drop across a section of known resistance such as a length of pipeline) or through a shunt by measuring the voltage drop across the shunt. Full-scale ranges from 1 to 10 or more volts are suited for measuring such quantities as structure-to-environment potential and voltage gradients in the earth resulting from the flow of stray current in the earth.

6. Where many soil resistivity measurements are to be made, one of the several available types of soil resistivity test instruments will save testing time.

7. Short jumper cables of insulated wire for interconnecting instruments, at least two hand reels with approximately 500 feet of insulated test wire, and small hand tools.

The equipment described in general terms is available through various cathodic protection equipment suppliers. Specific selections will depend on the particular requirements of the cathodic protection test program to be undertaken. Additional equipment items may be added as dictated by the requirements of these programs.

Most of the test instruments that will be used must be very sensitive in order to properly detect the small values of direct current and voltage encountered in cathodic protection test work. Accordingly, they are delicate and must be handled carefully. Rough usage can lead to failure or unreliable readings.

Situations may be encountered where the external resistance is so high that even high resistance voltmeters will have higher percentage reading errors than is acceptable. In such cases, potentiometer or potentiometer-voltmeter circuits (which draw no current from the external circuit when the unknown potential is being measured) or electronic voltmeters (which have extremely high internal resistance) may be used. Use of these instruments is discussed in Chapter 7 of the NACE book, "Control of Pipeline Corrosion," mentioned earlier.

Recording instruments are needed only when continuously varying potentials and currents are associated with the metallic system on which cathodic protection is applied. This may be caused by man-made stray direct current in the earth resulting from DC transit systems, DC mining operations, DC welding, and similar sources. Occasionally, in some areas, variable direct current will be induced in pipeline systems as a result of disturbances in the earth's magnetic field. Such currents are sometimes referred to as *telluric* currents. In marine work, structures such as steel piling and

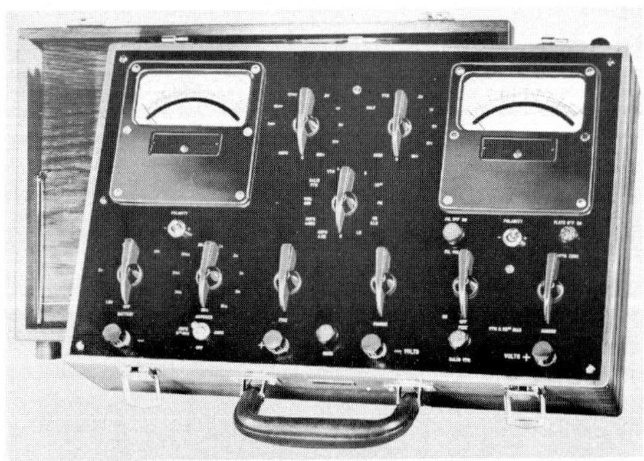

FIGURE 9.15 — One of the many types of multicombination meters used for measurements in underground cathodic protection system engineering. (SOURCE: photo courtesy of M. C. Miller Co.)

steel bulkheads will have varying current requirements for cathodic protection as the tide level changes. Recorders can be helpful in evaluating such situations.

Recorders suitable for cathodic protection work may be procured with ranges from one millivolt full-scale up to 100 volts full-scale. They can be self-powered (batteries or spring-wound mechanism) so that the chart advance device can be used where ordinary AC power is not available. With ranges within the limits indicated, they may be used for:

1. Measuring current flow on a structure (by the voltage drop method) using the millivolt ranges.
2. Measuring direct current (such as the output of rectifiers, galvanic anodes, or batteries) using appropriate shunts and matching millivolt range.
3. Measuring potentials between the structure under study and its environment, using an appropriate reference electrode, or measuring potentials between structures.
4. Measuring potential gradients in the earth caused by variable stray current (by recording voltage drop between suitable reference electrodes).

Factors Affecting Cathodic Protection Design

A number of things must be considered before a cathodic protection system is selected for any specific project. The following will serve as a checklist of some of the items that should be considered, together with a brief description of their significance. Some of the more important factors are discussed in greater detail elsewhere in this chapter.

Total Current Requirement

The total amount of current needed for cathodic protection of the project must be known. This may be determined by a current requirement test using a temporary setup. If the current requirement per cathodic protection installation is low (up to 1.5 or 2 amps), galvanic anodes may be the best choice, whereas large amounts of current may be more economically provided with impressed current systems using rectifiers or other sources of DC power. This is only a rough indication, and other factors explained later may govern the final decision.

Although an actual current requirement test will be needed to determine the amount of current that should be provided for a specific application, Table 9.5 can be used to get a rough idea of the magnitude of the test current that may be needed. Take particular note that the table is not to be used for design purposes. All current values shown are subject to variation, depending on environmental conditions. It will serve, however, to show the wide variation possible.

TABLE 9.5 — Rough Current Requirements for Cathodic Protection of Steel[1]

Conditions	Current Required, ma/m^2	(ma/ft^2)
Bare steel in moving seawater	100-160	(10 to 15)
Bare steel in quiet seawater	55-85	(5 to 8)
Bare steel in earth	10-30	(1 to 3)
Poorly coated steel in earth or water	1	(0.1)
Well-coated steel in earth or water	0.03	(0.003)
Very well-coated steel in earth or water	0.003 or less	(0.0003) or less

[1]Approximate only—not to be used for design.

Variations in Environment

For soils, the variations include heavy, poorly aerated soils (little oxygen access to protected surface) as well as those that are loose, sandy, and well aerated. Metal in soil which is poorly aerated tends to be relatively easy to polarize. More current may be needed to polarize a structure in soil having ready oxygen access to its surface. Acid soil conditions also may make it more difficult to attain polarization. Both oxygen and the acidic condition tend to have a depolarizing effect by removing hydrogen in the polarization films as it is formed by the cathodic protection process.

In water, the amount of water movement can have a pronounced effect. Structures in quiet waters tend to take the least current for protection. Highly turbulent waters produce a strong mechanical depolarizing effect from scrubbing by the moving water. Furthermore, the turbulent water is apt to be highly aerated and thus have an added depolarizing effect.

Variations in electrical resistivity of the environment should be known. Where other factors permit, the lowest available soil resistivity is usually the most suitable for anode locations for either galvanic anode or impressed current rectifier systems.

Protective Coatings

If the structure to be cathodically protected is coated, its type and in-place condition should be known. Whether or not the coating has good stability is important. *Stability* refers to the coating's ability to maintain a high degree of electrical resistance with minimum deterioration with time.

Measurement of the in-place resistance to earth of a coated structure made immediately after installation can be misleading, particularly in soils. Measurements made before the backfill has settled around the structure and become thoroughly water-saturated can indicate a much higher resistance than will be the case after conditions have stabilized.

Electrical Shielding

Where an underground system to be protected involves groups of closely spaced metallic components (such as banks of underground cable or conduits), a shielding effect can occur. Current from a remote cathodic protection source can collect readily on outermost layers in the bank, but relatively little current (and perhaps an amount insufficient for good protection) may reach interior components. The outer layer of structures, then, comprises an electrical shield. Such conditions may require placement of cathodic protection anodes within the shielded area to force protective current to flow where it is needed.

Economic Considerations

The first consideration is whether or not the expense of a cathodic protection system is justified at all. Structure replacement could be the better choice in some instances. In other cases, the structure may not be continued in service long enough to justify the cost of cathodic protection unless hazardous conditions are a factor.

If cathodic protection is an economic solution to the corrosion problem, the type of cathodic protection system selected should be that having the least cost. In addition to the initial cost of design and installation of the system, the cost of power (if required) and the cost of maintaining the system (inspections, repairs, and replacements) should be considered. Depreciation schedules, the value of money, and the salvage value also must be recognized.

Metal to be Protected

Although steel is not adversely affected by a reasonable degree of overprotection, certain metals that may be cathodically protected can be damaged if protective potentials are too high.

Aluminum, for example, is sensitive to alkali attack. The cathodic protection process creates an alkaline environment at the surface of the protected structure. If the protective potentials maintained on aluminum are too high, the environment can become so alkaline that the aluminum will corrode faster than if it is not cathodically protected at all. This type of attack is called *cathodic corrosion.* If cathodic protection is used on aluminum, it is suggested that protective potentials be maintained within the range of −1.0 to −1.1, as measured to a copper sulfate reference electrode, but not allowed to exceed −1.2 volts in any case.

Life Requirements

The expected useful life of the structure to be protected should be known. Where practical to do so, the design life of the cathodic protection system should be coordinated with that of the structure to be protected. Specifically, there is no point in designing a cathodic protection system that will last for a longer time than needed. The additional cost for the longer life is money wasted.

Maintenance Capabilities

The type of cathodic protection system selected can be influenced by the availability and reliability of maintenance facilities and personnel. If there is reason to believe that they will not be adequate, the simplest, most foolproof cathodic protection design may be the better choice, even though the initial cost of installation is higher. This, under some circumstances, tends to favor galvanic anodes over impressed current systems which have stopped supplying protective current for one reason or another (such as blown fuses, ground bed failure, accidental cutting of a cable, etc.) and have been allowed to remain in this condition for extended periods—sometimes years. During these periods, no benefit is derived from the installation, and corrosion damage to the supposedly protected structure can and generally does result.

Galvanic anode systems can, of course, be damaged as well. If the design provides for this, however, it can be arranged so that damage suffered will affect only a small part of the protection system without, necessarily, loss of protection potentials elsewhere. Damage to impressed current systems is much more apt to cause complete loss of protection for relatively large structure areas.

Stray Current Effects

Before preparing a cathodic protection design, it must be known whether or not stray direct currents are present. Currents which will most severely affect the structure to be protected are apt to be those of substantial magnitude that continually vary in quantity. These include stray currents from DC railway systems, mining operations using DC power, DC welding operations, and similar sources.

Depending on the severity of the stray current effect, the magnitude of the anodic potentials (at areas where stray current is discharging from the affected structure and causing corrosion) may be such that they cannot be counteracted readily with the usual cathodic protection system. This is particularly true where, at areas of current discharge, the structure may be a matter of several volts positive with respect to its environment. Such conditions necessitate special stray current control techniques involving metallic bonds from the affected structure to the source of the damaging current or other means of removing the stray current without harm to the affected structure.

The stray current sources described earlier can result in very rapid corrosion which is usually much more severe than the corrosion caused by other environmental factors (Figure 9.16). Another type of stray current which is variable in nature may be observed during periods of "magnetic storm" activity. Long structures such as pipelines or cables are most apt to be affected. During magnetic storms, the intensity of the earth's magnetic field can vary. When these variations occur, potentials are induced in the pipe or cable in much the same manner as potentials are induced in an electric generator.

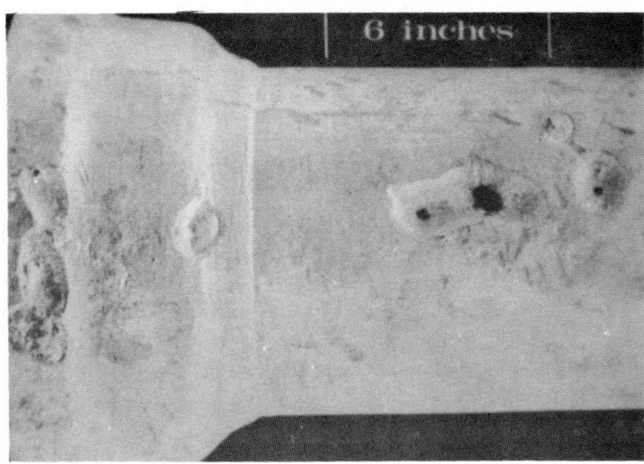

FIGURE 9.16 — Cast iron water main perforated by a large fault current.

Man-made variable stray current will usually demonstrate some pattern (as shown by recording instrument records), showing rush hour peaks on transit systems, shift changes in mining operations, etc. Variable stray currents resulting from magnetic disturbances, however, rarely, if ever, demonstrate any pattern. Further, they may be active in a given area for a time and then not appear again for a long time thereafter. Although sometimes intense for short periods, stray (telluric) current resulting from magnetic induction will seldom cause as much corrosion as uncontrolled man-made stray current because the currents are of relatively short duration and usually are not concentrated in a given area for an appreciable length of time.

In addition to the variable-type stray currents described, steady-state stray currents may be encountered. These may be caused by an impressed current cathodic protection system on an adjacent, but electrically separate structure if the ground bed is too close to the unprotected structure. Under such a condition, the affected structure passes through the potential gradient field around the ground bed in which the soil is positive with respect to the pipeline. This positive potential difference causes current to flow onto that part of the affected structure in the gradient field. Because there is no metallic path by which this current can flow back to the ground bed where the stray current originates, it must discharge into the earth from those parts of the affected structure that are outside the ground bed potential gradient field.

This situation may involve installation of metallic bonds between the affected structure and the structure to which the offending impressed current system is connected. In severe cases, moving the ground bed that is causing the stray current condition may be necessary.

When planning a cathodic protection system, the anode systems or ground beds should be located so that stray current from them will not be picked up by other structures and damage them. This is particularly important when using impressed current systems.

Figure 9.17 illustrates the general features of typical stray current conditions that may be encountered. Pipelines are used as a convenient means of illustration.

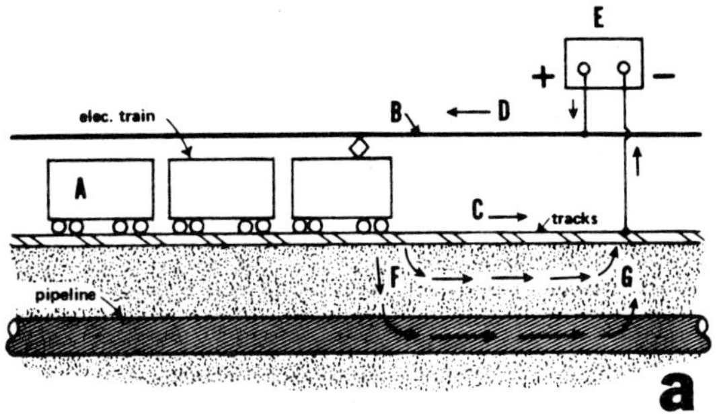

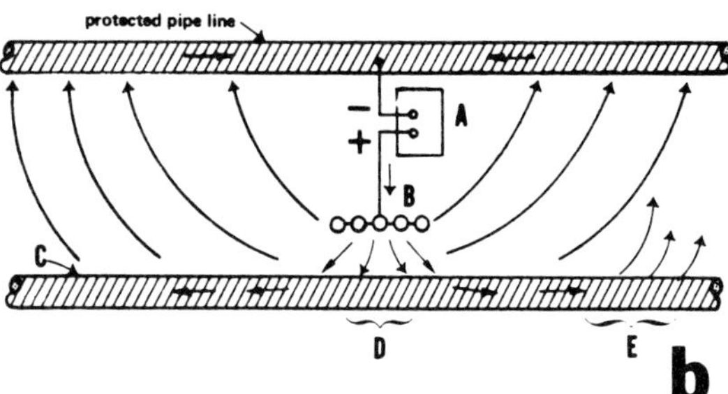

FIGURE 9.17 — Corrosion caused by stray current.

(a) Variable stray current from traction system: (A) train with electric motor prime movers; (B) overhead conductor; (C) tracks, return ground; (D) load current, variable; (E) DC substation supplying current to railroad; (F) part of load current flows from track to earth, metallic structures in earth path from train to substation will pick up current as shown and tend to be cathodically protected in pickup areas; (G) current discharging from pipeline to earth (corroding the pipe) in substation vicinity to return to negative terminal (ground) and complete the circuit.

(b) Steady-state current from nearby cathodic protection system: (A) rectifier for impressed current cathodic protection system; (B) ground bed; (C) pipeline electrically separated from cathodically protected pipeline; (D) pipeline picks up current from ground bed in this area and is protected; (E) current leaves line in this area to flow to protected pipeline to complete circuit. *Pipe corrodes where current leaves it.*

Temperature

The effect of temperature on cathodic protection design is primarily to change environment resistivity. Resistivities of soils and waters normally decrease as their temperature increases. The greatest change occurs with freezing. Resistivities of frozen soils or waters can be extremely high compared to those in the unfrozen condition. For this reason, measurements of the resistivity of frozen material would be meaningless for cathodic protection ground bed design purposes.

In marine work, the resistivity of seawater in tropical locations can be significantly lower than the same water in colder areas.

Ground beds should be installed below freezing depth in soils or waters where this may be a problem.

When measuring the resistivity of water samples from, for example, heat-exchanger water boxes which are to be protected, the measurement should be made at the normal water box operating temperature.

Anode Materials

Materials available to serve as galvanic or impressed current anodes have been described previously, as well as the sources of power for the impressed current system.

The selection of these basic elements in the cathodic protection design must be made on a sound economic basis, with proper attention to maintenance cost during the life of the system.

Wire and Cable

In impressed current systems, all underground or submerged cable from the positive DC terminal of the rectifier (or other power source) to the ground bed is at a positive voltage with respect to earth. For this reason, it must be very well insulated.

Remember that *all* buried parts of the ground bed assembly connected to the positive terminal of the power source may discharge current and corrode at any point where metal contacts the conducting environment in which the ground bed assembly is placed. This includes the cable from the rectifier to ground bed anodes and cable interconnecting anodes within the ground bed.

To prevent current discharge from the cables, wire having high-quality electrical insulation suitable for underground use must be provided, and all splices and connections must be insulated perfectly. At any defects in the cable insulation system, there will be current discharge that will corrode the wire until it separates, thus breaking the electrical connection between the rectifier and all or part of the ground bed installation.

By contrast, all buried wire connecting self-power-generating galvanic anodes to a protected structure are subject to current collection from the environment and are free of corrosion. Insulation on such wires is used to prevent their picking up unnecessary current. This also is true of the insulated cable extending from the negative terminal of the impressed current power source to the protected structure.

Cables from the negative terminal of impressed current systems to the protected structure, wires from galvanic anodes to the protected structure, and wires connected to the protected structure for testing purposes are all under protection and not subject to current discharge and corrosion if there are breaks in their insulation. It is a good practice, however, to use well-insulated cable and wire for such applications to prevent unnecessary consumption of cathodic protection current. The junction between a bare copper wire and a steel structure would create, unnecessarily, a strong dissimilar metal corrosion cell. Where such junctions are necessary, extremely careful and durable insulation from the environment is required.

Anode Backfill

Chemical backfill is used to surround galvanic anodes, in most cases, when they are buried in earth. There are several advantages to this. The special backfill provides a uniform environment surrounding the anode to promote uniform anode consumption and maximum efficiency. It isolates the anode material from direct contact with the earth, desirable to prevent action by soil chemicals which might otherwise cause high-resistance "passive" films to form at the anode surface and prevent discharge of useful amounts of current. It has a low electrical resistivity, so where the surrounding soil is of higher resistivity, this permits a lower anode-to-earth resistance and greater current output.

A typical backfill material used with magnesium anodes consists of 75% hydrated gypsum, 20% bentonite clay, and 5% sodium sulfate. A mixture of 50% molding plaster (plaster of paris) and 50% bentonite clay works well with zinc anodes. These backfill mixtures are available from suppliers.

For impressed current anode systems in earth, it is common practice to use a coke or graphite backfill to surround anodes. There are two advantages to this. One is that graphite or suitable coke has a very low electrical resistivity (lower than most soils) and helps to produce low resistance between anode and earth.

The second advantage is that if the coke or graphite is solidly packed around the impressed current anode, the greater part of the current discharged by the anode passes from the anode to the backfill through direct contact between the anode material and the backfill particles. This reduces the consumption rate of the anode. The current discharging from the outer surfaces of the backfill column will, however, consume the backfill material at a slow rate. The net result is a lower resistance, longer life anode installation than would exist if backfill were not used.

Suitable coke or graphite backfill materials are available from suppliers. Coke should be either coal coke or calcined (heat-treated) petroleum coke. Non-calcined petroleum coke can have high electrical resistivity and be unsuitable for ground bed backfill use.

Backfills customarily are not used with either galvanic or impressed current anodes in water immersion applications.

Miscellaneous

Electrical connections, particularly in galvanic anode systems, must be of very low electrical resistance to permit maximum current output. All underground or submerged connections should be welded, brazed, or soldered for permanent low resistance. This is also a good practice for impressed current systems and for test wire connections to protected structures.

Exposed metal at underground or submerged connections between impressed current anodes and cables to the power source must be perfectly insulated to prevent current discharge and accompanying corrosion.

Enclosed blocks suitable for terminating test wires from protected structures are available to facilitate periodic tests of the degree of protection attained.

Catalogs from suppliers of cathodic protection equipment and material give detailed information on the various components of cathodic protection installations.

Special Requirements for Specific Applications

Pipeline Applications

Most pipeline cathodic protection applications involve either galvanic anode or impressed current systems installed in earth for protection of external surfaces. Of the galvanic anode installations, most use magnesium as the anode material. Rectifiers are the most common source of DC power for impressed current systems.

For pipelines installed in the ocean bed (as for long harbor crossings and lines to offshore drilling operations), considerable use has been made of galvanic anode *bracelets*. These are essentially a ring of specially cast anodes encircling the pipe and attached directly to it. This permits having the anodes already attached to the pipe as it is laid. By so doing, the pipeline will be cathodically protected as soon as it becomes submerged. Used in conjunction with a good coating, sufficient anode material may be provided for long useful life. Zinc has been used most frequently for this type of installation.

Where surface soil conditions for pipelines on land are not suitable for ground beds installed near the surface, deep ground beds (vertical) may be installed if underlying earth resistivity is more favorable. Such ground beds usually are installed in a single hole which may be several hundred feet deep in some instances. Deep ground beds are used with impressed current systems. Particular care must be exercised during installation to avoid premature failures of anodes or anode leads that may not be repairable and may necessitate the installation of a complete new ground bed.

Other instances where deep ground beds are necessary include sites where right-of-way for surface ground beds cannot be obtained. For example, a deep bed can be installed on a pipeline right-of-way. They also are used in congested distribution systems where remote ground beds are needed, but where available sites for surface ground beds are not sufficiently remote from the pipes to be protected or from structures of other ownership.

In some congested areas, anodes (galvanic or impressed current) are distributed along the length of pipe to be protected. This permits placing the anodes close to the pipe, with each anode protecting a short length. The effect on other structures also may be controlled more readily. This type of installation may be more expensive than remote-type ground beds placed at much longer intervals, but may, nevertheless, be the best solution in some instances.

Where pipelines are banked, giving rise to severe shielding (as was illustrated by Figure 9.3), a continuous ribbon anode may be used within the bank and parallel to the pipelines to provide protective current within the bank. Such material is available in zinc or magnesium. For impressed current systems, platinum-coated wire or rod anodes are available if required.

Pipe Interiors

The interiors of large pipelines carrying corrosive liquids (such as seawater or industrial waste) may be lined with a suitable coating and protected with strip-type galvanic anode material. If the pipe interior is bare, relatively large amounts of current may be needed. In this case, an impressed current system may be used with platinum-coated anodes penetrating the pipe walls at intervals.

Fixed Structures in Seawater

Structures such as steel bulkheads, steel piles supporting piers or wharfs, offshore drilling platforms, and other similar structures may be protected with either sacrificial galvanic anode systems or impressed current systems.

Galvanic anode systems in seawater, for the most part, use much heavier anodes than those used in soil. This is because it is possible to obtain high current from them in a low-resistivity seawater environment, and because the weight is needed to provide reasonably long life.

For structures that can be polarized, a low-potential galvanic anode material (zinc or aluminum) is generally preferable to a high-potential material (magnesium). Magnesium will work perfectly well, but may discharge more current than is needed. The metal furnishing this wasted current is expensive.

Zinc or suitable aluminum alloy anodes can polarize a steel structure in seawater to within a few millivolts of the potential of the anode itself. Assume that potential is 50 millivolts, although it can be less. This means that the anode current will stabilize at a value sufficient to maintain polarization.

If magnesium anodes are used, however, the structure does not tend to polarize beyond about −1.2 volts to a copper sulfate electrode because the hydrogen overvoltage potential is reached and this results in the evolution of free hydrogen. This means that with a magnesium working voltage of about −1.45 volts to a copper sulfate electrode, there will be a driving voltage of approximately 0.25 volt (250 millivolts). Thus, on a comparable basis, the magnesium will discharge about five times as much current as is actually required to achieve polarization.

Because lower potential anodes will give full cathodic protection (*i.e.,* polarize the surface), the surplus current from the higher potential anode is, in effect, wasted. There is one compensation, however. The higher magnesium voltage tends to deposit a thicker calcareous coating (chemicals deposited from the seawater) on the structure surface than is obtained with the lower potential anodes. This thicker calcareous coating has protective value.

Special chemical backfills are not needed in the uniform seawater environment because galvanic anodes work satisfactorily without them.

Impressed current systems for fixed seawater structures may use suitable anode materials, without backfill, suspended in the seawater from the structure being protected or placed on the ocean floor. Treated graphite, high-silicon cast iron, platinized titanium, or lead-silver are used. Particular attention must be given to the design of the rectifier placement, header cable distribution system, and anode suspension or placement details. Abovewater parts of the system are subject to severe marine atmospheric attack, while other portions must be protected from (or designed to withstand) the mechanical forces exerted by moving seawater, as well as by water-carried debris or shipping traffic.

This chapter's bibliography contains a reference which gives the suggested current densities required to protect steel in various seawater exposures.

Marine Craft

Cathodic protection systems on hull exteriors must be capable of a wide range of current output. As a rule, a craft tied up at dockside will require much less current to keep the hull in a polarized condition than will be needed when the same craft is underway. This is because of the mechanical depolarizing effect of the rapidly moving seawater.

Furthermore, a craft which is fresh from drydocking after application of a new paint system initially will require relatively little current. As the paint deteriorates in service, current requirements may increase manyfold. This is particularly true with a craft that may encounter sand bars which remove paint rapidly through abrasive action.

Galvanic anodes (zinc and aluminum) are used to a considerable extent on ship hulls and are satisfactory if sufficient anode material is provided. Current output of the anodes automatically increases as the need for it increases. Design problems involve providing sufficient anodes to provide maximum-expected current demand. Enough anode weight must be provided to give reasonable life. This is usually matched to the normal interval between drydocking, or some multiple thereof.

Impressed current systems use hull-mounted anodes and some means of automatically adjusting the power source DC output to meet varying requirements. Present practice favors the use of potential-controlled automatic rectifiers. These devices are so arranged that the rectifier current output to hull anodes is automatically raised or lowered through control circuitry as the need changes. The control circuitry is guided by the potential between the hull and permanently installed through-hull reference electrodes placed at critical locations.

In tankers, tanks which are filled with salt water for ballast after discharging cargo are susceptible to severe corrosion. Galvanic anodes are used widely to prevent corrosion during the ballast period. Where there is an explosion hazard (as with tankers in petroleum products service), the use of impressed current cathodic protection systems would impose critical safety problems such as accidental short circuits and sparking. Although there is no probable spark hazard in galvanic anode circuits, an anode which broke away from its mounting and fell might (as it struck the steel tank bottom) create a spark with sufficient energy to ignite an explosive mixture. Zinc would appear to be the least likely to cause such sparking.

Marine Growth

The fouling of surfaces exposed to seawater can affect both corrosion and the application of cathodic protection. Very high applied current densities (much higher than needed for protection) will kill marine growth, while lower values seem to have only slight effects. The current can pass through the bodies of the organisms and protect the metal surface; it is even possible to deposit cathode scale beneath attached barnacles by this process.

Tanks

Tank exteriors normally may be handled by conventional cathodic protection systems. Tank interiors may be protected when they contain waters or other conductive fluids which are not strongly acidic. If tank interiors are well coated, galvanic anodes may be a good choice. If they are bare, impressed current systems may be used with anodes mounted inside the tank. Anode leads are normally brought through the tank wall to a rectifier or other source of DC power.

Industrial Plant Underground Networks

Many product manufacturing plants, refineries, electrical generating stations, etc., have complex networks of underground structures. Dissimilar metals are frequently involved, and where soil resistivity is low and corrosive, cathodic protection can provide substantial savings.

In complex networks, there usually is a considerable amount of shielding, so that a simple cathodic protection system using a remote ground bed may not permit sufficient current to reach structures within the shielded area. In such instances, cathodic protection systems (either galvanic anode or impressed current) using anodes distributed throughout the shielded area may be required.

Where impressed current systems are used in areas subject to explosion hazards (such as petroleum refineries), explosion-proof wiring systems and rectifiers must be used.

Piles and Well Casings

Corrosion on steel piles driven into earth tends, in most cases, to be concentrated near the upper end. This is particularly true where the piles are driven through fill or otherwise previously disturbed earth. Lower areas are generally much less susceptible to corrosion, as demonstrated by National Bureau of Standards work,[5] unless it can be shown that there are unusual subsurface conditions or a stray current situation which can justify cathodic protection.

If conditions are such that the upper ends require cathodic protection (and if many piles are involved), a distributed anode system (galvanic or impressed current) may be required. Such a system will concentrate the applied current at the upper pile ends and, if designed properly, will avoid shielding effects.

Piles used in marine structures are most susceptible to corrosion by the effects of the salt water to which they are exposed. Under normal conditions, areas where worst corrosion on such piles can be expected are as follows:

1. Just below the mudline (the point where the pile enters earth below the salt water).
2. Just below mean low water (in the continuously immersed portion of the pile).
3. In the splash zone (where wind and wave action wets the pile surface in the area above the mean high water level).

Usual practice is to concentrate cathodic protection current on those areas of the pilings contionuously immersed in salt water. This will take care of Conditions 1 and 2 above. Condition 3 is not helped much by cathodic protection because there is an insufficient amount of seawater electrolyte to carry protective current to the affected areas. Suitable protective coating systems are the best approach to corrosion control in splash zone areas.

Well casing exterior surfaces, being much longer than the usual pilings, frequently extend into subsurface zones where corrosion may be a problem. A technique which has been used widely for such well casings involves a conventional cathodic protection installation at the wellhead, with the current required for protection being determined by the *E-Log I technique*. This procedure involves applying test current, using a temporary ground bed, starting with a small amount of current, and increasing the test current by increments. At each increment of current, the well casing is allowed to

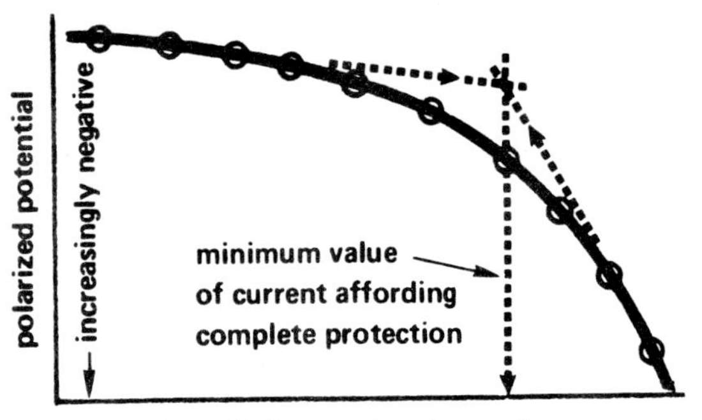

FIGURE 9.18 — Typical E-log I cathodic protection current requirement test.

polarize and the polarization potential and current are measured. Polarization potential (E) is plotted against the logarithm of the test current (log I) for each point, and the test is continued until a break in the curve is obtained. The current flow matching the break in the curve is the approximate current needed to stop corrosion, as indicated by this method.

The appearance of such a curve is illustrated by Figure 9.18. The current required to protect a well casing (which is, in effect, a vertical pipeline) usually proves to be considerably less than that needed for a horizontal pipeline of equivalent diameter and length buried at normal depth. This is because of the relative inaccessibility of the major portion of the well casing to oxygen compared to the accessibility of the near-surface pipeline.

Other Uses of Cathodic Protection

Limiting Damage by Cavitation

Numerous experiments have proven, and widespread practical application has confirmed, the fact that cathodic protection can help reduce the damage caused by cavitation. Cathodic protection of surfaces exposed to this effect (caused by the creation and collapse of microscopic vacancies in rapidly moving volumes of water, *e.g.,* on the surface of ship propellers) has been beneficial over a wide range of materials, velocities, and corrosive waters.

It is a common practice, for example, to affix anodes to the hulls of ships in the vicinity of propellers and to apply cathodic protection to the surfaces of the propellers themselves. It is assumed by most experimenters that cathodic protection helps under cavitation conditions by limiting the thickness of corrosion films on the affected surfaces.[6] These films are dislodged by the impact of water from collapsing voids, exposing fresh metal to corrode anew. When corrosion is reduced, less film is exposed to be eroded by the cavitation effect.

Crevice Corrosion Damage Reduced

Some materials susceptible to crevice corrosion, including Monel, certain stainless steels, mild steel, and others, display marked benefits from cathodic protection's effectiveness in limiting the extent and severity of this form of attack. Even at relatively high densities (30 mA/sq ft), crevice corrosion was reduced on Monel and Types 302 and 304 steels.[7] This could be important in such installations as desalting plants where large volumes of water are handled.

Cathodic protection also has proven beneficial in preventing or reducing the incidence of stress corrosion cracking in some materials. Some tests show that cathodic protection applied after stress corrosion cracking has started will bring the cracking to a halt.[8]

Cathodic Protection of "Tin" Cans

A considerable part of the protection afforded to food preserved in metal cans depends on the cathodic protection of the steel can provided by the tin coating. The protection conferred by the tin is effective in many acid and alkali environments and appears to be a function of thickness. Without the protection of the tin, the steel in the cans would be attacked by the contents of the can, ruining it either by contamination with corrosion products or by generation of hydrogen.

Domestic Hot Water Heaters

Magnesium (in the form of anodes) is used to protect the inside surfaces of glass-lined domestic water heaters. The anodes provide current to help polarize the defects in the glass coatings of the heaters and greatly extend their life.[9] Impressed currents are applied to these and to commercial hot water storage tanks to cathodically protect them.

Other Applications

Among numerous other installations of cathodic protection producing beneficial results are systems on the steel reinforcing wires of prestressed concrete pipe,[10] reinforcing steel in reinforced concrete structures,[11] and such things as brewery bottle washing machines[12] and the exposed metal underwater parts of wooden-hulled boats.[13] In each case, careful study of the corrosive environment, the areas and types of metals to be protected, and possible side effects need to be weighed.

Care needs to be taken, for instance, in applying cathodic protection to stressed steel under some conditions because it may cause hydrogen embrittlement (Chapter 6) and catastrophic failure. In the case of wooden-hulled boats, for example, it is necessary to avoid applying excessive current because the hydrogen generated by the excess current (as explained earlier) will create a highly alkaline condition near the metal parts which will destroy the wood.

An interesting use of the "decay" time delay of a cathodic protection current applied to a steel surface involves protection of a large paper making machine drum. Cathodic protection of this drum reduced its corrosion (which made it necessary to shut down the machine in order to resurface the drum) and extended the time the machine could operate without requiring resurfacing of the drum.[14]

References

1. Jakobs, J. A., A Comparison of Anodes for Impressed Current Systems, Materials Performance, Vol. 20, No. 5, p. 17 (1981).
2. NACE, Control of External Corrosion on Underground or Submerged Metallic Piping Systems, Recommended Practice RP-01-69, Houston, TX (1976).
3. Peabody, A. W., Control of Pipeline Corrosion, NACE, Houston, TX (1967).
4. Compton, K. G., Effect of A-C on Corrosion of Concentric Neutrals of URD Cable, Materials Performance, Vol. 21, No. 3, p. 31 (1982).
5. Romanoff, M., Corrosion of Steel Pilings in Soils, J. Res, N.B.S., Vol. 66C, No. 3, pp. 223-224 (1962).
6. Preiser, H. S. and Tytell, B. H., The Electrochemical Approach to Cavitation Damage and Its Prevention, Corrosion, Vol. 17, No. 11, pp. 535-549 (1961).
7. NACE, Effectiveness of Cathodic Currents in Reducing Crevice Corrosion and Pitting of Several Materials in Sea Water, Corrosion, Vol. 8, No. 2, pp. 50-56 (1952).
8. Logan, Hugh L., The Stress Corrosion of Metals, John Wiley and Sons, Inc., New York, N.Y., p. 16.
9. Fischer, H. C., Solving Design Problems for Cathodic Protection of Glass-Lined Domestic Water Heaters, Corrosion, Vol. 16, No. 9, p. 17 (1960).
10. Unz, M., Cathodic Protection of Prestressed Concrete Pipe, Corrosion, Vol. 16, No. 6, pp. 289-297 (1960).
11. Cornet, Israel and Bresler, Boris, Corrosion of Steel and Galvanized Steel in Concrete, Materials Protection, Vol. 5, No. 4, pp. 69-72 (1966).
12. Hale, D. W., Horton, C. C., and Michelson, J. A., Cathodic Protection Frequently Practical in Corrosion Control of Brewery Vessels, Corrosion, Vol. 13, No. 1, pp. 134-136 (1957).
13. NACE, Small Boat Industry Concerned About Corrosion, Materials Protection, Vol. 5, No. 5, p. 85 (1966).
14. Hamner, N. E., Applying Cathodic Protection to a Paper Making Machine, Materials Protection, Vol. 8, No. 2, pp. 48-49 (1969).

Bibliography

Deringer, W. A. and F. W. Nelson. A Field Investigation of Cathodic Protection in Glass-Lined and Galvanized Water Heaters. Corrosion, Vol. 8, No. 2, pp. 57-64 (1952).

NACE. Application of Organic Coatings to the External Surface of Steel Pipe for Underground Service. NACE Standard RP-02-75. Houston, TX (1975).

NACE. Current Density Requirements for Cathodic Protection of Steel in Sea Water. Materials Performance, Vol. 21, No. 2, p. 21 (1982).

Williams, Jean. Bibliography on Underground Corrosion. Part 1. Materials Performance, Vol. 21, No. 11, p. 40 (1982).

Chapter 10

Corrosion in Soils

CORROSION IN SOILS

Introduction

As was mentioned in previous chapters, various metals have varying tendencies to corrode, as illustrated by their relative positions in the Galvanic or EMF Series. Since this chapter will center largely around iron, it is well to note that iron is intermediate in these lists. Furthermore, it has been shown that pure iron, wrought iron, or mild steel behave about the same in underground situations. tions.

It also has been observed that iron pipe buried in bone dry soil suffers little or no corrosion. However, because of rain, natural springs, rivers, etc., soils are rarely dry, especially beneath the surface, so that real cases must be considered. Even in dealing with the simple case of moisture from natural rain, complex solutions similar to well water, river water, etc., could be encountered. All soils contain a variety of mineral matter, some of which is soluble to a greater or lesser degree; thus, we must deal with a variety of solutions.

If there were no minerals to dissolve in rain water, corrosion by soils would be almost nil. For instance, iron in pure water does not corrode. When oxygen and other substances are dissolved in it, however, the situation becomes much more complex.

An explanation of corrosion behavior under a variety of conditions is not easy to give; however, one important factor to consider is the electrical conductivity of the medium which surrounds the metal being corroded. Pure water is a very poor electrical conductor, but as substances dissolve in it, particularly those which ionize, conductivity rises sharply. The relationship between conductivity and corrosion, which will be discussed later, is a very logical one.

Electrolytes, Ions, and Dissociation Phenomena

Electrolytes

Let us review some of the principles learned in Chapter 2. It was stated earlier that pure water is not a good conductor of electricity, that water with certain substances dissolved in it is a good conductor, and that its conductivity varies over a wide range. It is thus important to examine the way in which the electric current flows through a water having good current carrying characteristics.

There are two principal ways in which electricity can flow; these are known as *metallic conduction* and *electrolytic conduction*. In metallic conduction, the energy is transferred from one metal atom to another—a process which takes place at a high speed. In electrolytic conduction, however, the energy is transferred by the actual movement through a liquid of charged particles called *ions*. Although the individual particles do not move very rapidly, the impulse travels through the water at a much higher speed—something like a wave traveling along a wire.

There are a number of differences between these two types of conduction. Metallic conduction, at low levels, does nothing to the metal except cause a slight heating effect. If this does not produce a high enough temperature to damage the metal, then there is no permanent change at all. On the other hand, the flow of current through an electrolyte produces many changes in the electrolyte itself and at the two terminals (which are called *electrodes*) by means of which the current enters and leaves the solution. One of these changes may be corrosion.

Ionization

The conduction of current by the movement of ions was referred to earlier. When any one of a large number of chemical compounds (but by no means all) is dissolved in water, the compound tends to break up, usually into two parts. Common salt, for example, is sodium chloride—each molecule of salt is made up of one atom of sodium and one atom of chlorine. In the undissolved form, this molecule is electrically neutral. When salt is dissolved in water, however, the two atoms separate and each atom develops an electric charge; the sodium atom will have a positive charge and the chlorine atom a negative charge. A charged atom is known as an *ion*.

The chemical formula for sodium chloride is NaCl; the Na stands for one sodium atom, the Cl stands for one chlorine atom. The sodium ion, on the other hand, is written as Na^+ and the chloride ion as Cl^-. (Note that the ion is called chloride instead of chlorine.) The chemical equation which shows the dissociation of sodium chloride can be written as:

$$\underset{\text{Sodium Chloride}}{NaCl} \rightleftharpoons \underset{\text{Sodium}}{Na^+} + \underset{\text{Chloride}}{Cl^-} \qquad (10.1)$$

The double head on the arrow signifies something important; the action is reversible, *i.e.*, when salt is dissolved in water, the molecules immediately start separating and recombining. At any given instant, in the case of salt, the great majority of molecules will be separated and will exist in the form of ions.

Copper sulfate, for example, has the formula $CuSO_4$. This means that a molecule is made up of one atom of copper, one atom of sulfur, and four atoms of oxygen. When this compound dissolves in water, it ionizes as does the salt, but it dissolves into two ions, as follows.

$$\underset{\text{Cupric Sulfate}}{CuSO_4} \rightleftharpoons \underset{\text{Copper}}{Cu^{++}} + \underset{\text{Sulfate}}{SO_4^{=}} \qquad (10.2)$$

This means that one ion of copper is formed, carrying a double positive charge, and one sulfate ion, carrying a double negative charge. This particular copper ion, with two positive charges, is known as the *cupric* ion. A similar one carrying only one such charge is known as the *cuprous* ion. Other compounds are even more complicated.

The compounds cited in these examples dissociate or ionize almost completely, *i.e.,* at any given moment there are only small percentages of whole molecules in the solution, the majority being in the form of ions. Compounds with this characteristic are known as good electrolytes. Others, like acetic acid, which dissociate to a much smaller extent, are considered to be poor electrolytes, while some, like sugar, which show almost no dissociation at all, are nonelectrolytes.

Water dissociates in this way:

$$\underset{\text{Water}}{H_2O} \rightleftharpoons \underset{\text{Hydrogen}}{H^+} + \underset{\text{Hydroxyl}}{OH^-} \qquad (10.3)$$

The first of these is known as the hydrogen ion and the second as the hydroxyl ion. Note that the hydrogen ion has a + charge, just like a metal ion. The dissociation of water, unlike that of the salts which have been mentioned, involves only a small number of the molecules present. At any instant, water is nearly all water. Very little of it is in the form of ions. This helps explain why pure water is a poor conductor, while a water solution is often a good conductor.

How Do Metals Corrode?

Insofar as corrosion in soils and waters is concerned, metals corrode largely by the formation of cells. A more complete descriptive term would be electrolytic cells. An *electrolytic cell* has four basic elements: an electrolyte, two electrodes, and a connecting circuit.

The *electrolyte* is a solution of a substance (which ionizes) in water. It does not need to be liquid. Ordinary soil is an electrolyte, unless it is absolutely free from water. Rock is also an electrolyte. Even granite has a small, but measureable, content of water and thus is considered to be an electrolyte. Concrete is an electrolyte, and sometimes it is a very good one. In other words, when nature needs an electrolyte in order to corrode a metal, it is seldom difficult to find.

The two electrodes a cell must have are usually both metal. Sometimes another substance forms an electrode or, very rarely, both electrodes. Carbon, for example, is a commonly encountered electrode, both in accidental cells and in those deliberately made by man. The two electrodes of a cell do not have to be of different metals. In some of the most important corroding cells, as will be discussed later, both electrodes are of the same metal.

The connecting circuit may take many forms. All that is necessary is a path by which current can flow from one electrode to the other, in addition to the path through the electrolyte. The many different kinds of cells differ in several respects, *e.g.,* electrolytes, electrodes, connecting circuits, etc. Some of the more important cells, from the corrosion viewpoint, are described as follows.

Galvanic Cells

Dissimilar Metal Cells

A *galvanic cell* may be one in which the two electrodes are of different metals. A familiar one is the common flashlight cell (Figure 10.1) where zinc and carbon are the electrodes. The electrolyte is an acid paste containing a number of compounds which serve to keep the cell active. Such a cell is the result of a number of years of intensive development and is not nearly as simple as it first appears.

For instance, where is the connecting circuit? As it exists on the shelf, there is none, and thus the flashlight cell is not really a functioning cell under these conditions. Even when it is put into your flashlight, there is still no available path for current to

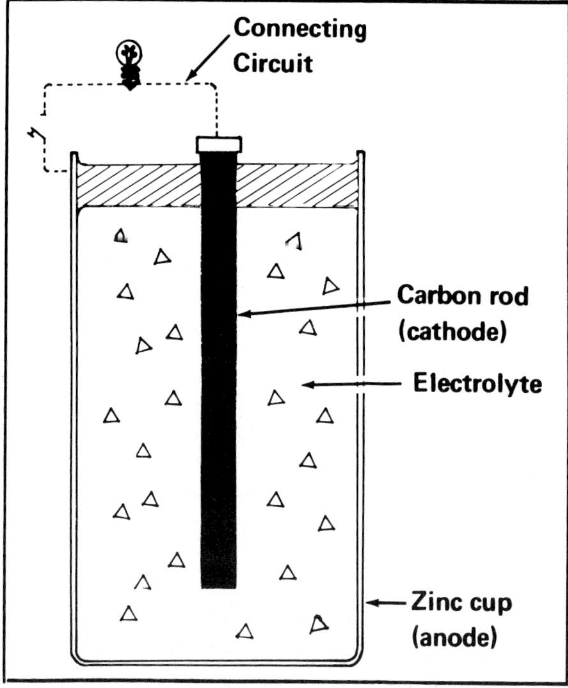

FIGURE 10.1 — Schematic of typical flashlight chemical cell.

flow until you turn the switch on. Then, with the establishment of a connecting circuit, not only does a working cell exist, but galvanic corrosion finally begins.

It is the zinc in this type of cell that corrodes. In time, in the old cells (and in some of the cheaper ones today) the zinc will develop a hole, and the acidic contents will leak out and cause damage. While today's cells are far less likely to do this, it does not alter the fact that the zinc corrodes.

In the case of a flashlight battery, the current runs from the carbon to the zinc through the connecting circuit. In any cell, the electrode from which current flows *to* the electrolyte is known as the *anode.* The electrode to which current flows *from* the electrolyte is known as the *cathode.* In a flashlight cell that is in use, the zinc is the anode and the carbon is the cathode. *The anode is the one which is corroded.* Only under a few unusual circumstances are there exceptions to this statement. The anode corrodes and the cathode usually does not.

Any two metals can be used to make a galvanic cell. In many cases, the voltage of such a cell can be estimated to some extent by the difference in the potentials of the two metals, as shown in the Electromotive Force Series of Chapter 2. Whether a metal will behave as an anode or a cathode in combination with another metal in the soil, can usually be determined by its relative position on the Galvanic Series in Seawater shown in Chapter 2. The metal which appears higher up on the list (more negative) will, in general, be the anode and will thus corrode. The metal lower down on the list (more positive) will be the cathode and thus will not corrode. Of course, this galvanic action will not take place under open-circuit conditions; there must be a connecting circuit.

Do galvanic cells occur often? Are they important as sources of corrosion? The answer to both questions is yes, although galvanic cells are not the *most* important kind of corroding cell. Whenever a copper pipe service line is directly connected to a cast iron gas or water main, a galvanic cell is formed (Figure 10.2). The soil is the electrolyte, the copper service line is the cathode, the iron (or steel) main is the anode, and the connecting circuit is completed by attaching the line to the main. Normally, such cells do not do any great amount of damage, because the anode (corroding) is so much larger than the cathode that the attack is spread out over a large area (refer to "Area Effects" in Chapters 2 and 15).

Another common galvanic corrosion cell, which also happens to involve copper and steel, is the one formed by the connection of ground rods and ground cables, traditionally of bare copper, with the steel piping, reinforcing rods, building columns, and similar structures around an industrial plant (Figure 10.3). Here again, the damage usually is not severe because the anode (steel pipe, etc.) usually is much larger than the cathode (copper cable and rods).

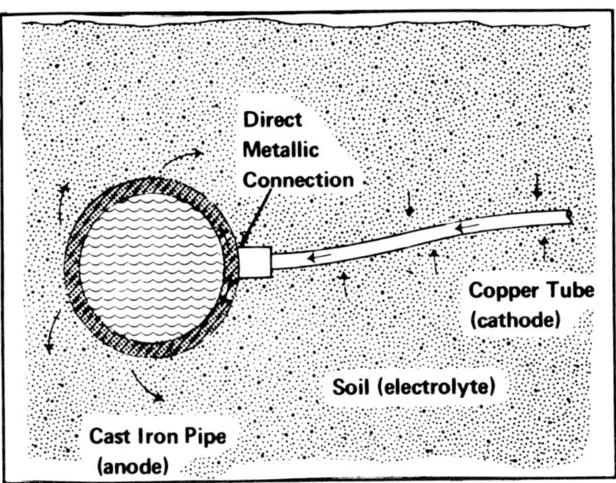

FIGURE 10.2 — Underground corrosion cell involving connection of dissimilar metals.

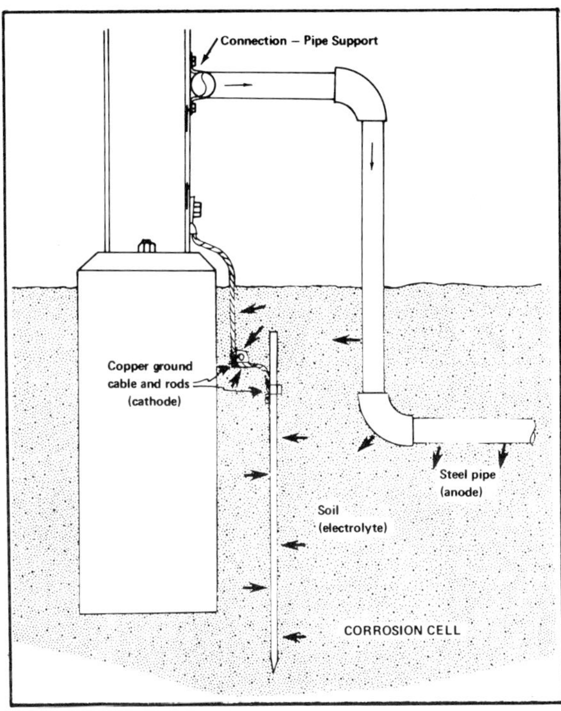

FIGURE 10.3 — Underground corrosion cell involving grounding rods and steel pipe.

Concentration Cells

A cell also may be formed when two electrodes of the same metal are connected while lying in electrolytes that contain different substances or the same substance in different amounts. For example, if one electrolyte is a dilute salt solution and the other a concentrated salt solution, a *concentration* cell may be formed. This takes place because one of the factors which determines an electrode potential is the electrolyte concentration. A similar cell is formed when the two electrolytes contain entirely different dissolved substances.

Predicting the behavior of such a cell is not always possible. When the salt involved is a simple one, and is not a salt of the metal which makes up the electrodes, the electrode which lies in the more concentrated solution may be the anode, and is thus corroded. When the salt involved is a salt of the

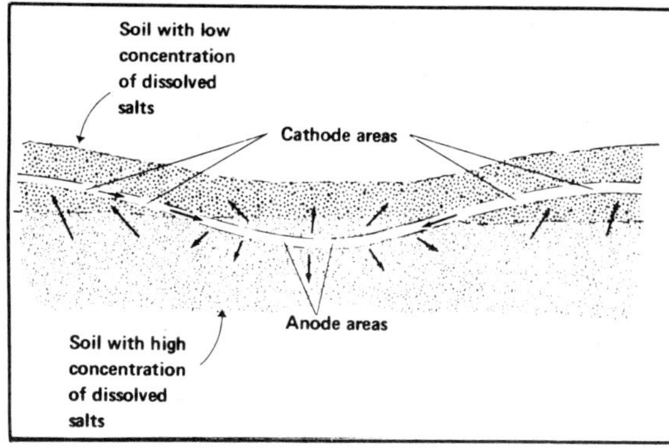

FIGURE 10.4 — Underground corrosion cell caused by different chemical concentrations.

metal, and no other salts are present, then the one in the dilute solution usually is the anode. Fortunately, we are seldom called upon to predict the behavior of such a cell in advance. Usually, we merely have to recognize it when it occurs.

Concentration cells are responsible for much of the corrosion which takes place in soils (Figure 10.4). Pipelines, for example, pass through different soils on their way across the land. Oil and gas well casings penetrate various strata of different compositions. In every case where two different soils are in contact with a single piece of metal, a concentration cell is possible.

In most cases in naturally occurring soils, that portion of pipe lying in the more conductive soil is the anode; that in the less conductive soil is the cathode; moist soils themselves act as the electrolyte (in this case, a compound electrolyte); and the pipe itself furnishes the connecting circuit. The current flows from the anodic area to the soil, then through the soil and from it to the cathodic area, and then along the pipe to the anodic area, as shown in Figure 10.4.

The dissolved salts in such a case are numerous and varied. They usually include compounds of aluminum, calcium, magnesium, and other metals, and they may be sulfates, chlorides, hydroxides, or any one of quite a variety of other compounds. Seldom do we need to know which salts are involved, because the activity is predictable from other indications and the remedies are the same no matter what the substances are.

Oxygen Concentration Cells

If two electrodes of the same metal are placed in a solution capable of acting as an electrolyte, and if there is a difference in the amount of oxygen reaching the two electrodes, a potential difference will develop between them. This is capable of operating a corroding cell. Such a cell is known as an *oxygen concentration cell,* a *differential oxygen cell,* or a *differential aeration cell.*

This type of cell can be made by putting two small pieces of copper in salt water, connecting them to a sensitive voltmeter, and then bubbling air or oxygen around one of them (Figure 10.5). In

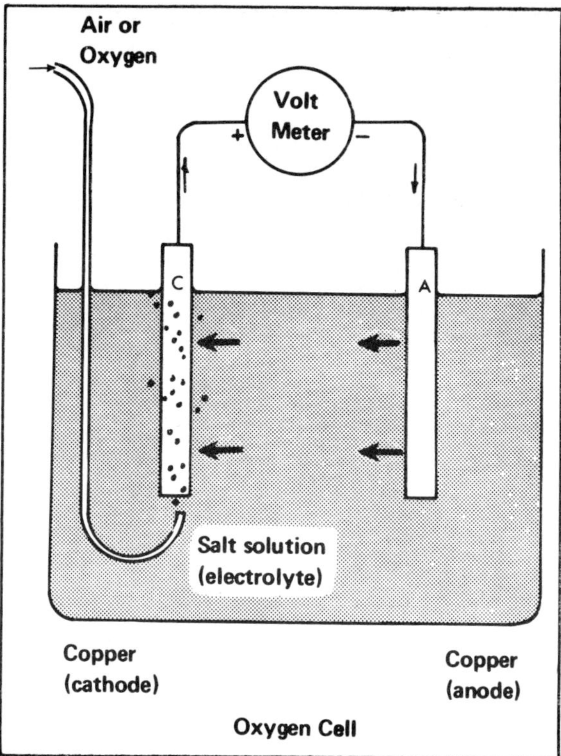

FIGURE 10.5 — Laboratory-type corrosion cell resulting from differential oxygen concentration.

almost all cases, such a cell will have for its anode the one which does not get oxygen, while the one with the plentiful oxygen supply becomes the cathode.

Such cells are very common on buried pipes, perhaps the most common cell found. For example, the pipe usually rests on undisturbed soil at the bottom of the ditch. Around the sides and on top of the pipe is relatively loose backfill which has been replaced in the ditch. Because the backfill is more permeable to oxygen (and the path is shorter) diffusing down from the surface, a cell is formed (Figure 10.6). The anode is the bottom surface of the pipe and the cathode is the rest of the surface. The elec-

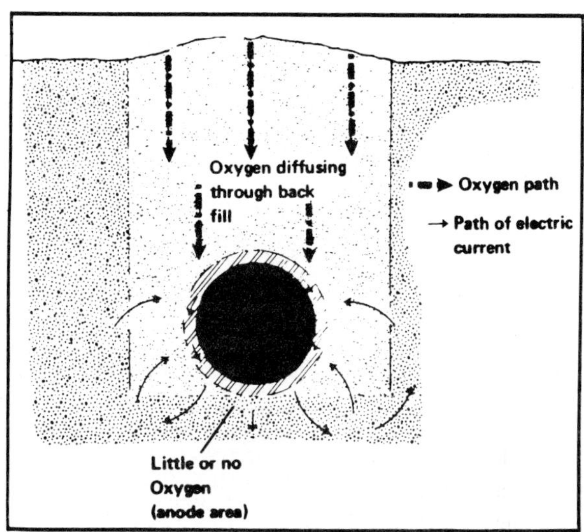

FIGURE 10.6 — Oxygen differential cell on pipeline in ditch.

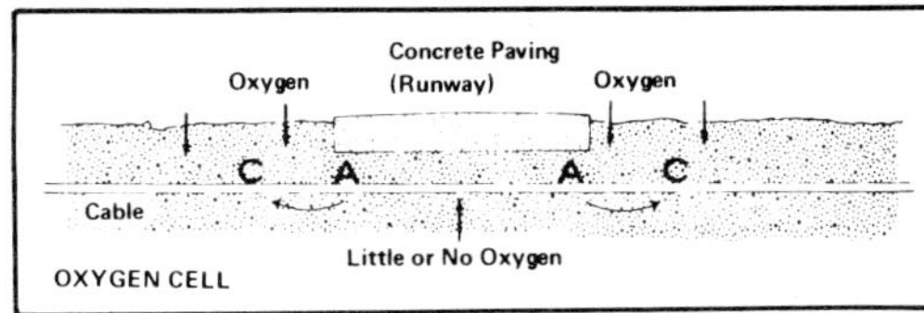

FIGURE 10.7 — Oxygen differential cell resulting from burial under paving: C = cathodic area; A = anodic area.

trolyte is the soil, and the connecting circuit is the pipe metal itself. This is the explanation for the fact that most corrosive attack on pipelines is on the bottom quarter of the pipe.

When a pipe or cable passes under paving, such as an airport runway, parking lot, or street (Figure 10.7), the portion under the paving has less access to oxygen than does that which lies under unpaved soil. Thus, a cell is formed: the anode is the pipe under the paving, the cathode is the pipe outside the paving, the electrolyte is the soil, and the connecting circuit is the pipe or cable. Although the entire length of pipe under the paving is anodic, most of the attack will take place not far from the edge. Because the path through the electrolyte is shorter to this part, most of the current takes this low-resistance path.

The casing of an oil or gas well usually is connected to a network of surface piping, *i.e.*, a gathering system which lies just below the surface. This can form an oxygen cell: the casing, at depth, forms the anode; the surface piping, with greater access to oxygen, becomes the cathode; again the soil is the electrolyte; and the piping and casing together make up the connecting circuit. The action of this cell can be controlled through insulating the surface piping from the casing by means of a special insulating union or other fitting. Note, however, that this will not stop all attack, due to the concentration cells described in the preceding section.

Oxygen concentration cells are responsible for much of the corrosion near the waterline on piling driven into the bed of the sea or in fresh water bodies. Wave action maintains a constant supply of oxygen to the metal just below the surface, while at greater depths there is little oxygen penetration.

Temperature Cells

The two cells just described, concentration and oxygen, are responsible for perhaps 90% of the corrosion in soils and much of that in natural waters. There is still a great deal of corrosion from other types of cells, however, so that additional types of cells cannot be ignored.

One such cell is the *temperature cell* (Figure 10.8). In it, the two electrodes are of the same metal, but one is maintained at a higher temperature than the other by some external means. In most cases, the electrode at the higher temperature becomes the anode.

A gas transmission line emerging from a compressor station serves as an example. The just-

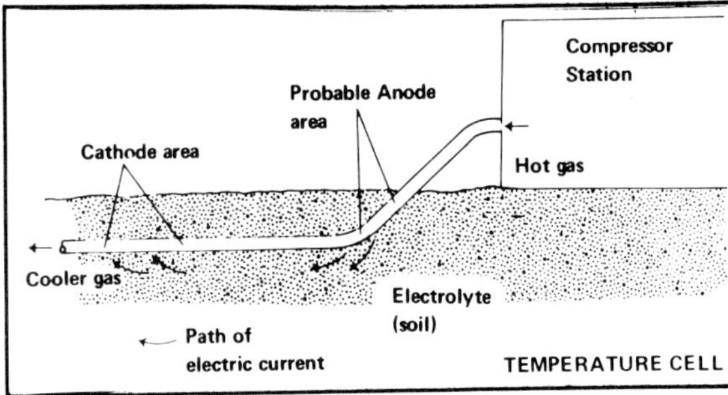

FIGURE 10.8 — Concentration cell resulting from heat differential.

compressed gas is hot, and, as it travels down the line, it loses heat by transfer to the surrounding soil (as well as losing some by expansion). The result creates a cell: the hot pipe near the compressor is the anode; the cooler pipe down the line is the cathode; the soil is the electrolyte; and the pipe itself is the connecting circuit. This is a particularly unfortunate type of cell because the high temperature just outside the compressor tends to damage the coating, so that the part of the line which is the anode also is likely to have the poorest coating.

Oil and gas well casings also experience similar cell attack. The pipe at some depth below the surface is at a higher temperature (this is a natural phenomenon; temperature increases with depth) and thus becomes the anode, the cooler pipe near the surface and the surface piping make up the cathode, the soil is the electrolyte, and the pipe is the connection. Note that this cell operates in the same direction as the oxygen cell described earlier; both tend to concentrate the corrosion on the casing at depth.

Impressed Current Cells

In all cells described so far, the source of the energy which makes the cell active has been within the cell. The source has been the solution potential of the anode (the amount by which it is greater than that of the cathode), some distribution of energy in the metal, or a concentration gradient in the electrolyte.

It is possible to have a cell in which the energy is supplied in the form of electrical current from an outside source. An electroplating cell is one example. In this case, energy comes from a generator supplying direct current. A storage battery which is being charged is another example. The power in this instance comes from the car's generator and actually forces current through the cell in a direction opposite to the driving voltage of the cell itself (the battery is really a set of three or six cells).

In a cell of this kind, the anode is still the electrode, from which current flows to the electrolyte, and the cathode is the one to which current flows from the electrolyte. In a storage battery, the "corrosion" which takes place at the anode is a reversible reaction: When the battery is discharging, this electrode becomes the cathode, and the chemical action is the reverse of that taking place during "charge."

Impressed current cells in which the soil is the electrolyte are of two kinds: accidental and deliberate. The deliberately impressed current system is the one which supplies cathodic protection to a threatened structure. Accidental systems, however, may exist under a variety of circumstances. Any direct current flowing in the soil from any source whatsoever can, if it finds a pipeline or other metal object in its path, collect on the pipe in one area and discharge from it in another. Where it collects becomes a cathode; where it discharges is an anode, and thus corrosion occurs.

The source of energy for such cells may be a distant generator, a direct current transmission line, a cathodic protection rectifier on some other line, a street car system, or an electric railway. Such a cell can result in troublesome problems if left uncontrolled (Figure 9.16).

Stress Cells

If an ordinary nail is dropped into a vessel of salt water, it will be attacked, and, after a time, rust will be observed discoloring the water. The initial points of attack will be, in almost every case, the point and the head. This is an example of a type of cell known as a *stress cell* (Figure 10.9). The two electrodes are of the same metal and the electrolyte is uniform. The difference lies in that one electrode is more highly stressed than the other. The area of high stress is always the anode of the cell, made that way by the extra energy supplied by the stress.

Stress cells can take on two basic forms. One, like the nail just described, has its anode established by residual internal stress or stress left because of something which has happened to the metal. In the case of the nail, the "happening" was the cold-forming of the head and point. If these stresses are "relieved" by heating the nail and letting it cool slowly, the stresses will disappear and this type of stress cell will be eliminated.

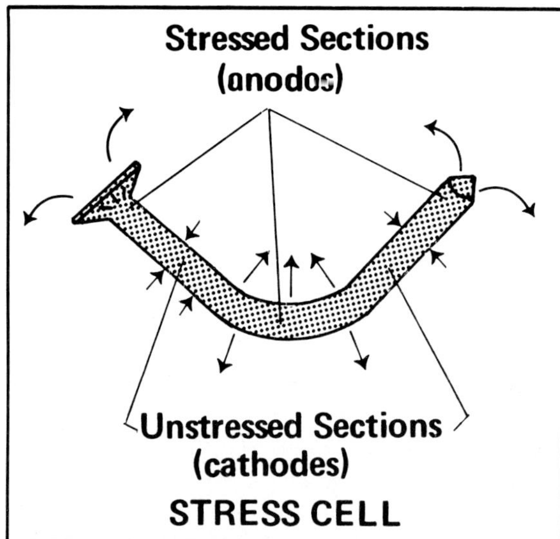

FIGURE 10.9 — Stress cell showing effect of stress in producing current discharge.

In the other type of stress cell, the metal is part of some kind of structure which is under stress. The most highly stressed part of the metal becomes the anode, with the less stressed or unstressed metal acting as the cathode.

Pipe which has been subjected to *cold bending* can set up such a cell. Other pipes, especially in manifolds and similar locations, that happen to have been forced into alignment and thus remain under large stresses, can also set up this type of cell. An improperly designed manifold may be relatively free from stress when built, but then develops some highly concentrated stresses while in service due to operating pressures, temperatures, or both. In addition to the hazard of simple mechanical failure, such a situation can intensify the stress corrosion problem, and is thus discussed further in Chapter 6.

Surface Film Cells

Some chemical or electrochemical reactions can take place on a metal surface, making it behave quite differently from a piece of clean metal. A film may be formed which is invisible, actually only a few molecules in thickness, but which may have a potential as much as 0.3 volt different from the unfilmed metal. Naturally, such a potential difference is enough to create an active corrosion cell.

Stainless steels are known for their ability to form such films, and thus owe much of their corrosion resistance to them. Aluminum, too, has this film-forming property, although its film represents a resistance barrier rather than a change in potential. Aluminum, in practice, is a much more durable metal than one would judge from its position in the EMF Series, made so by this ability to form a highly resistive film.

Steel in soil shows a similar change in potential with time. "Old" steel, *i.e.*, steel which has been in the ground for several years, is cathodic with respect to "new" steel, even when the two are identical in composition. It is thus strictly a surface phenomenon.

For example, suppose that a gas line has been in operation for ten years. During that time it has developed a number of leaks, and the rate of pit development is increasing. Before the leaking becomes critical, however, the gas demand increases to the point that the capacity of the line is not adequate. Accordingly, a new line is laid, parallel to the old one on the same right-of-way. Also, in order to give maximum flexibility in operation, crossovers are installed approximately every ten miles.

This situation forms an active corroding cell (Figure 10.10). The new line is the anode, the old line is the cathode, the soil is the electrolyte, and the crossovers form the connecting circuit. Total resistance of the cell is low because the lines are not far apart.

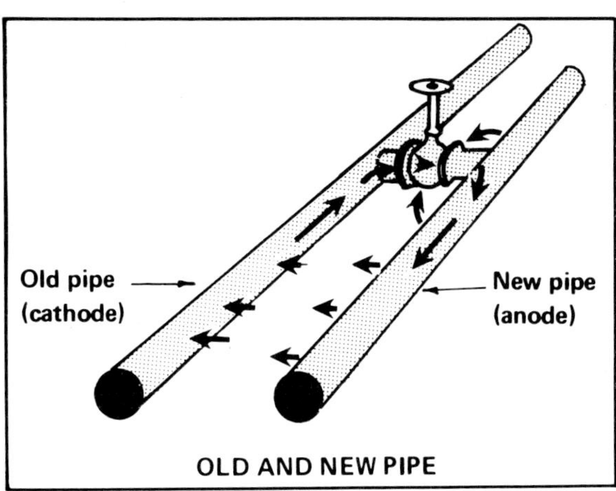

FIGURE 10.10 — Current flow resulting from interconnection of old and new piping.

The first effect is that corrosion is halted on the old line. Whereas it had developed a leak rate which was beginning to be troublesome, now it has no new leaks whatsoever. This seemingly happy state of affairs continues for several months, perhaps even a year or two. Then, suddenly, and to the great surprise and alarm of everyone concerned, the new line starts developing leaks, and at a greater rate than the old line had developed them. If the forces at work in this situation were not fully understood, it could well lead to the erroneous belief that the "new" steel was not as good as the "old" steel.

Miscellaneous Cells

Photoelectric Cell

A number of other differences can produce operating cells. There is, for example, the *photovoltaic cell* or *photoelectric cell.* One electrode is in the light, the other in the dark, and this sets up enough potential to make a cell operate. So far as is known, this is not important as a corroding cell, but it can be made to work in the laboratory.

Motion Cell

Motion of the electrolyte, with respect to one electrode, can also set up a cell. This may take place either from motion of the electrode or of the electrolyte. A river, by its flow alone, sets up a potential in the soil which can be picked up by a traversing pipeline. (This might be interpreted as an impressed current cell.) The motion of a ship through the water may generate small currents as well.

Pressure Cell

Since it is known that the potential of an electrode can be affected by variations in pressure, it is at least theoretically possible to set up a pressure cell. No actual example of corrosion from this cause is known, however.

Combination Cells

Any two, or more, of the cells described in preceding sections can operate at the same time. Sometimes they will work in the same direction, sometimes they will be opposed. For example, in a well casing, there is usually an oxygen cell and a temperature cell, both tending to make the deep pipe act as an anode. At the same time, there may be two or more ordinary concentration cells, as the casing passes through different strata of soil. Some of these will be in the same direction as the two mentioned, some will be in the opposite direction. Examples of combination cells could be multiplied almost indefinitely.

Pipeline Corrosion

By far the greater portion of the metal threatened by corrosion in soil and water is that which makes up pipelines. Thus, much of the remaining discussion in this chapter will be in terms of pipeline corrosion.

Almost any of the cells described earlier may be found on a pipeline, with the possible exception of the photoelectric cell. When a new line is laid in the earth, a vast number of cells come into being: the line passes through a variety of soils, giving rise to concentration cells; there are variations in oxygen access, setting up oxygen cells; there are impurities and differences on the surface of the pipe, creating galvanic cells; cold bends and welding stresses set up stress cells; and certainly, in the great complexity of the soil, there are the opportunities for any other kind of cell (perhaps currently unknown).

These cells are of various sizes and shapes. In some cases, anodes and cathodes are separated by an inch or two; in others, they may be miles apart. The cells also are of strengths varying from a few millivolts to perhaps half a volt. Anodic areas are of all sizes, tiny to large, as are the cathodes.

As soon as these areas become active, they begin changing. On cathodic areas, hydrogen may begin to form, as described in Chapter 2. Changes affect both cell resistance and potential. Anodes begin corroding, which puts new ions into solution at the surface. As these react with various components in the environment, concentrations change. Cell potentials also change; some increase, while others decrease.

Some areas which were initially anodic become cathodic. The reverse also may take place, although not as often. Generally, as time passes, the total anodic area on a line gets smaller, although total activity does not decrease at the same rate. The result is that the rate of attack at the worst locations tends to increase. Finally, among all of the various cells along the entire line, the most active anode loses enough metal by corrosion to penetrate the pipe wall completely. The line thus has its first leak and its corrosion history begins. What happens next?

Auxiliary Effects of Cell Operation

The principal effect of each of the cells described, and the effect which is of primary interest here, is the attack on the metal at the anode. Other reactions do occur, however, some of which are important in the corrosion process. One of these, as discussed in Chapter 2, may be the dissociation of water into its components, hydrogen and oxygen.

In many cells (depending upon the electrolyte), oxygen is formed at the anode, and in nearly all cells, hydrogen is formed at the cathode. If the chloride ion is present, chlorine gas may be formed at the anode. This generation of gas, either oxygen or chlorine, at the anode is not nearly as likely to occur in a natural corroding cell as it is in a cathodic protection system, especially where inert anodes are used.

Hydrogen

The formation of hydrogen at the cathode, however, can occur during both normal corrosion and cathodic protection. Initially, it may be formed as *nascent* hydrogen, *i.e.,* single atoms are produced. These may then combine with oxygen to form water, or with some other ion in the environment to form some other compound. It may dissolve in the metal (cathode area), or combine in pairs to form ordinary gaseous hydrogen. This latter reaction may occur only under exceptional conditions, because it requires a higher potential (known as the hydrogen overvoltage) than usually is present in naturally occurring corroding cells. It is a common event, however, in cathodic protection systems.

Under special circumstances, hydrogen formation can be damaging. Atomic hydrogen formed at the cathode may diffuse into the steel, and, at some depth below the metal surface, later combine in pairs to form the ordinary molecular gaseous hydrogen. Unlike the single atoms, these hydrogen molecules cannot migrate through the structure of the metal. Consequently, there is an accumulation of hydrogen gas within the metal, which can generate enough pressure to blister or split the solid steel (Figure 10.11). Only under special circumstances does this ever happen at the cathode of naturally occurring corroding cells. It can, however, occur in cathodic protection systems.

Electroendosmosis

Another side effect of the passage of electrical current through a porous medium like soil is *electroendosmosis.* Water is "carried along" with the current. Normally this effect is significant only with higher current densities than natural cells produce. Like hydrogen formation, it is more likely to be associated with cathodic protection. Its effect there is to carry water to the cathode and away from the anode. The latter effect, if the anode is above or not far below the water table, may result in an unwanted increase in anode resistance.

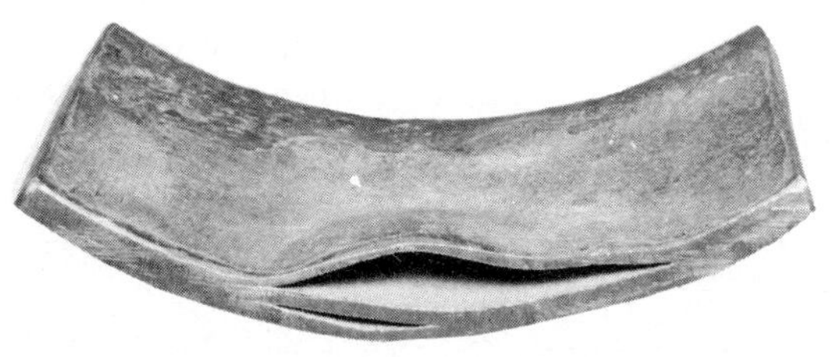

FIGURE 10.11 — Section of steel showing effect of accumulation of hydrogen at discontinuities to form blisters.

Cathode Scale

A third effect is the deposition of compounds from the soil on the pipe. This effect is extremely pronounced in seawater, where a cathode scale can be built up over a period of time that is like a hard, glossy enamel as much as an inch thick. This material is, when deposited in seawater, a mixture of calcium and magnesium oxides and hydroxides. It is a fairly complex substance, whose composition depends upon current density, among other factors.

In fresh or brackish water, cathode scale is formed only if the ions which compose it are present in the water; thus, its composition is even more variable than that formed in the sea.

This is also true in soils. Some soils will not form a visible scale, although there is enough calcium present in most soils for some scale formation. Often, when a line has been under cathodic protection, no scale will be visible until the pipe surface dries out. It then appears as a whitish discoloration. On a line which has not been under protection, cathode scale frequently is found in an irregular mottled pattern; this makes the actually active cathodic areas visible to the eye.

Pitting

When a pipeline that has been in corrosive soil without adequate protection for some time is examined, it is usually found that by far the greater part of the area is unaffected. Where corrosion has taken place, it is in the form of pits, which are relatively small areas where the attack has been deep (Figure 10.12). This characteristic of soil corrosion also is found in water, but not as universally.

In general, there always will be one spot on any pipeline where all of the conditions combine to give the highest rate of penetration. This is the spot where the first leak will occur. Since the underground pipe is invisible to the observer, it also will be the first place where corrosion will make its presence known.

On the rest of the pipeline, however, there will be many pits of varying depth and severity. In time, another one of these will perforate, and the time between the two may be much less than the time to the first one; then a third, and a fourth, and so on. That they will come faster and faster is known from experience, and the reasons are understandable from statistical analysis.

If the accumulating number of leaks which occur on a given pipeline is plotted on semilogarithmic

FIGURE 10.12 — Showing concentration in small areas of pits on underground pipeline. Arrows point to unrepaired pits. Other pitted areas repaired by welding. Circular patch is 1 in. in diameter. [SOURCE: Evaluating the Economy of Reconditioning and Coating Underground Steel Pipe, George W. Pemberton, Corrosion, Vol. 15, No. 5, pp. 279-281 (1959).]

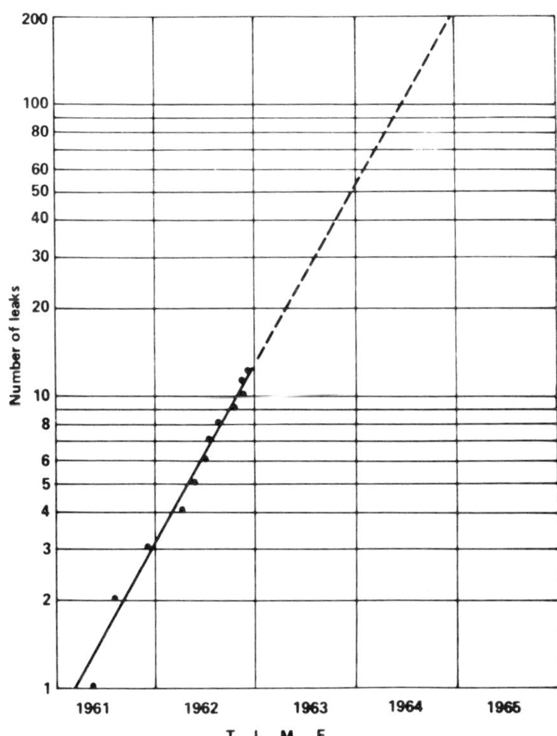

FIGURE 10.13 — Semilog plot of leak curve with projected rate.

paper, the resultant curve will tend to be a straight line. (Semilogarithmic paper, as shown in Figure 10.13, has its vertical scale logarithmic and its horizontal scale with uniform linear divisions.) It is a characteristic of a logarithmic scale that equal ratios occupy equal space, *i.e.,* the distance from 1 to 2 is the same as that from 2 to 4, or from 6 to 12, or from 1.75 to 3.5, or from any number to its double.

This can be verified by using a measuring device on the chart. The same thing is true of any ratio. The distance from 1 to 1.5 is the same as that from 2 to 3, or from 20 to 30, or from 150 to 225. It is this property, that of showing equal ratios as equal spaces, which makes it plot the ever-increasing rate of leak development as a straight line.

There is some theoretical ground for believing that the curve actually should be straight when plotted on full logarithmic paper, *i.e.,* that the time scale also should be logarithmic. This is not fully established, however, and the use of semilogarithmic paper is far more convenient. In an actual case, of course, the straight line is not perfect; statistical phenomena are seldom perfect, particularly when small numbers are involved.

Figure 10.13 shows such a curve for an actual pipeline. The line was built in November, 1958. For convenience, the curve starts at the beginning of the year in which the first leak occurred. This was May 16, 1961 when the line was about 30 months old. The second leak took place August 1st of the same year when the line was only three months older. It is interesting to note how steep the line would be if it were drawn on the basis of these two leaks alone. Successive leaks came to light as shown in Table 10.1.

On the last date in the leak history, the curve in Figure 10.13 was compiled. The dotted line beyond the last recorded leak shows the rate expected if nothing were done to prevent continuing leaks. This

TABLE 10.1 — Incidence of Leaks on Pipeline

Leak No.	Data
1	May 16, 1961
2	Aug. 1
3	Dec. 3
4	Apr. 4, 1962
5	May 18
6	Jun. 29
7	Jul. 20
8	Aug. 30
9	Oct. 10
10	Nov. 10
11	Nov. 15

line, when extended into the future, indicates that by the end of 1965 the line would have had approximately 200 leaks.

Unfortunately for our purpose of examining leak curves, but fortunately for owners of the line, the matter never came to a test. At the time the curve was drawn, work was already underway on the installation of a cathodic protection system. During installation, on December 14th, Leak No. 12 was found. The system went into operation in January. A leak occurred on April 14, 1964, over a year after the installation, and Leak No. 14 was found almost exactly one more year later, April 15, 1965. At the end of 1965, Figure 10.14 was drawn, showing the leak curve as affected by cathodic protection.

Historically, this has been the most useful function of the leak curve; to demonstrate to management the savings which can be had with cathodic protection. This is done by comparing the cost of the number of leaks to be expected with the cost of the cathodic protection which will prevent these leaks. When protection has been successfully applied to a system, the leak curve covering the periods of time before and after application (as in Figure 10.14) demonstrates the effectiveness of the method quite convincingly.

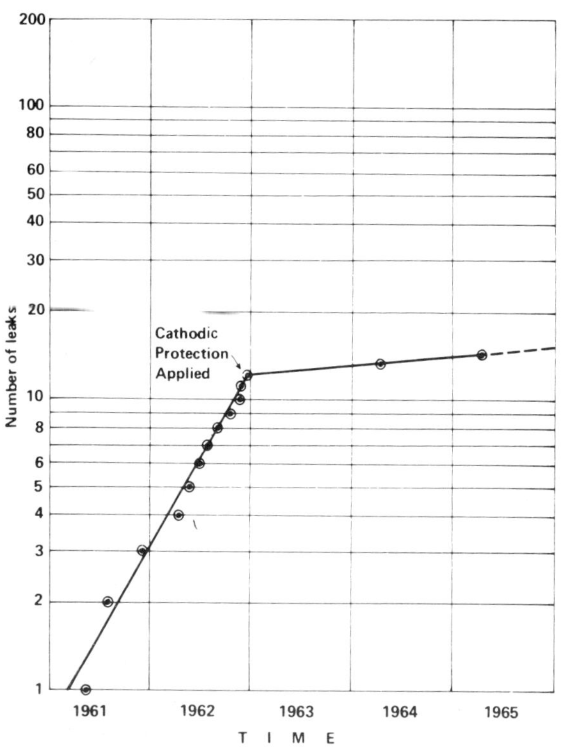

FIGURE 10.14 — Semilog plot of leak curve showing dramatic improvement in leak incidence following application of cathodic protection.

Of course, if cathodic protection is applied to a line when it is first built, before there are any leaks, no curve can ever be drawn. The "selling" job has to be done with other lines, older lines, or possibly lines of foreign ownership. As the examples become more and more remote in both time and space, the use of this method of demonstrating the usefulness of cathodic protection becomes less and less practical. It may almost be said that the better the corrosion engineer does his job, the harder it is to justify his position, and, if the job is done perfectly, it looks like he is not doing anything and is not needed at all!

Electrolyte Resistivity

What makes different lines have different leak curves? The greatest difference is in the soils in which lines lie. Probably the second largest factor, and it can easily be argued to be the first, is the coating. This point will be discussed later.

Pipe wall thickness, obviously, will also have an effect on leak rates. Other things being equal, a thin-walled pipe will develop a leak at a given location in less time than one with a thicker wall. Subsequent leaks will come at a faster rate, so the curve will be steeper. Oddly enough, though, the leak curve for a 50-mile line has the same shape as that for a 10-mile line, if everything else is the same. The longer line will simply have five times as many leaks at any stage of development, and on semilog paper, this merely means that the entire curve is shifted upward by a constant amount; hence, the slope is the same.

The same thing is almost true in comparing two pipelines which have different diameters but are otherwise similar. There are, however, some disturbing factors involved, and the statement is not entirely true.

It is when we come to the nature of the electrolyte that we find the factor which most profoundly affects the slope of the leak frequency curve. Soil comes in many varieties and has many different properties that can be measured. Soils may be classified by texture, organic content, dissolved chemicals, color, origin, etc., depending upon what one is interested in. The kind and amounts of the various dissolved salts, however, are of primary importance, and it turns out that the amount is more important than the kind.

An easy way to measure, approximately, the total amount of dissolved salts in soil, with particular importance given to those which ionize most readily, is by measuring the electrical resistivity of the soil. The lower the resistivity, the better the electrolyte; thus, soils provide electrolytes and in this way contribute to corrosion.

A careful distinction must be made between resistivity and resistance. *Resistance* is the property of a thing, a circuit element. For example, a piece of No. 10 copper wire 1,000 feet long has a resistance of 1 ohm. A piece of the same wire 100 feet long has a resistance of 0.1 ohm. A piece of No. 30 copper wire (very fine) only 10 feet long has a resistance of 1 ohm. Different circuit elements may have resistance ranging from less than 0.0001 ohm to several million ohms (or even more).

Resistivity, on the other hand, is the property of a substance. The material copper has a certain resistivity, silver has a different value, graphite still another, seawater another, etc. Any material capable of carrying current at all (and this includes almost every known substance) has a certain value of resistivity. For accurately defined pure substances, this is a precisely known and constant value available in handbooks. For variable substances such as brass, soil, or clay, the quantity itself is found to vary.

The resistance of a circuit element made of a uniform substance is determined by its resistivity and by its geometry, or its size and shape. Further details on how soil resistivities are measured is covered in Chapter 9.

In the control of pipeline corrosion, resistivity measurements are valuable in two respects: they help predict the probable corrosiveness of soils and waters, and they are needed for the proper design of cathodic protection systems.

Effects of Coating on Corrosion and on Cathodic Protection

An obvious method of controlling corrosion is that of interposing a barrier between the threatened metal surface and the corrosive medium, *i.e.,* some kind of coating. Since corrosion always requires the presence of an electrolyte (moisture) in contact with the metal, if a metal could be coated with a material which was absolutely waterproof and absolutely free from holes, all attack would be stopped. The coating, it should be noted, would not only need these two properties when it was applied, but the two properties would have to be permanent—the coating would have to remain perfect in both respects.

Unfortunately, a perfect coating does not exist. There are, however, coatings that approach this degree of perfection available at a high cost. We already know that corrosion can be controlled on a pipeline and on many other structures with a combination of a reasonably good coating and cathodic protection, and we know how much this combination costs. Thus, we are forced to acknowledge that no matter how good a new, improved coating might be, we cannot afford to pay more for it than we do for the older combination. Many questions therefore arise about the use of imperfect coatings: How good are they when used alone, what effect do they have on the progress of corrosion, and how do they affect the need for cathodic protection.

Without Cathodic Protection

Compared with a bare pipeline in the same location, a coated line will, in the long run, have fewer leaks. It may, however, have its first leak within a shorter time—simply because the coating, with its relatively few holes, may happen to concentrate corrosion activity at one or more of the small imperfections in it called *holidays.*

Under special circumstances, this effect can be even more pronounced. For example, the tendency of new steel to be anodic with respect to older steel can lead to a perplexing situation. Suppose that a bare line has been in operation for some time. At the worst spot, a leak develops. Soon others develop in the same general area and repairs are needed. So, a section of line is removed and replaced, thereby introducing new steel into the circuit.

At the same time, it is also decided to coat the new section. Since the coating is not perfect, there will be some holidays, or small areas not completely coated. This will have the effect of creating a cell with a small anode (the bare metal in the holidays) and a large cathode (the old steel pipe on either side of the new area). The potential of the cell will be as great as if the entire line were bare, but total activity will be concentrated on the small area of the holidays where an early failure is to be expected.

Any other circumstance in which the anodic section of a line is coated while the cathodic area remains bare may produce similar results. The total amount of corrosion may be reduced by the coating, but the speed of penetration (and hence the rapidity of appearance of the first leak) is increased.

In a stray current exposure area, it is extremely important that a good coating be applied to the line in the area of current pickup (which is cathodic, and not subject to attack) in order to increase electrical resistance and thus minimize the amount of current which will have to be disposed of elsewhere.

In summary, the net effect of coating without cathodic protection, as compared with the use of bare pipe, is to reduce the total amount of attack, but sometimes to increase local severity and, often, to affect distribution of the attack in novel and surprising ways.

With Cathodic Protection

The effect that coating has on cathodic protection requirements is simple and can be stated briefly: Application of a coating greatly reduces the amount of current required to obtain protection. The reduction may run from as much as 99.8% (for an extremely good coating) to as low as 50% (for a very poor, old, damaged coating.)

Effects of Corrosion and Cathodic Protection on a Coating

The opposing effects are also worth considering, *i.e.,* both corrosion and cathodic protection can adversely affect a coating. As stated earlier, it is essential that a coating, aside from being affordable, have the properties of being waterproof and free from holes or imperfections—not just at the time of installation, but permanently. It was also said, however, that neither of these properties can be had at a reasonable cost. This means that other coating properties which otherwise would not be necessary now become important.

First, it is necessary that a coating bond firmly to the metal surface. If the coating were perfect in all other respects, this would not be important. However, since the coating does have imperfections, it is necessary to keep moisture from penetrating indefinitely at each break. Realistically, then, a good bond is a definite requirement for any coating for pipelines.

It is also necessary, or at least extremely helpful, if the coating has a high electrical resistance. This is so that cathodic protection can be applied as cheaply as possible. The more conductive the coating, the larger the amount of current needed for cathodic protection. Thus, high electrical resistance is another coating requirement.

It is often said that a good pipeline coating should have high *dielectric strength,* which is the ability to withstand high voltage without failure. Although this is not needed underground, either with or without cathodic protection, all of the pipeline coatings which have proven useful do, in fact, have high dielectric strength. Therefore, any pipeline coating material with low dielectric strength is likely to be ineffective.

A high dielectric strength also makes it possible to use a high voltage holiday detector in the inspection of the finished coating: if the detection sparks through, then there is either a bubble, a piece of foreign matter, or a thin spot in the coating. This property, then, is a useful one, although it is not really necessary for a pipeline coating.

Two cathodic effects have already been described: the formation of cathode scale and the release of hydrogen. These can take place at the cathode of any type of cell, although the latter effect, the actual generation of hydrogen gas, takes place only at fairly high negative potentials—values not likely to be found in a corrosion cell, but common enough in cathodic protection systems or in stray current exposures.

Either of these effects is capable of damaging a coating. Given a small defect in the coating, the deposition of scale or of hydrogen under the edge of a sound coating can cause a progressive enlargement of the unbonded area. Under rare conditions, cathodic current can literally strip off a coating in a matter of hours. This can be one of the limitations of a cathodic protection system design.

Corrosion Surveys

Resistivity

A corrosion survey may be conducted to determine: (1) How severely is a pipeline or other structure attacked in a proposed location; (2) How severely is an existing structure being attacked; (3) How much damage has already been done; or (4) What steps can be taken to control the corrosive exposure? A survey of the resistivity along an existing or proposed route will give good answers to questions (1), (2), and (4), and will permit some fairly good estimations of question (3).

Except under what can only be described as research conditions, it is seldom profitable to conduct an extensive survey of resistivity along the route of a coated line, either existing or planned. There is really only one decision to make about a coated line, and that is whether or not to place it under cathodic protection. If the answer is negative, then the resistivity data serve no useful purpose. If it is affirmative, then all that is needed is information about possible anode bed sites, and these should be taken at a greater depth than those for the pipeline corrosion survey.

On a bare line, however, a resistivity survey makes a great deal of sense. It assembles data concerning two important problems: It tells where the line is subject to attack (hot spots), and it furnishes data needed for the selection of galvanic anodes of the proper sizes, or determines the amount of impressed current that is likely to be needed.

There are many adherents to the belief that any soil which is sufficiently corrosive to justify coating a pipeline, also justifies cathodic protection. In other words, no line should be coated unless it is also protected. Acceptance of this conclusion, of course, covers the first decision referred to above, thereby eliminating any justification (except as a research study) for a detailed resistivity survey along a coated line. Instead, a much more limited survey is conducted in conjunction with current requirement tests to obtain the data needed for the design of the cathodic protection system.

Potential Surveys

It is possible, unless stray current complicates the picture too much, to detect the pattern of corrosion currents on a pipeline by means of potential measurements made at the surface of the ground above the line. This type of survey was in vogue about a generation ago, but has since fallen into neglect, except as a research tool or in some exceptional cases. If it is a coated line, it is not really helpful to know where the line is being attacked because the remedy is the same regardless. In other words, locating anodic and cathodic areas on a coated pipeline costs money, and the information is not useful because if the line is to be protected, all of it will be protected.

Once the line has already been placed under protection, potential surveys, with readings usually at wide spacing using permanent test leads, are needed to ensure complete and continued protection. If trouble occurs and some sections of the line are found to be less than fully protected, then close potential measurements, *i.e.*, 25-ft spacing, may be useful in tracking down the difficulty.

A profile of the potential along the surface of the soil immediately above the pipeline can be run by any of several methods. Any sharp break in that profile must have an explanation. It may be an uninsulated lateral; it may be an accidentally shorted crossing line; it may be a shorted casing; it may be just a sharp change in soil resistivity; but it must have some reason for existence, and whatever is causing it may be what is robbing the line of protection.

Other Means of Protecting Steel

The preceding discussion has cited the use of cathodic protection as the prime method of protecting steel buried in soil. Other measures, however, also may be adopted in special cases.

When economically and mechanically attractive, a steel line or tank in the ground may be covered with an alkaline material such as limestone. Leachings from the stone will neutralize any acidic constituents in the soil and surround the steel with an alkaline environment which is noncorrosive. Calcium salts will eventually precipitate from the alkaline water to coat the steel. Any material with a reserve alkalinity could be used to encase the steel for this purpose.

There are also proprietary products on the market which claim to provide a protective sheath around buried steel in a wet soil.

Casing the exterior of a pipe underground is often required for mechanical and corrosion reasons. The working pipe is placed on insulating spacers inside a pipe of larger diameter. Only the exterior pipe is in contact with the soil and must be protected in the usual manner. The working pipe is now immune from soil corrosion. The annulus between the two pipes must be considered, however. Condensation of moisture, and perhaps corrosive gases, occurs in this annulus. An inhibited grease, oil, formulated aqueous mixture, or amine should be pumped into this annulus and maintained properly to ensure full protection of the working pipeline.

Structures Other Than Pipelines

Distribution Systems

Most of the previous discussion has been in direct reference to pipelines, for reasons given earlier. Other structures may be considered in terms of how they may differ from pipelines.

Distribution systems generally are composed of pipe of many different sizes and ages. The latter difference, in particular, can cause problems. Distribution systems are much more subject to accidental contacts with other lines, typically water lines. It is also common for distribution systems to be made of a greater variety of materials. For example, copper service lines may be used. Because of these differences, the approach to the protection of such systems is likely to differ in that more emphasis is placed on the use of small units and galvanic anodes, even when coated pipe is involved.

Gathering Systems

Gathering systems do not ordinarily differ much from pipelines. They offer more variety and often have more accidental contacts. Protection systems may encounter more interference problems in both directions. One common problem is that gathering lines, particularly in the early development of a field, may be laid in great haste, with little or no thought to the corrosion control problem. Although this is not, strictly speaking, a difference in corrosion behavior, it does account for some serious problems.

Plant Piping

The corrosion activity in a complex of piping in a refinery or other industrial plant has several unique features. First, there are always combinations of aboveground and underground piping; a considerable variety of sizes, materials, and functions; and almost invariably a great variety of coatings applied in original construction, creating a range of from bare to very well-coated pipe. The soil cannot tell what is inside the lines and thus treats them all alike, providing they are built alike.

Another factor present in almost every plant is the large amount of bare copper in the ground system, typically a collection of ground rods which are connected by a network of large diameter bare copper cable. This can cause much difficulty, both in causing corrosion when no protective measures are taken, and again as a complicating factor when protection is applied. Further problems may arise because of the presence of many paved and unpaved areas that set up troublesome aeration cells. Furthermore, the usual construction activity ultimately results in badly mixed soils. Soil resistivity maps, covering entire plant areas, are useful both in anticipating trouble areas and in planning control measures.

Well Casings

Although a well casing may be considered as merely a vertical pipeline, it also has several unique features. It costs much more per foot, so that leak repair is enormously expensive, if at all possible.

The corrosive exposure is complicated by the variety of soil strata through which it passes. Normally, these are much more varied than would be found on an equal length of horizontal pipeline. Casing is usually bare (by necessity or custom). Any investigation is severely hampered by the fact that only one end of the pipe is accessible, so that most conventional surveys cannot be conducted.

Underground Tanks

Underground tanks usually are large enough to contact more than one stratum of soil, so they may be subject to concentration cell action. They are almost always subject to oxygen cell attack, even when under pavement. Often, there are fittings of different metals, and seldom is the coating as good as that used on pipelines. One of the major problems in dealing with these structures is that it is difficult to justify an engineering study for each tank, yet they cannot all be treated alike.

Steel Piling

An important difference with steel piling is that a few pits or even holes have little effect on its structural strength. Consequently, much more corrosion can be tolerated than with pipelines. Piling is almost always bare, vertical, and hence subject to the same kinds of cells that attack oil well casings. Bonding often may be a problem because individual piles may not be interconnected electrically, a condition that makes both investigation and protection a problem.

Transmission Tower and Pole Footings, Guys, and Grounds

Electrical lines often have built-in problems because of the common use of extensive copper ground wire. The galvanic cells thus created can be ruinous. While this problem can be avoided by not interconnecting, there is still a problem with the corrosion of a large number of small units, which makes both study and protection difficult.

Corrosion of Materials Other Than Steel

Cast Iron

The corrosion rate of cast iron is comparable to that of steel in a soil. The iron is removed from the metal, leaving a network of carbon particles by the dealloying phenomenon termed *graphitization* (Chapter 5). The residual carbon retains the form of the object and, unless the weak structure is fractured, the graphitized form will convey liquids. Some cast iron water mains that are over 50 years old are in this condition and continue to adequately serve their purpose.

Once the cast iron is graphitized, the exterior becomes an extremely noble electrode in any galvanic couple. Thus, care should be used in attaching uncoated or unprotected cast iron to other metals in the soils.

Coating of the exterior of cast iron pipe with bituminous or lower oil coatings is often provided. Cathodic protection can then be applied if the mechanical joints used are bonded to provide one continuous electrical conductor.

Ductile iron behaves the same as cast iron in a soil.

Aluminum

Aluminum has been used underground with good results under certain circumstances. However, the metal is subject to attack by acids, alkalies, heavy metal salts, chlorides, and galvanic action. If attack does occur, either from soil corrosion, galvanic effect, or electrolysis, the area affected is normally small, but the pitting depth is severe.

As discussed previously, the metal may be cathodically protected in either the coated or bare conditions with appropriate precautions. The dense oxide film on the metal surface acts as a coating to a significant degree when protection is applied.

Zinc

As discussed previously, zinc is sometimes used as a reference half-cell in soils.

Galvanized steel is not normally installed underground. The thin zinc coating is quickly dissipated by galvanic action with any exposed steel.

Lead

At one time, lead was used extensively as sheathing for telephone cables, but has been supplanted by plastics in most areas. Although the metal is amphoteric (attacked by either acids or alkalies), a reasonable corrosion resistance was shown in most soils.

If cathodic protection of the metal is attempted, close control of the protection potentials is necessary to avoid alkaline attack on the lead.

Stainless Steels

It is sometimes necessary to use the austenitic stainless steels as underground piping. Very light gauge pipe is used, and any attack on the exterior can be serious. In many soils, the passive film on the metal remains continuous and no attack occurs. However, pitting can develop in heavy soils and in chloride environments. When in doubt, such as with the lack of prior experience, the metal should be protected in the usual manner.

Copper

It is well known that copper serves well in underground exposure, unless high sulfide concentrations exist. Burying copper in cinders, however, can cause severe pitting.

The brasses can dezincify in many soil exposures. Other copper alloys are resistant to essentially all natural soils.

Concrete

Concrete is extensively used underground as footings, piers, tanks, piping, etc. The material is normally used with a reinforcing agent such as steel bar of thin steel pipe, as in cylinder pipe. In most areas the concrete will be stable, but serious corrosion can occur from a number of sources.

The problems relating to the stability of concrete underground are comparable to those experienced in atmospheric or water services. The concrete may be attacked directly, or permeation of the concrete may occur with subsequent attack on the reinforcing steel.

When appraising the potential corrosivity of a soil, both the soil chemistry and the effect of groundwater must be considered. The soil must be judged aggressive if: (1) the pH is less than 6; (2) the pH is neutral, but hydrogen ion is available by exchange; (3) the sulfate or sulfide content is high; or (4) the magnesia content is high. Groundwater percolating into the soil to extract the aggressive ions is responsible for the attack. Dry soil is innocuous.

Acid attack on the alkaline concrete is to be expected. Thus, any acid spills in the vicinity of concrete structures should be neutralized and not allowed to penetrate deeply into the ground.

Sulfate and magnesium ions can be corrosive to the concrete in neutral solutions. Reactions occur with the calcium salts in the concrete to destroy the cohesion and produce a soft, porous mass. A Type V portland (ASTM Standard C150) will resist the sulfate more capably, but aluminate cements are used to combat magnesium salt attack.

Organic compounds, particularly esters, can rapidly degrade the usual concrete. Detergents can accelerate the rate of degradation. The freezing of absorbed water in the concrete can cause spalling or cracking.

Thus, when proposing the use of concrete underground, be attentive to water levels in the ground, the selection of the proper concrete (there are many types), the proper cure of the material, the density of the finished product, the cleanliness of the water and sand used, the depth of coverage of the reinforcing metal, the possible need to seal the exterior, and the possible need to maintain the reinforcing metal as an electrically continuous structure.

The surface can be treated chemically to densify and harden the exterior (fluoride treatments or sodium silicate washes) as a mode of corrosion protection. Bituminous coatings are often applied to seal the exterior. Epoxy coatings are most compatible with the concrete as a coating, patching compound, or adhesive.

It is imperative that the reinforcing steel be protected from corrosion, or cracking and disintegration of the concrete will occur as the corrosion products expand. Leachings from the concrete are alkaline and will protect the steel when only penetrating moisture reaches the metal. However, the rebar is often coated with epoxy or zinc-rich paints before casting the concrete.

Cathodic protection of rebar or cylinder steel in pipe can be applied if the metal structure is continuous. This means the certain joining of all rebar before casting the concrete and, in the case of cylinder pipe, the installation of metal-to-metal joint clips as the sections are placed in position. The careful use of cathodic protection of the steel is then possible.

Plastics

Plastic materials of construction have revolutionized many of the underground materials applications. Although certainly not immune to failure, the use of plastics negates the pitting, galvanic action, and other forms of localized attack experienced when using steel underground. The smooth interior will normally remain clean for the easy conveyance of liquids or gases. In diameters of six inches or less, the laying of pipe in the ground can be accomplished at the pace of a slow walk, which is an important economic advantage when combined with the savings from unneeded external protection.

The three major plastic materials used underground are types of thermoplastics: polyvinyl chloride (PVC), acrylonitrile-butadiene-styrene (ABS), and polyethylene (PE). These materials are not attacked by the concentration of acids, alkalies, or solvents encountered underground. Small amounts of moisture are absorbed by the plastics, but are not harmful. Certain specific organic compounds can cause stress corrosion cracking of the plastics. Of these, only the cracking of stressed PE by detergents in a contaminated soil could be envisioned as a potential problem.

Mechanical support of the materials by soil at a constant temperature overcomes two of the major problems associated with the use of plastics: the low modulus and strength, and the large thermal expansion property. When installed at sufficient depth to prevent collapse from heavy topsoil loadings (trucks and tractors), the materials will provide many years of trouble-free service.

When piping installations of these thermoplastics are made, care should be used in making up the joints to ensure their integrity, notches should not be made in the outer surface (notch sensitivity is common), and the pipe should be laid on a uniform base free of rocks or other hard, irregular pieces.

Other plastic materials, such as polypropylene, polybutylene, and glass-reinforced polyester or epoxy, are also used for specific services at greater expense. The glass-reinforced thermosetting plastics have been particularly successful as tank materials for the underground storage of a wide range of products (from water to gasoline). Long exposures in wet soils show only a superficial attack on the exterior of such thermoset plastics.

Bibliography

Biczok, I., Concrete Corrosion, Concrete Protection. Chem. Pub., New York, NY, 1967.

Romanoff, M., Underground Corrosion, Bureau of Standards, Circular 579, U.S. Govt. Printing Office, 1957.

NOTES

Chapter 11

Atmospheric Corrosion

ATMOSPHERIC CORROSION

Introduction

The specific area of corrosion in the atmosphere usually receives only minor attention in the writings of scientists and corrosion engineers who deal with the basic subject of corrosion. The corrosion phenomena encountered in chemical plants, underground structures, and, to a lesser degree, in seawater or at elevated temperatures, seem to offer more glamour and appear to be more spectacular. However, most corrosion damage to equipment and structures actually occurs in the atmosphere. The large segment of the paint industry committed to the manufacture and application of products for the protection of metals, as well as the large-scale operations of the galvanizing industry attest to the importance of controlling atmospheric corrosion.

Most all of the general types of corrosion occur in the atmosphere, and some appear to be largely restricted to it. Since the corroding metal is not bathed in large quantities of electrolyte, most atmospheric corrosion operates in highly localized corrosion cells. Thus, calculation of the electrode potentials on the basis of ion concentration, the determination of polarization characteristics, and other electrochemical operations are not possible in atmospheric corrosion as they are in liquid immersion types of corrosion. However, all of the electrochemical factors which are significant in corrosion processes do operate in the atmosphere. Thus, an understanding of electrochemical factors is still vital to an understanding of the corrosion mechanism.

The economic losses caused by atmospheric corrosion are tremendous and therefore account for the disappearance of a significant portion of metal produced. In the figures quoted by various estimators of national corrosion loss, specific items that require replacement as a result of the ravages of corrosion, as well as the cost of some corrosion control programs, are used as the basis of computation. Consider, for instance, agricultural machinery, steel structures, fences, exposed metals on buildings, automobile mufflers or bodies, and the myriad of other metal items which are discarded when they become unusable as a result of corrosion. These constitute direct losses from corrosion.

On the other hand, consider the major portion of a vast paint industry directed toward protecting metal from corrosion. Consider, too, the application costs of protective materials and the loss incurred when corrosion failure causes interruption of plant or process operation. These represent the intangible costs of corrosion which amount to even more than direct costs. F. L. LaQue once noted that without the existence of corrosion, nearly all metal objects and structures could be made from either iron or steel, so that there would be little need for any other type of metal. This thought tends to emphasize the tremendous impact of corrosion on our civilization.

Types of Atmospheric Corrosion

General or Uniform Corrosion

In general corrosion, surface layers of the metal are converted to corrosion products in such a way that the thickness of the section is uniformly decreased.

Pitting Corrosion

The metal is not corroded uniformly in pitting corrosion, but primarily at distinct spots where deep pits are produced. The bottoms of the pits are anodes in small corrosion cells, with the surrounding surfaces acting as cathodes. Metals that develop thin protective oxide films on their surfaces are particularly susceptible to this form of corrosion.

Dissimilar Metal Couple Corrosion

In the case of dissimilar metal couple corrosion, two dissimilar metals in the presence of small amounts of moisture form a small battery or corrosion cell (galvanic effect), with the less positive metal corroding at the exposed interface between the two metals. In the atmosphere, this corrosion does not extend very far from the junction of the metals and usually appears as a crevice-shaped pit. This type of attack is to be expected in the limited conductivity provided by the electrolyte. In certain instances, the accumulation of insoluble corrosion products may actually break the electrical continuity of the galvanic couple.

Intergranular Corrosion

Where segregation of some of the components or impurities of an alloy occur in a grain or crystal boundary, a potential difference may develop between the grain boundary material and the grain which will produce corrosion of the material that is less positive. This is referred to as intergranular corrosion since it follows the grain boundaries and acts like cracks in the metal.

Stress Corrosion Cracking

Under certain conditions, alloys under high tensile stress are susceptible to corrosion starting on the surface which will propagate through the lattice structure of the metal as cracks and lead to catastrophic failure. This is known as stress corrosion cracking. The stress may be residual, as from cold working and forming, or may result from external loading.

Corrosion Fatigue

Corrosion fatigue is the decrease in the fatigue resistance of a metal under cyclic stress. Any corrosion of a metal under a fatigue producing stress will cause reduced fatigue life. It is thought that corrosion produces pits, notches, or other starting points on the metal surface for the concentration of stress and the initiation of fatigue cracks.

Fretting Corrosion

If two faying surfaces exposed to the atmosphere under load are subjected to vibration, with respect to each other, which produces slight or incipient slip, fretting corrosion will develop at the areas of contact. It is presumed that protective corrosion films are destroyed by the vibration, thus repeatedly exposing fresh metal to the continuing corrosion process.

Crevice Corrosion

Retention of moisture and contaminants from the atmosphere in crevices may accelerate corrosion there, as compared with more boldly exposed surfaces resulting in crevice corrosion. This mechanism differs somewhat from the action of differential aeration in submerged crevices, but may produce an equally undesirable result.

Chapter 5 provides a more detailed description of these forms of localized attack.

Types of Corrosive Atmospheres

While atmospheres can be classified into four basic types, most of them are mixed and present no clear lines of demarcation. Furthermore, the type of atmosphere may vary with the wind pattern, particularly where corrosive pollutants are concerned.

Industrial

An industrial atmosphere is characterized by pollution composed mainly of sulfur compounds and nitrogen oxides. Sulfur dioxide from burning coal or other fossil fuels is picked up by moisture on dust particles as sulfurous acid. This is oxidized by some catalytic process on the dust particles to sulfuric acid which settles in microscopic droplets on exposed surfaces. Some sulfur dioxide and sulfurous acid are also present. The result is that contaminants in an industrial atmosphere, plus dew or fog, produce a highly corrosive, wet, acid film on exposed surfaces.

In addition to the normal industrial atmosphere in or near chemical plants, other corrosive pollutants may be present. These are usually various forms of chloride which may be much more corrosive than the acid sulfates. The reactivity of acid chlorides with most metals is more pronounced than that of other pollutants such as phosphates and nitrates.

Marine

A marine atmosphere is laden with fine particles of sea salt carried by the wind to settle on exposed surfaces. The quantity of salt contamination decreases rapidly with distance from the ocean, and is greatly affected by wind currents. The marine atmosphere also includes the space above the sea surfaces where splashing and heavy sea spray are encountered. Some splash zones could be classified as intermittent immersion.

Rural

A rural atmosphere does not contain strong chemical contaminants, but does contain organic and inorganic dusts. Its principal corrosive constituent is moisture and, of course, gaseous elements such as oxygen and carbon dioxide.

Arid or tropical atmospheres are special variations of the rural atmosphere. In arid climates there is little or no rainfall, but there may be a high relative humidity and occasional condensation. This situation is encountered along the desert coast of northern Africa. In the Tropics, in addition to the high average temperature, the daily cycle includes a high relative humidity, intense sunlight, and long periods of condensation during the night. In sheltered areas, the wetness from condensation may persist long after sunrise. Such conditions may produce a highly corrosive environment.

Indoor

The indoor atmosphere, while originally considered quite mild, actually can sometimes be quite severe. However, there is no typical contaminant or set of conditions associated with an indoor atmosphere. Any enclosed space which is not evacuated or filled with a liquid can be considered an indoor atmosphere. If not ventilated, it may contain fumes, which in the presence of condensation or high humidity could prove to be highly corrosive.

When considering atmospheres of any type, it must be understood that not only acidic contaminants are aggressive, but that alkaline materials (limestone, seashells, etc.) also can be corrosive when in contact with amphoteric metals (*e.g.*, aluminum, zinc, and lead). It also must be understood that the major contributor to the corrosion of all metals in the atmosphere is *oxygen*.

Factors Affecting Atmospheric Corrosion

Measurement of Atmospheric Factors

Various methods have been developed for measuring many of the factors that influence atmospheric corrosion. The quantity and composition of pollutants in the atmosphere, the amount collected on surfaces under a variety of conditions, and the variation of these with time have been determined. Temperature, relative humidity, wind direction and velocity, solar radiation, and amount of rainfall are easily recorded. Not so easily determined are dwelling time of wetness, and the quantity of sulfur dioxide and chloride contamination. However, methods for these determinations have been developed and are in use at various test stations. By monitoring these factors and relating them to corrosion rates, a better understanding of atmospheric corrosion has been obtained.

A method of measuring the time of wetness has been developed by Sereda[1] and correlated with the corrosion rates encountered in the atmosphere. A narrow strip of platinum is attached to, but insulated from, both sides of a zinc panel. When a film of moisture connects the platinum strip to the zinc plate, a galvanic cell or battery is set up. If the potential or the current through a shunt resistor is measured, the beginning and end of the "wetness period" can be determined. Sereda's work showed the effect of radiation from the panel on its temperature, the influence of contaminants, the difference between "groundward" and "skyward" surfaces of nonvertical surfaces, and the relation of temperature to corrosion rates.

Methods have been available for measuring the total amount of sulfur dioxide in a given volume of air, but this is related only indirectly to the effect of sulfur dioxide on corrosion. It is the actual amount of hydrated sulfur dioxide or sulfur trioxide deposited on metal surfaces that is important. The collection of sulfur dioxide by the lead peroxide cylinder method[2] has been employed and seems to correlate with corrosion rates when combined with the time of wetness. The presence of chloride contamination superimposed on the sulfur contamination greatly complicates the situation.

A number of methods of determining the contamination of the atmosphere by chlorides, principally in the form of sea salt, have been employed. The "wet candle method"[3] appears to be the simplest, but has the disadvantage that it also collects particles of dry salt that might not be deposited. In reality, it gives an indication of the salinity of the atmosphere rather than the contamination of exposed metal surfaces. A wick is wrapped around a glass rod and extends down into a container of water so that the wick stays wet. Particles of salt or spray are trapped by the wet wick and retained. At intervals, a quantitative determination of the chloride collected by the wick is made and a new wick is exposed. Other methods which measure the collection of particulate chlorides also have been developed.

Atmospheric Factors Other Than Contaminants

The most important factor in atmospheric corrosion, overriding pollution or lack of it, is moisture, either in the form of rain, dew, condensation, or high relative humidity. In the absence of moisture, most contaminants would have little or no corrosive effect.

Rain also may have a beneficial effect in washing away atmospheric pollutants that have settled on exposure surfaces. This effect has been particularly noticeable in marine atmospheres. On the other hand, if the rain collects in pockets or crevices, it may accelerate corrosion by supplying continued wetness in such areas.

Dew and condensation are undesirable from a corrosion standpoint if not accompanied by frequent rain washing which dilutes or eliminates contamination. A film of dew, saturated with sea salt or acid sulfates, and acid chlorides of an industrial atmosphere provide an aggressive electrolyte for the promotion of corrosion. Also, in the humid Tropics where nightly condensation appears on many surfaces, the stagnant moisture film either becomes alkaline from reaction with metal surfaces, or picks up carbon dioxide and becomes aggressive as a dilute acid.

Temperature plays an important role in atmospheric corrosion in two ways. First, there is the normal increase in corrosion activity which can theoretically double for each ten-degree increase in temperature. Secondly, a little-recognized effect is the temperature lag of metallic objects, due to their heat capacity, behind changes in the ambient temperature.

As the ambient temperature drops during the evening, metallic surfaces tend to remain warmer than the humid air surrounding them and do not begin to collect condensation until some time after the dew point has been reached. As the temperature begins to rise in the surrounding air, the lagging temperature of the metal structures will tend to make them act as condensers, maintaining a film of moisture on their surfaces.

The period of wetness is often much longer than the time the ambient air is at or below the dew point and varies with the section thickness of the metal structure, air currents, relative humidity, and direct radiation from the sun.

Cycling temperature has produced severe corrosion on metal objects in the Tropics, in unheated warehouses, and on metal tools, etc., stored in plastic bags. Also, the dew point of an atmosphere is the equilibrium condition of condensation and evaporation from a surface. To ensure that no corrosion will occur by condensation on a surface, the temperature should be maintained some 10 to 15 C above the dew point.

Some metals, particularly steel, are corroded in high relative humidities in the absence of temperature cycling. There is said to be a critical relative humidity at which corrosion rates suddenly increase. This is probably a function of the hygro-

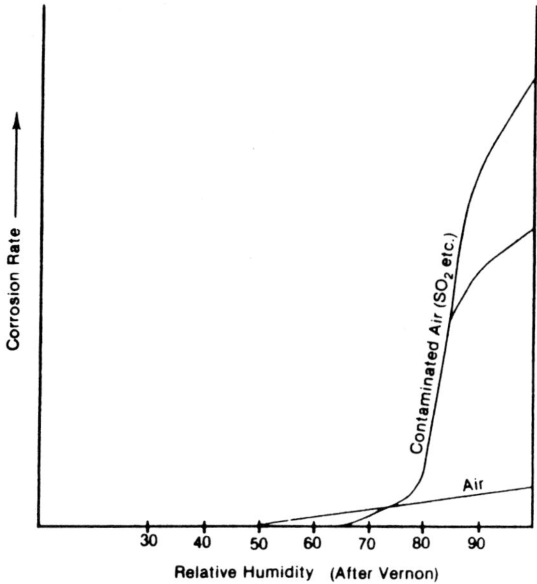

FIGURE 11.1 — Corrosion rate rises with relative humidity.

scopicity(1) of the corrosion products and of contaminants present. The danger of corrosion appears to rise rapidly as relative humidity passes 60% (Figure 11.1).

Organized Studies of Atmospheric Corrosion

The most comprehensive and extensive studies of atmospheric corrosion have been conducted by committees of ASTM (The American Society for Testing and Materials). Starting at the beginning of the century, various committees and task groups of this organization have exposed test specimens in carefully planned programs at test sites covering a wide variety of conditions. In the early stages of this test program, results were reported as changes in appearance, loss in physical and mechanical properties, or loss in weight. In recent years, a program has been pursued to correlate the atmospheric factors mentioned above with time of wetness and to correlate sulfur dioxide and chloride contamination with corrosion rates in several selected geographic locations.

Over the years, tests have been made on practically all of the common commercial wrought and cast metals and alloys, with and without protective coatings, in partially fabricated assemblies and in combinations. Small items of hardware and structural shapes have been exposed along with corrugated sheets such as those used for roofing. Several tests have been made of galvanic couples covering combinations of metals ranging from stainless steel to magnesium.

For many years, reports have been accumulated from various parts of the world on corrosion rates of a metal of nominal composition in a particular atmosphere. In the belief that loose terms like mild steel, aluminum, or brass did not provide any common denominator in the appraisal of relative corrosion rates at the sites from which the reports were obtained, one of the ASTM committees undertook a calibration program. In the initial phases, it used standard reference panels of zinc and steel whose composition was known and carefully controlled. More recently, it has added an aluminum alloy and pure copper to the calibrating metals.

A number of sites were selected in North America where standard test panels were exposed for varying periods of time, beginning the exposure at different times of the year. The changes in weight of these panels were used to establish a criterion against which corrosion in other areas could be compared. In order to compare the relative corrosivity of one site with another, the corrosion rate at the ASTM test plot located at State College, Pennsylvania, was arbitrarily chosen as 1.00, and all others were given as a ratio to this arbitrary base.

State College had been the scene of considerable amounts of atmospheric corrosion testing and is considered to be a mild rural atmosphere. The test site is located on the campus of Pennsylvania State University where the moderate climate varies in temperature from the upper nineties in the summer to below zero in the winter. The rainfall is typical of the rural Appalachian section of the United States.

Results of the first series of tests and calibration of nineteen test sites were reported by Larrabee and Ellis[4] in 1953 and 1959. (For details, refer to the original publications.) This work was so informative and useful that in 1960 ASTM decided to expand its program and include sites not covered in the earlier work. In the new program, instrumentation was provided in seven sites to gather data on atmospheric factors concurrent with the calibration tests. A total of 46 test sites were included in various parts of the free world. A preliminary report has been made by Coburn,[5] and extracted from his data are the rankings with two year losses for 45 of the sites, as shown in Table 11.1.

The results of other corrosion studies in the atmosphere which have been conducted by groups working within the framework of ASTM will be discussed in later sections of this chapter. Other data developed from tests conducted by the Naval Research Laboratories in Panama, British Ferrous and Nonferrous Associations, Ministry of Supply, and various producers of metal in the United States will be used as references.

Atmospheric Corrosion Tests

Effect of Conditions of Exposure on Atmospheric Corrosion

Copson,[6] in his paper on the design and interpretation of atmospheric corrosion tests, points out that in addition to the atmospheric conditions at the test site, other factors are important. Some of these

(1)Tendency to collect moisture from the atmosphere.

TABLE 11.1 — Atmospheric Corrosion: Comparative Order of Severity of 45 Locations Based on Steel and Zinc Losses

Ranking to State College Steel	Ranking to State College Zinc	Location	2-Year Exposure Grams Lost Steel	2-Year Exposure Grams Lost Zinc	Steel:Zinc Loss Ratio
1	1	Norman Wells, N.W.T., Canada	0.73	0.07	10.3
2	2	Phoenix, AZ	2.23	0.13	17.0
3	3	Saskatoon, Sask., Can.	2.77	0.13	21.0
4	4	Esquimalt, Vancouver Island, Can.	6.50	0.21	31.0
5	15	Detroit, MI	7.03	0.58	12.2
6	5	Fort Amidor Pier, Panama, C.Z.	7.10	0.28	25.2
7	11	Morenci, MI	7.03	0.53	18.0
8	7	Ottawa, Ontario, Can.	9.60	0.49	19.5
9	13	Potter County, PA	10.00	0.55	18.3
10	31	Waterbury, CT	11.00	1.12	9.8
11	10	State College, PA	11.17	0.51	22.0
12	28	Montreal, Quebec, Can.	11.44	1.05	10.9
13	6	Melbourne, Australia	12.70	0.34	37.4
14	20	Halifax (York Redoubt), NS	12.97	0.70	18.5
15	19	Durham, NH	13.30	0.70	19.0
16	12	Middletown, OH	14.00	0.54	26.0
17	30	Pittsburgh, PA	14.90	1.14	13.1
18	27	Columbus, OH	16.00	0.95	16.8
19	21	South Bend, PA	16.20	0.78	20.8
20	18	Trail, B. C., Can.	16.90	0.70	24.2
21	14	Bethlehem, PA	18.3	0.57	32.4
22	33	Cleveland, OH	19.0	1.21	15.7
23	8	Miraflores, Panama, C.Z.	20.9	0.50	41.8
24	29	London (Battersea), England	23.0	1.07	21.6
25	24	Monroeville, PA	23.8	0.84	28.4
26	35	Newark, NJ	24.7	1.63	15.1
27	16	Manila, Philippine Islands	26.2	0.66	39.8
28	32	Limon Bay, Panama, C.Z.	30.3	1.17	25.9
29	39	Bayonne, NJ	37.7	2.11	17.9
30	22	East Chicago, IN	41.1	0.79	52.1
31	9	Cape Kennedy, 1/2 mile from ocean	42.0	0.50	84.0
32	23	Brazos River, TX	45.4	0.81	56.0
33	40	Pilsey Island, England	50.0	2.50	20.0
34	42	London (Stratford) England	54.3	3.06	17.8
35	43	Halifax (Federal Building), NS	55.3	3.27	17.0
36	38	Cape Kennedy, 60 yds from ocean, 60-ft elevation	64.0	1.94	33.0
37	26	Kure Beach, NC, 800-ft lot	71.0	0.89	80.0
38	36	Cape Kennedy, 60 yds from ocean, 30-ft elevation	80.2	1.77	45.5
39	25	Daytona Beach, FL	144.0	0.88	164.0
40	44	Widness, England	174.0	4.48	39.0
41	37	Cape Kennedy, 60 yds from ocean, ground level	215.0	1.83	117.0
42	34	Dungeness, England	238.0	1.60	148.0
43	17	Point Reyes, CA	244.0	0.67	364.0
44	41	Kure Beach, NC, 80-ft lot	260.0	2.80	93.0
45	45	Galeta Point Beach, Panama, C.Z.	336.0	6.80	49.4

factors are the shape of the specimen; the direction it faces; the amount of shelter, drip, or runoff from other specimens; elevation; shading; and unusual contamination.

As an example of the influence of the shape[7] of the specimen, he cites tests of simulated service conditions for roofing materials which were assembled as shown in Figures 11.2 and 11.3. Where low-alloy steels were used in the test, perforation always occurred first in the rain troughs, next in the sheltered areas marked A, then at B; next in order were E, F, and G. Larrabee[8] found large differences in the corrosion of mild steel panels at one location under different conditions of shelter and the direction the panels faced. The difference in corrosion rate related to elevation at two nearby sites is shown in Figure 11.4.

Unusual contamination can be either detrimental or beneficial. One set of atmospheric exposure samples had to be moved because occasional visits to the area by a helicopter threw oil on the specimens and reduced the corrosion rate. In another instance, occasional contamination from the spray used by farmers on potatoes caused a serious chloride contamination that was unnatural for the area. Other tests have been located where contamination from unusual fumes is sometimes encountered, depending on wind direction and chance effluent from some local process.

For many years, different individuals and organizations have been putting specimens out in the atmosphere in all kinds of tests and in all kinds of shapes. There has been an attempt to standardize

FIGURE 11.2 — Simulated roofing racks. [SOURCE: Copson, H. R., Design and Interpretation of Atmospheric Corrosion Tests, Figure 17, Corrosion, Vol. 15, pp. 533-541 (1959)].

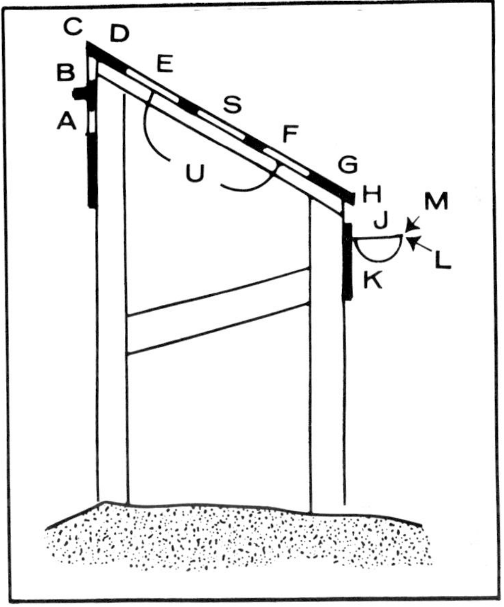

FIGURE 11.3 — Construction details of simulated roofing rack: (A) vertical surface, partly sheltered from direct rain impingement; (B and H) vertical surfaces, boldly exposed to rain; (C) sharp bend; (D and G) areas having lower surface in contact with wood; (E and F) plane inclined surfaces; (J and L) areas of water-line attack when trough is full of water; (K) area submerged when trough is full; (M) trough bead; (S) lock seam, half soldered, half not; and (U) fully sheltered under surfaces. [SOURCE: Copson, H. R., Design and Interpretation of Atmospheric Corrosion Tests, Figure 18, Corrosion, Vol. 15, pp. 533-541 (1959)].

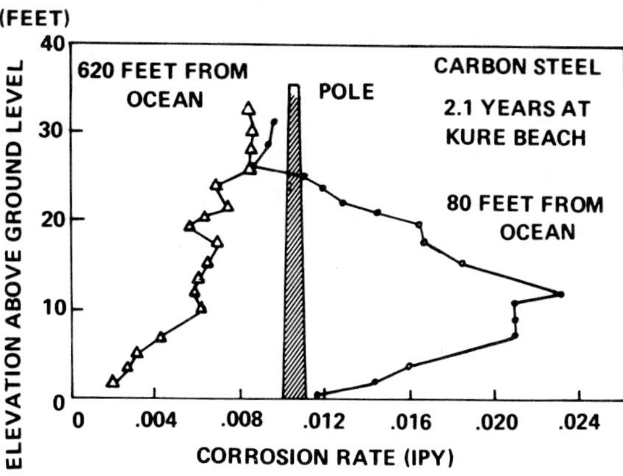

FIGURE 11.4 — Effect of height above ground at two adjacent marine sites. [SOURCE: Compton, K. G., Factors Influencing Measurement of Corrosion in Marine Atmospheres, Materials Protection, Vol. 4, No. 12, p. 13 (1965).

some of these tests, as well as the method of reporting the results. When initiating corrosion tests in the atmosphere, these standardized procedures should be consulted and used whenever possible.

There tend to be three general types of specimens: panels, tensile specimens, and stress corrosion specimens. The panels are usually in the form of sheets 4 inches wide, 6 inches long, and about 0.050 inch thick; the tensile specimens in a machined "dumb bell" shape to fit tensile testing machines; and the stress corrosion specimens in jigs to suit the type of stress being considered.

Panel specimens are usually placed in racks at a 30-degree angle to the horizontal, facing south or the source of major contamination. They are electrically insulated from the racks on which they are mounted and are arranged so that drip from their neighbors does not contaminate them. Cylindrical specimens are mounted horizontally, facing the same direction as the panels. They may be exposed fully to the weather or be partially sheltered, depending upon the requirements of the test. Several publications have covered the test fixtures and shape of specimens used for stress corrosion tests.

In most exposure tests, enough specimens are used so that removals may be made after periods of 1, 2, 7, and 20 years or 2, 5, 10, and 20 years. Such programs correct for materials that have changes in corrosion rate after the first one or two years. Very short-term tests usually are misleading in that the condition of the metal surfaces during the first few days of exposure may affect the initial corrosion rate, or "average weather conditions" may not be encountered during the initial exposure period.

Long exposures may be meaningless in that the contamination may change or the particular metals may no longer be of interest due to the development of new alloys. Older data should be interpreted in terms of the conditions existing at a given site during the time of the exposure. The control of atmospheric pollutants makes many of these locations less corrosive than in the previous period. For instance, the atmosphere of Pittsburgh, PA has become less contaminated, whereas that of State College increased for many years. Sandy Hook changed from a marine atmosphere to a mixed marine and industrial atmosphere.

Some tests are continued until failure, as in stress corrosion. However, in tests of protective coatings, as well as in others, periodic measurements of weight loss, pit depth, and change in tensile strength are made. All of these factors affect the duration of tests, but in most atmospheric exposures, a program covering a time span of ten years or more should be considered.

In many atmospheric corrosion tests, the specimens are observed every year for changes in appearance. At predetermined intervals, samples are removed, the weight loss is determined, tensile tests are made, and if pitting is significant, the deepest pit is measured, as well as a number of other pits, to determine average depth. In a few instances, composition of corrosion products is determined. Elaborate techniques such as X-ray diffraction, electron diffraction, electron microscopy, etc., are not often used in atmospheric corrosion tests. Some work has been done on measuring the thickness of oxide and other films by electrolytic reduction methods.

A number of testing laboratories are available which can be employed to conduct atmospheric corrosion tests.

Measurement of Corrosion

Results of corrosion tests are usually reported as loss in weight or reduction in thickness per unit of time; the change in tensile strength; time-to-fracture (stress corrosion cracking); time-to-perforation; time-to-10%-increase-in-electrical-resistance; time-to-initial-rust, 10% rust, or complete rust; or in some cases, changes in ductility. The reduction in thickness or penetration in microns per year (μm/y) or mils[2] per year (mpy) is the most general and useful measurement of corrosion rate. Weight loss in terms of milligrams per square decimeter per day (mdd), while useful, does not permit easy visualization. The use of metric expressions is now common, with μm (micron) per year (0.039 mpy) or mm/y (millimeter per year) being perferred.

Where pitting is the most important aspect of the corrosion behavior of the material, the maximum depth and average depth of several pits is a more useful expression than merely the average penetration. Pits may indicate perforation of a thin sheet of material, but in many cases the actual change in tensile strength is the most important observation. For this reason, the ASTM committees usually report all three changes in the specimen. A statistical treatment of the data may be made or the information may be presented in curves or tables.

Accelerated Tests

So far, reference has been made to natural exposures in the atmosphere. Many workers feel that evaluations are needed in times less than those required for on-site tests, and therefore have devised accelerated methods. The most important of these are the salt-fog or salt spray tests, the cycling humidity test, and various versions of the trade-marked Weatherometer[(3)] test. None of these reproduce the atmosphere, and no factor can be applied to the results of such tests to accurately predict behavior in a natural environment. The salt-fog tests have been standardized by ASTM and are useful in development work or in some lines of product control.

An interesting study has been conducted comparing salt-fog tests with natural exposure of zinc and steel. Panels of zinc and of steel were placed in the standard ASTM salt spray (fog) box and subjected to sprays of sodium chloride solution in concentrations of 3 and 5% and to sprays of the standard Navy synthetic seawater and natural seawater. Panels of the same material were immersed in natural seawater and were exposed to the atmosphere 80 and 800 feet from the ocean. The results are shown in Figures 11.5 and 11.6, and support the view that a salt spray test only determines the resistance of the panel to the salt spray test atmosphere and cannot be used reliably to predict behavior in any other atmosphere.

Corrosion Behavior and Resistance

Iron and Steel

Iron, in its various forms, is exposed to all kinds of environments. It tends to be highly reactive with most of them because of its natural tendency to form iron oxide. That it may resist as well as it does is sometimes due to the thin film of protective iron oxide formed on its surface by reaction with oxygen of the air. This film can prevent rusting in air at 99% relative humidity, but a specific contaminant (*e.g.*, 0.01% SO_2) may destroy the effectiveness of the film and permit continued corrosion. Thicker films of iron oxide may act as protective coatings, and after the first year or so, could reduce the corrosion rate, as shown in Figure 11.7 and Table 11.2.

While the corrosion rate of bare steel tends to decrease with time in most cases, the difference in corrosivity of different atmospheres for a particular product is tremendous. The relative corrosivity for open-hearth steel in atmospheres ranging from desert to the spray zone on an ocean beach is shown in Table 11.3. Similar ranges in corrosivity were determined by the ASTM calibration of test sites given in Table 11.1. In a few cases, the corrosion rates of ferrous metals have been reported as increasing with time, and careful analysis of the exposure conditions generally reveals that an ac-

(2)A mil is 0.001 inch

(3)Atlas Electric Devices Co., Chicago, IL.

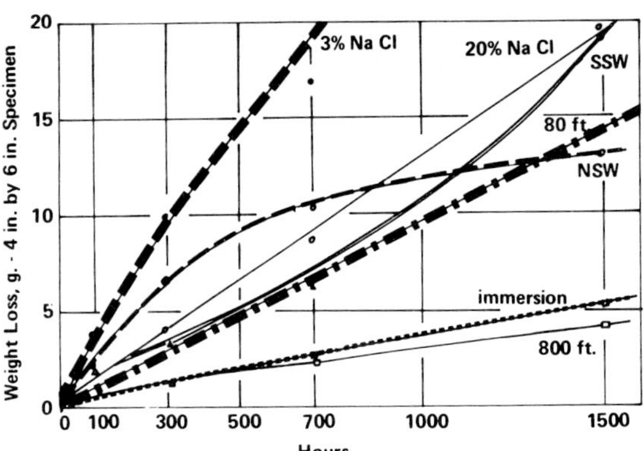

FIGURE 11.6 — Comparative tests on steel (0.05% copper) in spray of natural (NSW) and synthetic (SSW) seawater, sprays of 3 and 20% sodium chloride, in the atmosphere, 80 and 800 feet from the ocean at Key West, FL and in the ocean at Kure Beach, NC. [SOURCE: May, T. P. and Alexander, A. L., Spray Testing with Synthetic and Natural Seawater, Proc. ASTM, p. 50 (1950)].

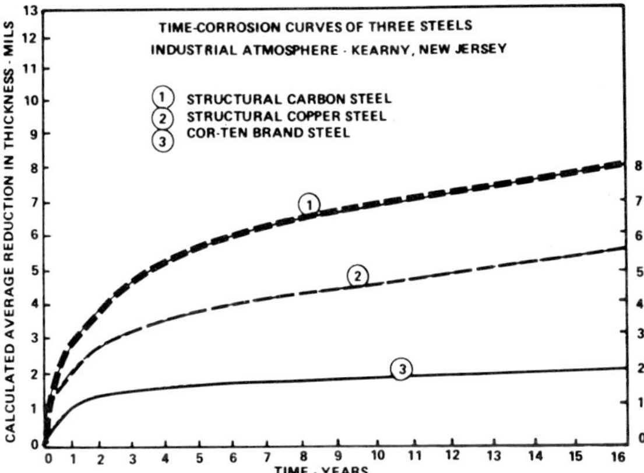

FIGURE 11.7 — Time-corrosion curves of three steels in industrial atmosphere, Kearny, NJ: (1) ordinary steel; (2) Cu-steel; and (3) Cor-ten., U.S. Steel Corp., Pittsburgh, PA. [SOURCE: Larrabee, C. P., Mechanisms of Atmospheric Corrosion of Ferrous Metals, Figure 5, Corrosion, Vol. 15, p. 526 (1959)].

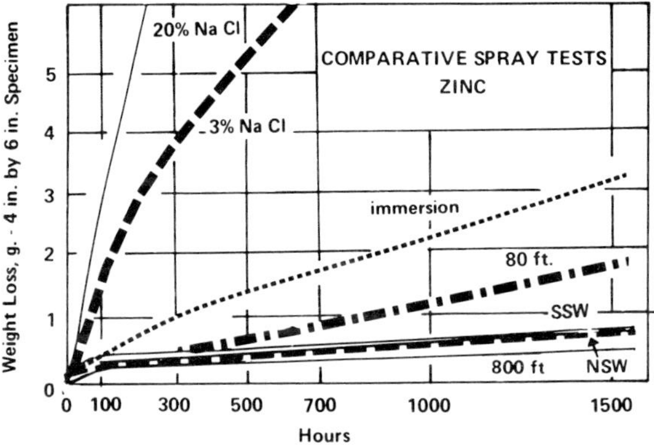

FIGURE 11.5 — Comparative tests on zinc of sprays of natural (NSW) and synthetic (SSW) seawater, sprays of 3 and 20% sodium chloride, in the atmosphere, 80 and 800 feet from the ocean and immersed in the ocean. [SOURCE: May, T. P. and Alexander, A. L., Spray Testing with Synthetic and Natural Seawater, Proc. ASTM, p. 50 (1950)].

TABLE 11.2 — Calculated Average Corrosion Rates of Low-Alloy Steels in Industrial Atmospheres

Year	Skyward Surface Mils Per Year	Ratio(1)
1st	0.5	60
2nd	0.06	8
3rd and 4th	0.045	6
5th, 6th, 7th, and 8th	0.025	3
9th to 16th	0.008	1

(1)Of the loss per year to the annual loss during the 9th through 16th years.
[SOURCE: Larrabee, C. P., Mechanism by which Ferrous Metals Corrode in the Atmosphere, Table 2, Corrosion, Vol. 15, No. 10, pp. 526-529 (1959).]

TABLE 11.3 — Relative Corrosivity of Open-Hearth Steel

Khartoum, Egypt	1
Abisco, No. Sweden	3
Singapore, Malaya	9
Daytona Beach, FL (Inland)	11
State College, PA	25
So. Bend, PA	29
Miraflores, Canal Zone, Panama	31
Kure Beach, NC, 800 ft. from ocean	38
Sandy Hook, NJ	50
Kearny, NJ	52
Vandegrift, PA	56
Pittsburgh, PA	65
Frodingham, U.K.	100
Daytona Beach, FL, near ocean	138
Kure Beach, NC, 80 ft. from ocean	475

cumulation of contaminating corrodents has occurred, thus changing the severity of the exposure.

It is generally conceded that steels containing low amounts of copper are particularly susceptible to severe atmospheric corrosion. In one test over a 3 1/2 year period in both a marine and an industrial atmosphere, a steel containing 0.01% copper corroded at a rate of 80 μm/y (3.1 mils per year), whereas increasing the copper content by a factor of five reduced the corrosion rate to only 35 μm/y (1.4 mils per year). Further additions of small amounts of nickel and chromium reduced the corrosion rate to the vicinity of a half-mil per year. This is illustrated in Figures 11.8 and 11.9 for Ni and Cr, respectively.

Other tests comparing gray cast iron, malleable iron, and these low-alloy steels indicated that their corrosion resistances were approximately the same. The influence of small alloying additions to steel has been discussed in detail and reported by various workers.

Frequently, tests on specific materials give a variety of results. A committee of ASTM found that

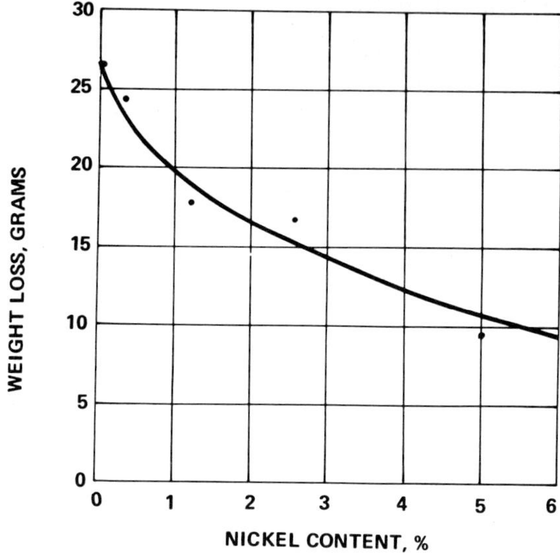

FIGURE 11.8 — Effect of nickel content on corrosion of steel in marine atmosphere.

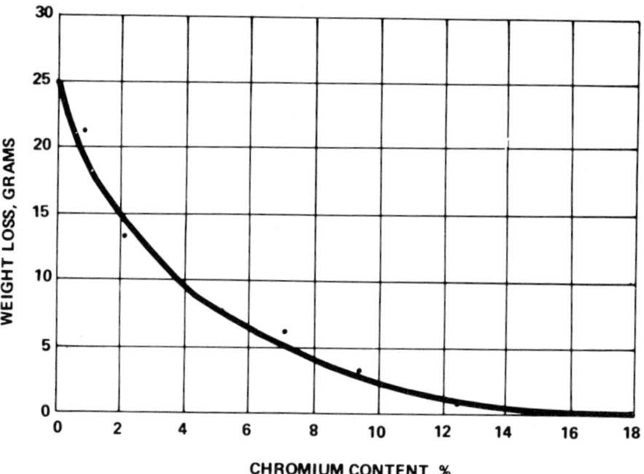

FIGURE 11.9 — Effect of chromium content on the corrosion of steel in 44 months in marine atmosphere 800 feet from ocean (specimens 4 × 6 inches).

the corrosion rate of malleable and nodular cast irons was about 60% that of mild steel (AISI 1020) at State College and 77% of that at an industrial site. This value dropped to 29% at marine sites. This indicates that markedly different rates will be found for various materials, depending on characteristics of the test site.

Plain cast iron appears to have a corrosion rate about one half that of 0.2% copper steel in a marine atmosphere. One has to be careful in citing such differences to stipulate the composition of the carbon steel because corrosion behavior of carbon steels is influenced so markedly by small variations in copper and phosphorus content. In an industrial atmosphere, a structural carbon steel showed a penetration of about 20 μm (0.8 mil), a copper structural steel about 10 μm (0.4 mil), and a low-alloy steel about 4 μm (0.15 mil) after five years of exposure.

In a study of the mechanism of rusting of low-alloy steel, Copson[7] reached the conclusion that the corrosion rate was controlled by the quantity and quality of the water reaching the surface. Contamination from sulfates and chlorides increased the solubility of the corrosion products and porosity of the rust. Copper and nickel tended to produce insoluble complex basic sulfates and lower porosity.

Larrabee thinks this explanation is oversimplified because small variations in alloying elements such as phosphorus and silicon, which do not form sulfates, have a significant effect on atmospheric corrosion. He found that an unusually low copper content in steel (0.0004% Cu) resulted in an increase of five times in corrosion rate over a steel containing the usual small amount of copper (0.20%). This was true in spite of a variation in sulfur content from 0.001 to 0.030%.

As shown in Table 11.1, it is impossible to give a corrosion rate for steel in the atmosphere without specifying the location, composition, and certain other factors. For one standard low-carbon steel, the loss in weight for the same sized specimen varied from 0.7 grams in two years at Norman Wells on Hudson Bay, to 336 grams on the beach in the Panama Canal Zone. This is a ratio of about 450:1! In between, we find some rural atmospheres producing a loss of about 10 grams, some industrial areas giving 20 to 40 gram losses, while mild marine environments cause losses between 50 and 100 grams.

If one can relate exposure conditions to those described in the literature, a fairly good estimate can be made of the probable corrosion behavior of a selected material. However, all aspects of the exposure of the metal surface must be considered. A high-strength, low-alloy (HSLA) steel may show an advantage in corrosion resistance of 12:1 over carbon steel when freely exposed in a mild environment. As the severity or the physical conditions of exposure change, the HSLA steel will show less superiority, until in crevices or the backside of structural forms in a corrosive atmosphere, it will be no better than carbon steel.

Very little needs to be said about the behavior of stainless steels (Types 200 and 300) which contain high percentages of nickel and chromium. The steels containing only chromium (Type 400) as the principal alloying constituent tend to rust superficially, but the others are relatively free from surface atmospheric corrosion. Many of them are susceptible to stress corrosion cracking in many common environments. Some crevice corrosion can be expected in insect screens and wire rope. (More will be said about the behavior of stainless steels in other chapters.)

Copper and Copper Alloys

Copper and its alloys are not exposed to the atmosphere in great quantities when compared with steel. However, the material brings aesthetic value to building construction, in addition to excellent corrosion resistance. The green aerugo (patina, verdigris, basic copper carbonate) formed on the surface provides an attractive decorative finish, while sealing the metal from further corrosion. As a consequence, some copper has been used for roofs, gutters, and as flashings on wood or composition-shingled roofs.

Extensive tests have been made of the corrosion resistance of copper and its alloys to various atmospheres. The most significant are those of the ASTM Committee on the Corrosion of Nonferrous Metals and of individual members of this committee. Various alloys were exposed in rural, industrial, and marine atmospheres for periods of up to twenty years. From data accumulated in these tests and the calibrations of relative corrosivity of the test sites, a fairly clear picture can be obtained of the corrosion behavior of copper.

In addition to the corrosion rates given in mpy, one must be mindful of dezincification of brasses and selective attack on some bronzes, as well as stress corrosion cracking illustrated by season cracking of brass. These types of corrosion contribute to the failure of the material in mechanical respects without significant weight changes or losses in thickness.

Where copper is used as flashing on roofs, corrosion has been encountered at the edge of the shingles as a continuous groove. The Copper and Brass Research Association's tests indicate that this effect is more pronounced when the atmosphere contains both chlorides and sulfides, and with wood shingles as compared with roofs of other composition. Tests indicate that all-copper roofs 20 mils or more in thickness would last several centuries in an industrial atmosphere.

If the green verdigris on copper alloys is desired for aesthetic reasons, pretreatment of the surface with appropriate acidic solutions is recommended. If oxidation by sulfur compounds precedes the desired reaction, the surface will present only a dark brown color for many years.

The exposure test by ASTM in four atmospheres during a period of seven years of exposure is given in Table 11.4. Results on all nonferrous metals in this test are reproduced here because they represent currently available products. In some of the older tests, a few of the metals are no longer being produced. Results of seven years of exposure are given in Table 11.4. The behavior of copper and copper alloys in three typical atmospheres from the two ASTM tests and from private tests is summarized in Table 11.5.

Nickel and Nickel Alloys

Nickel is widely used as a protective coating for atmospheric exposure, and some nickel alloys, while selected for other reasons, are also exposed to atmospheric corrosion. In ASTM tests, the results shown in Table 11.6 were obtained for several representative alloys.

As can be concluded from the data, nickel tends to be passive in the marine atmosphere. The

TABLE 11.4 — Seven-Year Data, Atmospheric Corrosion Test Program by Alloy and Exposure Site

Commercial Designation	Density g per cc	Panel Numbers	Days Exposed	Site[1]	Weight Change, mg per sq dm	CORROSION RATE mg per sq dm per day	CORROSION RATE mils per yr	PIT DEPTH, mils 4 Deepest Sky	PIT DEPTH, mils 4 Deepest Ground	PIT DEPTH, mils Avg.	PIT DEPTH, mils Maximum Sky	PIT DEPTH, mils Maximum Ground	Ultimate Strength ksi Control	Ultimate Strength ksi Exposed	Elongation in 2-in Percent Control	Elongation in 2-in Percent Exposed
ALUMINUM ALLOYS																
1188-H14	2.70	25-30	2557	A	34.25	0.0131	0.0070	1.5	2.4	2.0	4.4	6.7	11.7	11.9	12.0	11.2
		31-36	2558	B	62.64	0.0240	0.0127	1.7	1.5	1.6	4.7	3.3	—	11.8	—	11.[illegible]
		37-42	—	C	—[2]	—	—	—	—	—	—	—	—	—	—	--
		43-48	2538	D	10.43	0.0040	0.0021	NAA[3]	NAA[3]	NAA[3]	NAA[3]	NAA[3]	—	11.7	—	12.0
1135-H14	2.70	25-30	2557	A	52.15	0.0200	1.0106	1.6	1.1	1.4	4.5	2.7	16.4	16.3	11.8	12.4
		31-36	2558	B	72.36	0.0277	0.0147	1.6	1.4	1.5	2.8	3.0	—	16.6	—	11.6
		37-42	—	C	—[2]	—	—	—	—	—	—	—	—	—	—	—
		43-48	2538	D	11.97	0.0046	0.0025	NAA	NAA	NAA	NAA	NAA	—	16.5	—	12.1
3004-H34	2.70	25-30	2557	A	45.99	0.0176	0.0094	1.5	1.7	1.6	6.0	5.2	36.6	36.3	9.0	8.7
		31-36	2558	B	113.9	0.0436	0.0232	1.7	2.1	1.9	4.0	4.2	—	36.0	—	8.9
		37-42	—	C	—[2]	—	—	—	—	—	—	—	—	—	—	—
		43-48	2538	D	13.03	0.0050	0.0026	NAA	NAA	NAA	NAA	NAA	—	36.6	—	9.2
4043-H14	2.70	25-30	2557	A	62.18	0.0238	0.0126	1.2	1.2	1.2	5.1	3.3	22.8	22.2	10.7	9.4
		31-36	2558	B	67.76	0.0259	0.0138	1.5	1.4	1.5	4.5	3.6	—	22.2	—	10.3
		37-42	—	C	—[2]	—	—	—	—	—	—	—	—	—	—	—
		43-48	2538	D	11.43	0.0044	0.0023	NAA	NAA	NAA	NAA	NAA	—	23.0	—	10.4
X5454-O	2.68	25-30	2557	A	32.84	0.0125	0.0067	1.6	1.7	1.6	3.7	3.7	37.3	37.0	20.8	20.0
		31-36	2558	B	101.7	0.0385	0.0207	1.4	1.8	1.6	3.7	3.8	—	36.4	—	19.1
		41-42	2576	C	11.15	0.0043	0.0023	0.6	0.7	0.7	2.5	2.6	—	37.1	—	20.5
		43-48	2538	D	12.97	0.0050	0.0027	NAA	NAA	NAA	NAA	NAA	—	37.3	—	20.4
X5454-H34	2.68	25-30	2557	A	33.37	0.0126	0.0068	0.9	1.2	1.0	3.4	3.1	45.7	45.6	8.8	8.4
		31-36	2558	B	98.80	0.0374	0.0201	1.4	1.5	1.4	2.6	5.2	—	45.4	—	8.5
		37-42	2576	C	8.305	0.0031	0.0017	0.1	NAA	0.1	0.9	NAA	—	46.0	—	9.5
		43-48	2538	D	13.31	0.0051	0.0027	NAA	NAA	NAA	NAA	NAA	—	46.2	—	9.2
Alclad 5155 H34	2.66	25-30	2557	A	50.82	0.0191	0.0104	1.3	1.4	1.4	1.9	2.1	52.2	52.4	10.4	10.6
		31-36	2558	B	82.63	0.0310	0.0168	1.4	1.4	1.4	2.2	2.1	—	52.5	—	10.4
		37-42	2576	C	14.53	0.0054	0.0029	1.0	0.6	0.8	1.9	1.8	—	52.8	—	10.2
		43-48	2538	D	10.75	0.0041	0.0022	NAA	NAA	NAA	NAA	NAA	—	52.4	—	10.3
5456-O	2.64	25-30	2557	A	33.09	0.0123	0.0067	1.1	1.6	1.3	3.3	4.4	50.0	50.2	21.6	21.2
		31-36	2558	B	114.3	0.0425	0.0233	1.3	1.9	1.6	3.4	5.3	—	49.6	—	20.7
		37-42	2576	C	7.908	0.0029	0.0016	1.3	0.4	0.8	3.3	1.1	—	50.5	—	22.9
		43-48	2538	D	12.49	0.0047	0.0026	NAA	NAA	NAA	NAA	NAA	—	50.7	—	21.7
1100-H14	2.71	25-30	2557	A	58.56	0.0229	0.0122	1.5	2.1	1.8	3.5	4.4	18.5	18.7	8.5	7.0
		31-36	2558	B	72.07	0.0282	0.0150	0.7	0.4	0.5	1.6	1.5	—	18.5	—	8.0
		37-42	2568	C	24.77	0.0096	0.0051	0.7	0.4	0.6	2.9	2.1	—	18.5	—	7.5
		43-48	2538	D	11.75	0.0046	0.0025	NAA	0.2	0.1	NAA	1.3	—	18.7	—	8.5
2014-T4	2.80	25-30	2557	A	144.9	0.0425	0.0219	1.9[4]	1.4[4]	1.7[4]	2.7[4]	3.1	69.5	68.0	19.4	16.5
		31-36	2558	B	143.5	0.0561	0.0288	2.1[4]	2.2	2.2[4]	2.7	3.7	—	68.7	—	18.0
		37-42	2568	C	133.2	0.0519	0.0267	1.5[4]	3.7[4]	2.6[4]	3.1	6.6	—	65.9	—	13.0
		43-48	2538	D	43.11	0.0166	0.0085	0.2	0.1	0.2	1.5	1.2	—	69.7	—	19.0
Alclad 2014-T6	2.80	25-30	2557	A	48.46	0.0190	0.0098	1.1	0.6	0.9	1.9	1.3	66.5	69.2	8.9	9.0
		31-36	2558	B	91.30	0.0357	0.0184	1.5	1.0	1.3	2.1	1.4	—	69.6	—	9.0
		37-42	2568	C	13.83	0.0054	0.0028	0.8	0.3	0.6	1.2	1.2	—	69.3	—	9.5
		43-48	2538	D	13.10	0.0052	0.0027	0.3	<0.1	0.2	2.1	0.2	—	69.6	—	9.0
Alclad 2024-T3	2.77	25-30	2557	A	49.89	0.0162	0.0102	1.2	0.8	1.0	1.8	1.5	65.7	65.7	19.0	18.5
		31-36	2558	B	69.56	0.0204	0.0106	1.4	1.0	1.1	2.1	1.8	—	65.8	—	19.0
		37-42	2568	C	19.65	0.0076	0.0040	0.7	0.8	0.8	1.5	1.2	—	64.8	—	18.0
		43-48	2538	D	11.50	0.0034	0.0023	NAA	NAA	NAA	NAA	NAA	—	65.4	—	18.0
3003-H14	2.73	25-30	2557	A	55.92	0.0219	0.0015	1.7	1.4	1.6	2.4	3.2	22.2	22.0	7.9	7.5
		31-36	2558	B	120.2	0.0370	0.0248	1.4	3.1	2.5	3.1	4.8	—	21.9	—	8.0
		37-42	2568	C	26.68	0.0104	0.0055	0.6	1.6	1.1	1.8	3.2	—	22.2	—	8.0
		43-48	2538	D	13.56	0.0053	0.0020	<0.1	<0.1	<0.1	0.8	0.4	—	22.2	—	8.0
5005-H34	2.69	25-30	2557	A	63.36	0.0248	0.0133	1.4	1.4	1.4	3.8	3.3	22.5	22.5	7.9	7.0
		31-36	2558	B	77.06	0.0301	0.0161	0.5	0.8	0.7	1.9	3.0	—	22.1	—	7.0
		37-42	2568	C	18.70	0.0073	0.0029	0.6	0.4	0.5	4.5	2.5	—	22.6	—	7.0
		43-48	2538	D	11.60	0.0046	0.0025	NAA	NAA	NAA	NAA	NAA	—	22.5	—	7.5
5050-H34	2.69	25-30	2557	A	50.02	0.0196	0.0105	2.3	2.7	2.5	5.2	3.7	27.8	28.0	7.1	6.5
		31-36	2558	B	81.25	0.0317	0.0170	1.9	2.3	2.1	2.6	5.4	—	27.5	—	6.5
		37-42	2568	C	16.47	0.0064	0.0034	0.8	0.8	0.8	2.4	2.1	—	27.4	—	6.0
		43-48	2538	D	13.15	0.0052	0.0028	NAA	NAA	NAA	NAA	NAA	—	27.7	—	6.5
5052-H34	2.68	25-30	2557	A	40.36	0.0124	0.0085	1.5	1.7	1.6	2.6	2.8	35.6	35.2	10.3	10.0
		31-36	2558	B	95.01	0.0372	0.0200	1.8	1.7	1.8	2.2	2.2	—	35.4	—	10.0
		37-42	2568	C	11.98	0.0047	0.0025	1.0	0.3	0.7	2.2	2.0	—	35.5	—	9.0
		43-48	2538	D	12.87	0.0051	0.0027	0.1	NAA	0.1	0.5	NAA	—	35.5	—	10.0
5083-O	2.65	25-30	2557	A	56.34	0.0220	0.0120	2.2	1.4	1.8	5.1	3.5	44.0	43.5	20.5	19.0
		31-36	2558	B	117.1	0.0458	0.0249	1.8	2.8	2.3	3.0	4.4	—	42.9	—	19.0
		37-42	2568	C	29.75	0.0116	0.0063	0.6	NAA	0.3	2.1	NAA	—	43.8	—	19.5
		43-48	2538	D	21.14	0.0083	0.0045	0.1	NAA	<0.1	1.5	NAA	—	43.8	—	20.0
5083-H34	2.66	25-30	2557	A	49.06	0.0191	0.0104	3.3[4]	2.7[4]	3.0[4]	5.2	4.1	47.7	46.9	10.1	9.0
		31-36	2558	B	112.7	0.0441	0.0239	2.8[4]	3.2[4]	3.0[4]	3.4	3.9	—	46.7	—	9.5
		37-42	2568	C	14.06	0.0055	0.0030	1.0[4]	0.9[4]	1.0[4]	2.4	2.6	—	47.5	—	9.0
		43-48	2538	D	17.28	0.0068	0.0037	NAA	0.2[4]	0.1[4]	NAA	0.7	—	47.9	—	10.0

(continued)

TABLE 11.4 (continued)

Commercial Designation	Density g per cc	Panel Numbers	Days Exposed	Site[1]	Weight Change, mg per sq dm	CORROSION RATE mg per sq dm per day	mils per yr	PIT DEPTH, mils Sky	4 Deepest Ground	Avg.	Maximum Sky	Maximum Ground	Ultimate Strength ksi Control	Exposed	Elongation in 2-in Percent Control	Exposed
7075-T6	2.80	25-30	2557	A	145.0	0.0567	0.0292	4.2[4]	2.6[4]	3.1[4]	5.6	4.4	82.7	81.2	9.4	8.0
		31-36	2558	B	124.6	0.0490	0.0250	2.5[4]	2.5[4]	2.5[4]	4.0	4.8	—	81.4	—	8.0
		37-42	2568	C	118.6	0.0461	0.0237	2.0[4]	3.0[4]	2.8[4]	3.5	3.8	—	79.9	—	5.0
		43-48	2538	D	16.62	0.0066	0.0033	0.3	0.2	0.3	0.8	1.8	—	83.2	—	9.5
Alclad 7075-T6	2.80	25-30	2557	A	62.91	0.0246	0.0127	1.7	1.7	1.7	2.3	2.1	76.2	76.1	10.5	11.0
		31-36	2558	B	84.12	0.0329	0.0169	1.7	1.3	1.5	2.0	2.0	—	77.7	—	11.0
		37-42	2568	C	29.50	0.0115	0.0059	1.6	1.3	1.5	2.2	1.8	—	75.9	—	10.5
		43-48	2538	D	10.55	0.0042	0.0022	NAA	0.3	0.2	NAA	1.1	—	76.6	—	11.0
2024-T3	2.77	25-30	2557	A	96.55	0.0377	0.0196	3.1	2.4	2.8	3.7	2.1	70.3	68.7	19.4	16.8
		31-36	2558	B	109.5	0.0428	0.0222	2.7	2.3	2.5	3.3	2.8	—	69.0	—	16.7
		37-42	2576	C	76.55	0.0297	0.0154	3.2	5.2	4.2	3.8	6.7	—	64.7	—	9.2
		43-48	2538	D	13.37	0.0052	0.0027	0.3	0.3	0.3	0.7	2.7	—	69.6	—	19.7
2024-T36	2.77	25-30	2557	A	805.9	0.3152	0.1635	17.1[5]	6.3[5]	11.7[5]	34.2[5]	9.4[5]	75.1	58.8	13.3	3.5
		31-36	2558	B	143.5	0.0555	0.0288	1.8	1.9	1.9	2.3	3.0	—	67.7	—	7.6
		37-42	—	C	—[2]	—	—	—	—	—	—	—	—	—	—	—
		43-48	2538	D	14.06	0.0055	0.0029	0.4	1.6	0.7	0.8	2.5	—	74.7	—	13.8
2024-T81	2.77	25-30	2557	A	108.4	0.0423	0.0219	3.3	3.2	3.3	3.9	4.3	71.2	66.6	5.5	3.7
		31-36	2558	B	173.4	0.0678	0.0352	2.3	3.1	2.7	3.2	3.9	—	67.2	—	4.3
		37-42	2576	C	99.34	0.0386	0.0200	2.5	3.6	3.1	3.2	4.7	—	66.0	—	2.8
		43-48	2538	D	23.37	0.0092	0.0048	0.6	2.2	1.4	3.7	3.2	—	71.1	—	5.3
2024-T86	2.77	25-30	2557	A	127.4	0.0498	0.0258	3.4	2.2	2.8	4.5	3.0	78.2	72.5	5.0	3.2
		31-36	2558	B	185.7	0.0726	0.0377	1.6	2.0	1.8	2.2	2.4	—	74.2	—	3.9
		37-42	2576	C	135.0	0.0524	0.0272	2.2	3.3	2.8	2.6	4.1	—	71.6	—	1.5
		43-48	2538	D	23.43	0.0093	0.0048	0.4	2.3	1.4	3.0	3.7	—	77.5	—	4.9
Alclad 3003-H14	2.73	25-30	2557	A	47.32	0.0185	0.0097	4.9	4.7	4.8	5.3	5.1	21.9	21.9	6.3	5.8
		31-36	2558	B	84.86	0.0332	0.0175	4.3	4.4	4.4	4.9	4.9	—	22.0	—	5.7
		37-42	2576	C	17.75	0.0069	0.0036	4.0	4.5	4.3	4.5	4.9	—	22.0	—	5.7
		43-48	2538	D	11.58	0.0046	0.0024	NAA	NAA	NAA	NAA	NAA	—	22.1	—	5.5
5154-H34	2.66	25-30	2557	A	34.65	0.0135	0.0072	1.8	4.3	3.1	3.7	5.3	43.7	43.4	10.0	10.2
		31-36	2558	B	87.85	0.0344	0.0185	3.7	0.2	2.0	4.9	2.2	—	43.2	—	10.0
		37-42	2576	C	10.81	0.0042	0.0023	0.6	2.6	1.6	3.0	3.9	—	43.5	—	10.2
		43-48	2538	D	11.77	0.0047	0.0025	NAA	NAA	NAA	NAA	NAA	—	43.8	—	10.5
5357-H34	2.70	25-30	2557	A	31.40	0.0123	0.0065	3.1	4.9	4.0	4.5	6.5	24.8	24.8	7.2	7.2
		31-36	2558	B	79.35	0.0310	0.0165	3.3	4.6	4.0	4.7	6.1	—	24.6	—	7.0
		37-42	2576	C	8.821	0.0034	0.0018	0.7	3.2	2.0	3.0	4.3	—	24.7	—	7.2
		43-48	2538	D	12.22	0.0048	0.0026	NAA	NAA	NAA	NAA	NAA	—	24.9	—	7.8
6061-T4	2.70	25-30	2557	A	54.03	0.0211	0.0112	NAA	0.9	0.5	NAA	2.4	42.2	42.8	19.3	19.1
		31-36	2558	B	89.23	0.0349	0.0186	2.3	2.6	2.5	3.0	3.5	—	41.9	—	18.8
		37-42	2558	C	16.63	0.0065	0.0035	0.5	4.1	2.3	2.8	5.9	—	40.7	—	19.7
		43-48	2538	D	12.77	0.0050	0.0027	NAA	NAA	NAA	NAA	NAA	—	41.7	—	18.6
6061-T6	2.70	25-30	2557	A	64.21	0.0251	0.0134	2.3	1.5	1.9	4.5	3.2	46.7	46.5	11.0	9.4
		31-36	2558	B	95.13	0.0372	0.0198	2.2	0.6	1.4	5.1	2.6	—	46.2	—	10.0
		37-42	2558	C	28.83	0.0113	0.0060	0.2	2.7	1.5	2.8	3.9	—	46.6	—	10.3
		43-48	2538	D	12.86	0.0051	0.0027	NAA	NAA	NAA	NAA	NAA	—	47.1	—	10.9
Alclad 6061-T6	2.70	25-30	2557	A	54.19	0.0212	0.0113	0.5	1.5	1.0	2.4	2.2	45.0	44.7	11.9	12.0
		31-36	2558	B	80.28	0.0314	0.0167	1.2	0.6	0.9	2.4	2.2	—	44.7	—	12.0
		37-42	2558	C	33.35	0.0130	0.0070	NAA	2.0	1.0	NAA	2.2	—	44.9	—	12.0
		43-48	2538	D	10.85	0.0043	0.0023	0.2	NAA	0.1	1.6	NAA	—	44.9	—	12.0
HS30-WP	2.70	17-19	2207	S	1471.0	0.6667	0.3547	5.8	15.1	10.5	6.6	19.2	47.0	36.5	11.0	3.5
		29-31	2191	L	493.0	0.2250	0.1197	3.7	7.6	5.7	4.8	10.4	—	42.1	—	7.7
		41-42	2184	H	60.60	0.0280	0.0150	2.6	1.7	2.2	2.8	3.6	—	45.7	—	10.3
		53-55	2191	Ban	49.37	0.0227	0.0120	3.4	0.8	2.1	4.4	1.6	—	46.0	—	10.6
		127-129	2171	Ang	82.07	0.0377	0.0203	1.3	1.8	1.6	1.8	3.2	—	45.4	—	10.2
HC15-WP	2.79	17-19	2207	S	1488.0	0.6743	0.3473	2.5	18.0	10.0	—	—	67.2	58.0	9.8	3.6
		29-31	2191	L	537.6	0.2453	0.1263	2.5	10.0	6.0	—	—	—	65.1	—	6.3
		41-43	2184	H	62.50	0.0287	0.0147	0.9	0.9	0.9	1.0	1.2	—	67.1	—	9.1
		53-55	2191	Ban	44.10	0.0200	0.0103	0.8	0.8	0.8	1.0	1.0	—	67.4	—	9.6
		127-129	2171	Ang	96.37	0.0443	0.0230	1.0	1.0	1.0	1.2	1.4	—	66.6	—	9.2
LEAD ALLOYS																
Chemical Lead	11.34	25-30	2557	A	872.3	0.341	0.043	0.8	0.7	0.7	1.7	1.5	2.28	2.24	52.4	48.7
		31-36	2558	B	582.3	0.228	0.029	0.3	0.3	0.3	0.4	0.4	—	2.25	—	48.3
		A31-36	2213	B	349.1	0.158	0.020	0.4	0.5	0.5	0.9	0.8	—	2.37	—	47.6
		37-42	—	C	—	—	—	—	—	—	—	—	—	—	—	—
		43-48	2538	D	725.8	0.286	0.036	0.4	0.4	0.4	0.9	0.9	—	2.31	—	50.0
6% Antimonial Lead	10.88	25-30	2557	A	724.6	0.283	0.037	0.6	0.5	0.6	1.5	1.1	3.71	3.71	56.9	53.9
		31-37	2558	B	409.9	0.160	0.021	1.0	1.1	1.0	1.3	2.2	—	3.72	—	59.9
		A31-36	2213	B	279.2	0.126	0.017	0.8	0.6	0.7	1.6	1.1	—	3.74	—	47.6
		37-42	—	C	—	—	—	—	—	—	—	—	—	—	—	—
		43-48	2538	D	506.3	0.199	0.026	0.2	0.4	0.3	0.6	1.1	—	3.69	—	57.5

(continued)

TABLE 11.4 (continued)

Commercial Designation	Density g per cc	Panel Numbers	Days Exposed	Site[1]	Weight Change, mg per sq dm	CORROSION RATE mg per sq dm per day	CORROSION RATE mils per yr	PIT DEPTH, mils 4 Deepest Sky	4 Deepest Ground	Avg.	Maximum Sky	Maximum Ground	Ultimate Strength ksi Control	Ultimate Strength ksi Exposed	Elongation in 2-in Percent Control	Elongation in 2-in Percent Exposed
MAGNESIUM ALLOYS																
AZ31B-H24	1.77	27-30	2557	A	1829	0.716	0.582	14.5	5.7	10.1	15.0	10.0	41.3	33.2	17.6	16.3
		33-36	2558	B	3300	1.290	1.050	4.7	2.6	3.6	6.0	6.0	—	30.8	—	11.8
		37-42	2568	C	1425	0.555	0.451	10.3	8.0	9.2	13.0	10.0	—	35.8	—	9.1
		43-48	2538	D	1729	0.681	0.553	8.5	4.5	6.5	10.0	6.0	—	34.7	—	10.9
HK31A-H24	1.79	25-30	2557	A	1691	0.662	0.532	15.1	8.8	11.9	22.0	12.5	36.4	30.6	9.4	3.0
		31-36	2558	B	3953	1.550	1.250	26.9	6.1	16.5	28.0	11.0	—	20.7	—	0.8
		39-40	2568	C	1588	0.607	0.488	32.5	11.8	22.2	37.0	15.0	—	27.6	—	0.8
		43-48	2538	D	2015	0.793	0.637	26.7	23.3	25.0	31.0	31.0	—	23.6	—	0.8
HM21XA-T8	1.78	25-30	2557	A	1699	0.664	0.537	11.2	9.0	10.1	18.0	12.5	35.3	30.4	5.8	2.9
		31-36	2558	B	3998	1.560	1.270	26.2	5.9	16.2	28.5	8.5	—	19.6	—	0.6
		37-42	—	C	—(2)	—	—	—	—	—	—	—	—	—	—	—
		43-48	2538	D	2000	0.789	0.639	26.9	18.3	22.6	28.5	23.5	—	23.5	—	0.5
ZE10XA-H24	1.76	25-30	2557	A	3520	1.380	1.130	20.6	18.8	19.7	26.5	26.5	39.1	28.8	13.4	2.5
		31-36	2558	B	3682	1.440	1.170	13.8	9.4	11.6	22.0	11.0	—	28.7	—	6.1
		37-42	—	C	—(2)	—	—	—	—	—	—	—	—	—	—	—
		43-48	2538	D	1883	0.741	0.606	15.8	21.3	18.2	28.5	24.5	—	32.1	—	3.1
ZH11X1-H24	1.77	25-30	2557	A	2032	0.794	0.645	20.4	15.9	18.2	24.0	18.5	38.8	30.4	16.5	3.5
		31-36	2558	B	3930	1.550	1.260	25.8	7.0	16.5	33.5	11.5	—	23.7	—	1.3
		37-42	—	C	—(2)	—	—	—	—	—	—	—	—	—	—	—
		43-48	2538	D	1942	0.766	0.623	35.8	24.1	29.9	39.5	28.0	—	26.1	—	1.5
MOLYBDENUM																
Molybdenum	10.2	25-30	2557	A	1222	0.4782	0.0674	1.7	1.4	1.6	2.2	2.4	102.2	99.4	15.2	17.0
		31-36	2558	B	793.4	0.3101	0.0437	1.4	1.5	1.5	2.8	3.0	—	100.5	—	17.5
		37-42	2558	C	—(2)	—	—	—	—	—	—	—	—	—	—	—
		43-48	2538	D	320.1	0.1261	0.0178	1.7	1.6	1.7	3.7	3.3	—	99.4	—	15.4
NICKEL ALLOYS																
"A" Nickel	8.85	25-30	2557	A	150.8	0.0590	0.0095	0.8	0.6	0.7	1.4	1.0	62.5	62.0	44.0	42.5
		31-36	2558	B	1095	0.4282	0.0652	0.8	0.9	0.8	1.1	1.1	—	62.0	—	42.7
		37-42	—	C	—(2)	—	—	—	—	—	—	—	—	—	—	—
		43-48	2538	D	141.5	0.0058	0.0090	0.5	0.5	0.5	0.6	0.6	—	62.3	—	43.2
Incoloy	8.00	25-30	2557	A	33.60	0.0131	0.0023	0.6	0.4	0.5	1.3	0.7	85.0	84.8	44.2	44.3
		31-36	2558	B	11.72	0.0045	0.0008	nil	nil	nil	nil	nil	—	84.8	—	42.1
		37-42	—	C	—(2)	—	—	—	—	—	—	—	—	—	—	—
		43-48	2538	D	11.70	0.0046	0.0008	0.2	0.4	0.3	0.4	0.6	—	84.4	—	42.7
Inconel	8.54	25-30	2557	A	27.64	0.0108	0.0018	0.9	0.5	0.7	1.3	0.8	90.8	93.1	42.5	44.1
		31-36	2558	B	16.59	0.0064	0.0013	0.5	0.4	0.4	0.7	0.6	—	91.6	—	45.1
		37-42	—	C	—(2)	—	—	—	—	—	—	—	—	—	—	—
		43-48	2538	D	13.42	0.0053	0.0008	0.1	0.2	0.2	0.2	0.3	—	92.7	—	43.8
Monel	8.80	25-30	2557	A	224.6	0.0878	0.0143	0.7	0.6	0.6	0.7	0.7	75.0	75.1	39.5	40.0
		31-36	2558	B	505.5	0.1957	0.0318	0.9	0.9	0.9	1.4	1.4	—	74.5	—	40.0
		37-42	—	C	—(2)	—	—	—	—	—	—	—	—	—	—	—
		43-48	2538	D	125.4	0.0505	0.0082	0.6	0.7	0.6	0.8	0.8	—	75.7	—	40.0
Ni-O-Nel	8.00	25-30	2557	A	18.93	0.0074	0.0013	0.5	0.4	0.5	0.7	0.6	89.5	90.3	48.0	47.3
		31-36	2558	B	11.00	0.0043	0.0007	0.3	0.3	0.3	0.4	0.4	—	88.8	—	40.9
		37-42	—	C	—(2)	—	—	—	—	—	—	—	—	—	—	—
		43-48	2538	D	11.81	0.0046	0.0008	0.3	0.3	0.3	0.5	0.4	—	89.2	—	47.5
TITANIUM ALLOYS																
6Al-4V Titanium	4.42	25-30	2557	A	24.5	0.010	0.003	nil	nil	nil	nil	nil	133.0	133.2	9.9	9.5
		31-36	2558	B	25.0	0.010	0.003	—	—	—	—	—	133.7	133.7	—	9.5
		37-42	—	C	—(2)	—	—	—	—	—	—	—	—	—	—	—
		43-48	2538	D	24.3	0.010	0.003	—	—	—	—	—	134.2	134.2	—	9.1
8 Mn Titanium	4.76	25-30	2557	A	7.1	0.003	0.001	nil	nil	nil	nil	nil	145.8	145.8	17.2	17.6
		31-36	2558	B	4.0	0.002	0.001	—	—	—	—	—	—	146.0	—	17.4
		37-42	—	C	—(2)	—	—	—	—	—	—	—	—	—	—	—
		43-48	2538	D	10.2	0.004	0.001	—	—	—	—	—	—	144.8	—	17.1
2.5Al-16 Titanium	4.65	25-30	2557	A	27.7	0.011	0.003	nil	nil	nil	nil	nil	171.3	158.7	6.1	5.8
		31-36	2558	B	23.9	0.009	0.003	—	—	—	—	—	—	168.7	—	6.2
		37-42	—	C	—(2)	—	—	—	—	—	—	—	—	—	—	—
		43-48	2539	D	20.6	0.008	0.003	—	—	—	—	—	—	159.7	—	6.5
821	4.40	25-30	2557	A	15.1	0.006	0.002	nil	nil	nil	nil	nil	125.8	126.0	16.0	16.7
		31-36	2558	B	13.5	0.005	0.002	—	—	—	—	—	—	125.3	—	16.5
		37-42	—	C	—(2)	—	—	—	—	—	—	—	—	—	—	—
		43-48	2538	D	10.3	0.004	0.002	—	—	—	—	—	—	125.2	—	16.7
75A	4.54	25-30	2557	A	2.4	0.001	<0.001	nil	nil	nil	nil	nil	94.8	96.4	23.6	23.3
		31-36	2558	B	0.8	nil	nil	—	—	—	—	—	—	96.3	—	23.0
		37-43	—	C	—(2)	—	—	—	—	—	—	—	—	—	—	—
		43-48	2538	D	2.4	0.001	<0.001	—	—	—	—	—	—	94.7	—	25.3

(continued)

TABLE 11.4 (continued)

Commercial Designation	Density g per cc	Panel Numbers	Days Exposed	Site[1]	Weight Change, mg per sq dm	CORROSION RATE mg per sq dm per day	mils per yr	Sky	PIT DEPTH, mils 4 Deepest Ground	Avg.	Maximum Sky	Ground	Ultimate Strength ksi Control	Exposed	Elongation in 2-in Percent Control	Exposed
140A	4.65	25-30	2557	A	2.4	0.001	nil	nil	nil	nil	nil	nil	133.6	134.5	17.4	18.9
		31-36	2558	B	1.6	0.001	—	—	—	—	—	—	—	133.6	—	17.8
		37-42	—	C	—[2]	—	—	—	—	—	—	—	—	—	—	—
		43-48	2538	D	1.6	0.001	—	—	—	—	—	—	—	133.6	—	18.0
5Al-2.5Sn Titan	25-30	25-30	2557	A	2.4	0.001	nil	nil	nil	nil	nil	nil	124.8	123.5	15.4	15.2
		31-36	2558	B	1.6	0.001	—	—	—	—	—	—	—	124.5	—	14.3
		37-42	—	C	—[2]	—	—	—	—	—	—	—	—	—	—	—
		43-48	2538	D	1.6	0.001	—	—	—	—	—	—	—	124.3	—	14.3
4Al-3Mo 1V Titanium	4.54	25-30	2557	A	1.6	0.001	nil	nil	nil	nil	nil	nil	131.8	131.5	12.1	13.3
		31-36	2558	B	0	nil	—	—	—	—	—	—	—	131.9	—	13.5
		37-42	—	C	—[2]	—	—	—	—	—	—	—	—	—	—	—
		43-48	2538	D	0	nil	—	—	—	—	—	—	—	132.8	—	13.1
ZINC ALLOYS																
High Grade Zinc	7.13	25-30	2557	A	2048	0.801	0.161	2.2	1.1	1.7	2.9	1.7	19.0	18.7	86.3	66.8
		31-36	2558	B	2542	0.994	0.200	1.8	1.9	1.9	2.8	2.9	—	18.0	—	63.8
		37-42	—	C	—[2]	—	—	—	—	—	—	—	—	—	—	—
		43-48	2538	D	550.0	0.217	0.044	0.3	NAA	0.2	1.1	NAA	—	18.9	—	76.3
1 Cu Zinc	7.18	25-30	2557	A	2356	0.921	0.185	5.1	2.4	3.7	6.3	3.2	23.4	22.7	66.0	55.5
		31-36	2558	B	2459	0.961	0.963	5.7	2.4	4.1	7.8	3.0	—	22.2	—	50.1
		37-42	—	C	—[2]	—	—	—	—	—	—	—	—	—	—	—
		43-48	2538	D	584.7	0.230	0.046	1.6	1.3	1.5	2.3	2.2	—	22.8	—	55.5
TANTALUM																
Tantalum	16.6	25-30	2557	A		—	—	nil	nil	nil	nil	nil	—	80.2	—	2.9
			2558	—		—	—	nil	nil	nil	nil	nil	—	83.7	—	2.0
COPPER ALLOYS																
1.25Sn Phosphor Bronze	8.91	25-30	2557	A	1088	0.4254	0.0688	nil	nil	nil	nil	nil	42.0	42.1	46.9	47.5
		31-36	2558	B	993.3	0.3883	0.0628	—	—	—	—	—	—	41.5	—	47.1
		61-62	2558	C	265.8	0.1039	0.0168	—	—	—	—	—	—	41.8	—	48.0
		43-48	2538	D	402.8	0.1587	0.0257	—	—	—	—	—	—	41.4	—	48.3
90-10 Cupro-nickel	8.94	25-30	2557	A	1812	0.7085	0.1141	nil	nil	nil	nil	nil	50.4	49.2	32.2	32.5
		31-36	2558	B	2366	0.9251	0.1490	—	—	—	—	—	—	48.5	—	32.2
		37-42	2558	C	868.4	0.3395	0.0547	—	—	—	—	—	—	49.7	—	31.9
		43-48	2538	D	1288	0.5075	0.0818	—	—	—	—	—	—	48.8	—	32.1
5Sn Phosphor Bronze	8.90	25-30	2557	A	1562	0.6113	0.0993	nil	nil	nil	nil	nil	—	49.6	—	50.0
		31-36	2558	B	1187	0.4637	0.0750	—	—	—	—	—	—	49.4	—	46.0
		37-42	—	C	—	—	—	—	—	—	—	—	—	—	—	—
		43-48	2538	D	349.2	0.1377	0.0223	—	—	—	—	—	—	49.7	—	50.0
Tin Brass	8.78	25-30	2557	A	375.6	0.1470	0.0240	nil	nil	nil	nil	nil	—	46.3	—	52.0
		31-36	2558	B	1065	0.4163	0.0683	—	—	—	—	—	—	45.8	—	51.0
		36-42	—	C	—	—	—	—	—	—	—	—	—	—	—	—
		43-48	2538	D	470.7	0.1853	0.0303	—	—	—	—	—	—	46.0	—	51.0
2Si-7Al Bronze	7.78	25-30	2557	A	226.8	0.0890	0.0133	nil	nil	nil	nil	nil	—	75.8	—	46.0
		31-36	2558	B	689.3	0.2693	0.0507	—	—	—	—	—	—	73.8	—	45.0
		37-42	—	C	—	—	—	—	—	—	—	—	—	—	—	—
		43-48	2538	D	181.7	0.0717	0.0133	—	—	—	—	—	—	75.7	—	45.0
Red Brass	8.75	25-30	2557	A	516.7	0.20	0.0334	nil	nil	nil	nil	nil	42.4	41.9	42.0	40.8
		31-36	1558	B	1111	0.43	0.0711	—	—	—	—	—	—	41.4	—	41.0
		37-42	2558	C	371.9	0.15	0.0240	—	—	—	—	—	—	42.0	—	40.4
		43-48	2538	D	468.4	0.19	0.0306	—	—	—	—	—	—	42.3	—	40.0
Cartridge Brass	8.53	25-30	2557	A	457.2	0.18	0.0304	nil	nil	nil	nil	nil	49.1	48.8	59.0	57.5
		31-36	2558	B	1302	0.51	0.0862	—	—	—	—	—	—	47.9	—	58.3
		37-42	2558	C	248.8	0.10	0.0169	—	—	—	—	—	—	49.2	—	58.0
		43-48	2538	D	492.3	0.19	0.0321	—	—	—	—	—	—	48.8	—	57.0
10% Nickel Silver	8.69	25-30	2557	A	364.2	0.1424	0.0236	nil	nil	nil	nil	nil	56.9	55.6	46.7	46.7
		31-36	2558	B	1193	0.4665	0.0773	—	—	—	—	—	—	55.8	—	47.1
		37-42	2558	C	160.3	0.0627	0.0104	—	—	—	—	—	—	55.2	—	48.0
		43-48	2538	D	387.4	0.1527	0.0253	—	—	—	—	—	—	56.1	—	47.5
18% Nickel Silver	8.74	25-30	2557	A	329.4	0.1288	0.0212	nil	nil	nil	nil	nil	61.1	55.9	38.5	38.4
		31-36	2558	B	1233	0.4820	0.0794	—	—	—	—	—	—	54.1	—	38.7
		37-42	—	C	—[2]	—	—	—	—	—	—	—	—	—	—	—
		43-48	2538	D	370.1	0.1458	0.0240	—	—	—	—	—	—	55.7	—	38.9
Tough-pitch Copper	8.91	25-30	2557	A	1032	0.4037	0.0652	nil	nil	nil	nil	nil	35.4	32.8	44.0	44.8
		31-36	2558	B	901.4	0.3524	0.0570	—	—	—	—	—	—	33.4	—	45.2
		37-38	2558	C	396.3	0.1549	0.0250	—	—	—	—	—	—	33.7	—	44.5
		43-48	2538	D	462.3	0.1822	0.0294	—	—	—	—	—	—	34.0	—	44.5
No. 30	8.90	25-30	2557	A	689.0	0.2695	0.0436	nil	nil	nil	nil	nil	32.0	31.2	44.0	45.2
		31-36	2558	B	911.6	0.3564	0.0577	—	—	—	—	—	—	31.8	—	44.7
		37-42	2581	C	574.2	0.2225	0.0360	—	—	—	—	—	—	31.9	—	44.7
		43-48	2538	D	475.4	0.1873	0.0303	—	—	—	—	—	—	33.6	—	44.7

(continued)

TABLE 11.4 (continued)

Commercial Designation	Density g per cc	Panel Numbers	Days Exposed	Site[1]	Weight Change, mg per sq dm	CORROSION RATE mg per sq dm per day	mils per yr	PIT DEPTH, mils Sky	4 Deepest Ground	Avg.	Maximum Sky	Ground	Ultimate Strength ksi Control	Exposed	Elongation in 2-in Percent Control	Exposed
No. 60	8.90	25-30	2557	A	519.7	0.2032	0.0329	nil	nil	nil	nil	nil	33.9	33.4	43.5	44.7
		31-36	2558	B	897.0	0.3507	0.0567	—	—	—	—	—	—	33.6	—	44.2
		37-42	—	C	—[2]	—	—	—	—	—	—	—	—	—	—	—
		43-48	2538	D	390.1	0.1537	0.0249	—	—	—	—	—	—	34.3	—	44.3
No. 95	8.90	25-30	2557	A	603.5	0.23	0.0378	nil	nil	nil	nil	nil	39.5	39.2	39.5	38.3
		31-36	2558	B	969.5	0.38	0.0616	—	—	—	—	—	—	38.5	—	39.2
		37-42	2581	C	—[2]	—	—	—	—	—	—	—	—	—	—	—
		43-48	2538	D	511.1	0.20	0.0329	—	—	—	—	—	—	38.7	—	41.6
No. 180	8.90	25-30	2557	A	487.7	0.1910	0.0307	nil	nil	nil	nil	nil	—	46.9	—	—
		31-36	2558	B	979.9	0.3830	0.0613	—	—	—	—	—	—	47.6	—	—
		37-42	—	C	—[2]	—	—	—	—	—	—	—	—	—	—	—
		43-48	2538	D	209.2	0.1160	0.0187	—	—	—	—	—	—	47.7	—	—
Advance	8.90	25-30	2557	A	293.1	0.1147	0.0187	nil	nil	nil	nil	nil	—	57.6	—	40.0
		31-36	2558	B	964.6	0.3773	0.0610	—	—	—	—	—	—	57.1	—	40.0
		37-42	2558	C	78.98	0.0290	0.0050	—	—	—	—	—	—	57.3	—	40.0
		43-48	2538	D	159.3	0.0627	0.0100	—	—	—	—	—	—	57.5	—	40.0
No. 212	8.41	25-30	2557	A	542.1	0.2120	0.0363	nil	nil	nil	nil	nil	57.3	57.7	40.6	41.0
		31-36	2558	B	931.2	0.3641	0.0623	—	—	—	—	—	—	56.2	—	41.1
		37-42	—	C	—[2]	—	—	—	—	—	—	—	—	—	—	—
		43-48	2538	D	371.8	0.1465	0.0251	—	—	—	—	—	—	56.6	—	41.0
Beraloy A	8.26	25-30	2557	A	590.7	0.2310	0.0402	1.6	1.6	1.6	2.4	1.9	—	74.1	—	38.5
		31-36	2558	B	1015	0.3967	0.0690	1.8	1.6	1.7	2.2	2.2	—	74.4	—	37.9
		37-42	—	C	—[2]	—	—	—	—	—	—	—	—	—	—	—
		43-48	2538	D	434.3	0.1710	0.0298	1.2	1.0	1.1	1.4	1.1	—	75.2	—	33.1
Beraloy D	8.26	25-30	2557	A	552.3	0.2160	0.0376	1.4	1.4	1.4	1.6	1.7	—	67.9	—	49.5
		31-36	2558	B	1026	0.4010	0.0698	1.6	1.5	1.6	2.0	1.7	—	67.2	—	51.6
		37-42	—	C	—[2]	—	—	—	—	—	—	—	—	—	—	—
		43-48	2538	D	399.3	0.1577	0.0274	1.2	1.2	1.2	1.6	1.7	—	67.8	—	48.8
Beraloy C	8.75	25-30	2557	A	1036	0.4047	0.0704	1.3	1.1	1.2	1.6	1.3	—	50.1	—	35.2
		31-36	2558	B	1555	0.6077	0.1058	1.5	1.3	1.4	2.0	1.7	—	49.7	—	35.5
		37-42	—	C	—[2]	—	—	—	—	—	—	—	—	—	—	—
		43-48	2538	D	589.7	0.2323	0.0404	1.0	1.1	1.1	1.3	1.3	—	50.3	—	35.2
ALUMINUM ALLOYS																
5083-H34	2.65	25-30	2557	A	41.60	0.0132	0.0089	2.6[4]	1.9[4]	2.2[4]	4.3	3.1	53.1	52.2	12.0	10.5
		31-36	2558	B	129.1	0.0379	0.0206	2.5[4]	1.9[4]	2.2[4]	3.7	2.8	—	51.6	—	10.3
		37-42	—	C	—[2]	—	—	—	—	—	—	—	—	—	—	—
		43-48	2538	D	15.76	0.0062	0.0034	<0.1	0.4	0.2	0.3	1.6	—	52.1	—	11.0
7079-T6	2.80	25-30	2557	A	—[5]	—	—	1.5	1.4	1.5	3.1	2.2	78.2	78.1	11.0	10.6
		31-36	2558	B	123.0	0.0485	0.0250	1.4	1.3	1.4	2.4	2.6	—	77.5	—	10.5
		37-42	2850	C	—[5]	—	—	3.5	3.6	3.6	11.2	11.6	—	77.4	—	10.6
		43-48	2538	D	12.48	0.0049	0.0025	NAA	NAA	NAA	NAA	NAA	—	78.6	—	11.4
Alclad 7079-T6	2.80	25-30	2557	A	50.38	0.0198	0.0102	1.4	1.1	1.3	2.7	3.1	73.1	73.6	10.9	11.0
		31-36	2558	B	75.08	0.0296	0.0153	1.3	1.7	1.5	2.2	3.3	—	73.2	—	11.2
		37-42	—	C	—[2]	—	—	—	—	—	—	—	—	—	—	—
		43-48	2538	D	10.56	0.0043	0.0022	NAA	NAA	NAA	NAA	NAA	—	73.3	—	10.8
1199-H18	2.70	25-30	2557	A	25.17	0.0098	0.0052	4.5	5.3	4.9	6.3	8.9	16.7	16.1	4.8	5.0
		31-36	2558	B	52.47	0.0205	0.0109	NAA	NAA	NAA	NAA	NAA	—	16.0	—	6.0
		37-42	2333	C	—[2]	—	—	—	—	—	—	—	—	—	—	—
		43-48	2538	D	10.29	0.0041	0.0022	NAA	NAA	NAA	NAA	NAA	—	16.2	—	6.1
S1-H	2.70	17-19	2207	S	899.6	0.4080	0.2170	2.9	17.0	10.0	3.6	29.6	16.3	14.3	3.6	3.0
		29-31	2191	L	348.8	0.1590	0.0847	3.3	9.2	6.3	4.8	13.6	—	15.7	—	4.1
		41-43	2184	H	32.53	0.0147	0.0077	1.9	2.3	2.1	3.2	6.0	—	16.3	—	4.1
		53-55	2191	Ban	23.57	0.0107	0.0057	1.5	1.2	1.4	2.0	1.8	—	16.3	—	4.1
		127-129	2171	Ang	28.34	0.0130	0.0070	1.6	1.6	1.7	2.2	2.0	—	16.7	—	3.8
NS6-¼H	2.65	17-19	2207	S	1420	0.6480	0.3513	4.0	19.0	11.6	4.6	23.2	50.3	38.3	12.3	5.5
		29-31	2191	L	508.5	0.2317	0.1257	2.6	8.4	5.5	3.2	11.6	—	44.0	—	7.7
		41-43	2184	H	80.33	0.0367	0.0197	1.0	1.0	1.0	1.4	1.4	—	49.2	—	10.4
		53-55	2191	Ban	48.73	0.0220	0.0117	0.6	0.6	0.7	1.4	1.0	—	49.8	—	11.8
		127-129	2171	Ang	83.95	0.0387	0.0210	2.2	2.7	2.5	3.2	4.4	—	47.9	—	10.9

[1] A = Kure Beach (80-ft site), NC; B = New York area (Newark, NJ); C = Point Reyes, CA; D = State College, PA; S = Sheffield, England; L = London, England; H = Hayling Island, England; Ban = Banbury, England; Ang = Anglesey, England.
[2] Panels lost.
[3] No appreciable attack.
[4] Intergranular attack.
[5] Test data not considered reliable.

TABLE 11.5 — Corrosion of Copper and Copper Alloys in Mils/Yr (mpy)

Material[1]	New York Newark	State College, PA	Kure Beach La Jolla
T.P. Copper	0.054-0.057	0.017-0.029	0.050-0.065
Phosphor Bronze	0.075	0.022	0.099
Red Brass	0.071	0.030-0.017	0.033-0.013
Yellow Brass	0.084-0.086	0.030-0.017	0.030-0.006
90-10 Cupro Ni	0.149	0.082	0.114
Tin Brass	0.068	0.03	0.024
Admiralty Metal	0.099	0.02	0.013
Manganese Bronze	0.34	0.019	0.0795
Si Al Bronze	0.050	0.013	0.013
10% Ni Silver	0.077	0.025	0.023
18% Ni Silver	0.079	0.024	0.021
Advance	0.061	0.01	0.018
Be Copper	0.069	0.030	0.040
12 different Coppers	0.032	0.018	0.053

[1]Composition of these and other alloys is given in Chapter 4.

TABLE 11.6 — Corrosion in Mils/Yr of Nickel and Its Alloys in Various Atmospheres

Material	New York Newark	State College, PA	Kure Beach La Jolla
Nickel	0.0058	0.144	0.0085
	0.0095	0.065	0.009
Monel	0.0064	0.062	0.0067
	0.14	0.032	0.0082
Incoloy[1]	0.0023	0.0008	0.0008
Inconel[1]	0.0018	0.0013	0.0008
Nionel[1]	0.0013	0.0007	0.0008

[1]The International Nickel Co., Inc., New York, NY.

ratio between the corrosion rate for nickel exposed to the industrial atmosphere and that exposed to rural or marine atmospheres was 28:1. Nickel-base alloys usually are grouped into five classes, according to the principal alloying constituent: silicon, copper, chromium, chromium-iron and molybdenum-chromium, and chromium-iron and molybdenum-chromium. Some of the familiar trade names are Hastelloy C,[4] Monel, Nichrome,[5] and other Hastelloys.

Aluminum and Its Alloys

Aluminum, in its many forms, often is exposed directly to the atmosphere with no protective coating. Examples of such exposure are the facings or architectural trim on buildings, aircraft, household utensils, electrical conductors, small buildings, fences, outboard motors, and irrigation piping. It is exceeded only by steel in tonnage directly exposed to the elements. It is produced in the form of wrought products, extrusions, and castings with a large variety of alloying elements to impart desired mechanical properties. The atmospheric corrosion behavior of aluminum products fits into some fairly well-defined patterns that are related to composition.

[4]Stellite Division, Cabot Corp., Kokomo, IN.
[5]INCO Inc., New York, NY.

While pure aluminum has excellent atmospheric corrosion resistance, alloys containing copper and silicon as the principal alloying constituents are susceptible and should be used with care. A numbering system has been developed for wrought alloys which indicates the principal alloying elements by the first digit in the four-digit number designating the specific aluminum alloy.

1XXX Pure Aluminum
2XXX Copper
3XXX Manganese
4XXX Silicon
5XXX Magnesium
6XXX Magnesium-Silicon
7XXX Zinc

The casting alloys do not have a similar numbering system, so that the user or designer must check the composition to predict corrosion resistance. A listing is given in Chapter 4 of this text. Persons not fully acquainted with the corrosion characteristics of the various aluminum alloys should restrict their selection to the 1XXX, 3XXX, 5XXX, and 6XXX alloys. Expert opinion should be obtained when using the other alloys.

In a rural atmosphere, the corrosion rate for most alloys is about 2.5 "micro-inches" (0.06 μm/y) per year,[6] with those containing large amounts of copper about double this low rate. Changes in tensile strength because of corrosion vary from 0 to less than 1% for sheet material.

In a marine environment, the differences between alloys appear as a tenfold increase, from about 25 micro-inches (0.6 μm) for the less corrosion-resistant materials, to about 10 micro-inches (0.7 μm) per year for the better materials. Pitting also is about 10 times greater in marine atmospheres. In nearly all cases, the corrosion rate per year was greater in the severe industrial atmosphere of Newark, NJ than in the marine atmosphere. However, the pit depth appeared to be less in the industrial atmosphere, with the pit growth rate showing a definite decrease with time.

Some aluminum alloys develop severe pitting and a voluminous white corrosion product under some exposure conditions in a marine atmosphere. Aluminum roofs have been known to corrode severely at the overlaps. Some aluminum alloys also can be attacked intergranularly when exposed after certain metallurgical treatments (cold working or precipitation hardening). General intergranular attack or *exfoliation* can then occur (Chapter 5). The attack tends to start at sheared edges or punched holes, but is not restricted to these areas. Aircraft manufacturers, in particular, must guard against this type of corrosion.

In designing aluminum equipment, care must be exercised to avoid dissimilar metal couples and

[6]Micro-inch is one thousandth of a mil; or a millionth of an inch (0.000001 in.).

the attendant galvanic corrosion. Copper and rusty steel are particularly bad in contact with aluminum. Due to the passive film on stainless steel, it can be used in contact with aluminum in the atmosphere with little expectation of accelerated corrosion, despite the differences in potential. In addition, designers should be aware of the possibility that some aluminum alloys may be sensitized to intergranular corrosion by heat treatment.

For specific corrosion rates of various wrought aluminum alloys in four representative atmospheres, refer to Table 11.4. In this table, the atmospheres are: (A) Kure Beach, NC (marine—80 ft from the ocean); (B) Newark, NJ (industrial); (C) Point Reyes, CA (marine); and (D) State College, PA (rural). Commercial designations of these alloys are those indicated in the Aluminum Association Standard, 1966. Corrosion rates of sand cast and die cast alloys parallel those of the wrought alloys. High-copper alloys are most susceptible, with silicon next, while magnesium alloys appear to be the most resistant to atmospheric attack. Cylindrical tensile specimens with a minimum surface-to-volume ratio, while very rough from corrosion, did not show as large a loss in tensile strength as strips cut from sheets.

As would be expected, constant exposure to moisture with a limited supply of oxygen to the aluminum surface leads to the rapid corrosion of any aluminum apparatus or equipment component. This is due to the highly reactive nature of aluminum that leads to formation of oxides or hydroxides. In the presence of oxygen, a protective aluminum oxide film develops on any aluminum surface. This oxide film is substantially unreactive with the normal constituents of the atmosphere. If the film is removed by mechanical or chemical means and the aluminum exposed to water, a rapid reaction sets in and large quantities of the aluminum are converted to the hydroxide and subsequently to the oxide.

For instance, in aerated distilled water, aluminum is quite inert and corrosion resistant. In deaerated distilled water, aluminum is quite reactive, particularly if the protective oxide layer is initially removed by mechanical abrasion or by a scratch.

Stress corrosion cracking data appear to indicate that most of the 1000, 3000, and 6000 series of wrought alloys, the magnesium, and the silicon-magnesium casting alloys are relatively immune. Most of the other alloys may be susceptible under some conditions. This subject should be investigated carefully by the designer or user if stresses are high and the atmosphere is corrosive.

The usual corrosion behavior in the atmosphere involves pitting and roughening of the surface with a fairly large decrease in the corrosion rate after the first one to three years of exposure. In a marine atmosphere, the maximum pit depth might reach 6 mils within 2 years, yet not exceed 8 mils in 20 years. In an industrial atmosphere, initial pits might reach 4 mils in the first period and not exceed 5 mils at the end of 20 years. In terms of average corrosion penetration in mils per year, the initial rate would be about 0.025 mpy, dropping to less than 0.01 mpy by seven years.

Severe pitting has been encountered where aluminum surfaces were contaminated by either alkaline dust or coral dust containing chlorides, followed by condensation. On some of the South Pacific islands, dust collected from the surfaces of sheltered structures, such as those inside aircraft wings, contained 6% chloride by weight.

It must be realized that the oxide film on aluminum is exceptionally continuous and protective. As a result, corrosion can proceed beneath such a film without obvious external change. Both general and a type of *filiform* corrosion can develop beneath the oxide layer without a change in surface appearance. Probe the surface with a plastic, aluminum, or tungsten carbide gouge when such attack might be present.

In the design of aluminum structures, the usual precautions of avoiding crevices or pockets and coupling with dissimilar metals must be observed. An example of this is in the overlapping encountered between aluminum roofs and siding. In some marine or industrial atmospheres, the aluminum perforated at the laps within a few months. In addition, stress concentrations should be avoided, such as those found in some riveted structures, in the vicinity of welds, and at notches or inside corners. Where it is impractical to avoid dissimilar metals, the aluminum should be electrically insulated from the more noble metal by means of washers, sleeves, etc. In some instances, covering the noble metal with an organic finish is sufficient to greatly reduce galvanic couple corrosion.

To illustrate the problems encountered in designing aluminum structures, an amusing and fairly well-known, yet illustrative story comes to mind. A new, modern processing plant was built having an uninsulated aluminum roof attached directly to steel support frames. This was known to have worked well in several instances. However, the high relative humidity in the plant caused condensation on the cool roof during rainstorms, evenings, and cold spells in the winter.

Problem 1

Water dripped frequently, causing considerable annoyance to people working in the plant, and material was often damaged.

Partial Solution 1. Insulation of the cold roof seemed to be the obvious answer, so insulating material was sprayed on all interior surfaces. To ensure good adhesion of the insulation to the aluminum, an alkaline binder was used.

Problem 2

Moisture collected slowly and was retained in large quantities by the insulation. Now, chunks of dripping insulation "rained" down instead of water drops.

Partial Solution 2. The insulation was removed.

Problem 3

Galvanic corrosion aggravated by the alkaline environment resulted in numerous perforations of the aluminum roof along areas adjacent to the contacts with the steel supports.

Final Solution. The aluminum roof was not replaced, but the aluminum was electrically insulated from the steel supports. Nonalkaline insulation was applied and waterproofed by spraying a suitable compound to the exposed surfaces. This properly installed roof is now giving good service.

Zinc, Zinc Alloys, and Zinc Coatings

Zinc is exposed to the atmosphere in the form of galvanized sheet, as in flashings on roofs; as die castings, and as coatings on steel, either hot dipped or electroplated. Studies of the corrosion of zinc in the New York atmosphere indicated that there was less than a 10% difference between the corrosion rate of galvanized iron, zinc die castings, and three grades of rolled zinc. It was also found that the corrosion rate tended to be a linear function with time. In some instances, where the rate changed after a period of time, it was concluded that the amount of contamination in the atmosphere had changed.

The selection of zinc for calibration of test sites was based on experience that indicated this linear relation with time. The experimenters found that there was some difference in corrosion rate, depending upon the conditions immediately after exposure and from year to year, but that three successive two-year exposures established the corrosion index, and that the deviation with time was insignificant.

Corrosion rates in a large number of atmospheres were given in Table 11.1. Note the particularly low rates of attack on zinc as compared with steel in marine exposures. Such excellent resistance is acquired by the hard, dense, protective products of corrosion in a chloride atmosphere. Similar results cannot be obtained in a sulfurous atmosphere where the products are soft, voluminous, and nonprotective. A 20-year test gave the results listed in Table 11.7.

TABLE 11.7 — 20-Year Corrosion of Zinc

Type of Atmosphere	Average Penetration mpy	µm/y
Industrial	0.252	6.0
Seacoast	0.058	1.5
Rural	0.042	1.1
Arid	0.007	0.2

Zinc-base die castings usually are not exposed boldly to the outside atmosphere without a protective coating. When breaks or pits occur in coatings such as nickel and chromium, the corrosion of the die casting may be accelerated due to the dissimilar metal contact and thus give a false impression of the corrodibility of zinc. Many small parts of machinery, household appliances, and hardware are made of zinc-base die castings or "white metal"[(7)] and are exposed to an indoor atmosphere where their corrosion behavior is very good. In these cases where severe corrosion is encountered in this relatively mild atmosphere, the cause may be improper alloy selection or the use of material containing too high a percentage of impurities.

One of the most important atmospheric exposure uses of zinc is as coatings on steel. Most of these are hot-dipped galvanized coatings containing a small amount of aluminum. Recently, a large volume of sheet steel has been electroplated continuously with zinc as it comes from the rolling mills. Thickness of electroplated coatings is considerably lower than those applied by the hot dip process. Exposure testing of zinc-coated steel sheets has led to the parameters of life expectancy in years-to-failure, given in Table 11.8.

Experience has indicated that in severe industrial atmospheres, a 1 mil coating will last a little over 1 year and heavy galvanized coating will last about 2 years, and that almost the same results will occur in severe marine atmospheres. Supporting this view is the data supplied by Patterson on tests at Kure Beach, NC which have been modified to data shown in Table 11.9.

Note the wide range of zinc thicknesses indicated in the table. These range from the thin electroplate achieved by barrel plating, to the heavy plating afforded by hot dipping. When zinc plating is to be specified for truly corrosive conditions, such as industrial applications, the cost of moving large tonnages of steel often becomes the deciding factor in the use of the coating. Consequently, the relative cost of obtaining an incremental increase in the thickness of the zinc coating is minimal, and at least 2.0 ounces per square foot of steel should be specified.

[(7)]White metal is typically 93 to 96% Zn, 4% Al, 0.05% Mg, and sometimes 1 to 3% Cu.

TABLE 11.8 — Life Expectancy of Zinc-Coated Steel

Thickness mils	µm	Weight oz/sq ft	Years to Failure Rural	Marine	Heavy Industrial
3.4	86	2.0	48	32	13
2.1	53	1.25	30-35	20-25	8-10
1.7	43	1.00	20-25	15	6-8
1	25	0.6	8-10	5-8	3-5
0.3-0.5	10	0.2-0.3	3-4	2-3	1

TABLE 11.9 — Rusting Time of Zinc-Coated Steel

Weeks' Exposure	0.05-mil (1.3 μm) 800-ft lot Percent Rust	Weeks' Exposure	0.020-mil (0.5 μm) 80-ft lot Percent Rust
9	35	27	8
11	60	52	65
51	100	65	100

TIME TO 50% RUST

Zn, mils (μm)	Weeks 80-ft Lot	800-ft Lot
0.05 (1.3)	3	10
0.10 (2.5)	4	82
0.20 (5)	46[1]	162
0.5 (13)	104	—[2]
1.0 (25)	280[1]	—[2]
2.0 (50)	1000[1]	—[2]

TIME TO 100% RUST

Zn, mils (μm)	80-ft Lot	800-ft Lot
0.05 (1.3)	8	51
0.10 (2.5)	25	104
0.20 (5)	65	240
0.50 (13)	156	—[2]
1.0 (25)	400[1]	—[2]
2.0 (50)	2500[1]	—[2]

[1]Estimated.
[2]No rusting except at punched hole.

In the Tropics, whenever nightly condensation occurs, the corrosion of zinc is rapid, particularly in the absence of the washing effect of heavy rains. A heavy film of white corrosion product forms on the surface which tends to be hygroscopic and to retain a very reactive moisture film. In some cases, this film appears to be slightly alkaline from the reaction of water with the zinc to form zinc hydroxide. In other instances, it appears slightly acidic from the CO_2 and contaminants from perspiration in finger prints, traces of organic acids, etc. During World War II, much equipment was damaged in the Tropics by condensation on zinc-coated parts.

Miscellaneous Alloys

Many other alloys are exposed to the atmosphere under a variety of circumstances. The resulting corrosion problems tend to be specific. Lead and tin alloys tend to be inert in the atmosphere unless some specific type of contamination exists. As far as exposure to the atmosphere is concerned, tin appears mostly as a coating on steel containers where it can accelerate the corrosion of the more anodic steel by pitting. Lead is sometimes used on roofs, but more often is found indoors in chemical plants.

Due to their inertness, silver and gold are used as alloys to confer corrosion resistance to more reactive metals. Silver reacts with sulfur compounds in the atmosphere to produce heavy coatings of black silver sulfide. This corrosion product is soft and rubs off of electrical contacts.

Titanium alloys have very good corrosion resistance in the atmosphere, as do the iron-nickel-chromium alloys. There are many alloys that hold a specific place in the metal family, such as pewter, speculum, constantan, and the like, which possess certain atmospheric corrosion resistance properties, but are used so infrequently that they do not warrant general discussion.

Metallic Coatings

Any discussion of atmospheric corrosion should include consideration of the corrosion behavior of various metallic protective and decorative coatings. These are usually divided into two general categories: (1) those which confer protection to the basic metal; and (2) those that are used for decorative purposes.

As protective coatings, a distinction must be made between those such as zinc, cadmium, and aluminum on steel that protect by sacrificial behavior, and those that must provide a substantially continuous protective envelope around the protected metal. In the latter category are metals such as nickel, tin, silver, brass, and chromium.

Sacrificial metals corrode as coatings in much the same manner as they do as solid metals until the base or protected metal is exposed at pores or bared areas. The galvanic couple effect then begins to accelerate the corrosion of the protective coating. This galvanic couple effect tends to protect the base metal at the pores or bared spots. General corrosion of the sacrificial type of coating tends to follow a linear function that is peculiar to the particular site. Nominal rates can be obtained from the calibration data of Coburn, Ellis, and Larrabee in the ASTM studies.

This reaction is in contrast to what happens in the case of the more noble protective coatings that act as an envelope and resist the atmosphere due to their inertness, passivity, or protective films. As soon as a pore or bare spot appears, the corrosion of the base metal is accelerated. Die castings of a metal such as zinc covered with a copper-nickel-chromium coating will suffer severe corrosion at large pores or discontinuities in the coating.

While most metal coatings are applied by electroplating, some are produced by flame spraying, hot dipping, electrostatic sputtering, or vapor deposition. The corrosion rate of a metal coating is largely independent of the method of application, except where impurities play a role. Powders of various metals have been attached to the surface of a base metal by either organic or inorganic binders. Where these metal particles are in mutual contact, and in contact with the base metal, they can perform much like a metallic coating (Chapter 12).

In noble types of coating, porosity is very important and tends to decrease rapidly as the coating thickness increases beyond 0.5 mil. No clear cut information is available on the factors that influence porosity in electrodeposited coatings. Research on this subject has been conducted for many years by the American Electroplaters Society.

Plastics

Essentially all plastics freely exposed to the elements will change in some manner. The active rays of the sun become potent agents of change in the organic materials. Further polymerization of the resin can occur to produce embrittlement. Other types of new bonding can be triggered to make the plastic more crystalline. Any volatile component of the material, such as a plasticizer, can be evaporated. The polymer chains may be simply oxidized and broken up to destroy the product. Oxygen, ozone, and moisture act with the sunlight to degrade the plastics.

The external evidence of attack may be *blushing* (loss of gloss), chalking, or change in color of the product. This is often observed on the epoxy and polyester plastics that have not been coated. However, only mechanical tests will reveal the extent of degradation of either the thermoplastic or thermosetting resins.

The effect of high atmospheric temperatures or heating from direct exposure to the sun can be particularly severe on the thermoplasts. Creep or distension of the polyvinyl chloride and polyethylene plastics will occur readily unless provision is made to prevent overheating or stressing of the materials. Certain other thermoplasts can be dimensionally stable under all normal atmospheric temperatures. The strength of the thermosetting resins is not noticeably changed.

Plastics to be exposed freely in the atmosphere should be thoroughly tested. ASTM Recommended Practice D 1435 describes the appropriate conditions for such a test exposure and suggests tests that might be used to evaluate changes in the materials. Changes in mechanical and physical properties of the plastics are determined (Chapter 14) for definitive results. Weight gains or losses can be of interest, but do not provide substantive results.

Accelerated cabinet testing of the materials can be performed with essentially the same validity as when testing the metals. Poor materials can be eliminated, but extrapolation of the data to forecast the life of plastic parts in the atmosphere should not be attempted. ASTM D 1499, D 2565, and G 26 may be used to ensure the controlled cabinet testing of the plastics. ASTM D 750 should be used for the evaluation of the elastomers.

Preventive measures to ensure the integrity of plastic materials are similar to those for metals, *i.e.*, the exterior should be covered. The exterior surfaces are painted or metal plated to provide a barrier between the material and the atmosphere. Alloying of the basic plastic material with small amounts of other resins can often upgrade the stability of the material.

Common Forms of Atmospheric Corrosion

Stress Corrosion

While static stress in a metal lattice tends to shift its potential in the negative direction, this change is only a few millivolts, and only under exceptional circumstances does it accelerate general corrosion. Cracking as a result of combined stress and corrosion is, however, another matter, and one of great importance.

Although a vast amount of effort has been put into the study of stress corrosion cracking, a complete picture of the mechanism remains obscure. Some workers are of the opinion that several mechanisms are at work simultaneously and so mutually obscure each other. Under some circumstances, one mechanism seems to be predominant and leads the research worker to believe that the fundamental process has been discovered, while under other circumstances the same mechanism appears to be subordinate.

Stress corrosion cracking has been encountered in practically all types of alloys and is of the greatest economic importance. The vast amount of study and testing has produced alloys of low susceptibility and has established the thresholds of stress at which cracking is likely to be encountered. There is a general consensus that both an electrochemical and a mechanical mechanism are involved. Fortunately, each alloy is susceptible in only a few rather specific environments (refer to Chapter 6).

Stress corrosion cracking appears to progress in several stages. First, there appears to be a period of incubation during which it is thought that the electrochemical corrosion process produces conditions necessary for crack propagation. This is followed by the appearance of microcracks, and their growth into larger cracks.

The mechanism of crack growth is somewhat obscure, but in some cases hydrogen embrittlement or the precipitation of an embrittling phase at grain boundaries seems to be involved. Both the incubation period and the slow initial growth of cracks seriously hamper the study of the phenomenon. Specimens must be held in the stressed condition in the corrosive environment until the crack appears. This may occur in anywhere from a few minutes, to a few years. One of the frustrating features of this work is that it cannot be assumed that if a crack has not appeared after a certain lapse of time, it will not appear in a small additional period of time. This situation requires the testing of large numbers of specimens over long periods.

Unfortunately, the chloride ion is an almost universal accelerant in stress corrosion cracking in the atmosphere. It affects high-strength steels, stainless steels, and aluminum and magnesium alloys. It does not seem to induce cracking in most nickel-base alloys. Ammonia and nitrates have an adverse effect on copper-base alloys. The familiar season cracking of stressed brasses is greatly accelerated by nitrogen compounds.

It is advisable to investigate the stress corrosion cracking history of any alloy which is proposed for use in a stressed condition in the atmosphere. However, this does not guarantee that such a study will permit the choice of an alloy immune to particular exposure conditions. For instance, stainless steel messenger strand which supported telephone cables in a marine atmosphere had to be evaluated for the causes of severe stress corrosion cracking after many "experts" claimed there would be no problem in the proposed use of the material.

Corrosion Fatigue

Corrosion fatigue increases with the severity of the environment. Actually, corrosion fatigue can be considered as an increase in the fatigue rate over that in an inert atmosphere or vacuum. For all practical purposes, one should consider the increased aggressiveness of contaminated air against that of dry air. Most corrosion fatigue data are generated in a controlled laboratory-atmosphere.

Fretting

A related type of corrosion which occurs in the atmosphere is known as *fretting*. Where two metal surfaces are in contact and subjected to vibration so that slight slip occurs, there will be an accelerated corrosion or oxidation of the freshly bared surface of the metal at each cycle. Presumably, the slight slip removes the protective oxide film, exposing fresh metal to the atmosphere.

Another type of fretting is achieved when two closely mated surfaces slip under pressure. Welding of discrete areas on the two surfaces in the absence of an atmosphere can occur. Subsequent motion tears small particles of metal from one of the surfaces and exposes the particle to the atmosphere where it is quickly oxidized (Chapter 5). This type of corrosion has been particularly troublesome with ball bearings which are stationary but subjected to vibration.

Galvanic Corrosion by Dissimilar Metals

Dissimilar metal couples will be discussed in other sections of this text as a part of systems submerged in volume electrolyte. In atmospheric corrosion, galvanic corrosion is often found in fastenings. Several extensive tests have been made in various atmospheres to obtain quantitative data. In many early tests, electrical contact between the parts of the couple was lost as the result of the insulating effect of corrosion products, so that the data are misleading.

More recent work employed a threaded rod[9] on which was wound a small diameter wire of the more negative metal to increase the effective surface involved. A preferred configuration used a 4 × 6-inch plate of negative metal with a positive metal strip across the center, from which tensile specimens could be cut. This has given useful information on the subject. For details on these tests, refer to the yearly reports of ASTM Committee B-3.

As might be expected, these tests indicate that galvanic corrosion in the atmosphere is only significant where heavy contamination from chlorides and sulfates is encountered, and where there is a persistent film of moisture connecting the dissimilar metals. The data indicate that one needs to be concerned only with magnesium, aluminum, and zinc base alloys when coupled to copper alloys or rusty steel. Most of the other combinations pose no serious problems in atmospheric exposures.

Tarnish

Tarnish is an oxide or sulfide film on the surface of metals such as copper or silver. While these tarnish films usually only detract from the appearance of the metal, in the case of electrical contacts they may have a major effect of great consequence. The growth of oxide films on copper tends to follow a parabolic rate and is a function of temperature. Sulfide films behave similarly, but also are sensitive to the concentration of sulfur in the atmosphere.

The usual concept that tarnish is merely a surface discoloration is inadequate when applied to electrical contacts. Fogging of nickel and oxidation of steel at elevated temperatures in dry air can be considered a type of tarnish. Little or nothing can be done to prevent oxidation in air, but sulfur can be removed by suitable absorbent filters.

High-Temperature Corrosion

Corrosion at elevated temperatures in air or in special gases can be considered special cases of atmospheric corrosion. They will be covered in detail in Chapter 13 on high-temperature corrosion. However, consideration must be given to metals that must serve in corrosive atmospheres covering a wide range of temperatures. In what is usually considered to be high-temperature corrosion, water usually does not play an important role, and the mechanism of corrosion is somewhat different than when it does.

Many types of equipment must withstand ordinary atmospheric corrosion between exposures to high temperatures. Stacks fall into this category. While they are hot, there is little corrosion; when cool, they corrode badly. Since the high-temperature performance is controlling, ordinary protective measures may not be practical. Fortunately, many alloys designed for high-temperature operation are also resistant to moisture and atmospheric contaminants.

Metallizing followed by overcoating with a silicone is often used to protect hot stacks and breechings.

Preventive Measures

Atmospheric corrosion may be controlled by the use of protective coatings, by proper selection or modification of the material, and to a limited extent by controlling the atmosphere. The use of protective coatings has been touched upon in other sections of the chapter and will receive an extensive treatment in Chapter 12 on coatings. After one has selected an alloy with the proper mechanical properties and some possibility of meeting the corrosion resisting requirements in the atmosphere to which it may be subjected, small changes in composition or removal of impurities can produce marked improvement in its performance.

For instance, reduction of iron and nickel impurities greatly improves the corrosion resistance of certain magnesium alloys. Control of copper in aluminum has a similar result. The addition of small amounts of copper, nickel, or chromium can greatly increase the atmospheric corrosion resistance of carbon steel, as shown in Figures 11.8 and 11.9 and in this chapter's section on iron and steel.

Control of the atmosphere is possible only in cases of indoor or housed exposures. Corrosion and tarnish in telephone centrals and computer areas often is controlled by suitable air conditioning and filtering. The former keeps the relative humidity down, and the latter removes both particulate matter and sulfur compounds.

Organic coatings, discussed in the next chapter, prevent corrosion in the atmosphere primarily by acting as a barrier to deter contaminants and moisture from making contact with the metal surface. A few coatings perform a dual role. Inhibitive primers such as red lead and zinc chromate provide an oxidizing environment at the surface of the metal and tend to make it passive. The zinc-rich organic and inorganic coatings, because of the conductivity between zinc particles and base metal, tend to provide galvanic protection in addition to acting as barriers.

The corrosion protection afforded by barrier coatings depends upon their integrity and low moisture transmission. As soon as the integrity of the coating is destroyed by cracking, blistering, peeling, or excessive chalking, corrosion will set in. Coatings must be renewed before this happens, otherwise they will have to be removed and a whole new system applied to the clean metal. The life of a coating is a function of its composition, adhesion to the metal, and severity of exposure. Special treatments are applied to most metal surfaces to promote adhesion.

Surface reaction treatments which act as protective coatings are applied to aluminum and to zinc. Anodizing of aluminum provides a dense, hard protective coating which, when properly applied, gives improved results in all kinds of atmospheres. This coating is produced by an electrolytic treatment in either a sulfuric or chromic acid solution. It is sealed by prolonged immersion in boiling water or a dilute chromate salt. Zinc and cadmium are given a chromate treatment to increase corrosion resistance to condensation and mild contamination in indoor atmospheres.

Ceramic coatings (porcelain) are employed to give protection from atmospheric corrosion. They are usually used indoors and for sanitary reasons. Some silicate and silicone coatings are also used for special purposes.

Special Problems

Bolting and threaded fasteners pose a special problem because of the built-in crevice formed and the difficulty of obtaining a reliable coating on the sharp edges of the threads. Special types of coated bolts using polyimide coatings are now competing with the traditional galvanized, cadmium-plated, and hot-dipped aluminum-coated bolts. Selection of threaded fasteners for corrosive service warrants special consideration (*e.g.,* in and around cooling towers, marine locations, etc.).

Corrosion under wet insulation is a dangerous problem because it is out of sight. Severe damage by poultice corrosion can occur on aluminum, steel, and stainless steel, particularly under wet insulation, depending on the nature of the insulation and of the contaminants. Severe general corrosion, pitting, or even stress corrosion cracking can ensue, unknown to the casual observer, unless susceptible vessels and piping are suitably painted with an appropriate coating system *before* being insulated.

References

1. Sereda, P. J., Measurement of Surface Moisture, ASTM Bulletin, p. 53, Feb. (1955), p. 61, May (1959), and p. 47, May (1960).
2. Department of Scientific and Industrial Research, Atmospheric Pollution, 18th Report, H. M. Stationery Office, London, 1953.
3. Ambler, H. R. and Bain, A. A. J., Corrosion of Metals in the Tropics, J. Appl. Chem., Vol. 5, pp. 437-467 (1955).
4. Larrabee, C. P. and Ellis, O. B., Corrosiveness of Various Atmospheric Test Sites, Proc. ASTM, Vol. 53, p. 194 (1953) and Vol. 59, p. 183 (1959).
5. Coburn, S. K., Corrosiveness of Various Atmospheric Test Sites as Measured by Specimens of Steel and Zinc. Symposium on Atmospheric Corrosion, Am. Soc. Testing Mats., Boston, MA, June 25, 1967.
6. Copson, H. R., Design and Interpretation of Atmospheric Corrosion Tests, Corrosion, Vol. 15, pp. 533-541 (1959).
7. Copson, H. R., A Theory of the Mechanism of Rusting of Low Alloy Steels in the Atmosphere, Proc. ASTM, Vol. 45, p. 544 (1945).
8. Larrabee, C. P., Mechanisms of Atmospheric Corrosion of Ferrous Metals, Corrosion, Vol. 15, p. 526 (1959).
9. Compton, K. G., Mendizza, A., and Bradley, W. W., Atmospheric Galvanic Couple Corrosion, Corrosion, Vol. 11, pp. 35-44 (1955).

Bibliography

ASTM. Metal Corrosion in the Atmosphere. ASTM Pub. No. 435. ASTM, Philadelphia, PA (1968).

Rozenfeld, I. L. Atmospheric Corrosion of Metals. NACE, Houston, TX, 1972.

Chapter 12

Protective Coatings

PROTECTIVE COATINGS

Introduction

Putting a barrier between a corrosive environment and the material to be protected is a fundamental method of corrosion control. It is, in fact, the most widely used method of protecting steel and other metals. As a result, the industry which provides materials and services in this area is important to the economy of this country.

The concept of adding coatings to surfaces is so ancient that its beginnings are lost in the mists of history. Paleolithic cave drawings exist which are reputed to be tens of thousands of years old. They were applied by fingers, a splayed twig, or, as some believe, were sprayed by blowing pigments through a hollow reed. Until the Egyptians began using varnishes about 4000 B.C., there is no evidence that coatings were used for protection. Polychrome Greek statues became common by 300 B.C., and evidence exists that the Romans used coatings for both decorative and protective purposes. Coated Chinese artifacts are even more ancient than those in Western Europe. Thus, there is abundant evidence that man's interest in coatings extends far backwards into time.

Mechanisms by which coatings protect can be summarized as follows.

1. They prevent contact between the environment and the substrate, *e.g.*, nickel electroplating.
2. They restrict or limit contact between the environment and the substrate, as with most organic coatings.
3. They release substances which are protective or which inhibit attack, *e.g.*, chromate primers.
4. They produce an electrical current which is protective, *e.g.*, galvanizing.

It is useful to consider protective coatings as systems in which the component parts or stages of development are related and mutually reactive. While the concept that coatings should be considered as systems is not new, it is undoubtedly true that coatings often are applied in an offhanded manner and frequently without adequate consideration of factors that would help ensure success. The various elements of good coatings systems are described in this chapter.

Protective Coatings: The Systems Approach

A common, if not the most often encountered reason protective coatings do not perform well is that they have not been considered as systems. Most successful coating engineers approach a coatings project in much the same way that they approach any other engineering problem, beginning with the design of surfaces to be protected and ending with schedules for monitoring the completed work.

Elements usually considered in a systems approach include the following.

Design of Surfaces

Prefabrication decisions and post-erection corrections include elimination of sharp corners and hard to reach places, adoption of designs that will not collect or hold water or debris, provision for drainage of recessed zones, smoothing of rough surfaces and rounding of corners, removal of weld splatter, avoidance of skip welding, and caulking of crevices. Figure 12.1 illustrates some of these considerations.

Cylindrical or box shapes should be used whenever possible. If possible, the metallic materials selected should have the best resistance to corrosive attack or be alloyed to enhance coatings performance. Tests have shown that some alloying elements in steels, for instance, will lengthen the effective life of coatings. The HSLA (high-strength, low-alloy) steels are examples. They serve as an excellent base for a coating.

Designs also should be such that dwelling time of atmospheric condensation is minimized, and that surfaces shaded from the sun get extra attention. Large enclosed spaces that have not been totally sealed, such as the insides of box girders, should be ventilated, with fans if necessary.

Surface Preparation

Surface preparation is a critical part of the coatings operation and must provide a surface that is compatible with the coating material to be applied. The major considerations concern the cleanliness of the surface required and the surface area (roughness of surface; profile) to be obtained.

Material Selection

A coating combination must be specified after testing that will provide proper stability during the life of the project. The number of coats to be applied, the compatibility of the various coats, and the requirements for their maintenance must be considered. Obviously, the initial system and the maintenance program envisioned must protect the substrate during the life of the project.

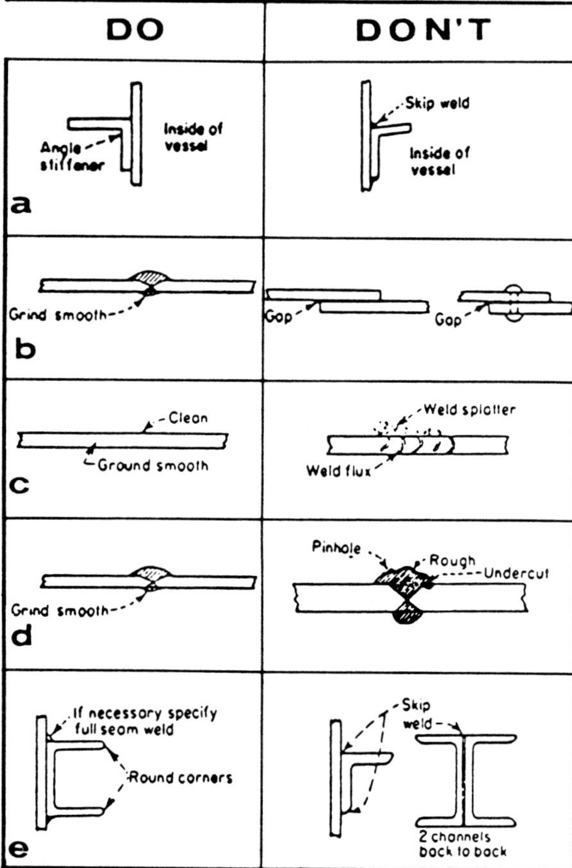

FIGURES 12.1 — (a) Stiffening members should be on outside surface of vessel or tank. (b) Butt welding should be used rather than lap welding or riveted construction. (c) All weld splatter should be removed. (d) All joints should be continuously and solidly welded. (e) Construction involving pockets or crevices that will not drain or cannot be cleaned and coated properly, should be avoided. [SOURCE: Figure copy adapted from material supplied by Metalweld, Inc. Also see Recommended Practice RP-01-78, Design, Fabrication, and Surface Finish of Metal Tanks and Vessels to be Lined for Chemical Immersion Service, NACE, Houston, TX.]

Application

The coating material must have application characteristics which would allow its proper application under all conditions existing during the coating process. Proper identification of the physical or other properties of the film expected from the application must be specified. Inspection procedures to verify the properties and to maintain uniformity of the system must also be stated. The safety of personnel involved in the work or of others in the vicinity must be considered.

Economics

System A versus system B must be appraised in true economic terms to yield the maximum protection for money spent, both initially and during maintenance of the project.

Surface Preparation

An essential preliminary to any coating operation is proper surface preparation. It is false economy to skimp on surface preparation in the belief that the coatings applied will make up for surface deficiencies. This is especially true of newer high-performance synthetic materials.

Even the most versatile oleoresinous material will give better service when applied to a properly prepared surface. Most engineers believe that, of the cost of a coating job, as much as one-half to two-thirds goes for surface preparation and labor. Consequently, a coating job that fails because of poor surface preparation represents a loss of at least two-thirds of the whole, whereas touch-up maintenance of a properly prepared and primed surface is a nominal maintenance expense.

It is essential to be practical when preparing a surface for coating. In some cases it is impossible to provide the best prepared surface because there is insufficient money to do the job, no time, or it is not permitted because of possible product contamination, fire hazard, or some other sufficient reason. Nevertheless, whenever and wherever it is feasible, the very best surface preparation consistent with good engineering practice is desirable. This assumes, for example, that it would be folly to apply a coating with a ten-year life span to a building destined to be razed in five years.

Some of the most commonly used methods of preparing surfaces for coatings, but not all, are discussed in the following sections.

Principles of Coating Adhesion

The cleaning of a metal surface is performed to achieve two basic goals: (1) to obtain a hard continuous surface which will remain stable as a coating substrate, and (2) to provide a cleaner, more metallic base to which the coating material will adhere more strongly.

Heavy brushing of the surface with wire brushes, carborundum, or other materials can attain some part of these goals. A firm oxide surface is produced in this way which is probably much less contaminated with atmospheric pollutants (chloride, sulfur, and nitrogenous and alkali compounds) than the original surface. If the remaining scale is firm, dry, and not excessive in thickness, a heavy coating can be applied to cover the rough substrate and keep active species from the atmosphere from diffusing to and through the scale. However, if all solid foreign material could be removed from the metal surface, a much improved adhesion could be achieved.

Consider, for instance, a metal surface (Chapter 3). If it were possible to go into space and remove all foreign material from the surface, a relatively high electron field would begin forming above the metal. A coating molecule of the proper configuration would form a bond with the metal atoms that would essentially be a chemical bond.

Unfortunately, when a piece of metal is cleaned in our atmosphere, the surface is badly contaminated under even the best of conditions. Ox-

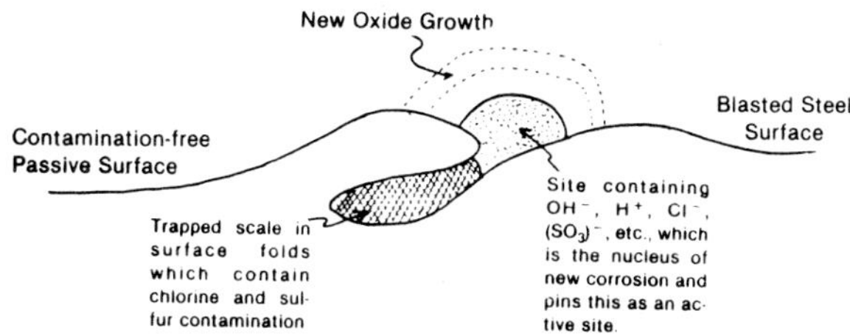

FIGURE 12.2 — Schematic which depicts the new surface corrosion processes of an industrially corroded, blasted, and environmentally tested steel.

ygen is immediately adsorbed. In addition, the cleaning procedure normally leaves some undesirable residue, and all original scale or other foreign material is not removed (Figure 12.2). Onto this surface we must apply a coating material that competes for the available electromagnetic field created by the electrons.

Thus, the cleaner we can make the surface and the greater the surface area, the greater the chance of tying the coating material to the metal, or actually the thin oxide on the metal, by the electromagnetic attraction between the two. These attractions are known as hydrogen, Keesom, London, Debye bond forces (all VanderWaal forces), and constitute the great majority of the bonding strength of the coating to the metal.[1] Mechanical attachment to the rough surface composes the remainder of the bonding strength.

Of the cleaning methods available to provide the cleanest and largest surface per unit area, the use of abrasive blasting produces a surface closest to the ideal. However, it must be remembered that the best of blasted surfaces is still badly contaminated with residual metal oxides, extraneous dirt, particles of the abrasive materials, and adsorbed gases on the metal. As a consequence, it is necessary to construct a coating molecule which will have the maximum capacity to compete for and lock into the available bonding sites. For this reason, coating materials containing OH, N, O, and particularly C = O functional groups provide the best adhesion to a metal substrate.

Any procedure which improves the cleanliness of a surface and/or increases the surface area of metal per square unit of area will improve the adhesion of a given coating material. Good wetting of the surface by the wet coating is obviously a requisite for good adhesion. However, the polar functional groups designed into the coating molecule to achieve good adhesion also provide good wetting properties.

The coating material must remain stable to maintain adequate adhesion. If embrittlement occurs through oxidation, cross-linking, or evaporation of a portion of the coating, the resultant stresses in the coating can pull it from the surface. Any loss of adhesion from such sources is undesirable.

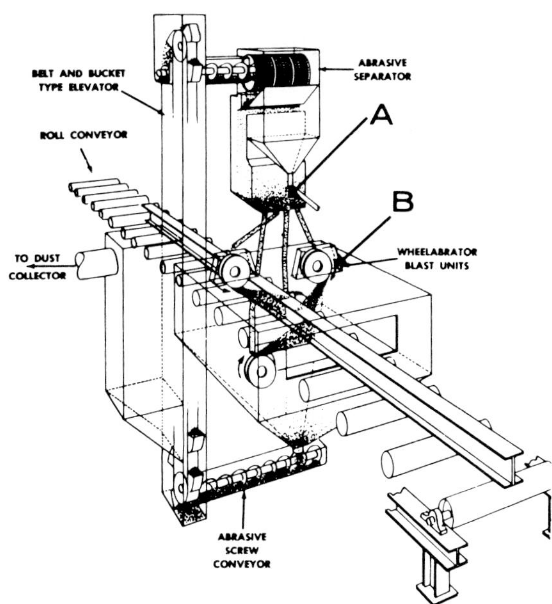

FIGURE 12.3 — Schematic of centrifugal blasting machine. Abrasive is fed into the wheel from the center (A), from whence it is impelled against the rapidly moving vanes (arrows show direction of rotation), and flung from the wheel perimeter (B) against the structural steel shape.

Abrasive Cleaning

The two principal methods of abrasive cleaning before application of coatings are centrifugal blasting and air pressure blasting. Other abrasive methods, such as barrel cleaning, will not be discussed here.

Centrifugal blasting, as shown in Figure 12.3, is accomplished by machines that propel abrasives against the surface to be cleaned by imparting velocity to them by means of a rapidly rotating wheel. The abrasive material (sand, cut wire, shot, glass beads, various hard oxides, or carbides) impinges against the surface, removing both the surface contamination and making a pattern of indentations, such as that shown in Figure 12.4. This pat-

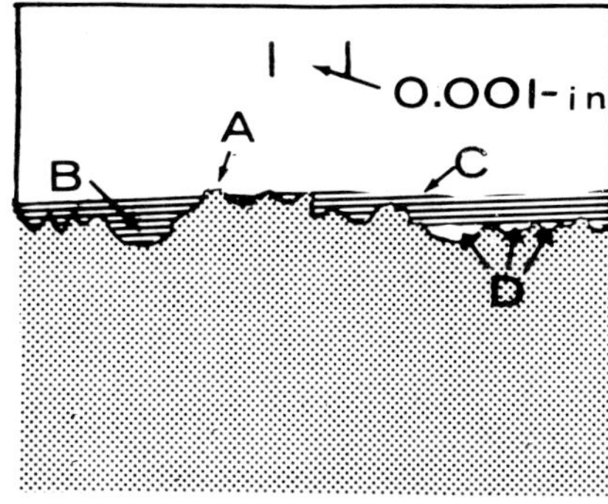

FIGURE 12.4 — Diagram showing cross section through a base which has been centrifugally blasted with G-50 grit (fine) and then coated. Note that peak A extends above coating surface, B indicates a valley, and C indicates coating surface. Coating fails first at peak A, while D indicates "bridging" of valleys where coating failure may start. [SOURCE: Bigos, Joseph, Anchor Pattern Profile and Its Effects on Paint Performance, Figure 5, Corrosion, Vol. 15, No. 8, pp. 428-432 (1959).]

tern, consisting of peaks and valleys, serves as a mechanical assist to the chemical bonding forces of materials applied to the surface. The total surface area exposed in a profile of this type obviously is much greater than that exposed by a smooth surface.

Abrasives used in centrifugal blasting systems are collected and recycled. They may or may not be cleaned during the cycle, but usually some particles of surface contamination recirculate with the abrasive.

Centrifugal blasting machines are of great value when continuous or intermittent shipments of steel are made to a site. All steel, as it is received, can be run through the machine and immediately primed to provide an advantageous start on the coating system to be used. Without this initial treatment of the steel, a more tedious, expensive, and difficult blasting operation must be conducted on the steel after erection to provide a surface in any way comparable to the centrifugally blasted metal. Even then, overlapping surfaces and many recesses in the erected structure cannot be properly cleaned and primed.

Cleaning of steel by blasting or pickling, followed immediately by a prime coat, can also be provided by many mills at an attractive price. The major problem then becomes one of handling the steel throughout transportation and handling from the mill to the field erection without removing excessive amounts of the prime coat.

Field blasting involves injecting a supply of abrasive into a rapidly moving air stream and expelling it through a nozzle of the proper configuration so that the solid particles impinge against the surface to be cleaned. Important considerations in good and efficient air blasting include:

1. The skill of the workman,
2. Good equipment producing the proper volume of abrasive at the proper velocity to do good work,
3. The quality of the abrasive, and
4. Contamination problems.

A typical rig for efficient pressure blasting is shown in Figure 12.5. This illustrates a standard commercial setup. Also, equipment in which the abrasives are stored in a small container fastened beneath the blasting gun may be used for small areas or inside shops for work on small pieces.

The skill of the workman in keeping the blast stream at the proper distance from and angle to the work, and his conscientiousness in being sure that all areas are properly cleaned are a prerequisite to a good job. He must have the right kind of equipment, including the proper size and type of nozzle for the job being done, the correct size and lengths of hose, proper air pressure, and abrasive of the right quality and size to produce the desired surface.[2] When skill and equipment are correct,

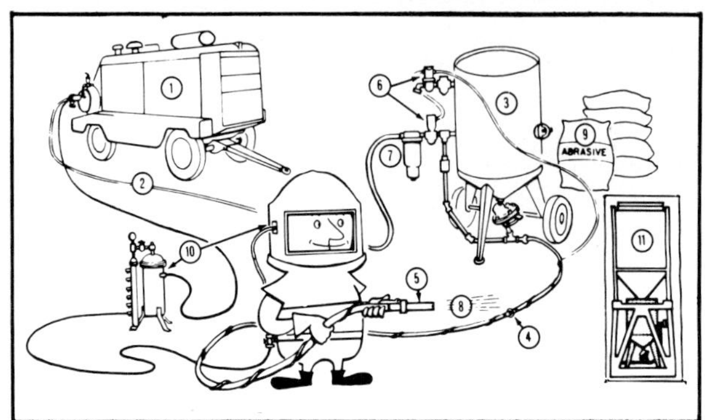

FIGURE 12.5 — Typical rig for compressed air abrasive blasting: (1) an adequate and efficient air supply (compressor); (2) air hose, couplings, and valves of ample size; (3) a portable, high-production blast machine; (4) the correct size of antistatic pressure hose with externally fitted quick couplings; (5) high-production venturi nozzle; (6) pneumatic remote control valves for safety and cost savings; (7) an effective moisture separator; (8) high nozzle air pressure; (9) the proper blasting abrasive; (10) a safe air-fed helmet and air purifier; (11) inset showing bulk abrasive storage hopper over pressure blast vessel equipped with remote control valves—when pop-up filling valve opens, abrasive from the hopper fills the machine. [SOURCE: Clementina, Ltd., San Francisco, California.]

achievement of a given grade of cleanliness is a function of time.

The National Association of Corrosion Engineers has adopted a schedule of four degrees of cleaning by blasting operations. These have approximately the same profiles for both sand and centrifugal blasting and are described briefly as follows.

1. White metal blast cleaned—produces a gray-white, uniform, metallic color with no obvious foreign material remaining.

2. Near-white blast cleaned—foreign matter is removed, but with differing shades of metallic gray surface allowable.

3. Commercial blast cleaned—rust and foreign matter is removed, except for tight specks of oxide or paint uniformly distributed over a minor portion of the surface; some residues in pits.

4. Brush-off blast cleaned—light mill scale and tightly adhered rust is allowed if properly distributed over the surface.

Actual wording of the four grades of surface preparation and detailed information concerning methods that may be used to achieve proper surface preparation are available from documents published by NACE. Similar specifications for surface preparations can be obtained from the Steel Structures Painting Council (SSPC).[3]

Also available from NACE are AISI 1020 (G10200) steel coupons which have been blasted either by sand or centrifugal methods to the four standards. These are encapsulated in transparent acrylic resin so that they can be compared with a blasted surface to determine the amount of contamination remaining.[4] Also available from NACE is the *Surface Preparation Handbook,* which contains information concerning surface preparation prior to coating.[5] The American Society for Testing and Materials can supply pictorial standards.[6]

Because time is the important factor insofar as economics is concerned, the importance of good and proper equipment and material cannot be overemphasized. Those who wish to investigate this aspect more deeply will find an excellent discussion in Chapter 3 of *Industrial Maintenance Painting* by Paul Weaver.[7]

In addition to sand (hard, crystalline variety), steel grit, slag (metal oxides), and other abrasives cited previously, aluminum oxide, walnut shells, ice, dry ice, and various other materials have been used as blasting agents to reduce the profile (surface roughness) obtained and contamination of the atmosphere.

Blasting of the inside of tanks, piping, and other difficult surfaces can be achieved by ingenious available methods. A variation of abrasive particle air blasting, useful in special cases, is abrasive wet blasting, which involves introducing abrasives into a rapidly moving stream of water and air. The combination obviously cuts down on the atmospheric pollution experienced when using the dry abrasive alone. Inhibitors are often added to the water to reduce rusting of the "clean" steel before the prime coat is applied. See Figure 12.6 for a sketch of a nozzle used for such purposes.

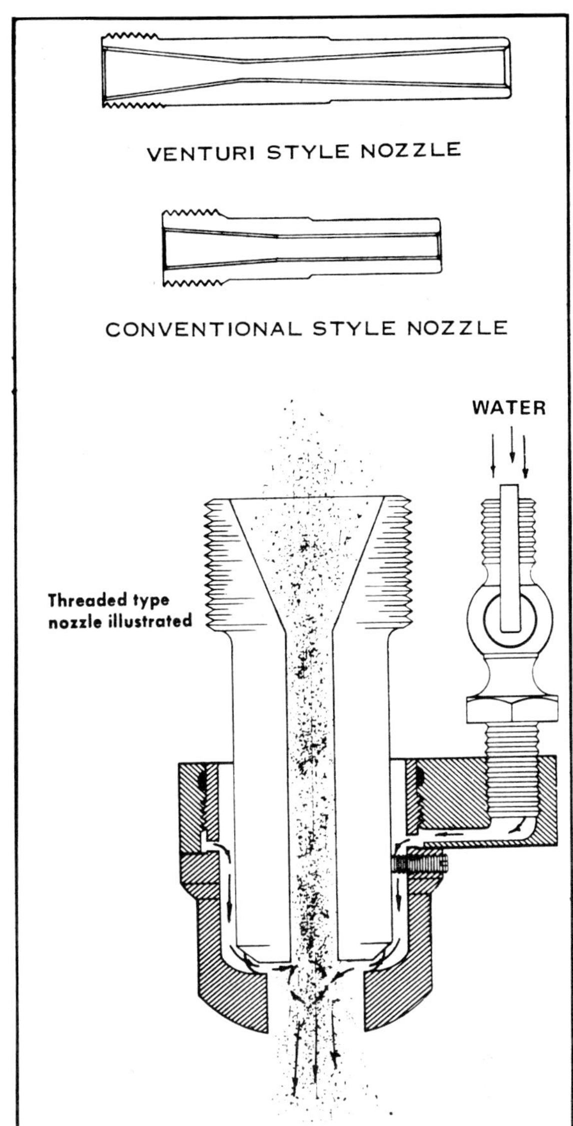

FIGURE 12.6 — (Top) High-performance venturi-type long nozzle, and (Middle) short conventional type. [SOURCE: Clementina, Ltd., San Francisco, California.] (Bottom) Attachment that makes "wetblast" possible by adding water as the abrasive leaves the nozzle. Widely used to clean stone, brick, concrete, and stucco. [SOURCE: Sanstorm Manufacturing Co., Fresno, California.]

Other Surface Preparation Methods

Although abrasive blasting is widely used by the process and oil industries, the following procedures are also frequently used. Pickling, for instance, is probably used to prepare more metal surfaces than all other procedures combined.

1. Solvent wipe-off.
2. Vapor or solution degreasing
3. Steam or hot water cleaning
4. Flame cleaning
5. Mechanical abrasion
6. Electrochemical cleaning
7. Pickling
8. Water blasting

Solvent Wipe. For certain surfaces, it is necessary only to wipe the surface with a selected solvent to remove oil, grease, and loose dirt. The use of heavy detergent solutions, isopropanol, ketones, and aliphatic and aromatic hydrocarbons for this purpose can be effective.

The major requirement for the process is that the wipe shall not leave a residual contaminant, but rather a clean, dry metal surface. Such a cleaning procedure is used on previously cleaned steel, stainless steel surfaces, and old paint films before overcoating, or on dirty steel prior to abrasive blasting.

Vapor or Solution Degreasing. A surface that is dry and free from oily residues can be obtained by suspending the item in a closed booth in which the vapors of a solvent are condensing. The vapors condense on the part, coalesce, dissolve the contaminant, and drip from the surface. Trichlor and perchloroethylene are common solvents used for this purpose. Many production line items are cleaned in this manner before coating.

Baths of cleaning solvents may also be prepared to remove oily or loose contaminants. Aqueous detergent systems, emulsion systems, and alkali baths are used with good mechanical or ultrasonic agitation to ensure scouring of the surface by the liquid.

Steam or Hot Water Degreasing. Hot water heaters or steam generators (*gennys*) are available for use in the field to clean metals, coatings, or concrete surfaces. Strong detergents or alkalies are added to the water to emulsify oils and similar organic contaminants on contact. Impingement of the hot, strong emulsifying agent on the surface removes even heavy soil at a rapid rate. This method can often be used with good results prior to abrasive blasting.

Flame Cleaning. The process of flame cleaning found greater favor in Europe than in America. A wide oxy-acetylene flame is played on the surface of steel parts to produce two effects. First, the heavier portions of rust scale or mill scale will pop from the surface because of the thermal expansion difference between the scale and the substrate. Secondly, the surface is freed of moisture if sufficient heat is applied to the substrate. Removal of loose rust particles after the treatment, followed by prompt coating can produce an adequate coating procedure for certain applications.

Mechanical Abrasion. Manual or power tool brushing of an oxidized surface obviously provides the poorest of cleaning procedures available. However, cost, location of the part, or the availability of tools often dictates the use of such a technique. Normally, only the top loose layers of rust on a piece of steel are removed through this method. Moisture and other contaminants remain in the residual scale. Chipping and grinding can remove the heavier scale, but will impress oxide and salts into the metal surface.

When using brushes or grinding wheels, it is important to consider the compatibility of the tool with the substrate. For instance, the use of steel or bronze brushes to clean aluminum can leave sufficient heavy metal (Cu or Fe) contamination on the surface to initiate severe pitting of the metal.

An air-actuated needle gun is also available to clean small areas by mechanical action. Barrel cleaning, where parts are placed in a rotating drum containing an abrasive, is also used for small parts.

Electrochemical Cleaning. Cathodic cleaning of a metal part probably provides the cleanest of all surfaces.

The metal to be cleaned is made the cathode in a low-resistance electrolyte bath, and some 6 to 12 volts are used to provide a cathodic current of 3 to 10 A/dm^2 on the metal surface. The hydrogen generated at the *surface* of the metal will remove oil, scales, paint, or any other foreign material (Figure 12.7).

A surface heavily coated with tenacious and thick deposits, such as those found on surfaces of fuel tanks in ships, may be partially cleaned by electrolytic descaling. This method involves filling the tank or vessel with low resistivity water (usually salt water) and making the vessel wall the cathode of an electrical circuit receiving current from an anode suspended in the liquid. Current densities sufficient to cause formation of hydrogen at the metal surface are imposed. This causes the coatings to separate from the metal and slough off into the electrolyte.

In this or any other electrochemical process using high current densities, the hazard of ac-

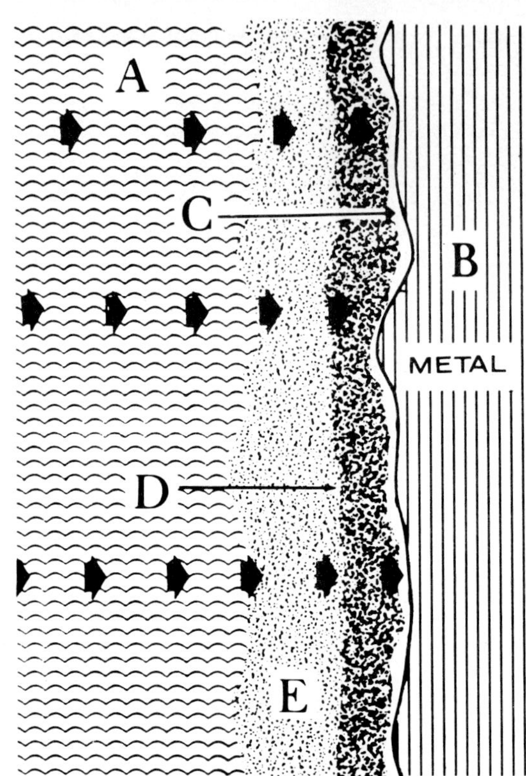

FIGURE 12.7 — Electrolytic descaling of steel surfaces is effected when an electrical current of sufficient density (black arrows) passes through an electrolyte (A), and reaches the surface of the base metal (B). (C) indicates a layer of hydrogen or of hydrated iron; (D) indicates a layer of iron corrosion products; and (E) indicates a layer of porous residue, such as may be deposited by fuel oil. Layer (C) is soft and thus permits washing away layers (D) and (E). [SOURCE: Cook, F. E. Preiser, H. S., and Mills, J. F., Corrosion, Vol. 11, No. 4, pp. 161-181 (1955).]

cumulating the explosive hydrogen gas must be recognized. Adequate venting of the gas in a safe manner must be provided.

One shipyard, at great expense, has devised a self-contained cathodic cleaning apparatus that can be used on the exterior of a ship hull to prepare it for recoating.

Although used less frequently, anodic cleaning of a metal surface is possible. As the metal ions are taken into solution at the anodic surface, all other extraneous material is forced into the electrolyte. Very smooth surfaces can be obtained by this process. Electrochemical milling (''machining'') of specific areas on a metal surface is used for this purpose.

Pickling. Great quantities of sheet, plate, coil stock, and other forms of metal are pickled in production mills or by reprocessors of the metals. On steel products, the goal is the removal of the mill scale (FeO-Fe_3O_4) formed on the surface during hot processing ($>$575 C). The bath is an aqueous hot hydrochloric, sulfuric, or less frequently phosphoric acid solution containing an inhibitor. The inhibitor must allow uniform attack on the metal, as little metal loss as possible, and when pickling steel, must leave a clean surface free of carbon smut.

Such an inexpensive cleaning procedure followed by thorough rinsing and drying provides one of the cleanest and most active surfaces for further processing. More steel is pickled, treated, and coated than by all other methods of preparation and protection.

The pickling process is conducted in large vats and does not lend itself readily to use in the field. Acid gels or washes based on phosphoric acid can be prepared (typified by MIL C-10578, Type C) which will lightly etch a steel surface when held on the surface once the heavy scale has been removed.[8] This type of "pickling" is restricted to smaller articles where the acid can be managed properly.

Water Blasting. Water blasting involves playing a stream of water moving at a high velocity (ca. 5000 psi) against a surface to be cleaned. The energy imparted to the water dislodges the scale or contamination from the surface, as shown in Figure 3.7 of Reference 7. This method is especially effective in removing resilient materials which may not be removed efficiently by abrasives, or in cleaning surfaces of complex cross section such as gratings or sieves. Water also may be used when sandblasting is a fire hazard. The procedure is used primarily in maintenance painting procedures.[9] NACE has a recommended practice for water blasting.[10]

Cementitious Surfaces

As with the application of coatings to metals, the important first step with concrete is that of surface preparation. When considering concrete, preparation problems are compounded because the character of the substrate may vary so widely. Concrete block construction may present a fairly uniform, hard surface, while some hydraulic concrete surfaces must be actually repaired before attempting coating of the structure. Somewhere between these two cases arises the problems of coating polymeric concretes, precast slabs, and gunite surfaces.

When present, any laitance (thin, cement-rich, weak surface) must be removed, cavities filled, other surface imperfections leveled, all dust and loose material removed, the surface neutralized if there exists alkali from salts of the cement, and the surface dried. To accomplish this end, hand brushing, acid etching, water blasting, steam cleaning, and sandblasting are often used.[11]

Cracks in concrete can sometimes be prepared by routing and applying an elastic epoxy, urethane, or other material that will expand and contract without rupturing as the concrete moves.

A large variety of coating materials are used on concrete, with the epoxies, acrylics, certain urethanes, asphalts, and vinyls heading the list.

If sealing of the concrete surface is desired, repeated treatment with a silicone wash may make the surface of concrete, granite, etc., satisfactorily water-resistant.

Another treatment that can be said to produce a conversion coating is repeated spraying with a dilute solution of magnesium fluosilicate. The chemical reacts with the outer portion of the concrete to produce a hard, water-resistant finish.

Substrate Properties Affecting Coatings Performance

Properties of the substrate to which coatings are applied will influence performance. Tests have shown advantages of certain alloying elements in steel which are conducive to improved service. It usually is conceded that most, if not all metals have oxides on their surfaces to which coatings are applied, the only exception being in certain exotic vacuum systems used for electronic equipment. Atmospheric oxide layers are usually thin on nonferrous metals.

As mentioned elsewhere, surfaces in compression often increase coating performance because they tend to be attacked less than surfaces that are not so stressed; on the other hand, surfaces in tension may cause coatings to fail earlier than they would otherwise. Differences in expansion rates between a coating and a substrate impose stresses on the coating which may lead to cracking and failure. Both inadequate and excessive coating thickness may lead to failure.

Metals high in the electrochemical series, such as magnesium and zinc, require special attention. Because organic coatings are permeable, corrosive solutions that reach the base metal will attack, causing generation of hydrogen which can cause blistering and ultimate failure. Galvanized surfaces are prone to reject organic coatings (Figure 12.8).

Wood substrates are porous and require sealers. Coatings applied to wood may suffer attack from water penetrating from an uncoated side. Some coatings will permit reverse transmission of vapor better than others. Examples are certain linseed oil formulations, some alkyds, and latex water-based materials.[9] Ample information on coatings for wood is available from other sources in addition to those cited here.[16]

Factors in Materials Selection

Voluminous information is available concerning selection of coating materials. Among thorough studies available are data published by NACE, including the *NACE Coatings and Linings Handbook,*[2] a looseleaf collection of NACE books and technical committee reports. *Coatings and Linings for Immersion Service*[12] is a similar NACE document. Numerous other books, reports, and studies may be identified in the references and bibliography for this chapter.

Additional information is available in the NACE monthly journals, *Materials Performance* and *Corrosion.* The NACE journal, *Corrosion Abstracts,* may have as many as 900 entries pertaining to coatings in its annual index.

FIGURE 12.8 — Because the galvanized surface of this circuit breaker box was not properly prepared, there was substantial loss of adhesion in less than a year.

Characterization of Coatings

A convenient way to characterize coatings is as follows.

1. Primers (first coatings applied to a surface).
2. Secondary or Intermediate (if required by some formulations).
3. Topcoats (selected for specific resistance to expected environment and compatible with second or primer coats).

Primers

The original term *primer* implied a coating first applied to a substrate because it had a singular adhesive affinity for it and/or because it provided better adhesion for a subsequent coat than the subsequent coat could achieve if applied directly. A primer also functioned as a carrier for an inhibitor or for an anodic metal loading such as zinc, both of which will be discussed further in later sections of this chapter.

In modern usage and for many applications, Primers usually are thin films (75 μm; 3 mils or less), and may be used on both metal and wood surfaces. On wood, their function is to seal the grain or to provide a smooth base for topcoats. On concrete, they should be compatible with the alkaline surface and help improve both adhesion and the life of subsequent coats.

References 13 and 14 describe developments in primer technology and protection mechanisms, among a wide range of other topics concerned with coating performance.

Intermediate Coats

A second coat may be either the top or final coat, or an intermediate coat added when multiple thin films are required. The intermediate coat must serve as a tie coat between the primer and the subsequent coats, and may be of a different composition than either of them. When the three are of different formulations, it is the intermediate coat that is used to provide the major film thickness for the coating.

Topcoat

A topcoat may be required to extend the life of the preceding coats. When designed as a topcoat, the film is normally more dense and hydrophobic than the remainder of the system to reduce the rate of moisture permeation to the underlying coats.

A topcoat also may be selected to confer reflectivity, reduce photodegradation,[15] or to effect color coding, among other reasons.

Glasses and other silica-based materials require coatings that can be fired (melted) on, as a rule. Limited adhesion of organic coatings can be achieved on etched surfaces.

Polymeric materials are seldom protected with organic coatings, although the coating of epoxy and polyester piping and vessels to prevent deterioration by ultraviolet rays is common.

The Coating Materials

Coating materials increase in complexity as time passes. More can be achieved by the use of a proper coating material today than by simply covering a surface to reduce contact of an environment with the substrate. Coatings of various types are applied to produce the following effects.

1. To provide a dense, inert barrier to shut off the environment.
2. To provide good adhesion to a metal surface for the application of other materials.
3. To serve as a better substrate for topcoats than a bare metal surface would afford.
4. To sacrificially protect the metal surface by cathodic protection.
5. To resist abrasion.
6. To cover a substrate while allowing the escape of gases from beneath.
7. For aesthetic reasons.
8. To provide a smoother, low-friction surface.
9. To prevent catalytic effect of a surface on a product in contact with the surface.
10. To hold an inhibitor at a metal surface for use when moisture permeates the film.
11. To provide surfaces that are more easily cleaned.

12. To prevent contamination of a contained product.

13. To protect the surface from high-temperature oxidation.

The solid product, to be residual on a metal surface, can be prepared separately as a resin (polymer) or can be formed on the surface after application. This resinous material, plus the pigments and fillers added, comprise the *solids* of the coating. These are carried in a *vehicle* (liquid solvent or dispersing medium) which may be varied in viscosity by the addition of a *thinner.* The vehicle also may ultimately be the *binder* of the coating.

Only the solids remain on a surface after the final coating has formed. However, a portion of, or the total vehicle may be reacted with the resin (100% solids coating) to produce the final film. *Pigments* are added to provide color or as inhibitors in a prime (first) coat. *Fillers* are selected to increase the bulk of the coating, improve the density, increase abrasion resistance, and increase the opacity of the film. Selection of the wrong filler or addition of excess filler can downgrade a coating by increasing the permeability or decreasing the cohesion in the film.

Lacquers are those coatings formed by the evaporation of the vehicle from a preformed polymer (coal tar resin dissolved in a hydrocarbon leaves a lacquer finish), while an *enamel* is normally described as a pigmented coating of greater thickness and hardness which may be formed from a resin that completes its "cure" by baking or oxidation in the atmosphere. *Curing* indicates the ultimate reaction of the film, whether effected by the atmosphere, catalysts, or baking. The process is normally associated with some cross-linking of the polymer chains of the coating resin, unless simple drying of the film occurs.

Obviously, the solids content of a liquid coating determines the final film thickness of the coating, depending on the amount applied over a specific area. Thus,

$$\frac{\text{Liters} \times \%\ \text{Vol. Solids} \times 1000}{\text{Sq. Meters}} = \mu\text{m}$$

or

$$\frac{\text{Gals.} \times \%\ \text{Volume Solids} \times 1604}{\text{Sq. Ft.}} = \text{Mils Thickness of Dry Film} \quad (12.1)$$

This theoretical coverage will not be precisely obtained because of the roughness of the surface and losses occurring during application of the wet film. Some 15% loss is to be anticipated for spray applications.

Resins used in the coating formulations can be natural or synthetic. Asphalts, copal, and tung are examples of the naturally occurring molecules, while vinyl, epoxy, and acrylics are examples of the man-made polymers prepared by the petrochemical industry.

If the resin product is not designed to provide a lacquer coating after evaporation of the vehicle and thinner, some reaction must take place to produce the desired polymeric film on the surface. When using linseed oil and certain other molecules, this curing is accomplished by reaction with oxygen from the air (*oxidation*). When a *catalyst* (accelerator; hardener) is added to a specific formulation, the additive actually reacts with the resinous solids to produce the hard, inert final film. Baking of certain types of coatings can produce a similar reaction between resin molecules to produce the coating.

Some of the more widely used coating materials are enumerated in the following paragraphs. It is most important to realize that each of the basic polymers listed does not produce a single coating. They serve as a base upon which it is possible to construct innumerable coating combinations. Although the base resin may impart some property of the coating that is fairly consistent (*e.g.*, epoxies resist alkalies), it is possible to alter the final polymer in such a manner that entirely new properties are obtained. Thus, one should not speak categorically about a coating group, but must specifically state the formulation or proprietary name of the material.

For example, consider the "urethane" coating of today. First, it must be stated which of the six categories of urethane formulations is being referred to. Then there are a wide range of properties that can be obtained from the coatings in each of those categories. These statements apply to the identification of all polymeric products (elastomers, plastics, and coatings).

Bituminous

Natural asphalts are the oldest of all known coating materials. These products were used alone or mixed with other materials before the time of written history. Today, the asphalts and the coal tars derived from petroleum or coal are important coatings. Unless exposed to direct sunlight, the materials are stable, inexpensive, heavy, protective coatings.

As with other coating materials, no single product exists. The asphalt coatings are formulated from a number of naturally occurring beds of bitumen or from petroleum residues. The coal tar product is recovered from a mix of molecules that allows production of coatings with varying properties. These can be applied as hot melts or as *cut backs* (diluted with hydrocarbon solvents) and have greater moisture resistance per unit thickness than do the asphalts.

The products are widely used for the coating of steel and concrete structures below ground, as roofing coatings, and in certain severely acidic atmospheric exposures. Heavy bituminous coatings are often reinforced with glass cloth, burlap, or other materials to improve lateral strength in the completed covering.

Oils

For many years, the formulation of oleoresinous protective paints was based on naturally occurring "oils" (long-chain, unsaturated acid and ester molecules) and rosins such as linseed from flax, tung, pine, castor, fish, and other sources which would provide a resinous film when the oxygen from the air reacted with the molecule to harden the film and produce some cross-linking. This atmospheric drying process was slow, with obvious disadvantages. Also, once the film was cured, the chemical resistance was not adequate to withstand many of the more severe exposures of an industrialized environment.

However, each material has its useful place, and the proper formulation of a "long" oil can produce a cured coating that is tough, ductile, dense, continuous, and very protective of steel exposed to rural and milder urban environments.

Phenolics

It is possible to synthesize a great diversity of phenolic resins. As a consequence, it is best to describe how the various resins are used rather than discuss specific chemical combinations.

First, there are a large group of substituted phenols that are added to oil, alkyd, and other coatings to upgrade their water resistance. Such a coating is phenolic *fortified,* but is not a phenolic coating.

Secondly, preparation of oil-soluble phenolic resins allowed the preparation of a group of true phenolic coatings to be used for atmospheric and fresh water immersion. These are phenol esters combined with oils, rosins, or newer synthetic resins. Certain of these phenolic resins are formulated into an inexpensive low baking finish.

Lastly, there is a class of phenolic resins applied as a lacquer and baked at high temperature (150 C or greater) to produce unique chemical resistance properties, such as excellent acid and solvent resistance. Alkali and strong oxidants are essentially the only environments in which they are not stable. The coatings are used extensively on the interior of cans, drums, tank cars, and storage vessels to prevent contamination of the contents by iron.

Alkyds

The alkyd resins (polyesters) replaced the oils to a major extent as the easiest material to apply at the lowest cost, beginning in the 1920s. These can be formulated for application under most any condition, including baking cycles. The enamel produced has good color stability with little chalking in the atmosphere. The excellent films obtained can be used in the milder industrial environments and most other atmospheric exposures, but should not be exposed to strong acid, alkali, or solvent conditions.

This category of coating should always be considered when assessing a job because of the ease of application.

Acrylics

As a category of coating resin, the same comments may be made for the acrylic resins as for the phenolic products; *i.e.,* the various resins available are used in three different manners.

Some of the acrylic resins are designed to be used as additions to upgrade other coating bases. Others are used in formulations that are applied on home appliances, office furniture, etc., and baked on production line systems. The product has high gloss, toughness, and aging stability.

Today, the greatest use of acrylic resins is in the preparation of latex paints. The original water dispersions were based on styrene-butadiene resins, but quickly yielded the field to the acrylic acid-vinyl polymers. Little need be said about the utility of the latex formulations available today.

The future holds promise of even better water-based coatings based on the acrylic and other modified resins (epoxies, phenolics, etc.). The environmental restrictions on the use of organic solvents in paint formulations makes the use of the water-base materials most attractive. Great strides have been made in producing nonrusting water-based primers and water-based electrophoretic coatings.

The materials do not have good solvent or alkali resistance. However, their durability in the atmosphere is excellent, although color retention can be a problem. They are normally compatible with previously applied oil, phenolic, or alkyd coatings. Acrylic latexes can be particularly beneficial in eliminating blistering problems of other coatings on wood where the more dense coatings prevented "breathing" (vapor transmission) through the paint.

Vinyls

Vinyls are one of the most widely used resin systems for painting in severe industrial environments, coastal regions, and for many water immersion applications. This can be attributed to the high density of the film, the hydrophobic character of the resin, the excellent resistance to oxidation, and the rapid formation of this lacquer-type film when applied. Vinyl films are resistant to acids, alkalies, and oxidants, but not to solvents, except straight-chain hydrocarbons and alcohols.

The vinyl coating resin is a mixture of vinyl chloride and ester molecules. The original coating formulations contained a low-solids content and had to be applied as multiple coats to achieve a reasonable thickness. Also, the primer, intermediate, and topcoats were of differing resin bases. High-build vinyls were eventually prepared which overcame these problems. However, the high-build materials do not have the density or chemical stability of the lower solids lacquers. Thus, in general, the vinyl coating system used for immersion service is not used as a maintenance paint, and vice versa.

Another unique group of coatings are the vinyl plastisols and organosols. These are heavy, tough coatings applied on such items as dishwasher trays and other production line items.

Epoxy

The epoxy molecule was prepared in Germany in the 1800s, but it was in the 1940s that use of the resins based on the molecule became commercially important. Their popularity increased rapidly because of the excellent adhesion attained by the resin and the broad spectrum of chemical resistance inherent in most of the formulations.

The epoxy molecule was combined with oils to produce the epoxy esters, a less costly, easily applied, less chemical-resistant coating which should not be confused with the *catalyzed* epoxies.

The two-package (catalyzed; activated) epoxies offer some of the best properties of any coating material available, once properly applied. However, care must be exercised in their application. Heavy films with fine adhesion to clean metal can be achieved in two coats. The film is resistant to acids, alkalies, and most solvents, but not to strong oxidants. Rather heavy chalking of the surface can occur, depending on the formulation chosen, and the surface can be difficult to overcoat at a later time due to continued curing and chemical resistance of the film.

The amine-cured epoxies are used extensively for the internal lining of storage vessels, hopper cars, and concrete structures. A hand layup of epoxy coating reinforced with glass cloth provides an exceptionally heavy coating for immersion service.

The combination of epoxy resin with coal tar polymer produces a less expensive coating with properties appropriate for use on the interior and exterior of piping, and many other severe exposure conditions.

Urethane

The urethane polymer is most familiar in the form of plastics, elastomers, foams, and coatings. Obviously, the different urethane resins must be capable of combining with many other types of molecules in order to produce this variety of end products. As a result, the category is divided into six types by ASTM. In each of these various types, a range of properties can be obtained for the coating, depending on the chemistry of the system. Elastic coatings, hard coatings, coatings with poor ultraviolet light stability, coatings with excellent ultraviolet light stability, coatings difficult to pigment, coatings readily pigmented with the usual additives, coatings with a chemical resistance no better than an alkyd, or coatings with very fine chemical resistance can all be obtained in this grouping. There is no such thing as a "urethane coating," so it is important to be specific.

The most unique property of this category of coating material is the superb abrasion resistance of the moisture-cured varieties. No other organic material with a hard, glossy finish can compare with a Type II urethane in abrasion resistance.

The materials are used widely for floor and other wood finishes, hopper car linings, and as maintenance coatings. Adhesion of the basic urethane resin to metal is not good, and the coatings are applied over primers of other resins such as epoxies.

Temporary Preservatives

One large area of coating use is that of the temporary preservatives. The term *temporary* can mean intervals from hours to months or years. Protection of metal during a period of days or months, depending on whether the exposure is indoors or outdoors, is normally expected. The coating materials may be divided into the following groups.

Oils, Greases, and Microcrystalline Waxes. Materials such as oils, greases, and microcrystalline waxes are identified and divided into various types by MIL (military) specifications. These range from thin oils applied to steel or aluminum sheets or to coil stock, to the heavy, cohesive greases or waxes that provide relatively permanent coverings over the metal substrate. The materials are often termed *slushing compounds* because they are applied by dipping, rough brushing, or spraying. Many are formulated with inhibitors to stabilize the surface of the metal.

Such materials may be used to preserve metals between periods of processing, for the protection of precision surfaces during shipment or storage, to maintain the interior surfaces of equipment during storage, or to prevent corrosion of steel surfaces when in contact with stagnant waters (*e.g.*, fire water tanks).[17]

Grease. Cloth impregnated with heavy, inhibited greases is also available for atmospheric exposures. Although too tacky for use in areas where there is personnel traffic, the materials provide an exceptionally heavy, stable covering over metal where undisturbed.

Strippable Coatings. Synthetic polymers can be used to "cocoon" equipment to prevent interchange of moisture with the atmosphere (Figure 12.9).

Antifouling Coatings

Because the life processes of living organisms often induce or accelerate deterioration of coatings, there is a wide range of additives designed to kill or

FIGURE 12.9 — Highly polished silver surfaces are subject to rapid oxidation or sulfidation, so a temporary, strippable vinyl coating is used. Sprayed-on silicones also are used for protection of surfaces such as these.

reduce the activity of bacteria and other micro and macroscopic living things. Substances used to impregnate wood to protect it against attack are not covered here, but they often influence the selection and performance of coatings applied over treated material.

Because steel is anodic to copper oxides in salt water, copper-containing boat or ship hull coatings require an electrically insulating coating on the steel. Copper compounds should not be used on aluminum, because the galvanic difference is even greater than it is for steel.

Biocides and/or biostats include various salts of metals, arsenic, tin, lead, and zinc, as well as organic poisons. Fungicides include many of those listed above, and such compounds as chlormethyl oxetane, chlorophenyl compounds, sulfuryl/sulfonyl pyridine and many others. They prevent attachment or growth of organisms.

Antifouling coatings have been in use for centuries on hulls of ships, and some copper formulations developed long ago are still in use. Newer compounds include variations of organic tin esters. Recent development of formulations in which the metal salts are bound into the organic molecule control and extend their dissolution rates into water, thus improving surface concentration levels vs time, and thereby extending effective life of the coating.

Fungicides are widely used in water-based organic coatings to control bacterial attack in containers and after the coating has been applied. A comprehensive guide[18] to the use of fungicides has been published in England.

Composite or Multiphase Coatings

Various combinations of primers, synthetic polymer coatings, inorganic fabrics, and thick film coatings are used for protection against corrosive fluids and gases, usually in tanks or vessels. A membrane of polymeric material will be laid over the inside surfaces of a steel tank to be lined with acid-resistant brick, such as that used for resistance to heat and impact in steel mill pickling tanks. Figure 12.10 shows details of a multilayer lining being put inside a gas scrubbing tower.

Composite coatings are typified by systems used on pipelines. As one example, a coating may consist of inhibited primers, followed by cross-linked butyl rubber-based adhesives, and polyethylene tape. They also may consist of a hot-sprayed thermoplastic, along with cut glass fibers or other additives.

Zinc-Rich Materials

Because of their singular merits in corrosion control in many environments and their worldwide use, special attention is necessary for zinc-loaded organic and inorganic materials. The materials have exhibited excellent protection of steel when used as primers, and in many instances, as a total coating system.

The zinc functions as an anodic material to provide cathodic protection of steel at those areas where moisture has penetrated the coating (pores, scratches, etc.). To achieve this end, the zinc particles must be in electrical contact with one another and the substrate to provide the metallic path of the cell (Chapter 2). Thus, the *loading* or metallic content of the coating must be very high. Coating thickness should be about 70 μm (3 mils).

The binder for the small zinc spheroids can be one of three kinds:

1. Inorganic silicate produced by evaporation of water,

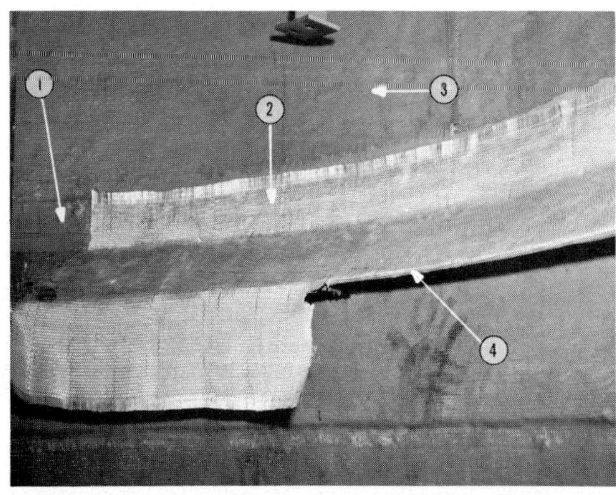

FIGURE 12.10 — Details of a liner being installed inside a gas scrubbing tower. The layers are: (1) trowel coat of epoxy filled with silica flour, (2) fiberglass cloth pressed into trowel coat, (3) fiberglass saturated with epoxy, and (4) edges are rolled over and sealed.

2. Inorganic silicate produced by reaction with water, or

3. Organic resin resulting from loss of solvent.

The original coating of this type used an aqueous sodium silicate solution as the vehicle (Type 1, above) with a final rinse of phosphoric acid to set the coating. Such a procedure is obviously not as simple to use as the straightforward mixing, application, and drying of the standard two-package (zinc and vehicle) coating which succeeded it. However, the coating had certain characteristics that have not been duplicated by the later products.

The two-package system of the first type dries by the evaporation of water from the vehicle to leave the silicate binder holding the zinc particles in place. A one-package version is also available.

The Type 2, two-package product has ethyl silicate and ethanol as the vehicle. The ethyl silicate then reacts with moisture from the air once the coating is applied. Ethanol is formed by the reaction and evaporates rapidly to leave the silicate binder. This product is more tolerant of moisture than is the Type 1 material, but should not be used where the relative humidity is very low.

The third category of zinc-rich coating utilizes customary organic polymers as vehicles which dry by solvent evaporation. Acrylic, chlorinated rubber, and catalyzed epoxy resins are used.

As indicated, these materials serve well as primers beneath organic topcoats. Underfilm creepage of corrosion is reduced where the substrate becomes exposed to the elements. Care must be taken in the selection and application of the topcoat to ensure that the resin is compatible with the zinc when moisture is present, and that adequate adhesion with good continuity is achieved.

All zinc-rich coatings require good surface preparation. The materials are the best primers available for use on galvanized metal.

The zinc-rich coatings are particularly good in chloride environments (marine areas, etc.) where the corrosion products form a hard, dense film over the zinc. The coatings should not be used in sulfurous atmospheres with or without topcoats.

In many instances, the zinc-rich coating has sufficient resistance to be used without a topcoat, or the application of a topcoat can be deferred for an appreciable period of time. Such a deferred system can save considerable money since the investment is delayed.

Inorganic (Nonmetallic) Coatings

Inorganics include numerous classes of materials, among them the hydraulic concretes (cements), ceramics and clays, glass, carbon, silicates, and others. On a volumetric basis, hydraulic concretes are the most common. So commonly used are these concretes that they often are not identified as coatings at all.

Hydraulic Concrete

Concretes are used to coat pipe inside and outside, especially that to be buried or submerged, as are water or sewer lines. Cast iron, ductile iron, or steel pipe may be coated at the mill by a process during which the pipe is spun on the center of its longitudinal axis while a mortar mixture is sprayed onto the inside surface in a uniform, dense layer. After proper curing, provided the pipe is handled carefully, this coating can protect the pipe interior against attack by water and many other liquid and gaseous corrosives.

The concrete coating can be obtained only on the exterior or on both sides of a thin steel pipe (cylinder pipe). Solid concrete pipe reinforced by heavy wire mesh or by prestressed wires is also available.

A concrete coating on steel of any configuration can be protective to the steel as long as it is intact because the alkaline reaction of hydraulic cement maintains a pH at the steel surface which effectively prevents corrosion. It does this when encasing reinforcing steel, for example, where it functions both as a strength and as a protective component.

Concrete is applied to external surfaces of pipe to protect them from soil or water in which the pipe is buried or submerged. In the case of underwater piping to transport gas or liquid hydrocarbons, for example, it may be mixed with barites or other heavy materials to confer negative buoyancy. It also may be used in mixtures of organic materials where its ability to maintain a noncorrosive pH at the surface of steel is exploited.

Ceramics

Ceramics also may be used in much the same way as hydraulic concretes, but they usually are applied to confer resistance to heat and/or attack by hot, high-velocity gases. Thus, they are used in exhaust passages emitting hot gases such as rocket nozzles, furnace linings, and other similar applications.

Also to be considered are the specialty concretes derived from the acid-resistant cements. These mortars can be applied to steel or concrete to provide remarkable acid, alkali, and heat resistance. They may be reinforced with wire mesh to sustain the heavy thicknesses normally applied. Coefficients of expansion close to that of steel have been attained to resolve one of the major problems when applying inorganic materials on steel. These can be used as floors, tank interiors, in scrubbers, and other process equipment of clean interior design.

Glass

Glass, a material now being produced in a variety of new and interesting formulations, usually is applied as a slurry, nearly always in a production line, but seldom for maintenance purposes. Once

TABLE 12.1 — Characteristics of Some Common Industrial Coatings[1]

Generic Name	Surface Prep.[2]	Thickness, mils	Application Method: Brush	Spray	Dip	Electrostatic	Fluidized Bed	Roller	Resistant To: Abrasion	Acids	Alkalis	Gases	Oil & Fats	Oxidizers	Solvents	Sunlight	Temp. Limit, F	Water: Fresh	Salt	Toxic	NACE Report Number
Alkyd	3	4.5-5	X	X	—	—	—	X	F	F	P	F	G	P	G	E	218	G	G	No	6B165
Asphalt	3	2.5-250	X	X	X	—	—	—	F	G	G	G	P	P	P	E	300	G	—	No	6A166
Chlorinated Rubber	3	>5	X	X	X	—	—	X	F	G	G	G	G	G	P	P	300 140[3]	G	G	No	6A356
Chlorosulfonated Polyethylene	1	8	—	X	—	—	—	—	E	F	E	G	G	E	P	E	212	E	E	No	6A166
Coal Tar (Epoxy, Urethane)	2	>30	X	X	—	—	—	X	E	E	E	E	E	P	F	E	250	E	E	Yes	6A162
Epoxies	2-3	7	X	X	X	X	X	X	G	E	G	F	G	P	E	G	250	F	F	No	6A256
Fluorocarbons	1	>7	—	X	—	—	—	—	—	E	E	E	E	E	E	E	390	—	G	—	6A163
Furfuryl Alcohol	1-2	>20	X	X	X	—	—	X	—	G	G	G	E	G	E	F	200	—	—	No	6A263
Neoprene	1	>20	X	X	—	—	—	—	E	G	G	G	E	P	P	E	200	G	E	No	6A363
Phenolics[5]	1	5-7	X	X	—	—	—	—	F	G	P	G	E	P	G	G	300	E	E	No	6A362
Polyesters[4]	1	—	—	X	—	—	—	—	—	G	F	G	E	G	E	G	175	E	E	Yes	6A462
Polyvinylchlor-Acetate	1	>250	X	X	X	—	—	X	G	F	G	P	E	F	P	G	150	G	E	No	6A154
Urethane	1	>12	X	X	—	—	—	—	E	P	P	G[5]	E	P	P	G	250	E	E	No	6A262
Vinyl	1	>8	X	X	X	—	—	X	G	G	E	E	E	E	P	E	150	E	E	No	6B163
Vinylidene Chloride	2	>10	X	X	X	—	—	X	E	E	G	F	G	F	F	G	150	E	E	No	6A157
Zinc-Filled Inorganic	1	>4	—	X	—	—	—	—	E	P	P	—	E	P	E	E	700	E	E	Yes	6B161

[1]Data are adapted from information available in the NACE Technical Committee reports referenced herein. Not all data match exactly the criteria in the reports. Refer to reports for exact and complete information.
[2]Numbers refer to specific surface preparation grades included in NACE Technical Committee Report No. 6G153.
[3]Wet temperature maximum.
[4]Random reinforced with chopped glass fibers.
[5]Baked formulations.

LEGEND: E = Excellent, G = Good, F = Fair, P = Poor. Recommendations are advisory only. Where no recommendations appear, data are lacking.

applied and as long as protected from mechanical abuse, glass surfaces are highly resistant to acidic and many mildly alkaline corrosives (especially the new types), are easily cleaned, but difficult to repair.

Glass lined and coated pipe, valves, pumps, and vessels of all kinds are widely used in the chemical, pharmaceutical, and food industries.

Glass coatings are available in two basic types and in various grades of integrity. As with the basic borosilicate glass from which the coatings are derived, the materials may be used up to about 175 C in most acid mixtures other than those containing fluorides.

Porcelain coatings are used extensively for their excellent atmospheric stability (signs, etc.), but are not normally used for immersion service of any type.

Specialty Coatings

The preparation of a coating system can be made from many other sources. A few additional coating bases are listed in Table 12.1.

As an example, coatings are used to "stop-off" areas of a metal when it is to be plated, etched, or treated at high temperatures.

Conversion Coatings

As the name implies, *conversion* coatings are produced by converting the surface of the base metal into a new, uniform, and normally more corrosion-resistant form. Some are used as prepared, but many others serve as a base for subsequent chemical treatments or overcoating. The more common conversion treatments are as follows.

Anodizing

Aluminum is always encased in a thin oxide film which provides protection of the metal itself. This film can be fortified or increased in thickness by making the metal the anode in an acid bath and discharging current from the metal.

As noted earlier, the anode environment is highly oxidizing. This anodizing condition produces a much heavier oxide film on the metal. Oxide film thicknesses of 1 to 25 μm (0.04 to 1 mil) are nominal. The films may be colored during formation or treated after formation to increase the stability. Greater resistance to permeation of contaminating ions on the surface is produced by this heavy film. Many aluminum building materials, car parts, and other structurals are treated in this manner.

Other metals, such as titanium, zirconium, and the stainless steels can also be "anodized," but the term is normally reserved for the treatment of aluminum.

Phosphatizing

The most successful treatment of steel to date has been by phosphatizing. The steel is pickled or otherwise well cleaned and immediately held in a bath of hot phosphoric acid containing zinc and perhaps manganese salts, plus other selected additives. A coating of iron-zinc phosphates is formed. The steel is removed, washed, and dried. The resulting film of some 1 to 10 μm (0.04 to 0.5 mil) thickness is then often given a chromate treatment to upgrade the corrosion resistance, because the film as formed has little resistance to attack.

The porous phosphate film can then be sealed with oils to resist corrosion, or serves as an excellent base for coatings. For many years, car bodies have been given this treatment prior to coating.

The requirements of a large vat to hold the hot acid salt solution limits the process to mills and other limited areas where large tonnages of steel are to be processed. Phosphatizing of zinc, cadmium, and aluminum can be conducted.

Cold phosphating of metal surfaces by spraying with a warm acid salt solution should not be compared with the phosphatizing treatment described above. Pickling of a metal surface can be obtained in this manner, and some residual phosphate formed, but not a full phosphate conversion coating.

Chromating

The oxidation of a clean metal surface greatly increases the corrosion resistance of the surface if a thin, continuous, dual metal oxide can be formed. The most satisfactory method of producing this film has been by treatment with chromic acid washes of one type or another. Aluminum, zinc, and cadmium are often treated in this manner, and the golden color left by the treatment can be seen on commercial items.

As with phosphatizing, the film produced serves as an excellent coating base. Indeed, it can be difficult to adhere most polymeric coatings to the nonferrous alloys without such a treatment. The thin films of chromium and substrate metal oxides are only some 0.01 to 0.2 μm in thickness and can be readily removed by abrasion.

Wash primers are a special type of chromate solution used to produce a conversion coating. A phosphoric acid solution containing chromate salts and a small amount of vinyl butyral resin is sprayed on the clean metal surface (steel, aluminum, etc.) and reacts with the alloy in the manner described earlier. Although the treatment does not serve as a primer, the film is an excellent base for most coatings.

Metal Oxides

The bluing of steel has long been used as a process for producing a protective oxide on steel. When oiled and maintained, the thin oxide film prevents the rusting of steel in the atmosphere. Such oxide coatings can be produced by exposing a clean steel surface to dry steam, strong alkalies, cyanides or a great number of other chemicals. The films must be very thin and continuous to have the mechanical and physical properties desired.

Oxide coatings have been produced on copper, nickel, tin, and gold for specific applications.

Pack Cementation (Diffusion)

Certain materials can be diffused into the surface of a metal to produce a surface product entirely different from the metal. Powders of carbonaceous materials, aluminides, chromium salts, borides, or silicides are packed around the object in a container. The mass is taken to an appropriate high temperature for sufficient time for the carbon, aluminum, chromium, or boron, etc., to diffuse into the metal surface. A hard, abrasion-resistant surface of some 20 to 50 μm (1 to 2 mils) in thickness is usually produced. These coatings are used to impart abrasion and/or high-temperature resistance to gas turbine blades and vanes, among other similar uses. Aluminum coatings frequently are used on heat exchanger surfaces subject to attack by sulfur gases (H_2S, SO_2, and SO_3).

The sherardizing process involves diffusion of a zinc coating into steel by tumbling steel parts with zinc dust at high temperatures.

Metal Coatings

For many rugged services, the choice of a metal coating is preferred. Where severe impact, abrasion, or high temperatures are a part of the materials selection consideration, the use of metallic coatings must be assessed. These are provided by a variety of techniques, as described in the following paragraphs.

Be aware that metallic coatings on a metal substrate may be electrochemically anodic, cathodic, or neutral. If complete coverage is achieved and maintained, this relationship is unimportant. However, if some electrolyte contacts the two entities, corrosion of either may be accelerated or accentuated, depending on the existing conditions (Figure 12.11).

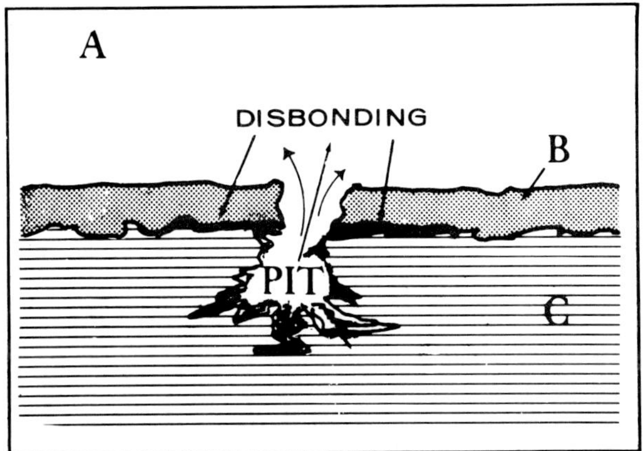

FIGURE 12.11 — Shows how a holiday (hole) in a coating cathodic to a base may cause a pit in the base metal. Arrows show approximate paths of ions leaving base metal: (A) electrolyte; (B) cathodic layer; and (C) base metal. The indicated disbonding may or may not take place. [SOURCE: Robert Draper, Nickel-Chromium Plating, Teddington, England, p. 10 (1961).]

Plating (Electrodeposition)

One of the five basic cathodic reactions can be used to good advantage. Metal plating on a cathodic surface can be achieved on a commercial basis to provide aesthetic and protective coatings.

The part to be coated is immersed as a cathode in a solution of the metal ions to be plated. Current is driven from an appropriate anode into the electrolyte, and the soluble metal plates out on the cathode. The current densities used are high, and the plating is done rapidly. Most any type of metal can be plated on a cathode. The more common plated metals are chromium, nickel, copper, cadmium, and zinc.

Chrome plate, as used on automobile parts and other items requiring protection under difficult conditions, is actually a three-ply coating. A *flash* of copper is first laid down, followed by a nickel coating which comprises most of the finished thickness. A thin, hard coating of bright chromium is then applied as the topcoat. The total thickness is normally 25 to 50 μm (1 to 2 mils).

The plated coating is designed to be pore-free. When the covering is punctured, accelerated corrosion of the steel substrate can occur because the steel is anodic to the topcoats. Severe attack can occur because of the adverse electrode area effects of the corrosion cell (Chapter 2).

Electroplated coatings normally fail by cracking from repeated expansion and contraction or impact.

One of the problems associated with plating is the presence of large quantities of hydrogen on the cathodic surface. Some of this hydrogen can enter the substrate metal (normally steel). For low-strength steels (hardness $< R_C22$), the hydrogen seldom poses a problem. However, the higher strength steels can be badly embrittled and fail prematurely. When poisoners for the dimerization of hydrogen (S, CN, As, Cd, etc.) are present in the bath, the amount of atomic hydrogen absorbed can be quite high. If immediate damage is not done, the parts may be baked at temperatures around 200 C (400 F) to drive most of the hydrogen from the metal.

For small, intricate shapes such as bolts and screws, plating metal may be applied by barrel plating. A rotating barrel containing the parts to be plated is turned slowly in the electrolyte, while current is discharged to the tumbling parts inside.

Brush plating is also performed, particularly for touch-up of small surfaces. In this case, an absorbent pad holding the electrolyte is wrapped around a metal anode and the brush is moved over the cathodic surface.

Today, chrome plating is also applied over plastic materials (primarily ABS) for many uses where a bright finish is desired.

Thin electroplated films of gold, platinum, and silver are used on electrical contacts and in electronic equipment. Because gold and platinum resist oxidation, they are important in contacts where small currents are being switched or passed through connections, because any oxides on the contact surfaces impose unwanted resistances in the circuits.

Platinum coatings on cathodic protection impressed current anodes permit imposition of higher voltages and current densities that other metals cannot sustain. Silver reflectance coatings must be protected by transparent topcoats against oxidation and sulfidation. Gold, rhodium, and osmium are widely used as coatings for small parts of operating equipment in industry, such as on meter and valve parts.

Gold leaf, not electroplated, has also been used for advertising and industrial signs and as coatings for roofs, especially those on government buildings, because of excellent resistance to atmospheric attack. For such uses, it was often the most economical coating.

Molten Bath

Galvanizing, tinning, and aluminizing are accomplished by cleaning the steel and immersing it in a controlled melt of the coating metal. Sufficient comment on the characteristics of the galvanized coating is provided in Chapter 11. The zinc coating produced is actually alloyed to the steel, as indicated in Figure 12.12.

Aluminum is applied to thin sheet and coil stock only in the mill. The coating is very continuous and very stable in marine atmospheres and high-temperature exposures. The aluminum is not to be considered as an anodic (sacrificial) coating on the steel.

Cladding

Metal layers of varying thicknesses can be applied to other metals by the following methods. These procedures are designed to obtain a more corrosion-resistant interior, while using the less expensive steel as a backing for strength.

Arc or Gas Welding. In arc or gas welding, relatively thick layers of weld metal are deposited either by manual or machine methods on surfaces. The interior of pulp digesters or other pressure

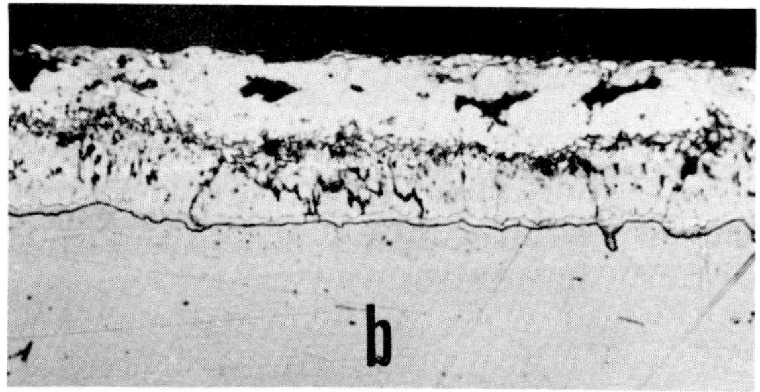

FIGURE 12.12 — Zones in heat-treated galvanized steel coatings: outer layer, essentially pure zinc; inner layer, zinc-iron alloy. [SOURCE; Daeson, 2nd Int. Cong. Proc., Figure 9b, NACE, p. 698, Houston, TX, 1963.]

vessels requiring an alloy composition to resist the chemical conditions can be constructed in this manner.

Roll Bonding. In roll bonding, layers of two metals are mated by heavy rolling in a mill after the surfaces have been thoroughly cleaned and treated. Clad thicknesses of 5 to 10% of the base steel thickness are common. Some small areas of unbonded metal will be present. Two metals may also be coextruded through a die.

Explosion Bonding. In explosion bonding, a base metal and a covering metal coating are contacted, placed in an appropriate enclosure with the coating material on top. A layer of explosive is placed over the coating metal. When the explosion is activated, the resulting shock wave merges the two materials, as shown in Figure 12.13.

Melting or Brazing. Lead, copper, and other low-melting alloys can be applied by the technique of melting or brazing. Figure 12.14 shows application of a lead coating to steel.

Soldering. A metal coating can be soldered to a base by flowing a solder layer on the mating surfaces, putting the metals in contact, and applying heat sufficient to melt the solder.

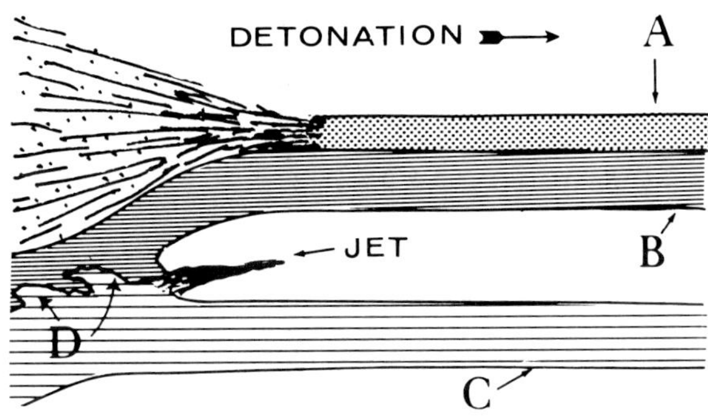

FIGURE 12.13 — Manner in which explosives bond a layer of cladding metal to a base: (A) explosive; (B) cladding metal; (C) base metal; and (D) shows how "jet" tends to secure cladding layer to base. [SOURCE: Pocalyko, Andrew, Explosion Clad Plate for Corrosion Service, Figures 2 and 5, Materials Protection, Vol. 4, No. 6, pp. 10-15 (1965).]

Metallizing

The metallizing technique for applying metal coatings is accomplished in several ways.

1. Feeding a wire into oxyacetylene or oxyhydrogen flames.

2. Feeding a wire into the heated zone created by electrodes arcing alternating current from which molten metal is impelled by a stream of compressed air toward the surface to be coated.

3. Blowing metallic particles suspended in a gas through a blow pipe flame.

Thorough surface preparation usually is required because the molten metal droplets must achieve adhesion before solidification. The coating achieved is somewhat oxidized and porous. Common applications are on steel surfaces exposed on offshore petroleum production platforms, chemical tanker compartments, stack breechings, and other structural steel locations. Metallized surfaces frequently are topcoated.

High Temperature

Ceramics, cermets (ceramic-metal mixtures), and various oxides are applied by a variety of methods, including plasma arc spraying, for high-temperature applications and for cutting edges. One application method is shown in Figure 12.15. This classification overlaps with the diffusion category and with coatings designed to provide hard, abrasion-resistant surfaces, such as those available with tungsten carbide, aluminum oxide, and nitrides. Developments are covered in a book edited by Holmes and Rahmel.[19]

Miscellaneous

Other application methods for metal coatings include the following.

Peen Plating. In peen plating, powders of soft metals, such as aluminum or zinc, are deposited on

FIGURE 12.14 — A machine is "burning" a lead coating onto a steel sheet. Similar claddings can be done by hand.

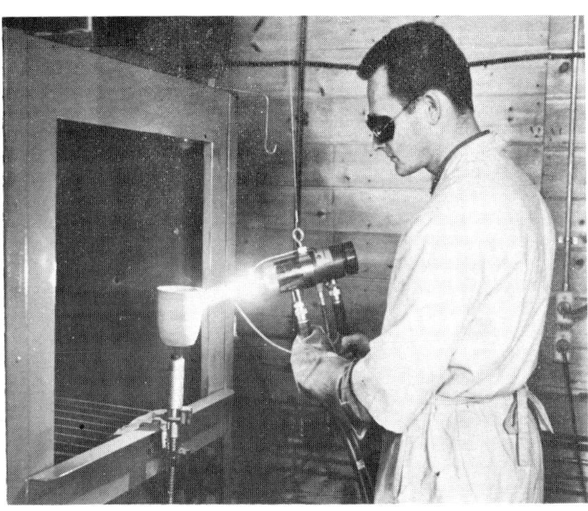

FIGURE 12.15 — A plasma flame spray gun is shown applying a high-temperature protective coating. [SOURCE: Photo Metco, Long Island City, N.Y.]

surfaces when objects are tumbled in a rotating barrel at higher temperatures.

Cathodic Sputtering. The metal to be deposited through cathodic sputtering is vaporized in an atmosphere of gas or in a vacuum from which it deposits on the object to be coated.

Chemical Reduction or Deposition. Particularly used for nickel, chemical reduction or deposition is conducted at 85 to 95 C and involves immersing the surface to be coated in a solution of nickel chloride and phosphoric acid, or borohydrides or alkyl amine boranes. A smooth, nonporous layer is produced which may contain up to 14% of phosphorus. The method covers surfaces in recesses and other irregularities, and is particularly suited for coating the insides of fabricated tanks, piping, valves, etc.

Salts can be obtained for the deposition of copper, tin, and certain other metals by the reduction process.

Pack Cementation. As discussed under the subject of conversion coatings, various selected materials may be diffused into a steel surface. In addition to the absorption of carbon, silicon, etc., new metal oxides and carbides can be welded to the steel matrix by pack cementation. The coatings produced are used to prevent high-temperature oxidation or resist severe abrasive conditions.

Application of Organic Coatings

Proper application of a coating can mean the difference between a successful job or a severe economic loss. In some instances, the choice of the application method can be the most important decision. At other times, more subtle influences determine the outcome of the work.

Atmospheric conditions are an important consideration. If excessive moisture is in the air, a coating must be applied very quickly after the surface is cleaned. Gross oxidation of the surface by moisture and oxygen in the air cannot be allowed.

Many coating resins are intolerant of moisture. Application of these materials on a dewy morning, under rainy conditions, or where rain is anticipated can be a waste of time and money. Also, the rheology of most coating materials is such that flowout of the film and proper solvent release will not occur below a given temperature. For many materials, this point is around 5 to 10 C (40 to 50 F).

The most difficult influence to control is that of contamination of the surface by some airborne species before the coating is applied. This can be a whiff of hydrogen sulfide, sulfur trioxide, or other chemical from a stack. It may be an alkaline material, such as fly ash, ground seashells, or limestone dust. When a coating is applied over such contaminants, a more rapid failure is certain to occur once moisture permeates the film.

Materials to be applied must be properly mixed. This process can be difficult for some materials, and adequate equipment to do the job should be available. When catalysts are to be added to the mixture, close attention to the proportions required must be given, and thorough post-mixing must be accomplished. The pot life of the mixture must be known and understood.

Many jobs have been downgraded because excessive or inadequate thinning of the coating material was made before the application. Overthinning to make the mixture more workable is the more common error.

Methods of Application

Many methods of applying a coating are available. The familiar techniques involve brushing, rolling, dipping, and palming of coatings of alkyd, acrylic latex, or varnish bases. Other procedures have been developed for more specific types of organic coatings and economy of operation. Some of the more important techniques are as follows.

Brushing and Rolling

Brushing and rolling rank first as the easiest and most commonly used application techniques.

Palming

Palming is sometimes referred to as a mitten application.

Dipping

Dipping is a technique that should be considered for intricate shapes such as grating, etc.

Air Spray

Air spray is an older spraying technique where a separate air stream was used to break up the coating stream.

Airless Spray

Airless spray is a more popular spraying technique where more pressure is used on the pot, and the tip of the gun is designed to disperse the coating. Less overspray and better control of the spray pattern are achieved.

Electrostatic Spray

With electrostatic spray, electrically charged particles of coating are directed through the air toward a surface whose charge is opposite that of the particles. The coating is thus attracted to the surface and will tend to deposit with little overspray. Good buildup on edges is also a positive factor. It is often used on production lines.

Electrophoretic

With the electrophoretic technique, electrically charged particles suspended in a conductive liquid are impelled toward a surface when a DC current is applied through the liquid. The particles tend to collect on surfaces of opposite polarity and will go into crevices and occluded areas in a relatively uniform thickness. Edges are particularly well coated. Automobile bodies are coated in this manner.

Fluidized Bed

When using the fluidized bed technique, thermoplastic polymers are powdered and suspended (levitated) in a tank through which air is pumped in such a way that the particles behave like a liquid. Surfaces to be coated are preheated and run through the bed. The hot surface melts the thermoplastic, fuses the coating, and subsequent cooling produces a uniform plastic sheathing. Thickness of the coating depends on the temperature of the part and the time of exposure.

Powder Spray

Parts to be coated by the powder spray method are heated to the appropriate temperature and then moved to an area where a fusible coating powder is sprayed onto the surface. Powder spray is a production line technique which can produce very uniform coatings.

Flame Spray

During the flame spray method, thermoplastic powders or rods are fed into an oxyacetylene or other flame and impelled toward the surface to be coated. The particles coalesce and solidify to form the coating. Metals are applied in the same manner at much higher temperatures.

Troweling

Concrete, mortars, epoxy, polyester, furane, and certain other polymers may be applied as heavy coatings by hand troweling. The organic materials are normally reinforced with glass cloth, while wire mesh might be used for the inorganic products (Figure 12.10).

Vacuum Deposition

With vacuum deposition, the coating material is placed on a heating source in the lower part of a container which is then put under vacuum. The item to be coated is suspended above the coating material. The coating material (organic or inorganic) is then heated until vaporization occurs. The vaporized coating molecules condense on the cold surface of the target object to produce the coating. Glass, plastics, and other nonmetallic bases may be coated in this manner.

Calendered or Sheet Lining

Prefabricated linings may be glued to surfaces, as shown in Figure 12.16. Applications are made inside tanks and large-diameter pipes. Polymeric linings can be designed to be inserted loose inside tanks. Large indoor or outdoor ponds and sumps designed to hold corrosives or ecologically restricted fluids may also be lined with polymerics.

FIGURE 12.16 — Application of calendered polyvinylidene chloride lining inside process piping. Linings are glued to substrate.

Testing

Coating operations of most any size should have some type of meaningful test program. The program should evaluate materials, surface preparation, application, and inspection procedures. Such programs are the only objective basis on which to make decisions relating to future work. These may be done in-house or contracted to the many consulting firms available to help those without the necessary personnel.

It is important that all parties having an interest in this work subscribe to the program to be instituted and agree on the basis by which the decisions are to be made. Beware of simply generating data. The evaluation procedures for coatings lend themselves to the accumulation of reams of meaningless data. Be sure that the tests conceived do not favor one type of material, and that they assess some important characteristic of the coating (water resistance, ultraviolet light resistance, etc.). Statistical design of the test program will provide more information for less money, show deficiencies in the program, and provide greater objectivity.

Whenever possible, use the standard test procedures and means of evaluation provided by ASTM, NACE, and others. Only in this manner can correlation with the results of other workers be established. New techniques of coating appraisal should be considered constantly, but not used as a daily procedure until evaluated by a variety of personnel working in the coatings field.

Coatings evaluation is conducted in: (a) the laboratory, and (b) the field.

Laboratory Testing

The number and types of tests that can be conducted in the laboratory is almost limitless. Because field testing of coatings is so costly in time, labor, and facilities, it has been the persistent aim of formulators, raw materials producers, and users of coatings to establish meaningful accelerated tests. These may involve a wide variety of comparatively simple, as well as highly sophisticated procedures.

None of them (so far as the literature shows) has consistently provided more than reliable means of eliminating the poorest among a group of competitive materials or correlation with some one specific exposure in the field.

The tests can define certain characteristics of a coating system. This discrimination usually is specific to the type of accelerated test, and frequently does not indicate in advance how a given coating will perform in service. However, the tests are invaluable when developing coatings or appraising new concepts in coatings application or use.

Some of the common accelerated tests are as follows.

General Immersion Tests

General immersion tests can be as simple as the standard saltwater immersion test for coatings to be used in the atmosphere, or as complicated as "cold wall" effect immersion tests for heavy coatings to be used constantly in immersion service. When coatings are to be used for immersion service, the NACE Standard on this subject should be followed.[20] Reproduce the anticipated conditions of the field exposure as closely as possible when evaluating coatings for immersion service. Reference 12 provides data relating to the properties of these lining materials.

Salt Spray Testing

Exposure of coatings to various formulations of salt spray or fog, with and without cyclic immersion and exposure to highly actinic light, heat, and to various wetting and drying cycles is a common practice. There was, for example, rough correlation between the nitric acid test[21] and salt fog exposures of 10 coatings systems. In another series of tests,[22] involving, in addition to salt fog, subjection to sulfuric acid, temperature shock, and saltwater immersion, results permitted selection of systems which apparently are giving good performance under service conditions.

Cathodic Protection Exposures

One accelerated test involves subjecting the material to which the coating has been applied to various voltages of impressed current, often in excess of that normally required to achieve cathodic protection, with the aim of demonstrating the ability of the coatings to resist the disbonding effect of the cathodic current or to resist electroendosmosis.

Impedance Measurements

Various approaches to the measurement of changes in the electrical resistance of a coating during exposure are available.

Condensation Apparatus

Relying on the known susceptibility of coatings to permeation by condensed water, a screening test is available in a laboratory-type cabinet which results in condensation of water on the painted surface. Disbonding of poorer coatings occurs.

Environmental Test Rooms

Environmental test rooms which permit various combinations of temperature, fogging, humidity levels, and shower effects to simulate rain can reproduce in an accelerated mode many of the factors present in an actual exposure.

In addition to the accelerated test methods just described, there are various specialized tests, such as those involving the use of barnacles to evaluate the toxic properties of antifouling coatings, equipment to produce high-intensity sound to test coatings exposed to sonar beams and cavitation, rotating disks to evaluate the effects of velocity of water impinging on surfaces, isotope-tagged liquids to measure the rate at which moisture penetrates a film, and others.

Coupons prepared for testing under laboratory conditions (for example, in areas not subjected to chemical fumes) may give results in accelerated tests which will not correlate with the performance of the same coating systems applied under field conditions.

Field or Service Testing

While the delays involved in field or service testing may seem interminably long in many cases (years may be required for completion of a full-scale field test), when coating expenditures run into the hundreds of thousands, and in some cases into millions of dollars, it pays to be sure about performance. Most field testing is done on-site under the supervision of the customer's testing staff. Ordinarily, coupons coated with the materials to be tested are exposed in multiples at several test sites in a plant as a first step.

After varying periods of exposure, depending on the design of the test program, coupons are removed from the exposure sites to laboratories where their residual properties are measured. At the end of such a program, it usually is possible, from among a group of coatings similarly exposed, to choose the one that performs best. This coating can be applied with some confidence that good service will be obtained.

As with other tests, select a field test panel and application procedure that is normally used for these exposures. The KTA panel (Figure 12.17) is a good example of a device widely used for this purpose.

Once the panel testing is completed, the usual procedure is to apply the most promising two or three candidate materials on tanks or other large metal areas in the plant site. Be aware when proceeding with this step that all surfaces in the field have far from the same exposure. The four quadrants of a tank are different in sunshine ex-

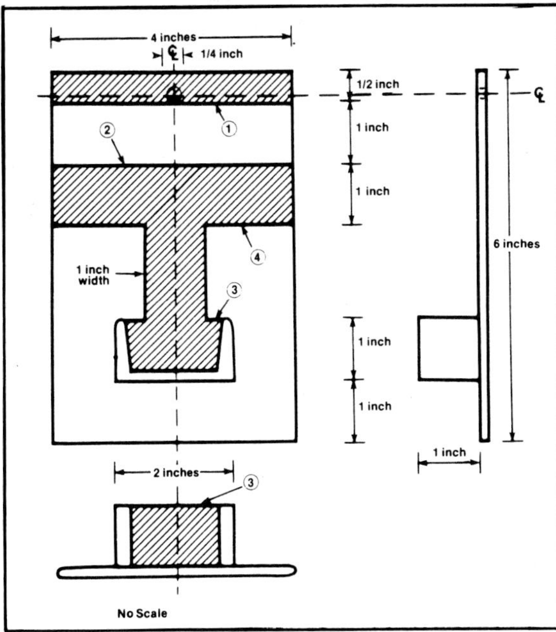

FIGURE 12.17 — Diagram of a repair test panel: (1) abrasive-blast strip on front, edges, and back; (2) abrasive-blast shaded area on front and edges; (3) abrasive-blast complete inside of channel, and (4) feather paint edge to bare metal.

posure, and the top and bottom halves of the tank are not the same because of the internal changes in liquid level.

It is worth reiterating here that in any large chemical or industrial plant, corrosivity of one plant location may be more or less severe by several orders of magnitude than the corrosivity at another location. Thus, exposure at a single location may not give results that can be applied with confidence everywhere in the plant. Furthermore, because some locations are less corrosive than others, they may not require systems as expensive as those used in the worst locations.

Specifications and Inspection

General Concepts of Specifications

It is a good engineering practice to have a clear goal in mind in the application of a coating. Much waste can be avoided by providing proper specifications applicable to a coating project, and planning in advance an inspection program which will ensure conformity with the specifications. It is not necessary, here, to go into minute detail with respect to what makes a good set of specifications. Ample guidance is provided in at least one readily available publication.[2]

The important admonition is to keep the specifications brief and understandable, and to be sure they contain only the requirements for the job. Specific properties of the coating to be measured should be identified, and the method of measuring the quantity should be stipulated. Establish acceptable parameters for the measurement. Then, during the inspection, enforce the specifications agreed to by all parties.

There is no substitute for a pre-job conference with the applicator, and perhaps the material supplier, to establish what is to be done, how it is to be done, what inspection procedures are to be used, and the acceptable limitations imposed by the specifications. At this meeting, the foreman designated to oversee the application should be present, as well as the officer of the applicator's company. Also, the inspector for the work should be present. Agreements by people having no oversight of the work normally have little meaning.

In most instances, do not try to tell the applicator how to do his job. A verbal or written description of such work cannot cover all the eventualities. However, be sure the applicator is properly informed as to any special requirements or limitations on work activities.

Some items that may be contained in any specification relate to such basics as surface preparation. Before a specification for surface preparation can be made, the type of coating system to be used must be established. If a thin film system is selected, the surface preparation method or surface profile characteristics should be known. Once this is known, the acceptable inspection standard should be identified.

Routines for surface preparation, time factors, acceptable intervals between preparation and coatings application (relevant to the environment present), and other pertinent details may be given.

Careful identification of the components of the system may be necessary. Characteristics of the primer, second, and third or final coats, their respective thicknesses and drying time allocations, and the range of permissible final coating thickness may be stated. Limits on addition of diluents, thinners, accelerators, or other changes in basic formulation can be established.

Inspection

The point at which the job is started is no time to decide on the techniques and thoroughness of the inspection to be made. As indicated, these factors should be succinctly stated in the specifications. There are many tests that may be specified which are essentially standardized. Consult these standard procedures.

Some of the tests which may be used relate to the following subjects.

Surface Preparation

The degree of cleanliness of a steel surface has been discussed under the subject of surface preparation. Agree on the use and interpretation of one of these standards. An alternative is to prepare a plate that has a minimum cleanliness finish agreed to by all parties. This plate can then be preserved and used as the standard.

Characteristics of a properly prepared surface may include specification of a surface profile in terms of mils depth, as well as other criteria. This is important in some instances, especially when a thin-film coating is to be applied. If the surface profile is such that an inadequate thickness of coating is deposited on the peaks, or the coating bridges the valley (Figure 12.4), a coating failure may follow, or, at best, a greater volume of coating will be required for adequate coverage of the peaks than would be the case if the profile were of proper depth and spacing.

Ensure that gauges and equipment are available able for checking surface profiles. In the hands of an experienced person, a good statistical estimate can be made in a short time of a properly prepared surface. Profilometers (useful laboratory instruments) give exact information on the profile, but it is difficult and usually impossible to use them in the field.

For the other techniques of inspection, in addition to methods described in NACE books and reports, a wide range of information on inspection tests for organic coatings is available.[23,24] These books cover most aspects of accelerated and on-site tests, including those for adhesion, hardness, abrasion, impact resistance, and others.

The following are brief descriptions of some of these inspection test methods for organic coatings.

Wet Film Gauging

Thickness of a wet film can be determined by any of several gauges which consist of a series of calibrated indentations along the edge of a hand-held ruler. Probable thickness of the dried film can be calculated if solids content and type of coating is known.

Dry Film Gauging

Electrical and magnetic nondestructive and visual destructive equipment for determining thickness are described in several places.[7,25] Additional information concerning these tests is available from NACE.

Coverage Completeness

Coverage completeness can be checked by using different colors of multiple coats or conductive and nonconductive coats.

Visual Inspection

Visual inspection can be conducted with or without magnification. The Tooke gauge,[7] which makes a diagonal cut through a coatings system to the substrate, permits measurement and identification of the coating layers under magnification.

Other Tests

Other tests include hardness, reflectance, adhesion,[14,26,27] and others.

The evaluation of the adhesion of a coating to a substrate is most difficult. If the coating has truly good tenacity, it is impossible. Many of the current tests will provide values of adhesion (if poor) or values of cohesion in the coating (if poorer than the adhesion). The evaluation of adhesion of polymeric coatings on wood is covered in an article by Kambanis and Chip.[28]

Inspection of Metal Coatings

Test methods for metal coatings are available from the American Electroplaters Society, ASTM, and other organizations. Tests are usually related to quality control procedures. Metal coatings applied on-site, such as those by metal spraying, should be tested by visual methods.

Miscellaneous Aspects of Coating Use

Training of Painters and Supervisors

Painters and supervisors should be sent to schools for coatings users held by manufacturers of materials and equipment, and conducted by such organizations as NACE, the Federation of Paint and Varnish Production Societies, and others. Safe and efficient use of equipment and materials is enhanced by these schools.

Legal Aspects

Liabilities for the safety of personnel and protection of plant buildings and equipment are spelled out in workmen's compensation statutes, insurance policies, etc. Anyone responsible for a coatings operation should be aware of personal as well as company liability in the case of an accident, fire, or other emergency situation.

Considerable information on bonding and liability practices is included in a NACE document.[2]

Maintenance

Maintenance of any engineered system is an obvious answer to excessive replacement costs. The maintenance of coating films, organic or metallic, is no different. If extensive deterioration has been allowed to occur, only complete removal of the coating with new surface preparation and total recoating is practical. The alternative is a planned maintenance program that provides touch-up of small areas of failure with tools and techniques that are economically acceptable.

Periodic inspection of coating projects should be made to determine the stability of the system and to establish the appropriate time for maintenance work.

Failures of Organic Coatings

A coating may fail as a result of a large number of potentially adverse conditions. Some of these can be defined as mechanical, as when abrasion or impact removes the coating. Thermal expansion of organic materials is high. Continued expansion and contraction of the coating over steel, and particularly concrete, requires a material which can resist significant internal stresses before failure occurs. The vast majority of failures, however, can be attributed to the chemistry of the coating and its environment.

All organic materials absorb gases and liquids to varying degrees, depending on the type of polymer exposed and the gas or liquid present. The rate at which this absorption occurs varies with the depth into the coating (or plastic) for a given system. Fick's laws of diffusion tell us in simple terms that the rate of this permeation (R) changes inversely with the square of the distance (d) into the material. Thus,

$$R = K\frac{1}{d^2} \tag{12.2}$$

expresses the diffusion rate of a liquid into a resin, with K being a constant for a specific system, *e.g.*, water in a particular vinyl coating.

The importance of thickness in a coating can thus be grasped quickly. The rate at which diffusion is occurring in the lower portion of the coating at a given time is greatly controlled by the distance from that point to the top of the coating. Based on empirical data, economics, and some theoretical considerations, the thickness of a coating necessary to resist moisture permeation from the atmosphere and otherwise resist deterioration is about 125 μm (5 mils).

What happens when moisture and oxygen penetrate the coating? Perhaps nothing, if a truly good bond has been achieved, but a number of reactions can occur. At the spot where the moisture contacts the steel, the initiation of further oxidation of the iron can occur. A corrosion cell comparable to that in a pit on freely exposed iron will develop. This can grow until the coating is opened by the mass of rust formed.

If some ionic contaminant existed at the interface between the coating and the steel, the electrolyte would be more conductive to implement the cell action. Many contaminants, such as chlorides, will actually enter into the chemistry of the cell to accentuate the attack. If mill scale exists in the area, a galvanic cell of 230 mV can be established, with the iron as the anodic electrode.

If conditions of humidity and oxygen permeation are right, the point of corrosion may begin to move in a random manner as the corrosion product reduces the oxygen content at the surface and the area becomes highly anodic to the surrounding cathode area of oxygen saturation. The worm track of corrosion which then occurs is termed *filiform corrosion.* Thin lacquers on food cans or on the exterior of automobile bodies can exhibit this type of failure.

The function of an inhibitor in the primer of a coating is to stop such reactions once moisture reaches the metal surface. The inhibitor should be sparingly soluble to produce an anion (*e.g.*, chromate) and cation (*e.g.*, zinc) which enter into the reaction at the anode and cathode to stop the corrosion (Chapter 7).

If the corrosion reaction is allowed to proceed, two of the cathodic reactions may disbond the coating beyond the minute point of corrosion. Hydrogen is produced at the cathode and will compete for the bonding forces which hold the coating to the surface. Molecular hydrogen (H_2) is a highly active gas which can simply pry the coating loose. In addition, hydroxyl ions (OH^-) are formed at the cathode. Coating materials not resistant to alkaline conditions can be attacked by the hydroxyl ion and lose adhesion. External cathodic currents increase the possibility of failure by the hydrogen and hydroxyl formation because much greater quantities of cathode products are formed.

In addition, external currents impressed on the steel as a cathode force more water through the coating than would be the case without the current. This forced diffusion of water through the coating is termed *electroendosmosis* (not to be confused with *osmosis* which occurs when water is drawn at a higher than normal rate through the film by a soluble salt lying beneath the coating). Thus, cathodic currents of sufficient magnitude can strip coatings from a steel surface. It is doubtful that any coating should be exposed to cathodic charges exceeding 2 volts (true differential).

The importance of adequate coating thickness has been emphasized. If good flowout is not obtained during application, thin spots (particularly at the edges) or actual *holidays* (holes) may exist in the film. At times, the profile pattern is too great for the coating, and peaks of metal protrude above the film. All of these obviously produce a more rapid failure of the coating.

Coatings overloaded with fillers or the wrong filler can be "cheesy" and porous. Rapid penetration of moisture to the surface will occur, with subsequent corrosion and rupturing of the weakly cohesive film. Such failures are often observed as a "star-burst" rupture in the film, rather than as a ductile tear observed in the better coatings.

Another source of failure may be from continued polymerization or oxidation of the coating once it is applied. The film becomes harder, less ductile, and changes in volume when these mechanisms prevail. Cracks in the coating then develop.

Esterification of the coating by foreign metallic cations to produce "soaps" of the coating can occur under the proper conditions when a coating containing ester groups is used. Zinc ions will produce these soaps with some coating materials.

Compatibility of the various coats must be considered. Even if the two coats are known to be compatible, application of the second coat too early or

FIGURE 12.18 — Cracking of a topcoat applied over a primer that had not completely dried or cured.

too late can be disastrous. Figure 12.18 shows an example of the former case where a topcoat was applied and dried over a primer coat that had not completely dried or cured. This example is probably of an alkyd over an oil primer. An example of the latter case can be the overcoating of a fully cured, catalyzed epoxy. The topcoat has little chance to "bite" into such a solvent-resistant surface.

The importance of cleanliness during application, uniform coverage, and good density and thickness in the coating should be evident when some of the more common modes of failure are understood.

Economic Aspects of Coatings Use

While not an absolute measure of importance, the magnitude of expenditures for protective coatings in the United States indicates how significant they are in controlling corrosion. A study made of NACE "user" members indicates that they spent $773 million in 1980 for coatings and related purposes. When this sum is extrapolated to the total NACE "user" membership, it rises to $4.12 billion. Projection of these figures through 1985, including inflation, results in a sum of almost $8 billion.[29] Because NACE membership represents only a fraction of the total number of persons using coatings for corrosion control, it is obvious that this activity is a major item of expenditure.

Methods of calculating the economics of coatings projects are presented elsewhere.[30,31] The important fact is to understand that simple, honest methods of evaluating the true cost of coatings work do exist.

When considering the true cost of a coating job, the application of economic principles must be the same as for any other investment or maintenance money. The life of System A must be compared with System B, the type of money being spent (depreciated capital vs maintenance) must be established, the tax rate known, and the value of money to the spender must be considered. System A can then be compared accurately with System B as an investment. Formulas allowing the rapid calculation of these factors have been prepared so that a course in economics is not required for the engineer to evaluate the alternatives.[31]

Expressions such as cost/ft.^{2}, cost/ft.^{2}/yr. (better), percent of maintenance costs, percent of capital costs, etc., are often manipulated as economic terms, but are unrelated to true economic comparisons necessary for a business decision.

Among numerous other cost data in a book by Banov,[9] the cost of coatings materials on metals ranged from 5 to 21% of the total, while surface preparation was constant at about 45% of the total. High-performance epoxy polyamide, vinyl, or urethane systems over 10 years represented a savings of 32 to 38% over "less expensive" systems in the exposures cited.

Among studies concerning systems for cost analysis and controls is one recently published in England which identifies the capital and operating cost implications when setting up or improving coatings procedures.[32]

Surface preparation of steel in a shop may cost 40% less than on-site, according to studies by Roebuck and McCage.[33] This same study shows surface preparation costs to be about 50% of the total cost of a coating job. A case history in this study of a 29 million sq. ft. (270,000 sq. m.) surface preparation job showed it to be economical to use on-site centrifugal blasting. Field blasting would have cost 2.79 times more than centrifugal cleaning.

These data underline the economic advantage of careful preliminary analysis of coatings problems and the savings that can be anticipated for high-performance systems.

Safety

In every industrial environment, the safety factor is an important consideration. In some environments it is of overriding importance. An exceptionally fine check-off list of safety factors to be considered on a job is available in Reference 2, with other comments in Reference 7. Some of the safety factors that need to be considered in coatings application are as follows.

Safety in Surface Preparation

Workmen using abrasive blasting equipment should be furnished good equipment in proper working order. They should be equipped with air-filter masks to prevent their breathing dust and scale.

Clothing should be adequate and safe. Safety shoes should be worn. Necessary goggles or safety glasses should be mandatory.

No equipment should be operated in areas where it will create sparks that might ignite explosive or flammable materials. It should not be operated where abrasives, scale, or overspray will damage or interfere with the operation of other equipment.

Safety in Materials

Coatings materials are frequently flammable, explosive, or poisonous, and sometimes all three. The characteristics of any material used should be known in advance, and any precautions required should be taken and rigidly maintained during the progress of the job.[2] Precautions on manufacturers' labels should be heeded.

Although it obviously is impossible in the scope of this chapter to fully cover all aspects of safety involved in handling explosive or toxic materials commonly found in coatings, some broad warnings can be given. In summary:

(a) Coatings incorporating flammable or explosive materials should not be used in the vicinity of open flames, sparks, or electrical equipment. Every precaution should be taken to prevent accidental fire or explosion by prohibiting smoking, requiring the use of nonsparking tools, or whatever other safety requirements are appropriate.

(b) When used in enclosed places, solvent concentration should be kept both below the explosive limit and below the acceptable toxicity level. Both limits vary among materials, so safety rules should be a function of materials used. Ventilation of enclosed places should be continuous during the operation and for three hours afterward when explosive or flammable solvents are used. Safety approved electrical equipment is mandatory.

Solvent vapors should be removed from tanks by suction because many vapors are heavier than air. Thus, the remotest and lowest ends of tanks should receive special attention. Workmen should wear approved compressed air masks. Shoes should have rubber soles and heels and no exposed steel nails.

Safety in Equipment

Whenever ventilation is a factor, it should be planned carefully and checked frequently. When necessary, automatic equipment should be used to make constant checks of air for poisons or explosive concentrations.

Diagrams showing the incorrect and correct ways to ventilate tanks are shown in Figure 12.19. Failure to ventilate according to the principles illustrated may lead to injury or death of workmen in enclosed places.

All rigging, lifts, platforms, hoses, or any other equipment used on the job should be inspected and maintained in safe order. Rigging should be done by experienced operators.

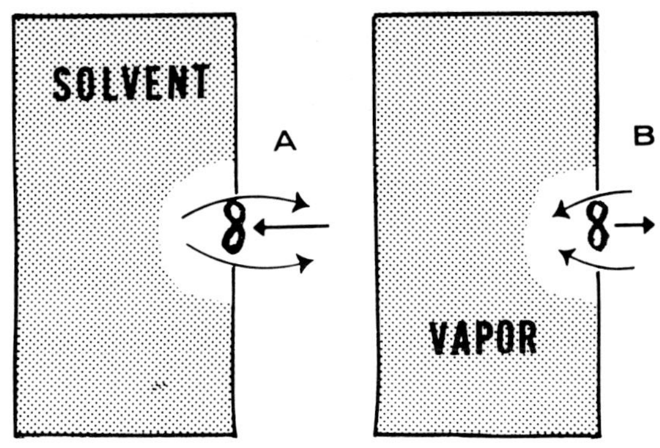

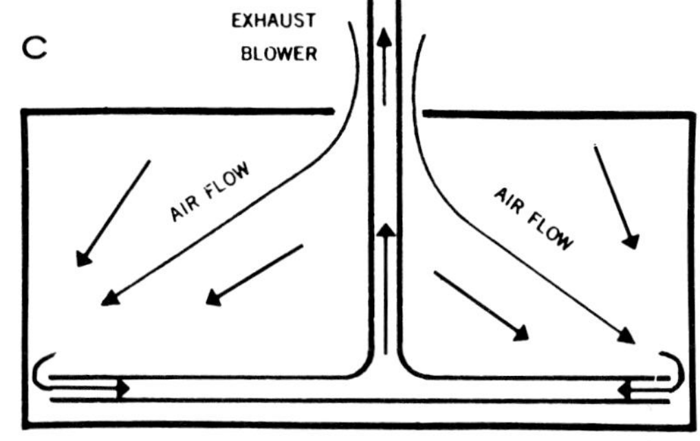

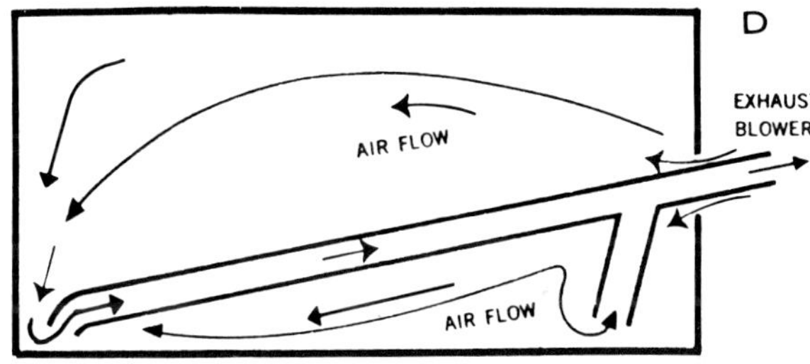

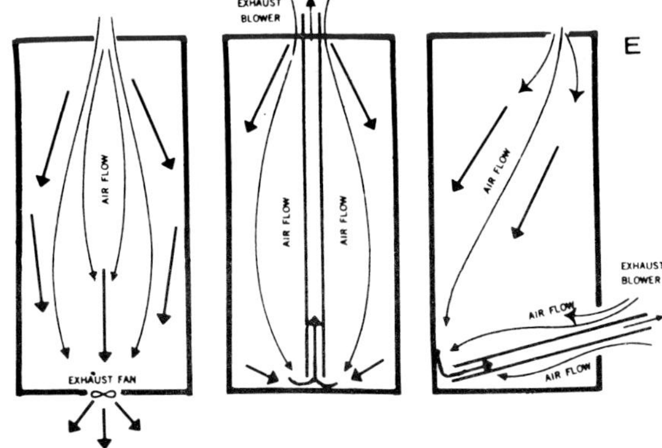

FIGURE 12.19 — Shows incorrect (A and B) and correct (C through E) methods of ventilating tanks where coatings are being applied. A and B show little or no ventilation of the enclosed volume. C through E show how gases displaced from remote or low areas are aspirated into the open, and how fresh air comes in to replace them. [SOURCE: Munger, C. G., Coatings and Their Safe Applications, Corrosion Control Reporter, Vol. 16, No. 4 (1965).]

Sources of Safety Information

Plant safety practices should be known and understood by plant paint crews or by outside crews doing contract work. In addition, the advice of plant safety engineers should be solicited and recommendations followed. Published information is available from trade associations, testing bureaus, and most important, is usually printed on labels and instruction sheets received with materials and equipment.

Those concerned with coating work should also be familiar with the U.S. Government safety standards regarding toxicity and other matters. There may also be additional state and local restrictions.

Conclusions

This discussion of protective coatings presents only a small fraction of the information that is available. Even though the list of references and accompanying bibliography is lengthy, many more sources other than those listed are accessible. The main message is that coating systems for the protection of materials need and probably demand more organized study than they have been given in many instances, and that the benefits attainable from proper preliminary planning, job execution, on-site inspection, and post-completion surveillance are well worth the effort and expense.

If anything is demonstrated by the facts set forth in this chapter concerning the problems associated with coatings, it is that decisions on coatings should not be left to the casual attention of inexperienced personnel. It is unmistakably clear that:

1. Decisions concerning effective protective coatings can be made properly only by experienced personnel.

2. The economic consequences of poor coating practices can be disastrous.

3. The number of coating materials available necessitates careful study if a proper material is to be selected.

4. Proper surface preparation is a prime economic factor for coatings in any environment.

5. Miscellaneous factors may determine the success or failure of an expensive coating job. These include such things as skill of the workmen, proper inspection of the job in progress and at its end, and provision of equipment to apply the coatings properly and promptly, among numerous others.

6. Continuous testing and evaluation of candidate systems and systems actually in service more than pays its way in providing a basis for proper selection.

References

1. Mullen, T. E., Chemical and Physical Forces of Adhesion of Coatings, Proceedings of NACE 24th Conference, p. 619 (1969).
2. NACE Coatings and Linings Handbook (Looseleaf), NACE, Houston, TX.
3. Steel Structures Painting Council, Surface Preparation Specifications (ANSI A159.1-1972), Pittsburgh, PA.
4. NACE, Visual Standard for Surfaces of New Steel Airblast Cleaned with Sand Abrasive, TM-01-70, and Visual Standard for Surfaces of New Steel Centrifugally Blast Cleaned with Steel Grit and Shot, TM-01-75, Houston, TX.
5. NACE Surface Preparation Handbook (Looseleaf), NACE, Houston, TX.
6. ASTM, Pictorial Surface Preparation Standards for Painting Steel Surfaces, ASTM D2200-63T, Philadelphia, PA.
7. Weaver, Paul L., Industrial Maintenance Painting, NACE, Houston, TX., 1967.
8. U.S. Government Printing Office, Paint Manual, 3rd Edition, U.S. Dept. Interior, Bureau of Reclamation, Superintendent of Documents, Washington, DC, 1976.
9. Banov, Abel, Paints and Coatings Handbook, 2nd Edition, Structures Pub. Co., Farmington, MI, 1978.
10. NACE, Surface Preparation of Steel and Other Hard Materials by Water Blasting Prior to Coating or Recoating, RP-01-72, Houston, TX.
11. NACE, Surface Preparation and Surfacing Materials for Cementitious Surfaces, NACE Report 6H175, Houston, TX.
12. NACE, Coatings and Linings for Immersion Service, Houston, TX.
13. Leidheiser, Henry, Jr., Ed., Corrosion Control by Coatings, Science Press, Princeton, NJ, 1979.
14. Leidheiser, Henry, Jr., Ed., Corrosion Control by Organic Coatings, NACE, Houston, TX., 1981.
15. Pappas, S. P. and Winslow, F. H., Eds., Photodegradation and Photostabilization of Coatings, ACS Symposium Series 151, American Chemical Society, Washington, DC, 1981.
16. Knofel, Dietbert, Corrosion of Building Materials, Van Nostrand-Reinhold Co., New York, NY, 1978.
17. Nathan, C. C., Ed., Corrosion Inhibitors, NACE, Houston, TX., 1973.
18. Springle, Robert, Guide to Paint Film Fungicides, Paint Research Association, Teddington, Middlesex, England, 1978.
19. Holmes, D. R. and Rahmel, A., Eds., Materials and Coatings to Resist High Temperature Corrosion, Applied Science Publishers, Ltd., Barking, Essex, England, 1978.
20. NACE, Laboratory Methods for the Evaluation of Protective Coatings Used as Lining Materials in Immersion Service, TM-01-74, Houston, TX.
21. NACE, Accelerated Testing of Marine Submerged and Topside Coatings, p. 8, Houston, TX.
22. Ibid., p. 11.
23. Ailor, W. H., Ed., Handbook of Corrosion Testing and Evaluation, John Wiley & Sons, New York, NY, 1971.
24. Gaynes, Norman I., Testing of Organic Coatings, Noyes Data Corp., Park Ridge, IL.
25. Gardner, H. A. and Sward, G. G., Physical and Chemical Examination of Paints, Varnishes, Lacquers, Colors, 11th Edition, Gardner Laboratories, Inc., Bethesda, MD, 1962.
26. Alle, K. M., Ed., Adhesion 5, Applied Science Pub. Ltd., Barking, Essex, England.
27. ASTM, Adhesion Measurement of Thin Films, Thick Films and Bulk Coatings, STP 640, Philadelphia, PA., 1978.
28. Kambanis, S. M. and Chip, G., Polymer and Paint Properties Affecting Wet Adhesion, J. Coatings Tech., Vol. 53, No. 682, pp. 57-63 (Nov., 1981).
29. NACE, A Survey of Estimated Annual Expenditures of the NACE "User" Membership, Order No. 52187, Houston, TX. (June, 1981).
30. Rodgers, J., Understanding Economic Aspects of Coatings is Vital to Engineering, Materials Protection, Vol. 9, No. 5, pp. 26-30 (1970).
31. NACE, Direct Calculation of Economic Appraisals of Corrosion Control Measures, NACE Standard RP-02-72, Houston, TX., 1972.
32. Paint Research Association, Costing for Industrial Paint Finishing, Teddington, Middlesex, England.
33. Roebuck, A. H. and McCage, D. L., Materials Performance, Vol. 15, No. 10, pp. 30-34 (1976).

Bibliography

American Concrete Institute. Applications of Polymer Concrete. Pub. 69. Detroit, MI.

Barton, Karel. Protection Against Atmospheric Corrosion. John Wiley & Sons, New York, NY, 1978.

Burns, R. M. and W. W. Bradley. Protective Coatings for Metals. Reinhold Publishing Corp., New York, NY, 1967.

Damusis, A., Ed. Sealants. Reinhold Pub. Corp., New York, NY, 1962.

Dismuke, T. D., S. K. Coburn, and C. M. Hirsch. Handbook of Corrosion Protection for Steel Pile Structures in Marine Environments. American Iron and Steel Inst., Washington, DC, 1981.

Edwards, J. D. and R. I. Wray. Aluminum Paint and Powders. 3rd Ed. Reinhold Pub. Corp., New York, NY, 1955.

Flick, E. W. Exterior Water-Based Trade Paint Formulations. Noyes Data Corp., Park Ridge, IL, 1980.

Gabe, D. R. Principles of Metal Surface Treatment and Protection. 2nd Ed. Pergamon Press, New York, NY, 1978.

Gillies, M. T. Water-Based Industrial Finishes. Noyes Data Corp., Park Ridge, IL, 1980.

Gov't Ptg. Office. Economic Effects of Metallic Corrosion in the United States. NBS Spec. Pub. 511-1. U.S. Sup'td. Doc., Table 8, Washington, DC.

Hammkind, H. S., *et al.* Surface Analysis of Interfacial Chemistry in Corrosion-Induced Paint Adhesion Loss. J. Coatings Tech., Vol. 51, No. 655, pp. 45-48 (1979).

Hamner, Norman E. Adhesion Fundamentals and Methods of Testing Organic Coatings. Materials Performance, Vol. 9, No. 5, pp. 31-36 (1970).

Ingham, H. S. and H. P. Shepard. The Metco Metallizing Handbook. 5th Ed. Metallizing Engineering Co., Inc., Long Island City, NY, 1951.

Mittal, K. L., Ed. Surface Contamination: Genesis, Detection and Control. Vol. 1 and 2. Plenum Publishing Corp., New York, NY.

Munger, C. G. and R. C. Robinson. Coatings and Cathodic Protection. Materials Performance, Vol. 20, No. 7, pp. 46-52 (1981).

NACE. Monolithic Organic Corrosion Resistant Floor Surfacing. RP-03-76. Houston, TX.

NACE. Protection of Motor Vehicles Against Tropical Corrosion. Materials Protection, Vol. 2, No. 4, pp. 61-63 (1963).

Powell, C. F., *et al.,* Eds. Vapor Deposition. John Wiley & Sons, Inc., New York, NY, 1966.

Silman, H., G. Isserlis, and A. F. Averill. Protective and Decorative Coatings for Metals. Finishing Pub., Ltd., Teddington, Middlesex, England, 1973.

Tator, K. B., *et al.* Influence of Surface Preparation Upon Performance of Protective Coatings in Various Atmospheres. Materials Performance, Vol. 22, No. 11, pp. 48-55 (1983).

Tatton, W. H. and E. W. Drew. Industrial Paint Application. Hart Publishing Corp., New York, NY, 1964.

Waindle, Roger. Automobile Corrosion Problems, Causes and Prevention. Materials Protection, Vol. 6, No. 12, pp. 31-32 (1967).

NOTES

Chapter 13

High-Temperature Corrosion

HIGH-TEMPERATURE CORROSION

Introduction

Understanding the behavior of metals at elevated temperatures, and especially their corrosion behavior, has only recently become an object of scientific investigation. Many techniques for studying reactions at high temperatures had to be developed. It is obviously difficult to observe the actual reaction between gases and metals at high temperatures and to watch the reaction products build up. It is easier to measure the change in weight after some interval, or even to measure the volume of gas consumed or the weight change continuously during the test. These techniques, however, are fairly recent developments. Formerly, one could only observe the appearance of the scale after the test was concluded and could only carefully examine it after it had cooled.

How a material had changed and how it actually had appeared at any given temperature was the subject of much speculation, but little accurate evidence was available. As has subsequently become clear, the appearance of a scale after cooling to room temperature may be quite different from the same scale at a high temperature. Many scales which are continuous and thus protective at a given temperature, will flake off or spall upon cooling, and hence create a misleading impression of their effectiveness during continuous service.

It is not surprising, therefore, that the first quantitative approach to oxidation behavior was made in the early 1920's with the postulation of the parabolic rate theory of oxidation by Tammann and, independently, by Pilling and Bedworth, nor that a more formal treatment of the problem would be delayed another decade, until the middle 1930's, when Wagner presented his theory of oxidation. Modifications and alternative theories continue to appear and presumably will continue to appear for a considerable time because there are still many gaps in the theory, particularly in its application to practical problems and in service experience involving complex systems.

In the following discussion, we will consider first the theory of metallic oxidation as exemplified by the behavior of pure metals under oxidizing conditions at elevated temperatures, and secondly, the effect of alloying additions on the corrosion resistance and mechanical properties of the alloys. The correlation with theory will be, at least qualitatively, shown. In this fashion it is hoped that an appreciation of the important factors controlling oxidation behavior will be gained.

We also will take a more detailed look at the behavior of engineering alloys at elevated temperatures in several different commonly encountered environments; namely, air and flue gas atmospheres, as well as in low pressure gases, vacuum, and molten metals and salts.

Although the temperatures above approximately 200 or 300 F are sometimes considered "high temperature," this chapter will be concerned primarily with temperatures above the "red-hot range," primarily 650 C (1200 F) and above.

Industrial areas where material stability at high temperature is important include utility operations (steam), cracking furnaces (ethylene, propylene, etc.), fertilizer operations (ammonia), reformer furnaces (CO and H_2 from methane), and various turbine operations.

The Materials

The most common alloys used for exposure at high temperatures are the iron-nickel-chromium alloys, sometimes referred to as the stainless steels. The compositions of the more commonly used stainless steels were given previously in Tables 4.3 and 4.4 (Chapter 4). The materials used may be cast or wrought. For most furnace operations, the cast metals described in Table 4.4 are used, with the HK-40 (J94204) currently being the most popular.[1]

Ordinary iron has been used successfully from room temperature up to temperatures of about 480 C (900 F), and for short periods of time up to 590 C (1100 F). At temperatures beyond this, ordinary steels tend to corrode rather heavily. However, there is a trend to resort to *cementation* (surface diffusion treatments) on carbon and low-alloy steels to render them more resistant to corrosion. Depending upon the corrosiveness of the medium involved and the exposure temperature, one or several of the following may be used: siliconizing, aluminizing, chromizing, or a somewhat different method of surface treatment such as plating, plasma spraying, cladding, etc.

If medium- or high-alloy steels are used, chromium-bearing stainless steels are generally recommended. Both the straight chromium steels (400 series) or the chromium-nickel steels (300 series) are used successfully up to approximately 870 C (1600 F). At still higher temperatures, the higher nickel and higher chromium alloys of the 300 and 400 series are used, and at temperatures somewhat in excess of 1093 C (2000 F), only the alloys with more than 20% Cr can be used with relative safety and particularly for short or intermediate periods of time.

During long-term use at high temperatures, a phenomenon known as *creep* is experienced, during which relatively low loads may cause a metal to deform very slowly and possibly rupture. Although this is not a corrosion phenomenon, it is nevertheless a factor that must be considered by anyone considering or recommending alloys for use at very high temperatures. It is on the basis of the creep (% per unit time) to be expected at the operating temperature that furnace parts are designed to withstand the mechanical stress.

At still higher temperatures, it may be necessary to use *refractory metals*—meaning those that have good high-temperature mechanical properties, plus melting temperatures well above 1650 C (3000 F), as shown in Figure 13.1. Thus, we will be considering alloys of chromium, columbium (sometimes called pelopium or niobium), molybdenum, tantalum, rhenium, or tungsten. Although titanium and zirconium melt above 1650 C (3000 F), their high-temperature mechanical properties are rather unsatisfactory, so they are not generally considered in the category of refractory metals. Similarly, vanadium is not a popular contender for high-temperature applications for reasons that will soon become obvious.

Unfortunately, all of the above-named refractory metals, with the exception of chromium, have poor high-temperature corrosion resistance in air or other oxidizing atmospheres. Some success has been obtained in developing useful life at high temperatures with protective coatings on the other refractory metals. In a few cases, some encouraging improvement in corrosion resistance has been developed by alloying.

Basis of Corrosion Resistance

Corrosion resistance at high temperatures stems from one of two basic factors; one deals with thermodynamics, the other with kinetics.

Thermodynamics

In the case of high-temperature corrosion resistance relating to thermodynamics, the corrosion reaction may be impossible for reasons of metallic nobility.

The term *nobility* refers to a quality in metals which indicates a preference to remain in the metallic state (like gold, as we already know), rather than the combined state in the form of oxide or sulfide, as most metals do. This is why non-noble iron keeps rusting to revert back to its preferred state, iron oxide or iron ore.

A study of metal preference comes under the heading of thermodynamics, and is therefore an important area in corrosion studies. The term *thermodynamics* comes from the fact that energy is lost (or gained) in the form of heat as the metal reacts to

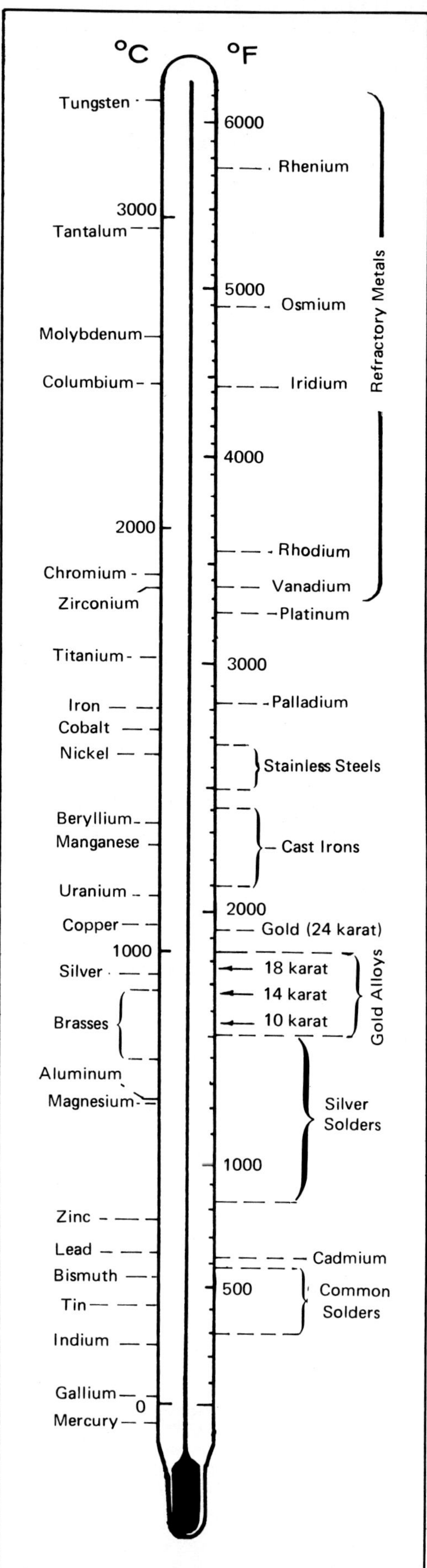

FIGURE 13.1 — Melting points of some metals.

form oxides or other compounds. The stability of the various oxides differs, as indicated earlier.

Two important points are involved. First, the more energy released when forming the compound, such as an oxide, the more stable the compound. Secondly, as the temperature is raised, more of the oxidizing species is required to maintain the oxide or other film.

As an illustration, let us consider a bar of copper and a bar of iron at 980 C (1800 F) in an atmosphere made up of 20% H_2 and 80% H_2O. This atmosphere is partly oxidizing and partly reducing. Under these conditions, copper will not corrode and will remain bright. Iron, on the other hand, will become coated with a layer of iron oxide.

In the case of copper, the reaction

Cu	+ H_2O	→	Cu_2O	+	H_2	
Copper	Water Vapor		Copper Oxide		Hydrogen	(13.1)

is thermodynamically impossible. In fact, the reverse reaction would occur if some copper oxide were on the copper; *i.e.,* the copper oxide would be reduced by the hydrogen to form H_2O and metallic Cu.

In the case of iron, however, the reaction

Fe	+ H_2O	→	FeO	+	H_2	
Iron	Water Vapor		Iron Oxide		Hydrogen	(13.2)

is possible and would occur. The 80% H_2O-20% H_2 mixture would not permit H_2 to reduce FeO to form Fe and H_2O, as in the case of copper.

Thus, we may say that copper is corrosion-resistant under these circumstances, or is more noble than iron. Iron, however, is thermodynamically subject to corrosive attack.

All of this can be plotted to produce the curves shown in Figure 13.2 for the metal-oxide systems. Note that environments other than oxygen (water, CO_2) produce metallic oxides because of their oxidizing capacity. Table 13.1 lists a number of common gases encountered in high-temperature works, and the type of reaction to be expected when they predominate.

TABLE 13.1 — Oxidation-Reduction Conditions Existing at High Temperature

If these species predominate:	
Oxygen	Hydrogen
Sulfur Dioxide	Ammonia
Sulfur Trioxide	Hydrogen Sulfide
Water (Steam)	Sulfur
Chlorine	Carbon
Carbon Dioxide	Carbon Monoxide
Molten Salts	
the system is:	
Oxidizing	**Reducing**

Referring back to our original statements regarding copper and iron at 980 C (1800 F) in a 20% H_2-80% H_2O gas atmosphere, Figure 13.2 indicates that although pure H_2O would oxidize copper (use point "C" on left vertical axis as a fulcrum), as little as about 0.1% H_2 would be adequate to suppress copper corrosion.

This is important to remember and is in accord with the relatively high nobility of copper. In the case of iron, however, about 70% H_2 would be necessary to suppress iron corrosion at 980 C.

It can be determined from Figure 13.2 that the $2Fe + O_2 = 2FeO$ line at 980 C indicates an equilibrium gas mixture of about 2:1 for H_2:H_2O ratio. Since a gas made up of 2 parts H_2 and 1 part H_2O is a 67% H_2 gas, 70% H_2 or more would favor oxide reduction at 980 C, and iron would remain bright in this atmosphere. If the iron were to be cooled in this atmosphere, it would oxidize because a "drier" hydrogen is necessary at lower temperatures.

Attempt to determine the H_2:H_2O ratio for iron at 150 C by using a straight-edge on Figure 13.2, and then check the answer with that given at the end of the chapter (Self-Check Test No. 1). If you can use the chart to determine this, you will have learned a very useful technique.

By using the heavy line shown from fulcrum "B," we can determine that 10^{-7} or more volume percent of oxygen also will sustain the oxide on iron at 1800 F.

One might also consider the possible behavior of an alloy such as an iron-nickel chromium stainless steel in moist hydrogen at high temperature. One should be able to deduce with the aid of Figure 13.2 that only Cr_2O_3 would be expected to form, and that the iron and nickel components in the alloy would behave in a noble fashion in such an atmosphere. Thus, a degree of control can be exerted on the type of oxide formed initially, which in turn would affect subsequent behavior.

By using Figure 13.2 in an appropriate fashion, one could determine the partial pressure of oxygen that would cause any of the listed metals to corrode, as well as the CO:CO_2 ratios or the H_2:H_2O ratios to produce a similar oxidation. Of course, if a mixture of all these gases were present simultaneously, the situation would become more complicated and its discussion would be beyond the intended scope of this chapter.

One can further add to the complexity of a situation by introducing other gases such as sulfur dioxide, hydrogen sulfide, chlorine gas, or other "contaminants" which, in addition to oxides, tend to form sulfides, chlorides, etc.

A later section dealing with the corrosion of alloys by hot flue gases involves some of the above-

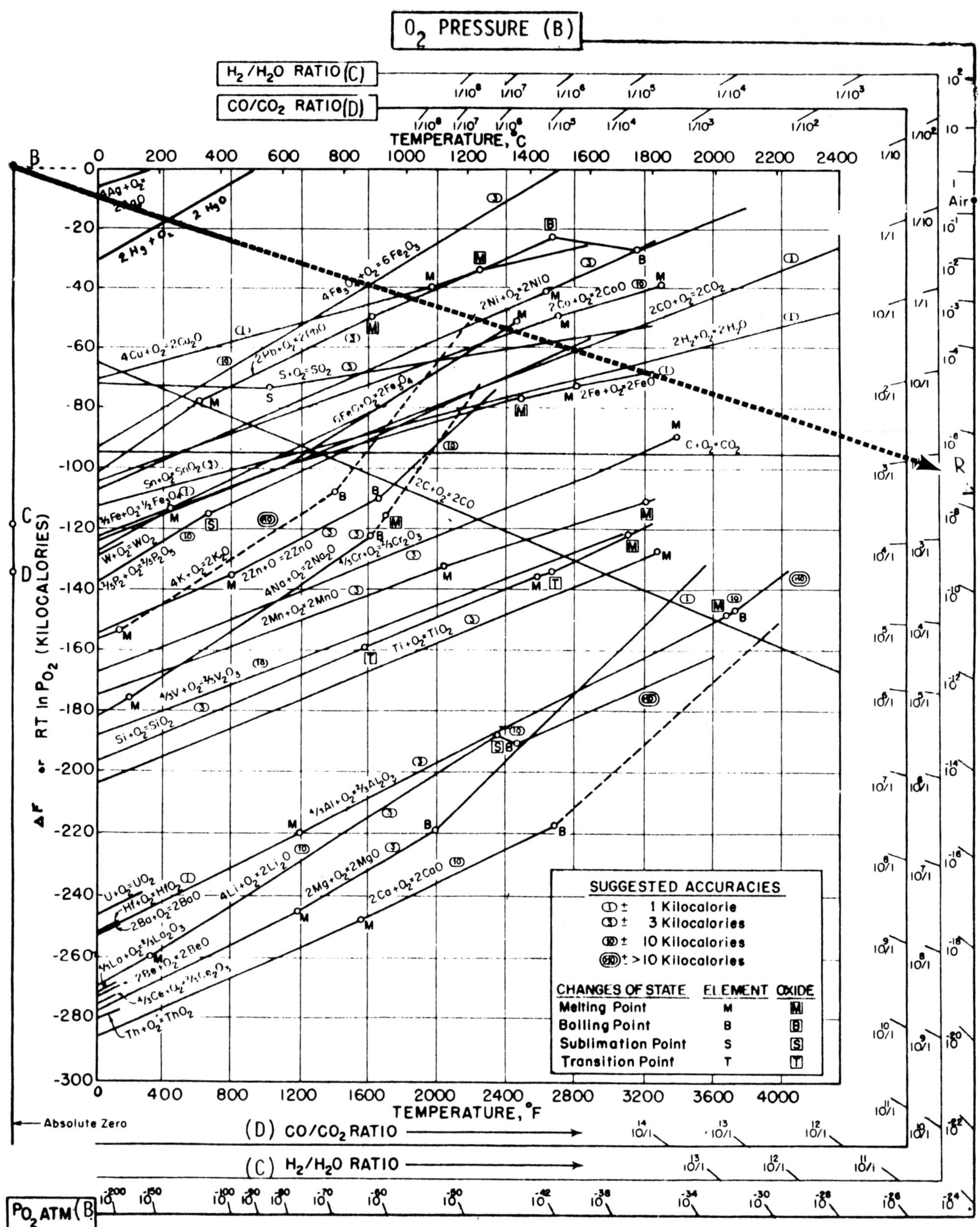

FIGURE 13.2 — Nobility of metals is illustrated in this chart. This permits determining equilibrium oxygen pressures or hydrogen:water vapor and carbon monoxide:carbon dioxide gas ratios that tend to develop at various temperatures.

mentioned complexities in which several oxides and sulfides are known to form. Although a sulfide chart similar to the oxide chart has also been published,[2] its application to alloys in mixed atmospheres is quite complex and will not be attempted here. Instead, an empirical approach will be used to note the gradual changes in corrosion behavior of alloys as composition is varied in an orderly manner.

Kinetics

In the case of kinetics, whether or not a metal is considered corrosion-resistant is dependent not on "if it corrodes," but on "how fast it corrodes." Practically all real situations come under this category. That is, over 99% of cases involve situations where corrosion is thermodynamically possible. The determination of corrosion resistance lies simply in how fast corrosion occurs. If it is extremely slow, we conclude that the condition of corrosion resistance has developed.

On the other hand, if corrosion is fast or intermediate in speed, we are likely to conclude that the metal in question does not have corrosion resistance and that a corrosion failure can soon occur. The meaning of "fast" or "slow" is usually dependent on the expected life of the object in question.

As to what factors determine whether a high-temperature corrosion reaction will occur slowly or rapidly, we must consider the characteristics of the corrosion product layer which forms. Herein lies the possibility of diffusion resistance or effective separation of metal from corrosive environment, which is the other basis of corrosion resistance.

Corrosion Products

When a metal is oxidized at elevated temperatures, a stable oxide or other compound generally covers and "coats" the exposed metal surfaces. Thus, the corrosion product layer may act as a barrier between the underlying metal and the corrosive environment (air, flue gas, molten salt, or any other corrodent). The compound may be a solid, liquid, or gas. For instance, if chromium, vanadium, and molybdenum were heated in air at 1100 C (2000 F):

1. Chromium would form solid Cr_2O_3,
2. Vanadium would form liquid V_2O_5, and
3. Molybdenum would form gaseous MoO_3.

Needless to say, only chromium would be severely corroded. If the oxide formed serves as an effective barrier, corrosion can be strongly retarded and only a thin oxide layer will form. We may then say that a protective (diffusion-resistant) scale had formed, or that the metal was corrosion-resistant under the conditions of exposure. An example of thin oxide film formation is shown in Figure 13.3.

On the other hand, if the barrier is not very effective, corrosion can and will continue at a moderate rate. We would then conclude that the oxide scale was only slightly protective and that the metal was not corrosion-resistant under these conditions of exposure. In other words, the diffusion of reactants was not appreciably reduced and the reaction was able to continue at a significant rate, forming thick oxide layers, as shown on other specimens.

The most common observation on iron-base alloys is a multiple layer oxide scale similar to that shown in Figure 13.4. With ordinary iron, the innermost layer is wustite, FeO; the intermediate layer is magnetite, a spinel form of oxide, Fe_3O_4; and the outermost layer is hematite, Fe_2O_3. There are cases where only one or two oxide layers will form, while in other instances as many as seven layers have been observed.

Another unusual observation of representative nonprotective oxide formation is a corrosion prod-

FIGURE 13.4 — Three oxide layers often are observed on low-alloy iron above 570 C. In this case, the inner layer was 55% of total layer thickness and the outer layer 30%. X2.7.

FIGURE 13.3 — Thin oxide formed at 1800 F (982 C) in 100 h on specimens at right in 2, 5, and 11 o'clock positions. Specimens at left shown before exposure.

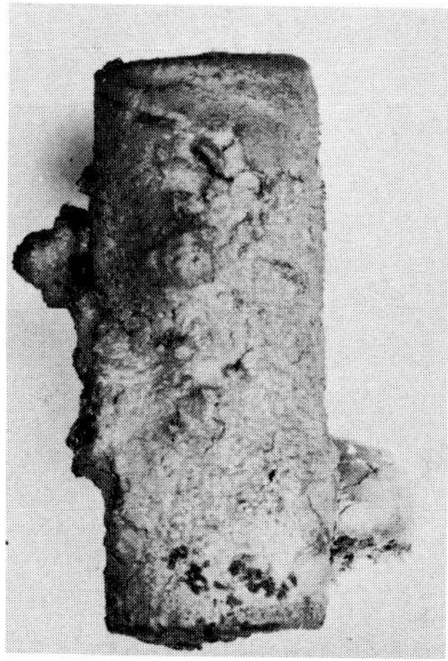

FIGURE 13.5 — Molten phase forms sometimes as it has on the high-nickel alloy specimen shown after exposure to reducing flue gas containing less than 1% hydrogen sulfide. X2.

uct with a low melting point. This would be expected to be accompanied by severe corrosive attack. An illustration of this phenomenon showing partial melting of the scale layer is shown in Figure 13.5.

How Corrosion Proceeds

Once a corrosion product layer is formed, continued reaction must involve the diffusion of at least one of the reactants through the corrosion product layer so that the corrosion reaction will continue. This will be true regardless of whether single or multiple scale layers exist.

Let us assume, as an example, that nickel is to be exposed to air at high temperature. Corrosion can theoretically continue through the nickel oxide layer by means of diffusion in either direction, alone or by counter-current diffusion, as illustrated in Figure 13.6.

From the atomic viewpoint, diffusion through the oxide occurs by atomic movement which is

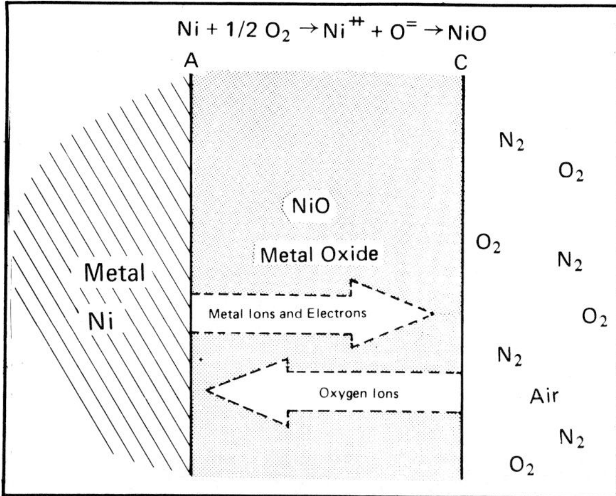

FIGURE 13.6 — Schematic diagram showing counter-current diffusion, *i.e.*, diffusion of oxygen inward and metal ions plus electrons outward. Thus, the anode is at the metal/metal oxide interface A, where oxidation ($M \rightarrow M^{++} + 2e$) occurs. The cathode is at C, the oxide/air interface where reduction ($\frac{1}{2}O_2 + 2e \rightarrow O^{=}$) occurs.

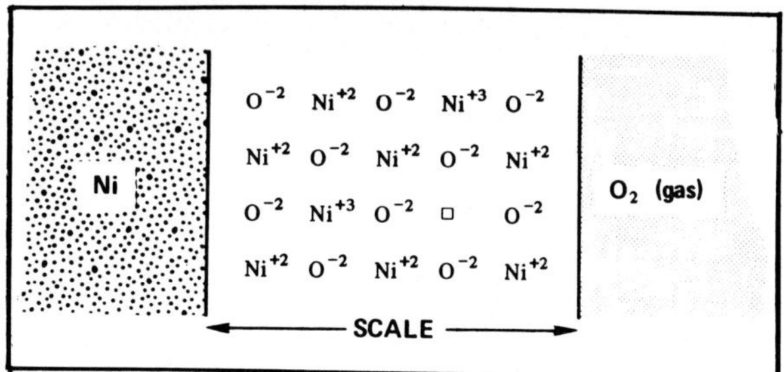

FIGURE 13.7 — Nickel oxide as an example of metal-deficit scale. Schematic of NiO lattice shows a few Ni^{+3} sites among the Ni^{+2}. For every two Ni^{+3} sites, there is a vacant □ Ni^{+2} site.

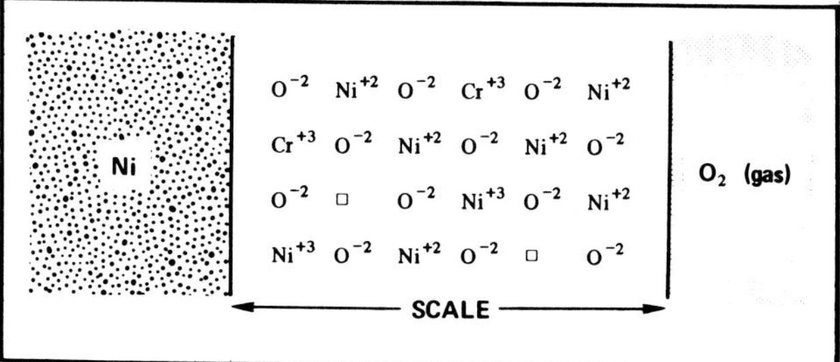

FIGURE 13.8 — Nickel oxide containing small amounts of chromic oxide. Note that this increases the density of vacant □ Ni^{+2} sites.

made easier by existing atomic vacancies, as illustrated in Figures 13.7 and 13.8. The schematic diagram in Figure 13.7 represents the type of oxide that forms on nickel (and on a number of other metals, including copper, cobalt, manganese, and chromium). Nickel forms only one oxide NiO, but, as the figure shows, it is a *metal-deficit* oxide; *i.e.*, a slight excess of oxygen atoms over nickel atoms will be found under normal conditions, and the charge balance (the scale must remain electrically neutral) will be maintained by a few of the nickel atoms, assuming a higher valence state.

For this type of oxide, the controlling step in increasing scale thickness is the ability of the nickel ions to migrate outward toward the surface. If a small quantity of a higher valence ion (Cr^{+3}, for example) is added to the nickel, as in Figure 13.8, charge balance will require a greater number of nickel vacancies and hence an easier, quicker path for nickel ions to migrate to the surface. Conversely, if a lower valence metal ion (Li^{+} is the best known example) is added, charge balance requirements will reduce the number of defects or vacancies, and the oxidation rate, conforming to the nickel ion diffusion rate, will decrease.

The opposite situation, a *metal-excess* scale, is also possible (zinc is an example). In this case, extra metal ions are believed to occupy intermediate positions in the scale. The introduction of a higher valence metal ion will reduce the number of zinc ions in the scale and thereby lower the oxidation rate. A lower valence metal ion will, of course, have the opposite effect, causing the metal-excess scale to respond to the addition of other metal ions in a manner just the reverse of that in metal-deficit scales.

Some metals form more than one oxide. In the case of iron, for example, Fe_2O_3, Fe_3O_4, and FeO can form, depending upon the quantity (usually measured as a partial pressure) of oxygen available. Generally, all three types will form concurrently on the metal surface. Fe_2O_3, which requires the most oxygen, will be found on the outside, and FeO, which requires the least, on the inside, with Fe_3O_4 between them. The boundaries between the various oxides usually are quite well defined and, provided the several layers remain intact, each will grow according to the parabolic rate law (as discussed in a later section).

In most instances, the predominant diffusion is that of metal ions plus electrons outward to the cathode (*i.e.,* to the oxide-air interface shown at "C" in Figure 13.6).

One should not overlook the migration of the tiny electron. If electron mobility was low, as in the case of oxides which behave as electrical insulators (Al_2O_3, BeO, Cr_2O_3, etc.), the metals developing these compounds as surface films could be expected to have good corrosion resistance.

In actual practice, diffusion, as illustrated in Figure 13.6, occurs in both directions, but keep in mind that corrosion can continue even if diffusion occurs in only one of the two directions.

Because nickel at high temperatures does not form an oxide layer with a good diffusion-resistant coating, alloying elements are generally added to provide added imperviousness to the oxide layer. Chromium is an excellent alloying element, and the 80 Ni-20 Cr alloy is one of several compositions commercially used for heating elements in high-quality domestic electric toasters or electric irons. This alloy generally develops a suitable $NiO\text{-}Cr_2O_3$ spinel (oxide) on the surface, but under suitable conditions, Cr_2O_3 forms. This oxide is even more diffusion-resistant than the spinel.

Linear Behavior

If the oxide film or scale cracks or is porous, *i.e.,* if the corrosive gas can continue to penetrate readily and react with the base metal, no protection will be afforded and attack will proceed at a rate determined essentially by the availability of the corrosive gas. In this case, the rate will not sensibly change with time, and, as is apparent from Figure 13.9, the weight change or depth of penetration from oxidation is a straight line or linear function of time and may be expressed:

$$y = kt \qquad (13.3)$$

where y = scale thickness or weight change, t = time, and k = a constant, which will, however, be dependent upon temperature, the material being tested, and other conditions of the test.

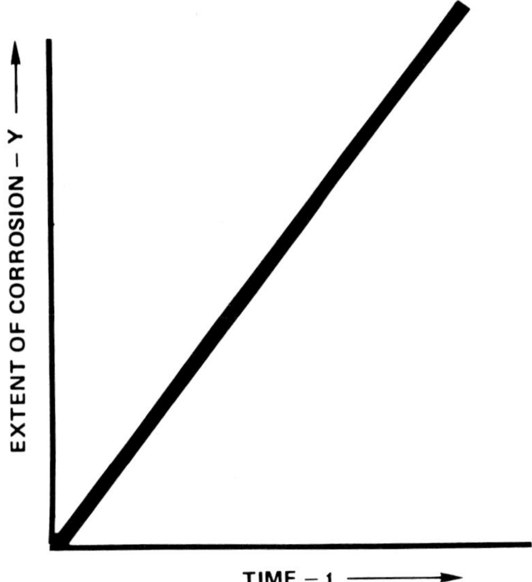

FIGURE 13.9 — Linear relationships between time and extent of corrosion which may be measured as depth of penetration or weight change (gain or loss).

Parabolic Behavior

If, on the other hand, the scale formed is continuous, adherent, and prevents easy access of the corrosive gas to the underlying base metal, a considerable measure of protection may occur, and the extent of protection will increase as the scale thickens. In this case, the availability of the corrosive gas will not determine the reaction rate.

Diffusion through the scale will be the slowest and hence, the rate-controlling step. The diffusing element that controls the reaction rate may be either the oxygen moving inward or the metal moving outward. The latter is more likely in alloys possessing good corrosion resistance. As the scale thickens, everything else remaining the same, the diffusion rate will decrease, so the rate of scale growth will decrease and be inversely proportional to its actual thickness at any point in time. The solution to this, as illustrated in Figure 13.10, is a parabola having

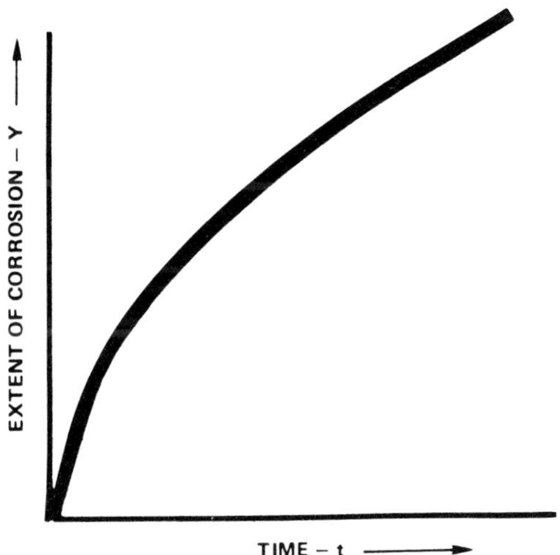

FIGURE 13.10 — Parabolic relationship between corrosion and time.

the equation:

$$y = k'\sqrt{t} \qquad (13.4)$$

Quite clearly, after some reasonable time has elapsed, the rate of scale growth will decrease to a low level; *i.e.,* a considerable measure of oxidation resistance will be obtained and, provided the scale remains intact, will be maintained for very long times.

Under special conditions, particularly at intermediate temperatures, it appears that after the initial scale formation, the oxidation process essentially stops. No change in scale thickness or weight is observed after very long intervals. Presumably this occurs because, when the scale reaches some minimum thickness, the diffusion rate is so low as to be essentially zero, thus creating a condition under which the metal should last almost indefinitely without further deterioration. Some investigators have chosen to call this type of behavior *tarnishing* and thereby to distinguish it from the higher temperature phenomenon of *scaling*. The latter will be of greater concern in the remainder of this discussion.

The parabolic rate law, although obviously an oversimplification of the actual state of affairs during oxidation, nevertheless provides a relatively accurate approximation of the behavior of the majority of metals and alloys used at high temperatures.

For example, Figure 13.11 gives actual test data on several pure metals.[3] Although the rate varied, all of the specimens oxidized according to the parabolic law. Figure 13.12 illustrates the behavior of a nickel-chromium alloy, which is basic for most of the heat-resistant and superalloys in use today. It can be seen again that, although the rate of scale formation is reduced from that of pure chromium or nickel at the same temperatures, parabolic behavior continues.[4]

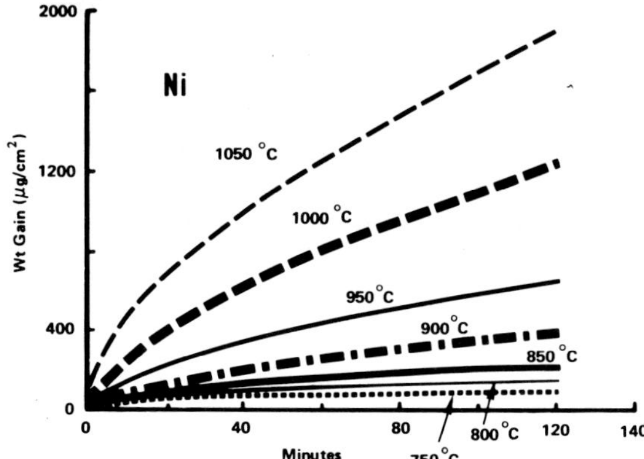

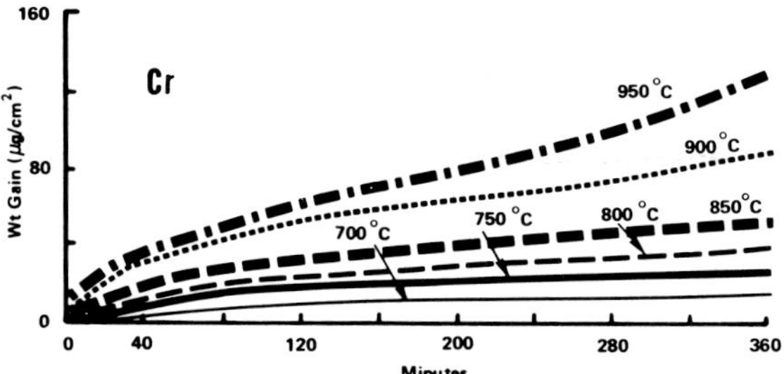

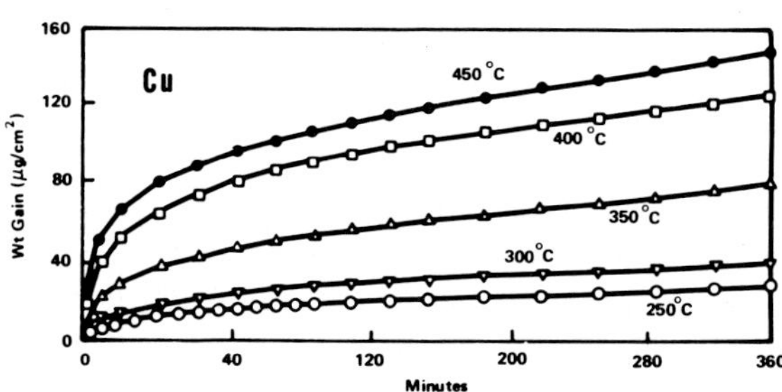

FIGURE 13.11 — Effect of temperature on corrosion-time curves for Ni, Cr, and Cu in 0.1 atm oxygen.

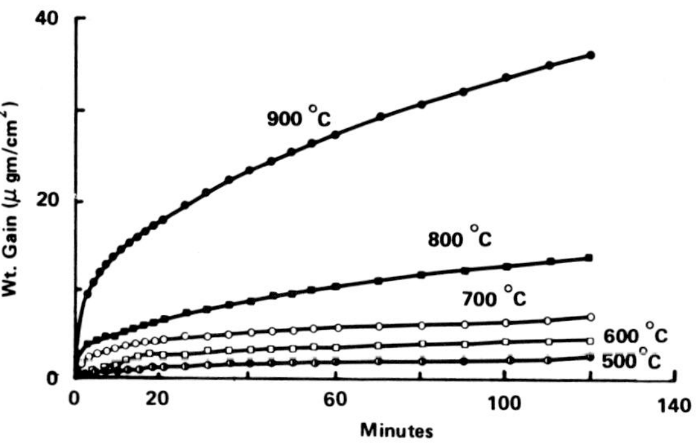

FIGURE 13.12 — Effect of temperature on oxidation of NiCr alloy 493.

Characteristics of Scales

Thus far, we have been concerned principally with oxide scale formation, and have assumed that once formed, the scale would remain intact and protective. With most oxidation-resistant alloys, this is in fact the case. But, on many more alloys, the scale is protective for a short time only, and with others it offers no significant protection.

Scale thickness may be an important factor. In early stages, the scale is thin and reasonably elastic. It also will have little intrinsic strength and usually will remain tightly fixed on the base metal. As it thickens, however, its own less desirable properties may become manifest. Heavy scales tend to be brittle and therefore to rupture and spall from the surface.

Figure 13.13 presents data on a stainless steel that reflects this type of behavior. At the lower temperatures, the scale thickens slowly and the parabolic rate law is followed. At the highest temperature, the rate also is initially parabolic, but at some point the scale breaks away locally and there is an immediate increase in the scaling rate as a new protective scale is formed.

The volume of scale formed, relative to the volume of the metal reacting, often has been cited as a principal factor controlling the scale continuity. It has been postulated that for maximum performance, the scale volume should exceed that of the

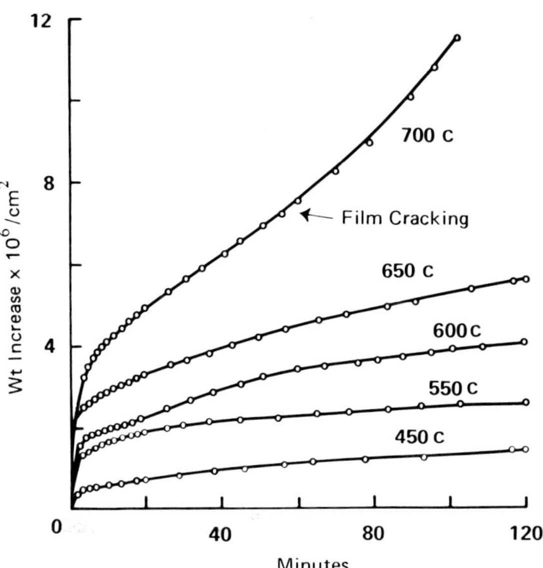

FIGURE 13.13 — Oxidation of stainless steel at various temperatures in 0.1 atm oxygen.

metal consumed. In this case, the surface scale would be under some compressive stress, desirable for most brittle materials, and the tendency to crack would be minimized. Certainly, if the scale volume were considerably less and tensile stresses existed, the opportunity for rupture would be much greater. Figure 13.14 represents, in a schematic fashion, typical ways in which scales have been observed to rupture.[5]

The matter of scale adherence is also important, as Figure 13.14 shows. Particularly if temperature variations occur, a weakly adherent scale is likely to separate locally from the base metal, a phenomenon sometimes termed *blistering*, and, because it has little intrinsic tensile strength, eventually rupture and spall. The relative thermal expansivity between scale and metal also has been cited as an influential factor when temperature fluctuations are involved. Probably of greater significance is the nature of the interface between the metal and the scale. If the interface is planar, it is easy to conceive of a shear crack initiating at the interface and propagating rapidly across the surface.

If, on the other hand, the interface is rough, where incipient grain boundary attack beneath the scale is occurring, a keying action may prevent rapid crack propagation and thereby improve scale adherence. It is thought by some that the major effect of reactive rare earth additions to heat-resistant alloys is not to enter the scale and lower diffusion rates, but to accumulate in grain boundaries and thus ensure a small amount of grain boundary attack to provide the keying action.[6]

A considerable effort is often made in industry to produce an initial scale on the interior that has the optimum properties. Centrifugally cast tubing is often bored to remove porosity near the inner surface, honed to provide a smooth surface profile, and then steamed heavily to produce a thin, continuous,

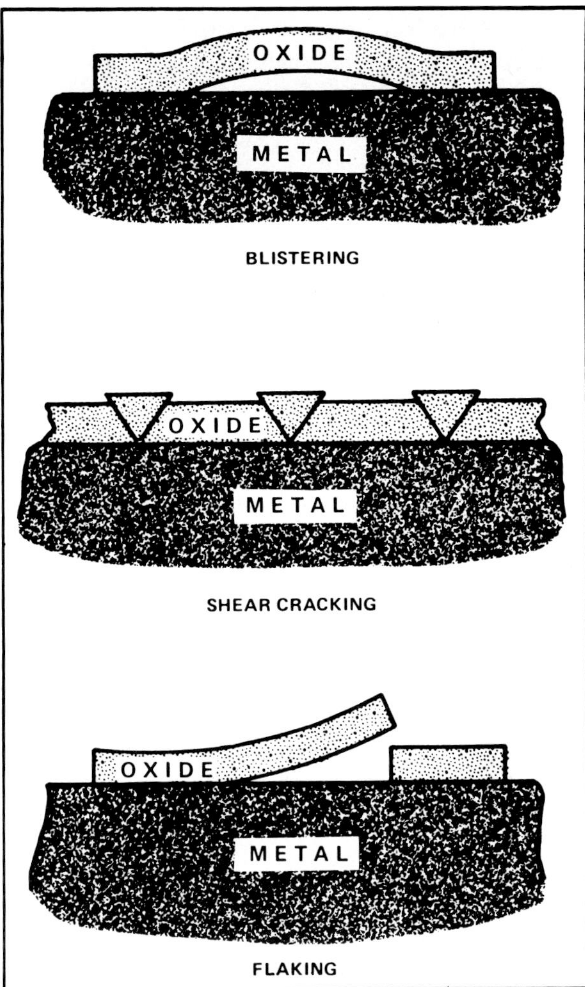

FIGURE 13.14 — Schematic of some of the hypothetical ways the integrity of film or scale is impaired.

dense, resistant oxide layer on the interior. Such a procedure has remarkably increased the life of tubing sections in some services.

Internal Oxidation

Thus far, we have been concerned with the development of external scales and their effect upon continuing oxidation. It is possible, however, for oxidation to proceed without an external scale, provided certain conditions exist. It is most important that the alloy contain, as a minor ingredient, an element which is considerably more reactive than the major constituent. Silicon in copper is one example, chromium in nickel is another. It is also essential that the rate at which the minor element diffuses toward the surface be less than that at which oxygen, sulfur, or nitrogen, etc., diffuses inward. It is possible, however, to stabilize the minor element by formation of a compound, *e.g.*, chromium carbide.

Internal oxidation is controlled by the concentration of oxygen in surface layers. This quantity must be insufficient to oxidize the nobler major constituent, but sufficient to oxidize the reactive minor constituent. External scale will form only when the effective pressure of oxygen at the metal-gas interface exceeds this range. Conversely, failure of surface scale to form indicates that the oxygen partial pressure at the interface is at such a level that it will not react with the major constituent.

An early example of destructive internal oxidation was found in failures of the chromel segments of chromel-alumel thermocouples used to measure the temperature of a hydrogen-rich gas. The hydrogen was able to penetrate the protection tube and set up an atmosphere within the tube of 90% nickel-10% chromium chromel alloy which was reducing to the nickel but oxidizing to the chromium. No surface scale formed, but the fracture surfaces showed the characteristic green color of chromium oxide. One method to correct situations of this sort is to include in the thermocouple assembly an active *getter* for oxygen, such as a strip of titanium, which, by removing the oxygen, will leave an atmosphere that is reducing to both constituents of the alloy.

Corrosion in Air

Of all gaseous, high-temperature environments, air is no doubt the most common. It is, therefore, given considerable attention here. Although we consider air to be roughly 80% nitrogen (N_2) and 20% oxygen (O_2), it also contains small but sometimes significant amounts of water vapor (H_2O), carbon dioxide (CO_2), carbon monoxide (CO), sulfur dioxide (SO_2), hydrogen sulfide (H_2S), and various other chemical fumes and rare gases.

This discussion of air corrosion will be limited to iron alloys, especially those containing chromium and nickel (stainless steels and heat-resistant alloys).

Adequate resistance to corrosion by air at elevated temperatures [above 650 C (1200 F)] implies the use of appropriate alloying elements. These elements generally include chromium, and also may contain one or several of the following beneficial alloying elements: nickel, silicon, aluminum, and possibly titanium. Many papers written on the effects of alloying with these and other elements to reduce corrosion will be summarized.

The relative amounts of alloying elements are determined generally by alloy composition, temperatures and environment to be encountered, degree of corrosion resistance required, and anticipated life, as well as the physical or mechanical properties required. However, the discussion to follow will be restricted to a consideration of the influence of various alloy additions on corrosion resistance.

Influence of Alloy Additions

The influence of alloy additions on corrosion resistance is largely dependent on the composition of the metal prior to the addition. That is, the effect of a certain alloy addition would be entirely different in degree, and sometimes in nature, for a carbon steel when compared with the effect of the same addition to a high-alloy steel. However, the following discussion will give the essential corrosion characteristics of an alloy containing the element under discussion.

Nickel and chromium, the two main alloying elements in heat-resistant alloys, will be treated in greater detail than the others. Silicon and aluminum are also quite effective, but must be used carefully because they can have a harmful effect on mechanical properties and other alloy characteristics. Still other elements also are sometimes added to the iron-nickel-chromium base to produce specific characteristics in fabrication or performance requirements.

Effects of Chromium

Chromium (Cr) is considered the most highly desirable addition to heat-resistant steels. In addition to its ability to provide effective corrosion resistance, it is further capable of improving certain high-temperature mechanical properties.

Although additions of chromium to iron are described here, it seems logical that chromium additions to other metals also will have a similar beneficial effect. It is the formation of a thin chromium-rich oxide layer on the metal surface that lends resistance to further corrosion.

Figure 13.15 shows the sharp decrease in iron's corrosion (oxidation) by air at a temperature of 1000 C produced by additions of Cr. Note that additions of chromium beyond 20% have little effect, particularly at the lower temperatures. As will be described later, these iron-chromium alloys are ferritic (*i.e.,* they have a body-centered cubic structure) or have notoriously poor high-temperature strength. Nickel additions to iron-chromium alloys not only improve air corrosion resistance, as described in the following section, but mechanical strength as well.

Effects of Nickel

Nickel (Ni) is not alloyed to ordinary iron for the development of improved high-temperature corrosion properties. It is effective, however, as il-

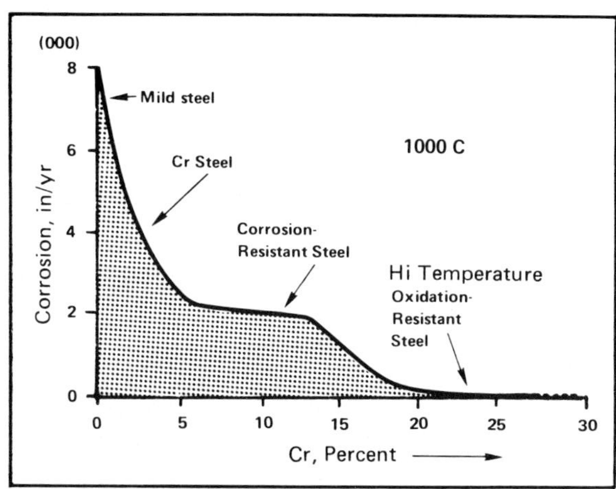

FIGURE 13.15 — Effect of chromium on reducing corrosion in air at 1000 C (1830 F) in FeCr alloys.

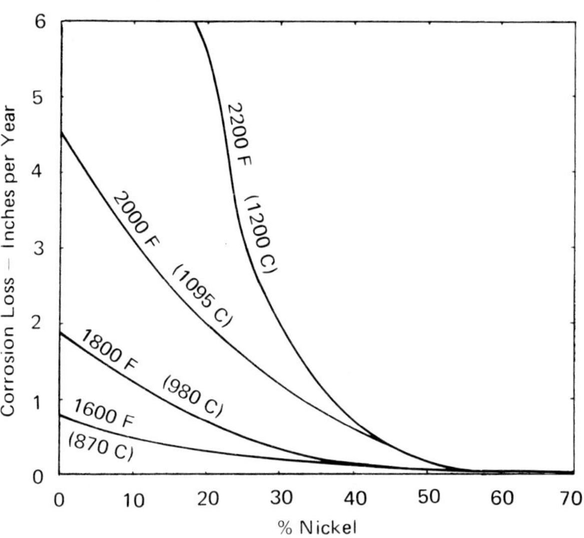

FIGURE 13.16 — Effect of adding nickel to an Fe11Cr alloy in air at 870, 980, 1095, and 1200 C.

lustrated in Figure 13.16, in alloys having 11% Cr. Note that its beneficial effects, like chromium, tend to fade after reaching some specific amount. However, the amount of nickel needed to achieve a low corrosion rate is greater than that of chromium, and the needed amount also increases for higher exposure temperatures.

Thus, from the corrosion viewpoint, nickel of up to about 20% is effective at 870 C (1600 F); of up to about 30% is effective at 980 C (1800 F); and of up to about 50% would be desirable at 1200 C (2200 F).

At higher chromium levels, the optimum nickel level for maximum air corrosion resistance decreases. This effect can be shown clearly by means of isocorrosion lines (compositions with the same weight losses) on a *ternary* diagram; *i.e.*, one which shows the changing composition of an alloy made up of three alloying elements; in this case, iron, nickel, and chromium.

Figure 13.17 shows the relative amounts of Ni (nickel) and Cr (chromium) necessary as alloying additions to Fe (iron) to reduce very heavy corrosion to lesser amounts. Note that at 1200 C in air, the addition of about 28% Cr is about equivalent to the combined addition of about 50% nickel plus 10% Cr, since all compositions on a given isocorrosion line yield the same corrosion rate.

The combined effect of Ni and Cr additions to iron in reducing corrosion by air at 1800 F is more clearly portrayed in the *space* diagram (three-dimensional), shown as Figure 13.18. This diagram has been drawn from the voluminous data reported[7] from corrosion studies at Battelle Memorial Institute.

Figure 13.18 should emphasize that the data reported to this point are based on 100-hour tests, and because of the deceleration of rates encountered in most high-temperature tests, the reported corro-

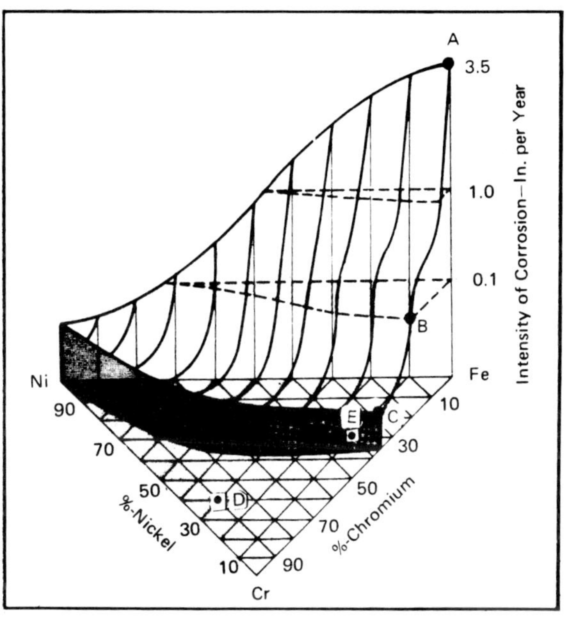

FIGURE 13.18 — Space diagram illustrating relative effects of nickel and chromium contents on intensity of iron alloy corrosion in air at 1800 F (980 C): C = 18Cr8Ni74Fe; D = 60Cr30Ni10Fe; and E = 30Cr10Ni60Fe.

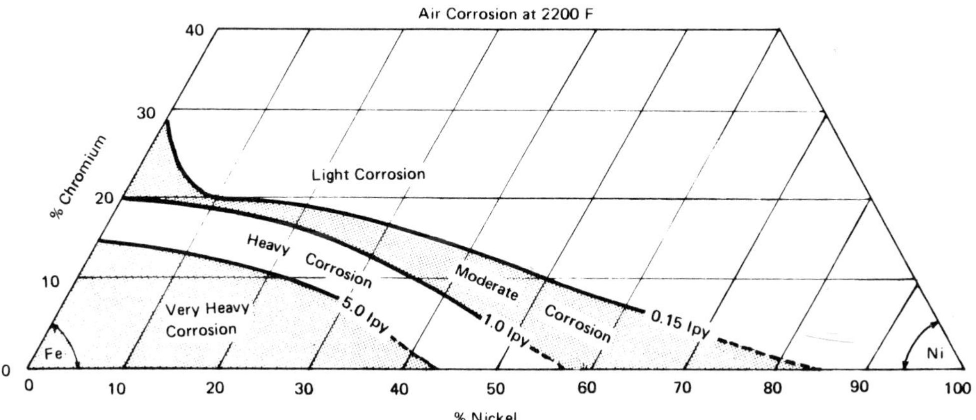

FIGURE 13.17 — Ternary diagram showing alloy composition areas of relatively heavy and light corrosion in air at 1200 C (2200 F). Rates are based on 100-h tests and expressed in ipy average metal loss. [SOURCE: Brasunas, A. deS., Gow, J. T., and Harder, O. E., Proceedings of ASTM, Vol. 46, p. 129 (1946).]

sion rates are conservative, as far as long-term projections are concerned. In other words, the average corrosion rate actually will be less in a 200-hour or 1000-hour exposure test.

Effects of Aluminum

The beneficial effect of aluminum additions in suppressing air corrosion of iron alloys is surprisingly strong. When amounts of up to 12% Al are added, a considerable reduction in high-temperature corrosion is achieved. However, mechanical properties suffer quite strongly, thus making this approach to corrosion resistance less attractive.

The marked beneficial effects of Al additions are clearly shown in Figure 13.19. Where low mechanical properties can be tolerated, the use of limited amounts of aluminum should be considered.

One approach to the aluminum-rich use with relatively little effect on mechanical properties is surface impregnation or *aluminizing*. This high-temperature cementation process is the diffusion of Al into iron, and has been known for many years to produce an effective surface for exposure in air or sulfur dioxide fumes to temperatures in the neighborhood of 1095 C (2000 F). The commercial process of *calorizing* (aluminizing) has been in use for over 60 years[8] and has definite advantages for certain high-temperature applications.

Effects of Silicon

Although silicon additions to steels are beneficial both in high and ordinary temperatures, its use is restricted because of marked embrittlement of the alloy when the amount exceeds 2 or 3%. Thus, the

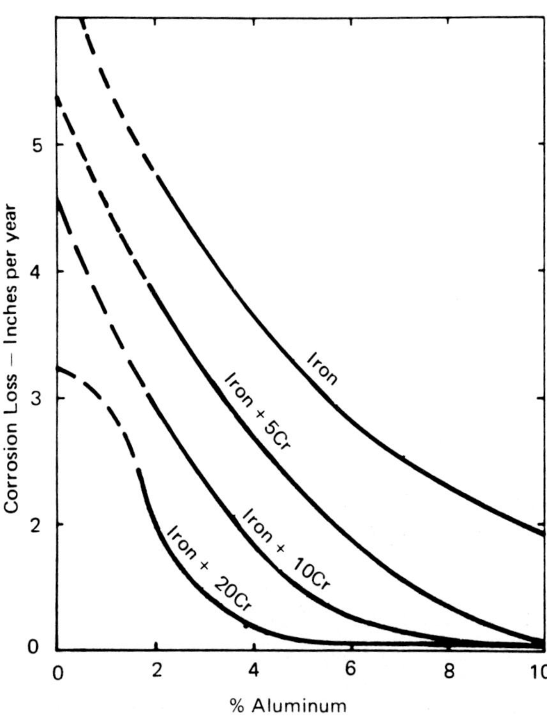

FIGURE 13.19 — Effect of adding aluminum to iron and to FeCr alloys on air corrosion at 2000 F (1095 C).

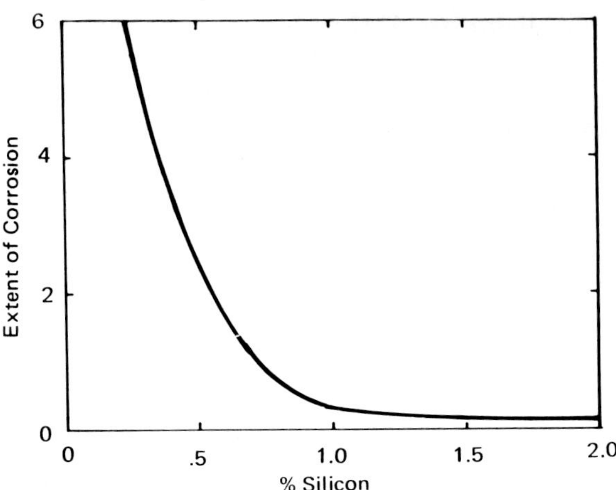

FIGURE 13.20 — Effect of increasing silicon content in a 5Cr0.5Mo steel on resistance to air corrosion at 815 C (1500 F).

actual use of silicon is generally limited to about 2%.

When used in combination with other beneficial alloying elements, silicon can be quite effective. This is illustrated in Figure 13.20. The commercial steel containing 5% Cr and 0.5% Mo can have its corrosion resistance improved by raising its silicon content from 0.2% to about 1%. There appears to be no further advantage to increasing silicon beyond 1% for applications limited to 815 C (1500 F) in an air environment.

For instance, a test made with a 5% Cr steel, with and without Si and Al additions, gave the results shown in Table 13.2 in a 1-day test in air at 732 C (1350 F) and at 1024 C (1875 F).[9]

Effects of Other Elements

A simple and direct statement concerning the effects of various alloying elements on high-temperature corrosion cannot be made. Effects necessarily depend on the base metal composition to which the alloying addition is being made, the temperature of an exposure, and the nature of the corrosive environment itself.

When very strong effects result, as from Cr, Ni, Si, and Al, mentioned previously, clear-cut statements can be made. However, if the effects are weaker, they are difficult to describe and may be masked by small variations in other factors. Nevertheless, with the risk of possible error, the following statements are made.

TABLE 13.2 — Effects of Si and Al on Performance of 5Cr Steel up to 1024 C

	Wt Loss mg/sq cm	
Percent Alloys	1350 F (732 C)	1875 F (1024 C)
5 Cr (plain)	56.0	164.1
+ 2 Al	23.2	146.1
+ 2 Si	0.4	68.1
+ 2 Si + 2 Al	0.2	4.0

Manganese (Mn). Small additions of manganese have *very little effect.* Although few data are available, Mn is not known to have a pronounced effect on high-temperature corrosion rates. In several actual tests,[8] a mild detrimental effect has been observed. Significant amounts of Mn are used in 200 series steels to replace part of the nickel in the analogous 300 steels.

Cobalt (Co). Cobalt additions have *very little effect.* Cobalt additions beyond 50% have been made at 850 to 1100 C, and no significant effects have been noted in air or flue gas atmospheres on alloys containing 22 to 29% Cr.

Molybdenum (Mo). Molybdenum additions are generally considered *detrimental.* Although Mo additions to stainless steels may be beneficial for room temperature corrosion tests in specific environments, their use at high temperatures is inadvisable. Small amounts (less than 4%) may have little effect on high-alloy materials. Larger additions to high alloys or smaller amounts in lower alloy steels can result in *catastrophic attack,* which will be described later.

Copper (Cu). Copper additions can be *somewhat harmful.* Ordinarily, copper is not added to alloys for high-temperature service, although it is sometimes advantageous in stainless steels used to resist chemical attack. It is deleterious at high temperature. As shown in Table 13.3, two heat-resistant alloys were prepared: one with 0.0% Cu, the other with 1.38% Cu (both alloys also contained 0.45 C, 15.3 Cr, 35 Ni). Corrosion rates (in mm per year) for the exposures indicated show copper to be harmful in heat-resistant alloys.

Arsenic (As). Additions of arsenic have *very little effect.* There is evidence that arsenic is beneficial to atmospheric corrosion resistance of mild steel, but this effect does not appear to exist at elevated temperatures. Two 35Ni-15Cr alloys were prepared: one with 0.0% As, the other with 0.086% As. Corrosion results (in mm per year) are shown in Table 13.4. One may conclude that this amount of arsenic in the 35Ni-15Cr alloy is neither beneficial nor harmful.

Vanadium (V). Vanadium additions are considered *harmful.* All evidence points to vanadium as a very harmful alloying element to heat-resistant alloys, as far as corrosion resistance is concerned. This is attributable to the very low melting point of its oxide, V_2O_5, which is well known to promote more rapid corrosion, even to the extent of catastrophic rates.[10]

TABLE 13.3 — Influence of Cu on 0.45C15.3Cr35Ni Alloys

(millimeter per year or mm/α)

	Air Corrosion		Reducing Flue Gas at 1800 F	
Cu, %	1800 F	2000 F	Low Sulfur	High Sulfur
0.00	0.46	1.22	0.46	1.30
1.38	0.99	1.37	1.09	6.9

[SOURCE: Brasunas, A. deS., Unpublished research.]

TABLE 13.4 — Influence of Arsenic on 35Ni15Cr Alloys

(millimeter per year)

	Air Corrosion		Sulfur-Containing Flue Gas	
As, %	1800 F	2000 F	Oxidizing-1800 F	Reducing-2000 F
0.00	0.28	0.79	0.18	0.43
0.086	0.28	0.99	0.28	0.30

[SOURCE: Brasunas, A. deS., Unpublished research.]

Carbon (C). Additions of carbon are generally only *slightly harmful.* To persons quite familiar with the harmful effect of carbon on sensitization, weld decay, and other forms of room temperature corrosion, the statement that carbon is tolerable for high-temperature applications may come as a surprise.

Carbon is recognized as being highly undesirable in 12% Cr steels, 18-8 stainless steels, etc. Principally, its deleterious influence is twofold. When present in steels in excess of 0.1%, a substantial amount of carbides is generally formed, especially when the metal is in an aged or sensitized condition. This is the usual condition of the heat-resistant steels, since their application is generally at elevated temperatures. In forming a carbide, some of the chromium which ordinarily would be present in the matrix becomes associated with the carbon particles as chromium carbide. Hence, the matrix is depleted of some of its chromium, and the resulting steel could have the corrosion characteristics of a chromium-bearing steel containing less chromium than a chemical analysis would indicate.

Since the approximate chemical formula for chromium carbide is Cr_4C, and recognizing that the atomic weights for chromium and carbon are 52.01 and 12.00, respectively, we can calculate that for every unit of weight of carbon,

$$\frac{4Cr}{C} = \frac{4(52.01)}{12.00} = \frac{208.04}{12} = 17.3 \qquad (13.5)$$

weight units of chromium are required. However, because some carbon is soluble in the metal matrix, the ratio 16:1 is more suitable. Thus, in calculating the probable chromium content of the matrix (background metal) of the low (0.30% C) and high (1.01% C) carbon 26Cr-20Ni alloy of Figure 13.21, the following information is obtainable:

$$\text{Matrix Cr} = \text{Total Cr} - 16\,C. \qquad (13.6)$$

For the 0.3% carbon alloy,

$$\text{Matrix Chromium} = 26 - 16(0.30) = 21.2\%\ Cr. \qquad (13.7)$$

For the 1.01% C alloy,

$$\text{Matrix Chromium} = 26 - 16(1.01) = 9.8\%\ Cr. \qquad (13.8)$$

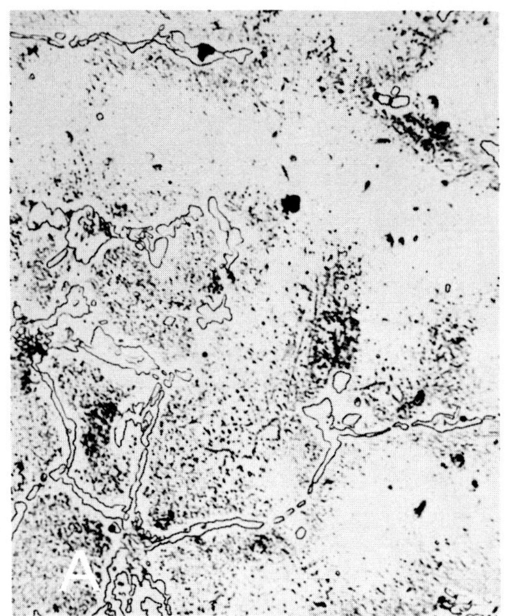

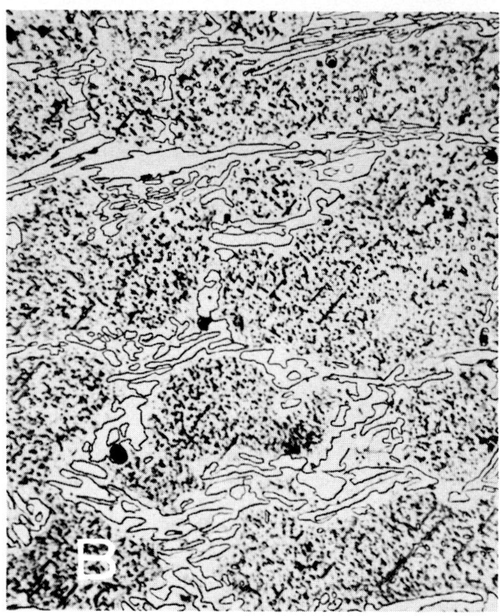

FIGURE 13.21 — Microstructure of aged medium- (A—0.32% C) and high-carbon (B—1.01% C) 26Cr20Ni alloys. X250.

This would seem to imply a marked susceptibility of the high-carbon alloy to corrosion. However, high-temperature air corrosion tests made on these alloys indicate that the chromium gets into the oxide layer to supply corrosion protection equally well when obtained from the carbide phase as when derived from the matrix.

One method by which carbon may act as a detrimental element is its formation of a continual network of *interdendritic* carbides (also referred to as *primary* or *eutectic* carbides) in cast alloys. These particles are often extremely susceptible to attack. This phenomenon is a serious problem in the liquid corrosion field and, to a lesser extent, to the hot gas corrosion field. Subsequent discussions will point out that not all iron-chromium carbides are equally susceptible. As the composition of the matrix varies according to the general composition of the steel, the composition of the carbide is also altered. Therefore, a different behavior should be anticipated.

Carbide-forming elements, such as titanium and columbium, are often added to alloys, particularly to those of the 18Cr-8Ni type. These elements act as carbide stabilizers, as was described in Chapter 3. These elements have a greater affinity for carbon than chromium does, and consequently they form carbides of their own, leaving all the chromium as a matrix constituent. The carbides formed by the stabilizing elements are found as very small particles scattered throughout the grains themselves, rather than in grain boundaries, thereby reducing appreciably the prevalence of intergranular corrosion.

The influence of moderate carbon variations (0.3% C) on the high-temperature corrosion resistance of alloys containing substantial amounts of alloying materials is essentially nil. Although the depletion of approximately 5% matrix chromium from a 12% Cr steel or an 18-8 steel might be a serious matter, with richer alloys of the 25Cr-20Ni type, the effect of removing 5% chromium would be almost imperceptible.

Furthermore, we should be more concerned with the composition of protective oxide layers than matrix metal composition. In other words, during high-temperature corrosive attack, a high-chromium protective oxide layer cares not whether its chromium came from the metal matrix or from a carbide.

In all likelihood, since decarburization (Figure 13.22) generally occurs at the metal surface preceding oxide formation, it releases chromium to the matrix. Thus, chromium can serve equally well in low- or high-carbon alloys at elevated temperatures.

$$Cr_4C + O_2 \rightarrow CO_2 + 4Cr \qquad (13.9)$$

$$4Cr + 3O_2 \rightarrow \underset{\text{Protective Oxide}}{2Cr_2O_3} \qquad (13.10)$$

Other Alloying Elements. Specimens of the 35Ni-15Cr type were also prepared with 2 and 4% additions of various other alloying elements. These included: tungsten (W), columbium (Cb), tantalum (Ta), titanium (Ti), zirconium (Zr), beryllium (Be), as well as vanadium and molybdenum, which were discussed earlier.

In a very long-term laboratory test of 1000 hours, specimens were exposed to air at 1095 C (2000 F), cooled, weighed, and handled lightly to remove loose oxide scale daily, then re-exposed for additional 24-hour periods until a total of 1000

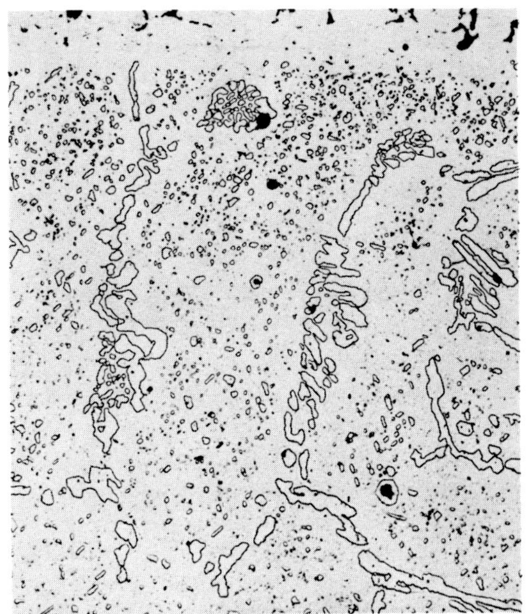

FIGURE 13.22 — Photomicrograph of cross section of heat-resistant alloy after mild corrosive attack. Although primary and secondary carbides appear throughout, they do not appear in the surface zone because of decarburization.

TABLE 13.5 — Effect of Alloy Additions on Air Corrosion of 35Ni15Cr Alloy at 1095 C (2000 F)

Behavior	% Alloy Additions 35Ni15Cr	Corrosion Rate Influence
Slightly Beneficial	2 and 4 Mo, 2 and 4 Zr, 2 and 4 Be, 2 W	Reduces Corr.
Slightly Harmful	4 W, 2 Ti	~1.5X
Harmful	4 Ti, 2 and 4 Cb, 2 and 4 Ta	2X or 3X
Extremely Harmful	2 and 4 V	~20X

[SOURCE: Brasunas, A. deS., Unpublished research.]

hours (42 days) had elapsed. Results are summarized in Table 13.5.

Summary

The relative corrosion resistance of metals in air at high temperatures is determined largely by their ability to form a protective scale layer. The extent of protection by the scale will depend on scale composition and its characteristics. The following factors determine the nature of the scale.

1. Composition of metal
2. Composition of the surrounding environment
3. Temperature
4. Time

The important characteristics of adherence, expansion, diffusion resistance, volatility, etc. are all determined by the above four parameters.

Extraneous factors such as variations in atmospheric composition, temperature fluctuations, presence of small amounts of minor elements, foreign contaminants, external stresses or abrasion, etc., all can have an important effect on the nature and extent of attack. All too frequently, factors are not reported or even known, and test data often seem to be erratic or even conflicting.

The behavior of alloys can be explained when all the facts are known, but since this is not always possible, duplicate or related tests should be performed. When several hundred specimens have been involved in a corrosion survey, the reliability of test results increases. Such was the work referred to earlier with iron-nickel-chromium alloys where ternary data charts summarized air corrosion at several temperatures.

An even more compact chart can be drawn to summarize this research study in nomographic form. Figure 13.23 summarizes the air corrosion tests over a wide temperature range. This chart permits estimation of probable corrosion results at other temperatures and for a broad range of composition.

As an example exercise, use this chart to determine the probable rate for an alloy composed of 14% Cr and 30% Ni (remainder iron) when exposed to air for 100 hours at 1900 F. Plot your lines on Figure 13.23 and compare your answer with Self-Check Quiz No. 2 at the end of the chapter.

Other Gaseous Media

We have discussed oxygen reactions at great length for several reasons. It is, of course, the most common high-temperature corrosion reaction. Consequently, it has been more widely studied and is better understood than less common forms of high-temperature corrosion.

A more compelling reason, however, is that what one learns about oxygen is directly applicable to most other gas-metal, high-temperature reactions in which a scale is formed. In fact, all of these reactions should be classified as oxidation, whether the scale is an oxide, a sulfide, or a halide, because the valence charge of the metal has been increased. In the following sections, then, we shall not go into as much detail, but in many instances will merely point out differences between the reactions under consideration and those already discussed.

Corrosion by Flue Gases

Corrosion by flue gases has many elements in common with air corrosion at high temperatures. Oxygen is the major element to consider, largely in the form of uncombusted gaseous oxygen (O_2) or its combination in the form of carbon dioxide (CO_2), water vapor (H_2O), or possibly even sulfur dioxide (SO_2). These gases, in addition to nitrogen (N_2), are the major constituents in *oxidizing* flue gases which contain an excess of O_2.

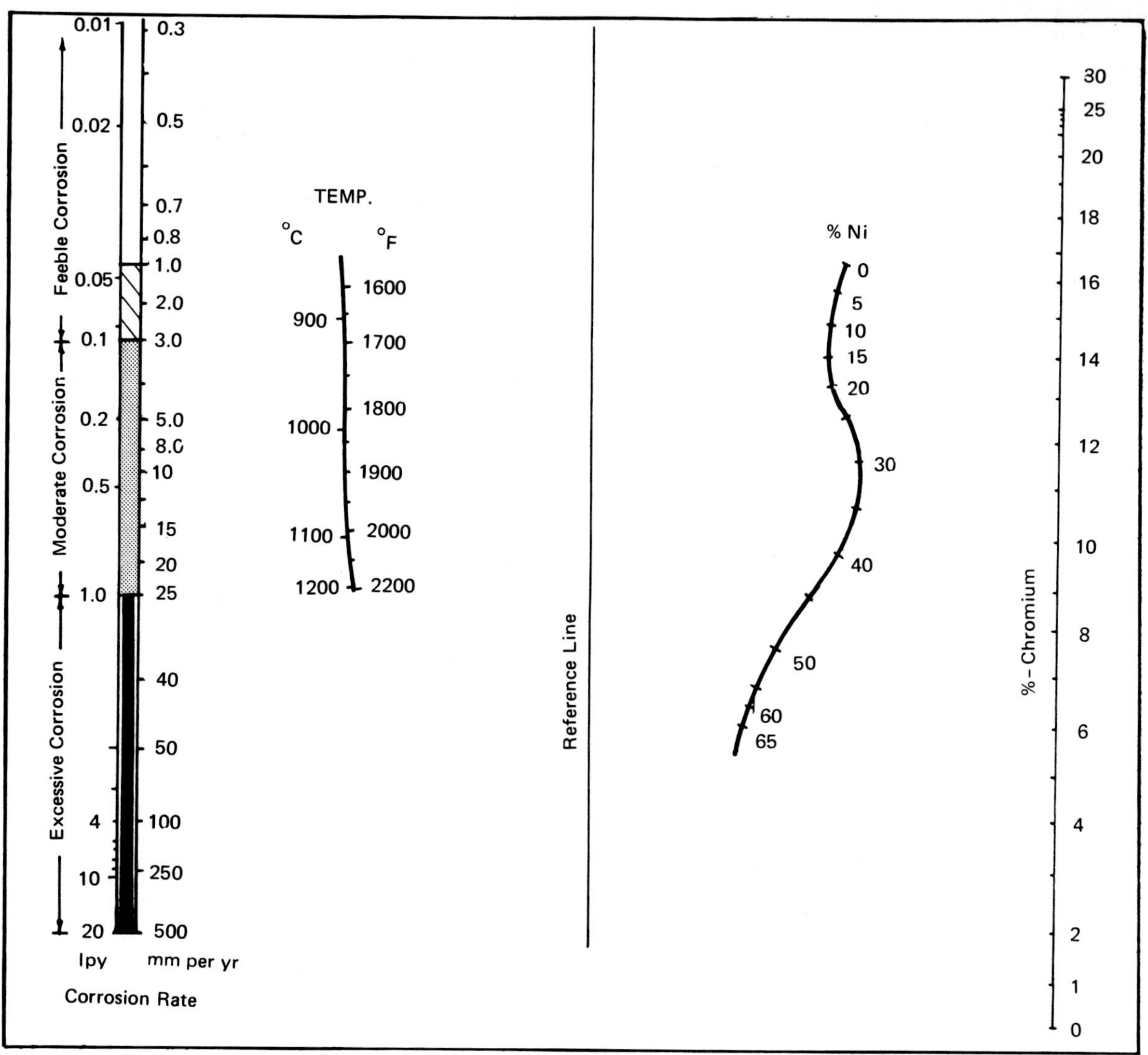

FIGURE 13.23 — **Nomograph illustrating corrosion in air of FeNiCr alloys. Based on 100-h continuous exposures.**

If there were insufficient oxygen for complete fuel combustion, the above oxidizing gas constituents would be present in smaller amounts and would be partially replaced by carbon monoxide (CO), hydrogen (H_2), and hydrogen sulfide (H_2S), respectively. Such gases are referred to as *reducing* flue gases.

Of course, there sometimes exists an in-between condition when the appropriate amount of air (or oxygen) is used for fuel combustion and the flue gas cannot be appropriately referred to as oxidizing or reducing. In such cases, the term *neutral* may be used.

Sulfur Content

The major constituent in flue gas corrosion which differentiates it from air corrosion, other than the oxygen referred to above, is the sulfur content. Although it originally may be present in the form of complex organic compounds, after combustion it occurs as SO_2 or H_2S, depending on the oxygen availability, because either can be converted to the other by either of the two following equations:

$$H_2S + O_2 \rightleftarrows H_2 + SO_2 \quad (13.11)$$

$$H_2S + 2CO_2 \rightleftarrows H_2 + SO_2 + 2CO \quad (13.12)$$

To repeat, reducing flue gases tend to have significant amounts of H_2, H_2S, and CO whereas oxidizing flue gases are relatively rich in H_2O, SO_2, and CO_2 (Table 13.1). One should keep in mind that no flue gas is completely or totally reducing or oxidizing. When CO_2 is largely present, some CO can also be found; when H_2 is present, some H_2O will be there, etc.

Some fuels are known to be quite high in sulfur, while others tend to be low. As a rule, bituminous coal is quite heavy (Table 13.6) and is a major "villain" in today's city air pollution problems. Fuel oils vary somewhat in sulfur, and sometimes also contain trace amounts of organic compounds of vanadium that can be much more corrosive, as will be mentioned in the next section dealing with catastrophic corrosion.

Among the lowest sulfur fuels are the natural gases. Some have significant traces of sulfur, but many are extremely low. Economics and avail-

TABLE 13.6 — Sulfur Content in Some Fuels and Their Combustion Products

Fuel	Sulfur Content of Gas Uncombusted	Combusted
Good Natural Gas	Trace	—(2)
Coke Oven Gas		
Purified	30 gS(1)	6 gS
Unpurified	600 gS	125 gS
Fuel Oil		
Low S	0.7% S	25 gS
High S	2.5% S	100 gS
Bituminous Coal		
Low S	1% S	55 gS
High S	5% S	300 gS

(1)gS = 1 grain S per 100 cu ft.
(2)Much less than 1 gS.

ability usually dictate which fuel a given industry will use, and the corrosion engineer is expected to keep the plant relatively free of serious operating problems.

Two principal methods are used to report sulfur content of gases. One is by volume percent and the other by referring to its weight in 100 cubic feet of gas. The former method is easier; *i.e.,* a 2% SO_2 content means that for every hundred gas molecules, two are likely to be SO_2 molecules. This system is less suitable when trace amounts are present, and references must be made to very small amounts, such as 0.00164% SO_2. In such cases, the weight in *grains* of sulfur per 100 cu ft of gas is used.

A grain is one seven thousandth of a pound. This is an ancient unit of weight which represents the average weight of a grain of wheat. The above amount, 0.00164% SO_2, represents one grain of sulfur per 100 cu ft of gas, and will be abbreviated in this section of the text as 1 gS.

Thus, one can compare fuels more easily in regard to sulfur content. A coal with a 4% sulfur content would generate a flue gas under ideal combustion with the proper amount of air, containing about 250 gS. Similarly, other typical flue gas comparisons are shown in Table 13.6.

Sulfidation

In general, oxidation by sulfur or sulfidation is a considerably more destructive form of high-temperature corrosion than oxidation by oxygen. The reasons are not hard to find. Sulfide scales tend to crack and spall more readily than oxides. In some cases, depending on the form in which sulfur is present in the atmosphere, continuous sulfide scales cannot form, so attack will proceed linearly; *i.e.,* the scale will afford no protection. The melting points of metallic sulfides usually are lower than those of the corresponding oxides.

Table 13.7 compares the melting points among a few metal sulfide-metal eutectics with the base metal melting points. Note that most sulfide eutectics are molten at temperatures well below the melting points of the base metals. Oxide eutectics usually melt at temperatures much closer to the melting points of the metals on which they form. Thus, the sulfide eutectic melting temperature frequently will constitute a limit to the upper service temperature of the metal or alloy, while only rarely does this occur with an oxide. As will be discussed later, the presence of a molten phase generally will cause a vast increase in the corrosion rate of any metal.

TABLE 13.7 — Differences Between Metal and Sulfide Eutectic Melting Points

	Metal °F	Metal °C	Metal-Sulfide Eutectic °F	Metal-Sulfide Eutectic °C
Co	2723	1495	1610	877
Cr	3362	1850	2462	1350
Cu	1981	1083	1958	1070
Fe	2802	1539	1805	985
Mn	2300	1260	—	—
Ni	2651	1455	1194	645

Nevertheless, the basic mechanism of sulfidation is closely akin to air oxidation and, to a considerable degree, the same alloying elements are useful in providing resistance to attack. As shown in Figure 13.24, the behavior of chromium follows the same pattern in sulfidation as in oxidation.[11] It is particularly interesting to note that, when sufficient chromium is present, the relative amount of nickel or iron is of secondary importance, despite the fact that pure nickel is much more susceptible to sulfur attack than pure iron. Chromium, however, is

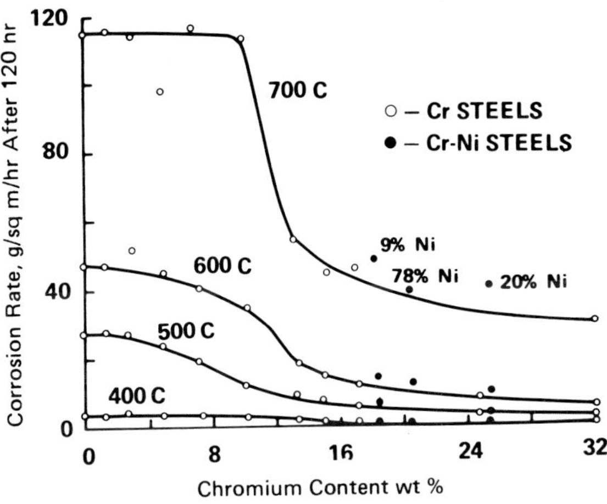

FIGURE 13.24 — Effect of Cr content of steels on corrosion resistance in hydrogen sulfide. Curves were drawn through open circles representing binary CrFe steels. Data for CrFeNi alloys were plotted as solid circles and were found close to curves for binary alloys, despite the great differences in FeNi content. [SOURCE: Schafmeister, P. and Naumann, F. K., DieChem Fab., Vol. 8 (1953).]

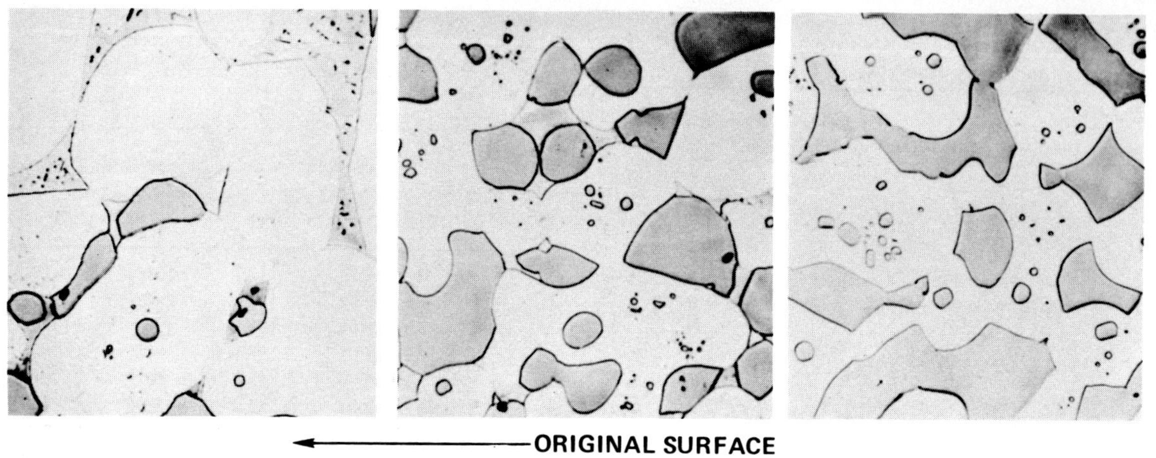

FIGURE 13.25 — Duplex sulfides formed in a 78Ni16Cr alloy. Gray areas are sulfides; white is austenite matrix. X750.

much more attracted to sulfur than either iron or nickel, and the scale formed is essentially a chromium sulfide, when sufficient chromium is present in the alloy, whatever the balance of the composition. Eventually, of course, when the chromium is depleted, the remaining base metal will be attacked rapidly.

Figure 13.25 illustrates a nickel-chromium alloy wherein, near the surface, sulfidation has progressed to the point where both chromium and nickel sulfides coexist. This condition occurred only because so much sulfur had been introduced at the surface that the available chromium was almost totally consumed. Under these extreme conditions, the remaining nickel-rich matrix will react to form nickel sulfide. More often, only chromium sulfides will be found, as in the center portion of the specimen.

The form in which sulfur is present will affect the rate of attack. When oxygen is also present, as would be the case with SO_2 or SO_3, a mixed oxide-sulfide scale frequently forms and offers a greater degree of protection than the sulfide scale produced by H_2S or organics (which in most cases decompose to give H_2S) and, of course, sulfur vapor. Figure 13.26 shows some data on the relative performance of a nickel-chromium alloy exposed to either pure SO_2 or pure H_2S.[12] Note that parabolic behavior is indicated, although nonprotective scale is evident at an early stage.

As with oxygen attack, aluminum may be a useful addition to improve sulfidation resistance. Table 13.8 gives some data on the performance of a number of iron-nickel-chromium alloys with and without aluminum.[13] In every case, the presence of aluminum provided an additional degree of resistance. Similar improvement could be expected when cementation techniques are used to develop siliconized, chromized, or aluminized coatings on ferrous alloys.

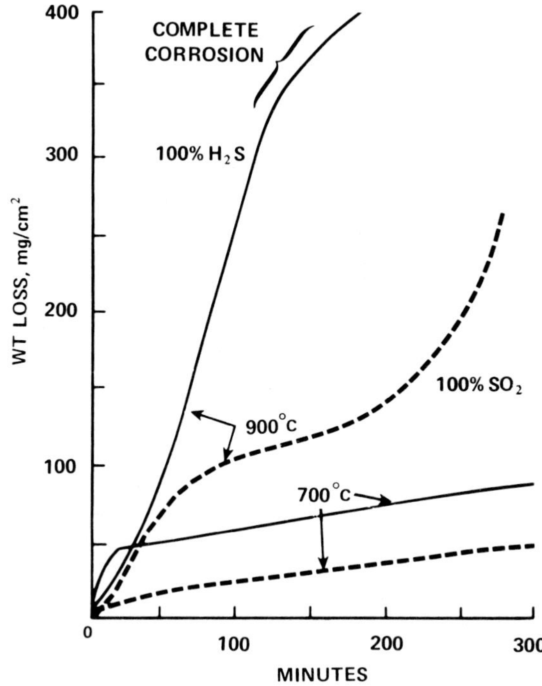

FIGURE 13.26 — Corrosion of NiCr alloy in sulfur-bearing gases. [SOURCE: Bradbury, E. J., Hancock, P., and Lewis, H., Metallurgia, Jan. (1963).]

TABLE 13.8 — Influence of Al on Sulfidation Resistance of Alloys(1)

Materials	mpy No Al	mpy with Al Additions, 3%	4%
18 Cr, 8 Ni(2)	44.6	17.1	3.1
20 Cr, 32 Ni	16.3	7.1	—
16 Cr, 78 Ni	98.7	12.1	—

(1)Test conditions: 985 F (530 C), 480 psi H_2S in H_2. Vol %—1.4 to 1.7

(2)Type 304.

Rate of Attack

The attack on iron-nickel-chromium alloys by oxidizing and reducing flue gases has shown strong similarities to air corrosion, particularly to the oxidizing gases. Two major differences are:

1. Oxidizing flue gases are less oxidizing than air; *i.e.,* they contain less free oxygen, and although they also contain CO_2, H_2O, and SO_2, the oxygen supply is smaller. In reducing flue gases, the available oxygen supply is still smaller.

2. They contain sulfur-bearing gases which may not attack high-chromium, nickel-free alloys excessively, but which will be corrosive to high-nickel alloys because of the instability and low melting points of the nickel-sulfur corrosion products (Figure 13.5).

Corrosion in oxidizing flue gases has been found to be similar to that encountered in air, especially if no harmful solid particles are entrapped and the sulfur content is low. High-sulfur oxidizing flue gases cause only slightly more corrosion.

Reducing flue gases, however, are very sensitive to the sulfur content. If the sulfur content is very low (below 5 grains per 100 cu ft of gas), corrosion is generally less severe than in air. This is attributed to the relatively small oxidizing power of the gas, and may be likened to air that is somewhat depleted in oxygen. However, when the sulfur content is appreciable, exceeding 100 grains per 100 cu ft of gas, corrosion due to intergranular sulfide penetration can be quite severe. Weight change indications may not properly describe the severity of metal damage in such instances. High-nickel alloys (above 35% Ni) with less than 15% Cr are especially susceptible to this form of corrosion.

Whenever low grade (Bunker C-type) fuel oils are used for heating boilers, etc., entrained tiny particles of vanadium pentoxide and other impurity particles can be deposited on hot metal surfaces and result in extremely severe corrosion attack. This subject is discussed under a later section concerning molten phases.

Isocorrosion lines on ternary diagrams similar to those shown in Figure 13.17 for air, have been reported[14] from the work performed at Battelle Memorial Institute. The major difference between diagrams representing corrosion under oxidizing as compared with those depicting reducing conditions is that the high-nickel alloys are heavily attacked in sulfur-bearing flue gases, particularly under reducing conditions.

Nitriding

Nitriding has been employed for years to provide hard, wear-resistant surfaces on certain low-alloy steels, and probably the first corrosion problems involving nitriding related to the retorts used to produce nitrided surfaces. Ammonia is the source of active nitrogen in the hard surfacing process and is the principal source of nitrogen as a corrodent. The nitrogen molecule is relatively stable and, except at very high temperatures over very long periods, has not created serious corrosion problems. Active nitrogen, readily produced by the decomposition of ammonia, is another matter, and rapid nitride formation will occur at temperatures below 540 C (1000 F). Because ammonia is an important component in many chemical processes and is essential in the burgeoning fertilizer industry, the need to cope with nitriding is readily evident.

To some extent, nitriding differs from the gas-metal reactions considered before. Although the ultimate reaction product, as shown in Figure 13.27, may well be a nitride scale, frequently the damaging effects become manifest as a loss of aqueous corrosion resistance or reduced ductility from nitrogen which, although it has diffused into the metal, has done so in an amount insufficient to produce a discrete scale. When the scale does form, it is quite brittle and offers little protection to the underlying metal.

If a plentiful supply of active nitrogen is available at the surface of the metal (its passage into the metal can be followed by the etching response of the alloy), a *smooth front* parallel to the surface will be found—evidence that preferential grain boundary diffusion is *not* a significant mode of entry. This is clearly illustrated in Figure 13.27. If, however, a limited supply of nitrogen is available, the continuous diffusion front will not be observed. Nitride needles frequently can be seen not only near grain boundaries, but within the grains as well.

It may generally be concluded from the latter that diffusion of nitride-forming elements, such as chromium, aluminum, and titanium, occurred over

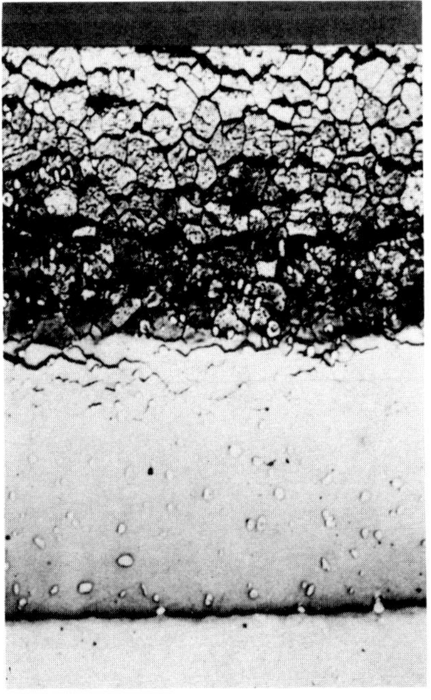

FIGURE 13.27 — Type 446 steel (28% Cr) after 1540 h at 260 C (500 F) in anhydrous ammonia. Note cracking in upper areas. 100X.

a fairly long period of time in order to concentrate sufficiently to permit precipitation. It appears that the locations where these elements concentrate, rather than the nitrogen (which may be assumed to be uniformly distributed throughout the structure) determine the locations of the precipitates.

As was noted above, such metals as iron, chromium, aluminum, and titanium readily form nitrides and are not generally suitable if the nitriding potential is high. Nickel and copper, on the other hand, do not form stable nitrides at elevated temperatures and might be expected to confer some resistance to nitriding. In fact, both nickel and copper are successfully used to "stop off" portions of steel parts during commercial nitriding when only a partial hard surface is desired. Neither element is entirely satisfactory for prolonged high-temperature service, because they are relatively sensitive to contaminants, such as sulfur, which may be present, nor do they have outstanding oxidation resistance.

It has been found that high nickel-chromium alloys provide the best engineering compromise. Figure 13.28 gives the results of a three month test in a chemical plant ammonia line operating at 510 C (935 F).[15] Pre-oxidation of the specimens provided at least temporary improvement for the more susceptible stainless steels, but maximum resistance was obtained with the nickel-base alloys.

The anomalous behavior of pure nickel should be mentioned. The attack on these specimens was markedly different from that on the balance of the alloys in that it was entirely intergranular, whereas the others showed uniform corrosion penetration. An X-ray diffraction analysis of the former revealed that the only nitride in the grain boundaries was manganese nitride (a small amount of this element is present in commercially pure wrought nickel), and that some graphite (perhaps because a small quantity of hydrocarbon was found in the ammonia stream) was also present.

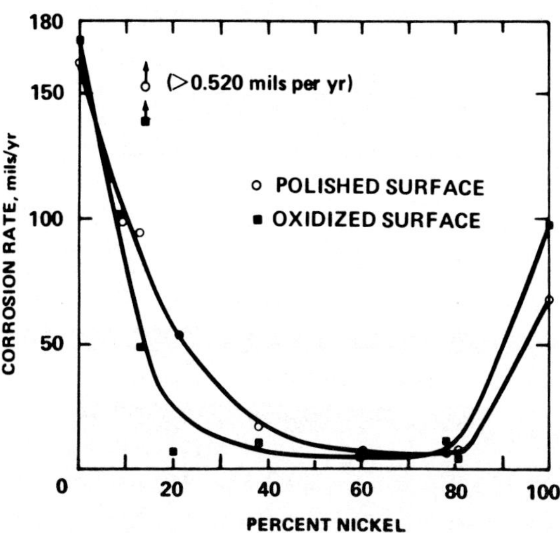

FIGURE 13.28 — Corrosion behavior of polished and oxidized surfaces of nickel alloys in plant ammonia line.

Carburization

Carburization is not, in the sense we have used the term previously, a form of high-temperature corrosion. Scale formation and metal waste are not direct products of the carburizing reaction, although loss of ductility can occur. Nevertheless, it is an important gas-metal reaction, and is so intimately connected with high-temperature corrosion that it must be studied and understood by anyone seriously interested in the factors affecting the behavior of metals at elevated temperatures.

Like nitriding, carburization is employed commercially to develop hard, wear-resistant surfaces on certain low-alloy steels. Steels, when exposed to a suitable carbonaceous gas or liquid at high temperatures, may absorb some portion of the carbon atoms which reach the metal surface. The higher carbon surface will respond better to heat treatment and develop a higher surface hardness, while the lower carbon interior will be softer but tougher, a desirable characteristic for many applications.

Again, some of the earlier difficulties with carburization undoubtedly originated because of the materials used to contain the reaction for carburizing steels. Currently, because of the enormous use of high-temperature treatment to alter or crack the molecular structure of hydrocarbons derived from petroleum products, the problem of carburization and the selection of materials to withstand it have great industrial significance.

As already mentioned, carburization is largely a gas-metal reaction. Although under some very critical conditions it may be possible to introduce carbon into a metal via a solid-state reaction (that is directly from carbon or graphite into the metal), the incidence of such an event is so infrequent as to be of no importance. Generally, a gas containing carbon monoxide or hydrocarbons such as methane, propane, etc., will be the source of the carbon.

At high temperatures, carbon will be released at the surface of the metal and, if the conditions are right, will diffuse inward. As in the case of nitriding, the diffusion occurs uniformly across the grain, in most instances, and is not intergranular (Figure 13.29).

Most metals have a significant solubility for carbon, *i.e.*, they can absorb a quantity without undergoing any drastic physical change, although some properties (such as response to heat treatment) may be altered measurably. Eventually, if carbon continues to be driven into the metal, its solubility can be exceeded and more dramatic changes will occur. In pure nickel, which forms no stable carbide at high temperatures, embrittlement from graphite precipitation may occur, and in the more commonly used heat-resistant alloys of the iron-nickel-chromium system, precipitation of chromium-rich carbides will occur.

Initially, such carbides will have little effect on the chemical or physical behavior of the alloy. In

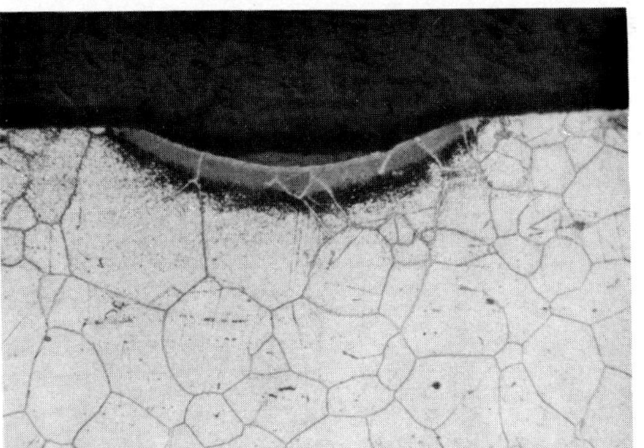

FIGURE 13.29 — Photomicrograph of cross section of a very small pit. Intense carburization is clearly evident in zones being corroded. X130.

fact, most commercial, chromium-containing alloys such as the stainless steels have a small quantity of chromium carbides present. However, as carburization continues and the quantity of carbides increases, the situation will change. Since chromium forms carbides more readily than the other constituents of stainless steels, the matrix of the alloy will be depleted of chromium.

As discussed previously, chromium is the element which contributes most to a resistance to oxidation and sulfidation. The chromium-depleted matrix will be unable to develop adequate protective scales and will become quite susceptible to these and other types of attack [Equation (13.5)]. Furthermore, the presence of continuous grain boundary carbides, as illustrated in Figure 13.30, will both increase the rate of intergranular attack (because the carbides are more reactive than the matrix) and harmfully affect the mechanical properties because cracks can propagate more readily along the brittle cracks.

In chemical reactions of the type involved in carburization, for example

$$2CO \rightarrow \underline{C} + CO_2, \qquad (13.13)$$

where $\underline{C}$ indicates carbon in solution in the metal, there is some equilibrium quantity of carbon in solution in the metal for each ratio of $CO:CO_2$. The precise amount will be strongly influenced by alloying elements. If the ratio of $CO:CO_2$ in Equation (13.13) is high, the driving force to cause carbon to enter the metal is great. On the other hand, if the ratio is low, the driving force is reduced. If reduced sufficiently, carbon may actually be removed from the metal, *i.e.,* not carburization but decarburization would be the result. At some intermediate ratio, of course, the carbon content of the metal would not change because the quantity already present in the alloy would be in equilibrium with the amount available from the gas.

Carburization may be minimized, then, by selecting an alloy that has a carbon content close to the equilibrium amount for the gas in question. Such a selection will reduce the driving force, and even if perfect equilibrium is not attained, harmful changes in the alloy should proceed at a relatively slow rate. For example, consider the photomicrographs in Figure 13.30. These alloys of the same base composition were exposed in the same carburizing environment, yet one was carburized a great deal and the other almost not at all. It is evident that at fixed chromium and nickel contents, increases in silicon improve the carburization resistance of iron-nickel-chromium alloys.

Figure 13.31 illustrates in a different way the important effect of chromium on carburization resistance.[16] It is significant that additions of both Si and Cr to iron, at least over the ranges covered by the alloys tested, lower the equilibrium quantity of carbon that can be kept in solution. Further, chromium is a strong carbide former. Nickel also lowers the carbon solubility limit in iron-nickel-chromium alloys and helps to improve carburiza-

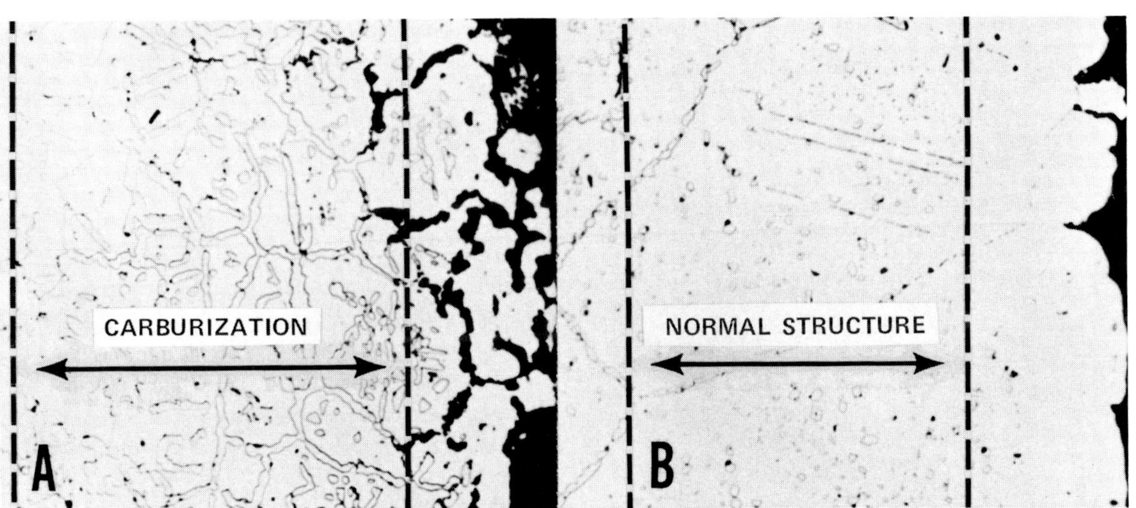

FIGURE 13.30 — Effect of variations of silicon content on carbon pickup after 7340 h at 1675 F (913 C) in a carburizing atmosphere: (A) 0.47% Si, % C = 0.96 (significant carbon pickup); (B) 1.00% Si, % C = 0.14 (no carbon pickup). Atm: Hydrogen 34%; CO 14, CH_4 12.4, CO_2 0.4, N_2 balance. Dew point +8 to 25F. X250.

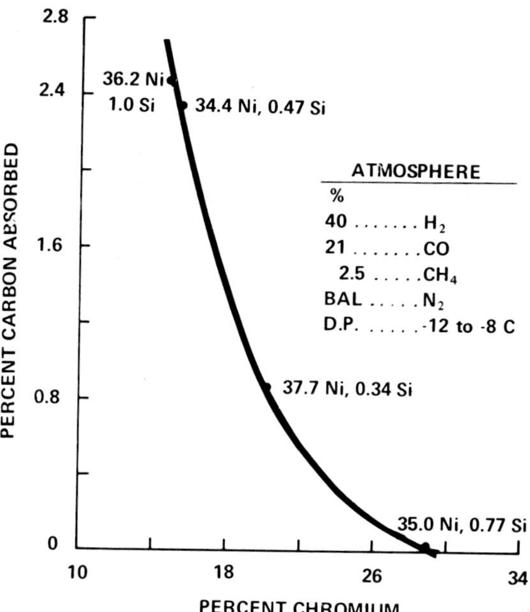

FIGURE 13.31 — Cr content vs carbon absorption at ~35% Ni level. 1500 h gas-carburizing at 1785 F (974 C).

tion resistance, although not as effectively as chromium or silicon. However, when the carbon potential is very high, it is impossible, with alloys currently available, to eliminate carburization completely.

Another ultimate effect of carburization can be *metal dusting.* This phenomenon occurs in process operations where oxidizing and reducing conditions alternate. When the environment is on the reducing side (CO predominant), carburization of the metal to a shallow depth can occur at breaks in the protective oxide film. When the exposure then changes to oxidizing, the high-carbon area of the metal is burned out and the metal reacted to the oxide. A depression is left in the metal surface where the carburized area existed (Figure 13.32), and the metal oxide is swept downstream in the process as metal dust.

FIGURE 13.32 — Hemispherical pits in stainless steel caused by metal dusting attack. X9.7.

Hydrogen Attack

Hydrogen attack is also a high-temperature phenomenon. It should be clearly distinguished from hydrogen embrittlement which occurs at temperatures near ambient and is usually associated with aqueous corrosion as the source of the hydrogen. At higher temperatures, above the dew point, high-pressure hydrogen, frequently a component of some process gas, is the more usual source. Ammonia production, mentioned previously in connection with nitriding, is an industrially important example. In order to join nitrogen with hydrogen to form NH_3, the two gases must be brought together in the presence of a catalyst under thousands of pounds of pressure, and resistance to hydrogen attack is an important consideration in selection of materials to contain this reaction.

Hydrogen attack does not involve corrosion in the usual sense. Although a reaction to form hydrides may occur with some metals, this is not an important consideration with respect to any material that has engineering significance, except titanium.

The attack with which we are concerned relates to the reaction of hydrogen with readily reducible carbides, or in some cases oxides, within the alloy to form methane or steam. These gaseous products, under high pressure, will cause small local ruptures which, of course, will impair the structural integrity of the metal and lead to early failure. In some cases, if structural discontinuities are present, hydrogen under pressure will accumulate at these points, and even without a further reaction may cause damage or blistering when the external pressure is reduced, as during a shutdown. In this case, the internal pressure in the defect will be maintained at a high level and will attempt to relieve itself by expanding against the wall, now no longer supported by the pressure of the contained gases.

Hydrogen is a difficult gas to cope with at elevated temperatures because of its ability to diffuse rapidly through metals in atomic form and because of its extreme reactivity. If the hydrogen, during its passage through the metal, encounters readily reducible compounds such as the carbides previously mentioned, a reaction will occur, and once having occurred, cannot be reversed.

Since the movement of hydrogen through the metals cannot be prevented, the principal recourse is to make certain that the carbides formed are sufficiently stable to resist reaction with hydrogen. Such elements as chromium and molybdenum form carbides considerably more stable than iron carbides, so steels containing additions of these elements have been quite successful in resisting hydrogen attack. Note, however, that the problem of blistering results from defects such as seams and inclusions and will not be eliminated merely by selecting a stabilized alloy.

Figure 13.33 contains a summary of service

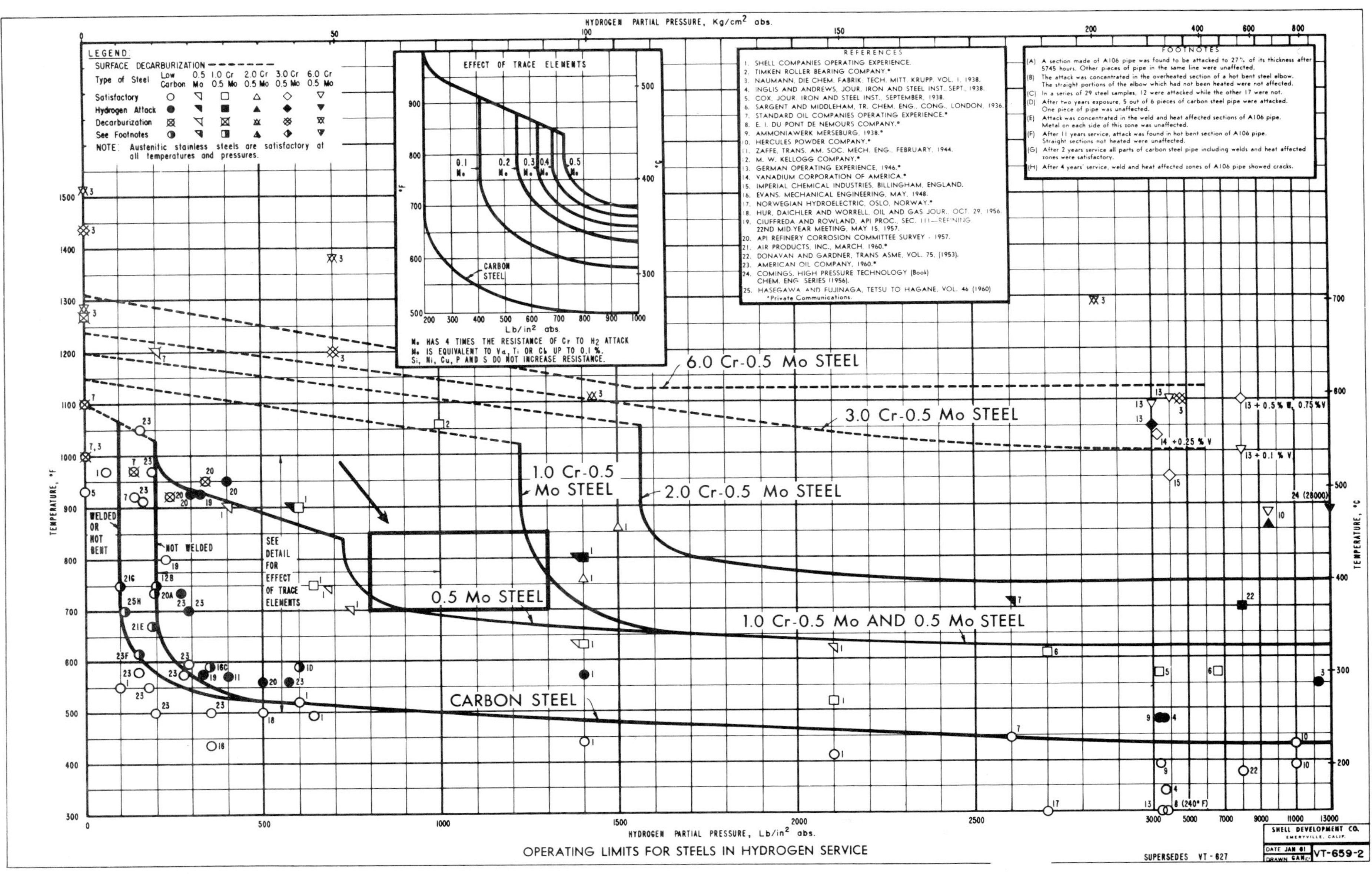

FIGURE 13.33 — Temperature-pressure combinations where several types of steels react with hydrogen to cause hydrogen attack and decarburization.

experiences relating the behavior of a vast number of alloys to hydrogen pressure and temperature.[17] The curves attempt to delineate the maximum conditions under which a given alloy may safely be used for prolonged service.

It is important to repeat that atomic hydrogen goes *right through* common metals. In a *clad* vessel (*e.g.*, copper, nickel, or stainless-clad steel), the *backup* steel must be alloyed to resist the process-side conditions of temperature and partial pressure of hydrogen. Otherwise, the steel substrate will be attacked by atomic hydrogen which passes through the corrosion-resistant cladding.

Halide Attack

From the industrial viewpoint, the principal halides of interest are chlorine and fluorine, together with HCl and HF vapors. Bromine and iodine are of secondary importance, but relatively few data exist even for the gases of major significance.

Despite the small amount of actual data, it is of interest to consider halide attack because it introduces a new mode of scale breakdown. In principle, corrosion by halogens proceeds in essentially the same fashion as in oxygen, sulfur, or nitrogen attack. The gas, upon contacting the surface, will oxidize the metal atoms there and form a halide layer. The scale, while it remains in place, will offer some measure of protection against continuing attack. The major difference is that metal halide compounds all have very high vapor pressures, so that when some critical temperature is exceeded, the scale will volatilize and leave the metal surface exposed for further rapid reaction.

Although a few oxides, such as MoO_3, are volatile at relatively low temperatures, and others, such as Cr_2O_3, may reach significant vapor pressures at very high temperatures, volatilization of oxides can, in most cases, be ignored, whereas it cannot for halides. Table 13.9 gives some values for the maximum recommended temperatures for some typical engineering alloys in chlorine and hydrogen chloride, and Table 13.10 gives corrosion data for fluorine. These results are based on very short-term tests, as noted, and are believed to be conservative. Similar behavior would be expected with the other halides.

Chromium additions can make the metal less resistant to halide attack.

TABLE 13.9 — Corrosion by Dry Chlorine and Hydrogen Chloride at High Temperatures[(1)]

Alloy	°F at Which Given Rate is Exceeded in Short-Term Lab Tests, in./mo 0.0025	0.005	0.01	0.05	0.1	Upper Limit for Continuous Service, °F
Corrosion in Dry Chlorine						
Nickel	950	1000	1100	1200	1250	1000
Inconel	950	1000	1050	1200	1250	1000
Hastelloy B	950	1000	1100	1200	—	1000
Hastelloy C	900	1000	1050	1200	—	950
Hastelloy A	900	1000	1100	1200	1250	900
Magnesium	850	900	950	1000	1050	850
Chromel A	800	900	1000	1150	—	850
Monel	750	850	900	1000	1000	800
18-8-Mo	600	650	750	850	900	650
18-8	550	600	650	750	850	600
Platinum	900	950	1000	1050	1050	500
Hastelloy D	400	450	550	—	—	400
Deoxidized Copper	350	450	500	500	550[(2)]	400
Carbon Steel	250	350	400	450	450[(3)]	400
Cast Iron	200	250	350	450	450[(3)]	350
1100 Aluminum	250	300	300	350	350[(4)]	250
Gold	250	300	350	400	400	—
Silver	100	150	250	450	500	—
Corrosion in Dry Hydrogen Chloride						
Platinum	2300	—	—	—	—	2200
Gold	1800	—	—	—	—	1600
Nickel	850	950	1050	1250	1300	950
Inconel	800	900	1000	1250	1350	900
Hastelloy B	700	800	900	1200	1300	850
SHA-1	700	800	900	1150	1250	850
Hastelloy C	700	800	900	1150	1250	850
Hastelloy A	650	750	900	1150	1200	800
Hastelloy D	550	700	850	1200	1150	800
18-8-Mo	700	700	900	1100	1200	800
25-12-Cb	650	750	850	1050	1150	800
18-8	650	750	850	1100	1200	750
Carbon Steel	500	600	750	1050	1150	500
Ni-Resist (Type 1)	500	600	700	1000	1100	500
Monel	450	500	650	900	1050	450
Silver	450	550	650	850	—	450
Cast Iron	400	500	600	850	950	400
Durichlor	350	400	500	650	750	350
Duriron	350	400	500	650	700	350
Copper	200	300	400	600	700	200

(1)Values based on short-time lab tests under controlled conditions. Data are indicative only and not applicable to estimates for equipment service life (°F − 32) × 5/9 = °C.
(2)Ignites about 600 F (316 C).
(3)Ignites at 232 to 260 C (450 to 500 F).
(4)Ignites at 204 to 232 C (400 to 450 F).
[SOURCE: Brown, M. H., DeLong, W. B., and Ault, J. P., Ind. Eng. Chem., Vol. 39 (1947).]

Molten Phases

In all environments considered up to this point, we have ignored the possibility of corrosion product fusion and have limited the discussion to gas-metal reactions with solid or, in the case of halides, gaseous reaction products. However, there are many situations where a clear distinction cannot be made. The corrosion product itself may melt, or contaminants may lower the melting point into the service temperature range. Moreover, important industrial processes are based upon the ability to contain static or flowing systems of fused salts or liquid metals for long times without harmful corrosion. It will be necessary, then, to consider the effect of the presence of molten phases on gas-metal reactions, and subsequently to outline the important factors relating to molten salt and liquid metal corrosion.

The numbers of contaminants which may interfere with the normal development of a protective scale during high-temperature service is virtually impossible to enumerate. The combustion of leaded gasoline contributes to deposits of lead oxide and other compounds on exhaust valves which can shorten their lives.

Much effort has also been expended on the problems resulting from the combustion of residual oils, because the ash contains, among other things, vanadium pentoxide, a compound which in the pure form melts at 690 C (1274 F), and whose melting temperature may be lowered further by the presence of other compounds.

TABLE 13.10 — Corrosion of Commercial Metals and Alloys by Fluorine

(Rate: in/mo)

Material	°C °F	200 392	250 482	300 572	350 662	400 752	450 842	500 932	600 1112	650 1201	700 1292
Nickel		—	—	—	—	0.0007	0.0019	0.0051	0.029	0.016	0.034
Monel[1]		—	—	—	—	0.0005	0.0015	0.002	0.060	0.080	0.15
Inconel[1]		—	—	—	—	0.038	0.096	0.062	0.17	0.13	0.51
Deoxidized Copper		—	—	—	—	0.16	—	0.12	0.99	—	2.9
Aluminum (1100)		—	—	—	—	Nil	Nil	0.013	0.018	—	—
Magnesium (Dow Metal G)		Nil	Nil	Nil	—	—	—	—	—	—	—
Type 430		0.0007	Nil	0.255	0.078	0.078	—	—	—	—	—
Type 347		Nil	0.145	0.213	0.517	0.795	—	—	—	—	—
Type 304-Cb		Nil	Nil	0.075	0.462	0.665	—	—	—	—	—
Type 310		Nil	Nil	0.031	0.354	0.561	—	—	—	—	—
Armco Iron		Nil	0.002	0.009	0.008	0.024	0.3	11.6	—	—	—
Sheet Steel (0.007% Si)		Nil	0.016	0.004	0.002	0.012	0.30	7.4	—	—	—
SAE 1020 (0.22% Si)		0.038	0.48	0.66	0.147	0.54	1.52	—	—	—	—
SAE 1030 (Trace Si)		0.002	0.008	0.009	Nil	0.015	0.54	19.8	—	—	—
SAE 1030 (0.18% Si)		—	—	0.75	—	—	—	—	—	—	—
SAE 1015 (0.07% Si)		—	—	0.83	—	—	—	—	—	—	—
Music Wire (0.13% Si)		—	—	0.40	—	—	—	—	—	—	—

[1]The International Nickel Co., Inc.

[SOURCE: Myers, W. R. and DeLong, W. B., Chem. Eng. Prog., Vol. 44 (1948).]

The small amount of sea salt which may enter a reactor or gas turbine with the combustion air can also initiate a sequence of events which culminate in catastrophically high corrosion rates.

Welding slag, improperly or incompletely removed, also will cause difficulty and frequently premature failure. However, no attempt will be made to discuss even a representative sampling of such contaminants. We shall merely try to show how they affect corrosion behavior.

The harmful contaminants or reaction products have one thing in common: whether they themselves are solid, liquid, or gaseous, they react with the metal or corrosion products to form a fused phase. This phase will destroy the integrity of the protective scale by fluxing or dissolving it and leave the underlying metal available for further corrosion. Frequently, the fused phase will attract and concentrate potent corrodents present in the atmosphere at exactly the place where they do the maximum amount of damage, the place where the scale is disrupted.

Ideally, the contaminating phase would be removed and the problem eliminated. However, this is frequently not possible. Molybdenum is an important and necessary component of many heat-resistant alloys. The cost of removing vanadium from residual oil, or removing salt from sea air is excessive.

In some cases, careful alloy selection can improve performance. However, other necessary characteristics often limit this approach. Alloys of 50 to 60% chromium and 30 to 40% nickel, respectively, have excellent resistance to residual oil ash. However, their rather poor high-temperature mechanical properties limit their range of applications, although not their usefulness when they are mechanically suitable.

Protective coatings are used frequently, and, at least where small, readily inspected parts such as turbine blades are involved, appear to be effective. In some cases, additives which prevent the fused phase from forming offer a solution. For example, the addition of magnesium sulfate (common epsom salts) to the fuel has permitted gas turbines to operate on residual oils without excessively high corrosion rates. A general solution to the problem does not exist, and the presence of any low melting contaminants should be viewed with suspicion.

Hydroxides

When molten hydroxides are present, similar results may be expected. In cases where the air atmosphere is excluded, however, reaction with oxygen can still occur since oxygen is a component of hydroxides. In the case of a metal in sodium hydroxide:

$$\text{Metal} + 2NaOH \rightarrow \text{Metal Oxide} + Na_2O + H_2 \quad (13.14)$$

Sometimes the reaction can also form metallic sodium, as:

$$\text{Metal} + 2NaOH \rightarrow \text{Metal Oxide} + 2Na + H_2 \qquad (13.15)$$

depending on the metal involved and other conditions of temperature and pressure.

Nickel is expected to behave in a noble fashion, but an iron-nickel-chromium alloy should have a mixed reaction—and it does.

Cyanides

Molten sodium cyanide is used commercially both to carburize and nitride steels. High-temperature containers for molten cyanide are generally made of high-alloy materials to resist this reaction. Tests at 843 C (1550 F) in a mixture of sodium carbonate, cyanide, and chloride have shown that carburization resistance increases with higher chromium, nickel, or higher silicon content. However, although Cr up to 30% was desirable, and specimens with nickel contents beyond 50% were among the best, silicon content seemed to be optimum between 1.5 and 2.0%. Lesser amounts of Si caused higher carbon pickup, and larger amounts resulted in oxidation phenomena, as was described under chloride attack when Cr was oxidized.

An interesting observation was that nickel-free ferritic steels of more than 20% Cr showed no carburization, presumably because of the combined effects of high chromium and the low carbon solubility in the ferritic (bcc) structure.

Corrosion in Liquid Metals

Solid metals in contact with liquid metals are frequently encountered in high heat-flux heat exchangers, mercury boilers, hot dip tinning or galvanizing operations, die casting processes, etc.

A certain degree of potential interaction can be predicted from binary phase diagrams, and these interactions may be considered chemical in nature. That is, intermetallic phase formation would be indicated and the extent of mutual solubility could be noted. However, as was pointed out in the previous chapter, very low solubility is no indication of negligible interaction. The mercury-iron phase diagram shows practically no solubility of one for the other (less than one part per million), yet damaging interaction has occurred because of *solution deposition,* or mass transfer caused by thermal gradients.

Corrosion in liquid metals can result from direct interaction of the metals, or, if alloys are involved, selective dissolution of certain constituents by the liquid metal. Furthermore, if impurities are present in the solid or liquid metal, even in trace amounts, the situation can become more complex.

Tests for compatibility have included static and dynamic tests. Usually numerous liquid metal-solid metal combinations are screened in the isothermal static test, and fewer of the more promising materials are tested in the expensive and time-consuming thermal gradient dynamic tests. Some of the numerous varieties of dynamic tests are as follows.

1. Forced circulation loops using electromagnetic or mechanical pumps.
2. Thermal convection loops of various designs.
3. See-saw-type tests.
4. Gyrated toroid loops.
5. Standing thermal convection pipe (Figure 13.34).

In Figure 13.34, molten metal is introduced up to the 13-inch level and placed vertically in a furnace to a depth of 8 inches. The upper portion is cooler, but is well above the melting point of the molten metal contained. A temperature difference of about 93 C (200 F) was measured in this particular test.

Let us assume that the interaction between the solid metal, "A" (container), and liquid metal, "B," was shown by a static constant and uniform temperature test to have negligible or very slight mutual solubility, and therefore to show practically no corrosion.

Here, in the thermal gradient test, there would obviously be a very slight difference in solubility of metal A in metal B at 1500 F vs 1300 F, as is illustrated in Figure 13.35. Therefore, metal A would dissolve slightly at 1500 F up to the saturation point. Because of thermal convection, this hot, saturated liquid then rises to the top, where at 1300 F it becomes supersaturated and precipitates a metal on the cooler wall in the form of dendrite-like A crystals. Then the cooled liquid metal sinks to the bottom to dissolve more solid metal. A close-up of the cold wall crystals is shown in Figure 13.36.

Continuation of this simple process causes a problem at both hot and cold ends of a closed cycle system. At the hot end, the metal wall gradually becomes thinner and could perforate. At the cold end, crystals could build up to the point where liquid metal could no longer flow through.

There is no known molten metal or salt in which mass transfer caused by temperature differences (of even a few degrees) can be totally ignored. Extraneous impurities in molten metals often may have a profound effect on mass transfer, in either direction, *i.e.,* they may inhibit or speed it up. Needless to say, the problem is unique, disturbing, interesting, and challenging.

Corrosion in Vacuum

A perfect vacuum is a space devoid of gaseous particles. Yet one of the best manmade vacuums (about 10^{-9} mm pressure) still contains about 800,000,000 molecules per cubic inch (or about fifty million molecules per cubic centimeter).

The more noble metals would not react with the remaining "few" molecules of O_2, CO_2, or H_2O

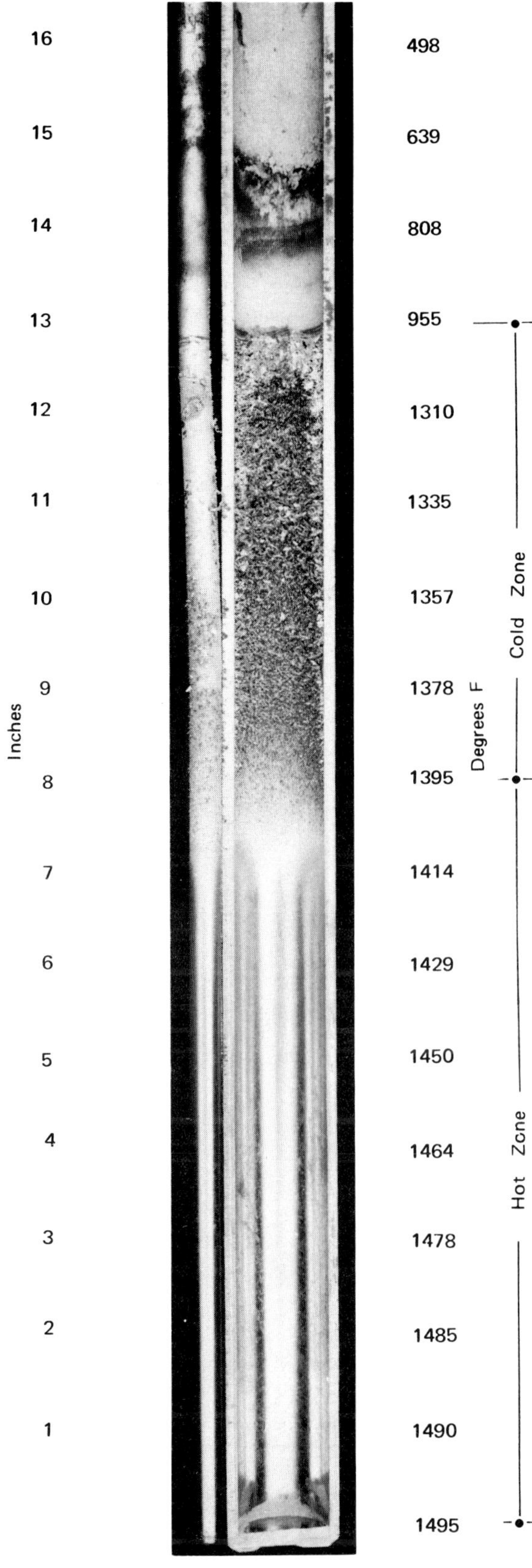

FIGURE 13.34 — A standing thermal convection liquid metal corrosion test. The thermal gradient between the 1 and 13-in levels causes natural convection and mass transfer from the hot zone (polished surface identifies dissolution zone from which metal dissolved) to the cold zone where solid metal crystals build up. There was no overall weight loss or gain during this 8-day test. X0.5.

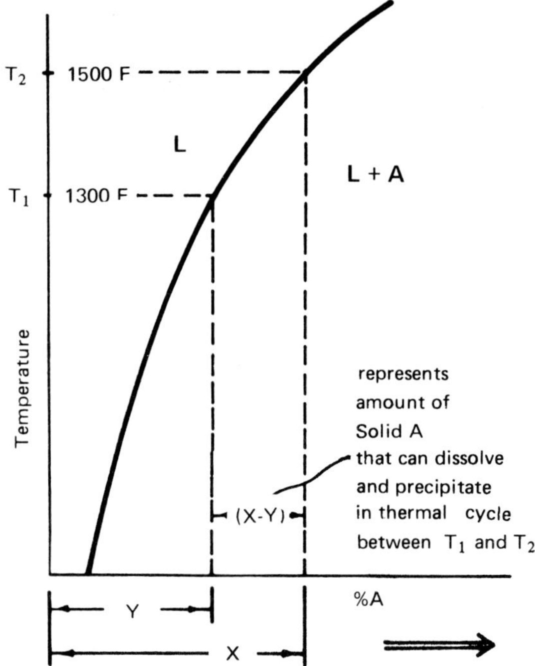

FIGURE 13.35 — Solubility of solid metal A in liquid metal B as a function of temperature. (This line is known in phase diagram studies as a liquidus.) An amount, X, is soluble at T_2 (1500 F), and Y is the amount of A soluble at T_1 (1300 F). Thus, X-Y may be precipitated during a thermal cycle between T_2 and T_1.

FIGURE 13.36 — Cold zone crystals which have precipitated during a solution-deposition reaction caused by thermal cycling. X24.

to form any oxide, but the more active metals (like Cr, Ti, Al, Be, or Mg) would react, probably superficially, to form a thin oxide layer, or in some cases, a nitride layer, depending on the metal and nature of residual gases. This is why purging a "vacuum" with hydrogen or argon can be very helpful in making the residual molecules more inert; that is, allowing fewer of the molecules to contact the metal surface.

Perhaps a more important form of high-temperature deterioration in vacuum has to do with evaporation or volatilization.

All substances have a vapor pressure, which may be extremely low even at elevated temperature. However, changes can occur over a period of time that should be given at least passing attention.

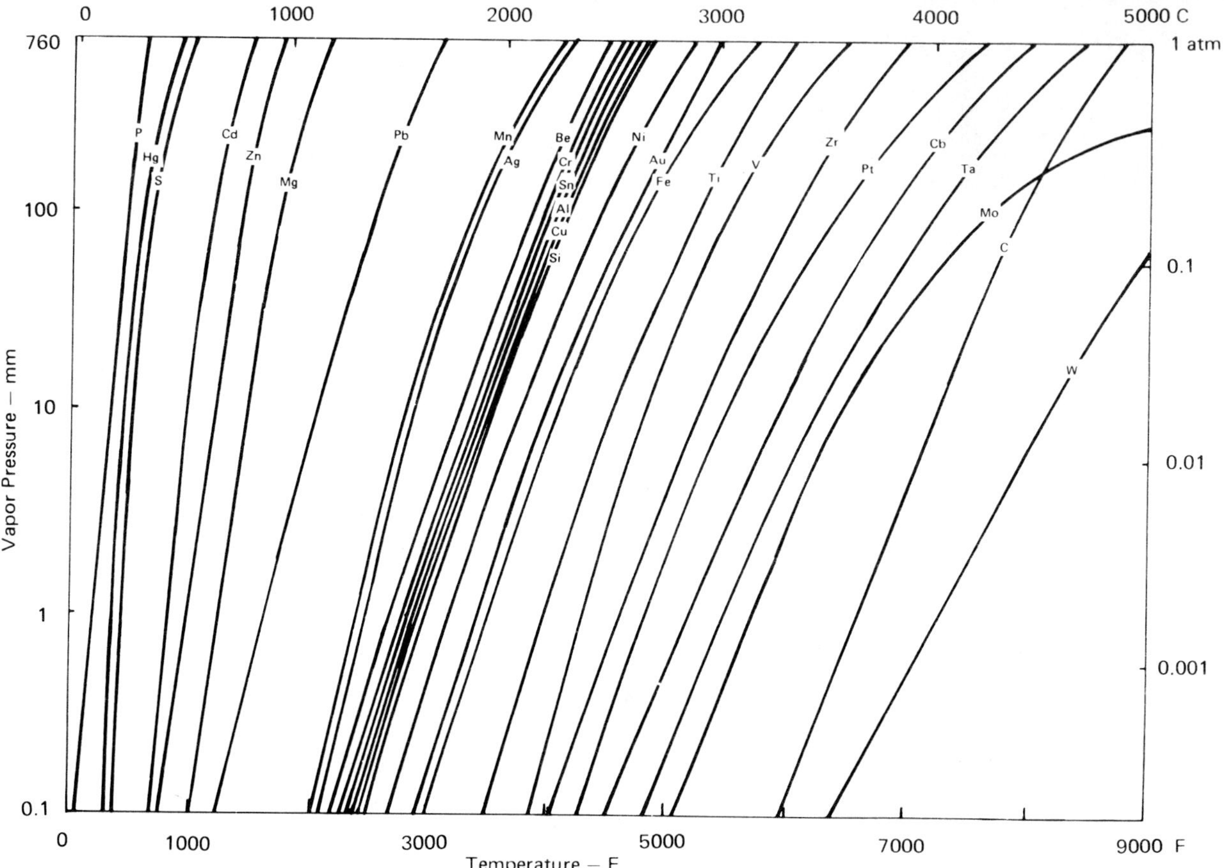

FIGURE 13.37 — Vapor pressures of some important elements.

Brass, an alloy of copper and zinc, can undergo a form of corrosion (even in perfect vacuum) which can be termed *dezincification.* Zinc has a relatively high vapor pressure, whereas that of copper is much lower. Thus, zinc would vaporize preferentially, leaving a copper-rich surface. While in aqueous corrosion, zinc often goes into solution as ions, in a vacuum it goes into space as atoms. The effect on the remaining metal is much the same.

Figure 13.37 illustrates the different volatilization tendencies of a number of common metals. Phosphorus and sulfur are important from a metallurgical viewpoint and are also included. Advantage is taken of the very high vapor pressures of Zn and Mg in purifying these metals on a commercial scale.

Note the relatively high vapor pressure of chromium when compared with iron or nickel.

If an iron-nickel-chromium alloy were heat treated in vacuum, there is a likelihood of superficial Cr loss due simply to evaporation. If a trace of O_2 or H_2O were present, a chromium oxide could form to tarnish the surface. This has often been observed to occur in some commercial vacuum heat treating operations.

Some metals, like Mo, that have very low vapor pressures, could nevertheless suffer high vaporization losses in poor vacuum because some oxygen or H_2S usually is present and the oxide is highly volatile.

Mechanical Behavior

A serious problem facing the engineer-designer in the realm of high-temperature applications is that relating to the mechanical properties of alloys at high temperatures. Loss of strength and ductility, poor creep characteristics at elevated temperatures, and phase transformations under service conditions are said to have caused more failures than corrosion. Therefore, a brief consideration of these properties and the effects of corrosion on mechanical properties is essential to the corrosion engineer who is concerned with metals at temperatures above 540 C (1000 F).

Mechanical Property Testing at Elevated Temperatures

Tensile tests at elevated temperatures have been made to determine the extent of decrease in strength as influenced by an increase in temperature. These tests are generally made in the conventional manner, *i.e.,* by increasing the load at a given rate until fracture occurs, which generally requires a few minutes. These short-term tensile tests may be of fundamental interest, but have no direct correlation with actual service conditions where a constant low load is usually encountered and creep of the alloy occurs.

Creep is the gradual stretch of an alloy under low load at high temperature. Creep strength is the present-day basis to determine the relative merits of various heat-resistant alloys for high-temperature service. Creep strength tests may be likened to low constant-load tensile strength tests at elevated temperatures. In this type of test, nothing seems to happen under low loads, but accurate measurements show that something is going on all the time. It may produce fracture in time periods up to 10,000 hours and provide engineering design data for high-temperature applications.

This type of test closely duplicates actual service conditions. To illustrate this method of testing, the following example is given.

A common high-temperature alloy is one containing about 35 Ni, 15 Cr, 0.5 C, and 1% each of Mn and Si. If creep tested at 870 C (1600 F), it would slowly deform under load. With greater tensile loads, of course, the stretching rate would increase. Figure 13.38 illustrates four different loads ranging from 83 MPa (12,000 psi) down to 35 MPa (5000 psi). The creep (stretching) rate at the high load was 5000 times greater than the low load. Fracture occurred in 12.5 hours at 12,000 psi, but did not occur in 50 days with the lower load, at which time the test was suspended. Note that ductility decreases as rupture time increases; this is typical in creep-rupture testing.

The four creep curves shown in Figure 13.38 would be similar for increasing temperature, as well as for increasing load; *i.e.*, for a given load, the higher the temperature, the higher the creep rate.

A summary of numerous creep tests similar to those just described is generally plotted in the manner illustrated in Figure 13.39. Creep rate and rupture time can be roughly estimated for any load by extending these curves. As an approximation, one may note that for each 90 C (200 F) rise in temperature, the creep strength decreases about 50%, a fact that could be useful for on-the-spot "guesstimations." However, this estimation procedure should be used carefully and limited to the range of about 650 to 1100 C (1200 to 2000 F). Higher temperatures (for Fe-Ni-Cr alloys) may approach the melting range and conditions of severe corrosion; lower temperatures can involve precipitation or transformation phenomena of a strengthening nature.

As composition varies within normal chemical specification allowances, differences in strength properties should be expected. Furthermore, if other alloying elements are introduced, changes can be expected. The nature of change will depend on base composition, temperature, and environment.

Metallurgical changes such as grain growth, precipitation phenomena, transformations, and even corrosion will affect mechanical properties. As a rule, the above changes in structure have a weakening effect. Any notches (stress raisers) created are particularly detrimental, *e.g.*, as in intergranular attack.

C	Ni	Cr	Si	Mn
0.40	38.52	14.10	1.27	0.85

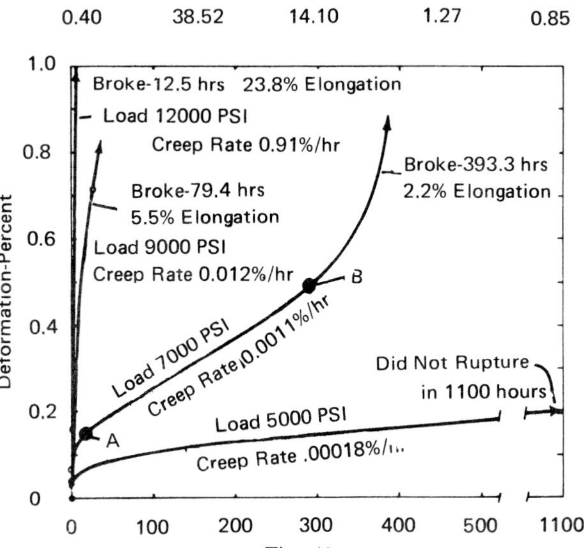

FIGURE 13.38 — Time-deformation curves for HT-type cast alloy for four loads at 870 C (1600 F). Three important mechanical properties are assessable from creep-rupture tests: creep rate, rupture time, and percent elongation.

Effect of Composition Variations on Strength

At the risk of being only about 75% correct, the following generalizations regarding alloy additions on strength are presented.

1. *Carbon* is added to strengthen steel, and it does actually increase the creep strength below 480 C (900 F). Above 600 C (1100 F), the effect diminishes, and the influence of carbon on steel is definitely weak above this temperature.

2. *Titanium* is beneficial below 815 C (1500 F), and is of questionable advantage above this temperature.

3. *Molybdenum* definitely strengthens low- and high-alloy steels.

4. *Chromium* is a feeble strengthener, and is used mainly to impart corrosion resistance.

5. *Manganese* has been found to increase strength when added in amounts up to 1.3%. Additional amounts of manganese cause a very slight decrease in strength.

6. *Silicon* has a very weak effect and is added chiefly for corrosion resistance in amounts up to 1.5%. It is a ferrite promoter, being about three times as effective as chromium. It has been known to facilitate the transformation of alpha to sigma in high-chromium alloys at about 1300 F (740 C). Sigma is a very brittle phase.

7. *Aluminum* generally weakens metals at high temperatures and is added (other than in deoxidation amounts) primarily for corrosion resistance or to control grain size.

8. *Nickel* has no perceptible effect when added

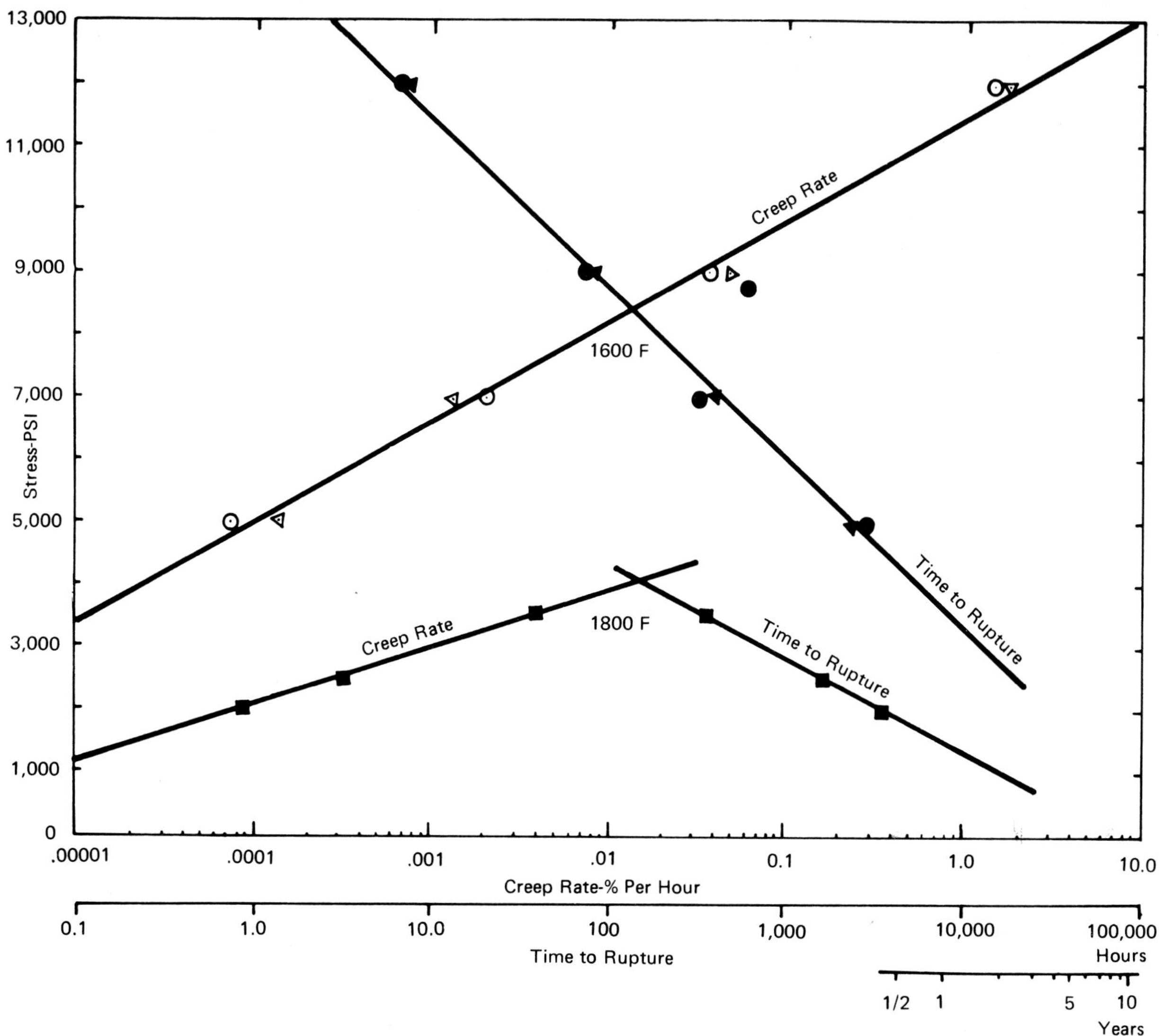

FIGURE 13.39 — Average minimum rate and rupture time at 870 C (1600 F) and 980 C (1800 F) for HT-type cast alloys.

in amounts up to a few percent. Greater amounts increase the strength somewhat.

9. *Cobalt* is a feeble strengthener for iron, but in the case of certain highly alloyed materials, it appears to be a definite strengthener.

10. *Tungsten* is positively a strengthener, but only half as effective as molybdenum; nevertheless, it is quite good. Tungsten is a ferrite promoter, increases the hot-hardness of heat-resistant steels, improves creep strength, and tends to reduce the ductility.

11. *Vanadium,* when added in amounts of 0.5 to 1.5%, is quite effective in strengthening iron, especially in the 1200 F range. It is a renowned carbide and nitride former.

12. *Zirconium* (Zr), *Titanium* (Ti), and *Columbium* (Cb) are elements which form stable carbides and nitrides and are considered to be ferrite promoters. The addition of small amounts of these elements are known to render low-carbon, 12% Cr steels nonhardenable.

13. *Phosphorus,* surprisingly enough, in amounts of 0.2 and 0.4% does strengthen iron, but greater amounts embrittle it.

Effects of Microstructure

Ferritic heat-resistant alloys are very stable when subjected to aging or cyclic heating through a wide range of temperatures. These and the martensitic alloys all have relatively low creep strengths at high temperatures. These alloys are strongly magnetic at room temperatures.

Austenitic alloys are characterized by their superior creep strengths and retained ductility over long periods of time. Fully austenitic alloys retain their stability under any thermal treatment. A slight microstructural change is effected, however, by the agglomeration of secondary carbides from the supersaturated austenite, resulting in a slight decrease in strength and ductility. Austenitic alloys can be nonmagnetic, weakly magnetic, or strongly magnetic, depending upon their composition and prior mechanical treatment.

Alloys containing more than one phase are rather susceptible to wide variations in properties and characteristics, depending upon the thermal history of the alloy. Possible high-temperature phases (other than carbide) are ferrite, nitride, austenite, and sigma.

The partially austenitic alloys are considerably inferior to the fully austenitic alloy in strength characteristics. Therefore, care should be exercised in the selection of alloy types for high-temperature service, or in balancing the composition of marginal alloys in the multiphase vicinity (such as 18Cr-8Ni, 25Cr-12Ni, and 25Cr-20Ni) by proper control of carbon, nitrogen, nickel, manganese, and other austenite-promoting alloying elements. These alloys and others were listed in Chapter 4.

Summary

It has been demonstrated that corrosion can be serious at high temperatures because of general metal loss (metal thickness reduction) or concentrated local attack (pitting), which is expected to have a harmful effect on the mechanical properties.

Intergranular attack has been shown to be severe in some cases, thereby reducing the effective metallic cross-sectional area. This results in a weak structural member. Furthermore, intergranular attack tends to act as a stress-raiser in stressed members. The detrimental effects of notches and cracks on mechanical properties are well recognized, and the danger of deep corrosion penetration via the cracks cannot be overemphasized.

Another deleterious effect resulting from corrosion is that associated with decarburization. This phenomenon was illustrated in Figure 13.23. Since it has been pointed out that the strength of an alloy is affected by the carbon content, it is obvious that decarburization of the surface results in a weakening of that part of the metal. The alloy would then contain a core of average strength surrounded by a case or ''skin'' of low strength. This reduces the effective working limits of the alloy. In a similar fashion, an alloy might become carburized if the conditions were favorable. This could result in an embrittled surface.

The need for a careful study of the properties of a heat-resistant alloy and its behavior in the anticipated environment is of considerable importance in the selection of a suitable alloy for a particular service application. New alloys and nonmetallic materials which are continually being made available to industry are making it possible to make better selections and to establish safe working limits within which the material can be expected to give satisfactory performance over a reasonable length of time.

References

1. Moller, G. E. and Warren, C. W., Survey of Tube Experience in Ethylene and Olefin Pyrolysis Furnaces, Materials Performance, Vol. 20, No. 10 (1981).
2. Richardson, F. D. and Jeffes, J. H. C., J. Iron & Steel Inst: Oxide chart, Vol. 160, p. 261 (1948); Sulfide chart, Vol. 171, p. 167 (1952).
3. Gulbransen, E. A. and Andrew, K. F., J. Electrochem. Soc., Vol. 104, Nos. 6 and 7 (1957); Gulbransen, E. A., Copan, T. P., and Andrew, K. F., J. Electrochem. Soc., Vol. 108, No. 2 (1961).
4. Gulbransen, E. A. and Andrew, K. F., J. Electrochem. Soc., Vol. 106, No. 11 (1959).
5. Evans, U. R., Trans. Electrochem. Soc., Vol. 91 (1947).
6. Lustman, B., Trans. AIME, Vol. 188 (1950).
7. Brasunas, A. deS., Gow, J. T., and Harder, O. E., Proceeding of ASTM, Vol. 46, p. 129 (1946).
8. Ruder, W. E., Calorizing Metals, Trans. Electrochem. Soc., Vol. 27, p. 253 (1915).
9. Brasunas, A. deS., Unpublished research.
10. Brasunas, A. deS. and Grant, N. J., Trans. ASM, Vol. 44, p. 1117 (1948).
11. Schafmeister, P. and Naumann, F. K., Die Chem. Fab., Vol. 8 (1953).
12. Bradbury, E. J., Hancock, P., and Lewis, H., Metallurgia, January (1963).
13. Backensto, E. B., Prior, J. E., Sjoberg, J. W., and Manuel, R. W., Corrosion, Vol. 18, No. 7 (1962).
14. Hunter, M. A., Corrosion Handbook, Uhlig, H. H., ed., p. 694 (1948).
15. Moran, J. J., Mihalisin, J. R., and Skinner, E. N., Corrosion, Vol. 17, No. 4 (1961).
16. Moran, J. J. and Mihalisin, J. R., SAE, Paper 500A, March, 1962.
17. Nelson, G. A. and Effinger, R. T., Weld. Res. Supp., January (1955).

Bibliography

Campbell, Ivon E. and Edwin M. Sherwood. High Temperature Materials and Technology. J. Wiley & Sons, New York, N.Y., 1967.

Clark, Frances H. Metals at High Temperatures. Reinhold, New York, N.Y., 1950.

Draley, J. E. and J. R. Weeks. Corrosion by Liquid Metals. Plenum Press, New York, N.Y., 1970.

Hauffe, K. Oxidation of Metals. Plenum Press, New York, N.Y., 1965.

Kofstad, P. High Temperature Oxidation of Metals. J. Wiley & Sons, New York, N.Y., 1966.

Kubaschewski, O. and B. E. Hopkins. Oxidation of Metals and Alloys. Academic Press, New York, N.Y., 1962.

Smithells, C. J. Met. Ref. Book. Butterworth and Co., 1962.

Answer to Self-Check Quiz No. 1

Between $10^2/L$ and $10^3/L$
(about 300 to 1) H_2/H_2O ratio

Answer to Self-Check Quiz No. 2

0.1 ipy (100 mpy) or 2500 $\mu m/y$

APPENDIX A — F to C Temperature Conversion Chart

°F	°C	°F	°C	°F	°C
-40	-40	800	427	1832	1000
0	-18	900	482	1900	1038
32	0	932	500	2000	1093
100	38	1000	538	2100	1149
212	100	1200	649	2200	1204
300	149	1400	760	2300	1260
400	204	1500	816	2400	1316
500	260	1600	871	2500	1371
554	300	1652	900	3000	1649
600	316	1700	927	5000	2760
700	371	1800	982		

Chapter 14

Testing and Inspection

TESTING AND INSPECTION

The Value of Corrosion Testing

The corrosion of metals is governed by fundamental laws. As we understand these laws better, we become better able to predict the performance of a particular metal, how it reacts under a given set of conditions, and how its performance can be improved. At present, however, it is necessary to develop experimentally most of the information we need concerning the corrosion of metals.

Metals are used under countless and changing conditions, so that the unexpected is not uncommon. It is because of this limited predictability of metal performance that corrosion tests are so important.

Properly conducted, corrosion tests can mean the savings of millions of dollars. They are the means by which we can avoid using a material under unsuitable conditions or of using a more expensive material than is required. Corrosion tests also help in the development of new alloys that perform more inexpensively, efficiently, longer, or more safely than the alloys now in use. Also, quality control corrosion tests are a means of ensuring that the alloys we make and purchase have the capabilities expected of them.

Scope of this Chapter

Because of the large number of metals, nonmetallic materials, and applications, it is impossible to cover all phases of corrosion testing in this chapter. The intent, therefore, is to cover the fundamentals of commonly used materials testing and to develop guidelines for proper test methods. Because the content of this book covers basics, emphasis is given to those responsibilities frequently assigned to a laboratory technician.

Corrosion testing programs can be simple ones which are completed in a few minutes or hours, or they can be complex ones which require the combined work of a number of investigators over a period of years.

This chapter emphasizes laboratory-type corrosion testing of metals. However, field and in-plant testing is often required to solve process problems and to determine the condition of equipment.

Types of Corrosion Tests

Corrosion tests are divided into two broad categories: (1) tests made in the laboratory under controlled conditions, and (2) tests made in the field under natural or service conditions.

Laboratory Controlled Tests

Tests in the laboratory are made with pure chemicals under closely controlled conditions.[1] Thus, each test should be reproducible in repeat runs of fixed duration. This reproducibility is especially important when a new test method is used or when a new alloy or fabricated item is evaluated. If reproducible conditions can be achieved, then differences in data truly reflect a difference in the resistance to corrosion of the materials being tested, and the investigator knows if an improvement has been made.

Purpose

Laboratory tests give us an early indication of what will happen in actual practice. However, the time required for an "indication" is dependent upon the purpose and nature of the tests. Laboratory corrosion tests generally are conducted over a period of time ranging from a few minutes to as long as a year or more.

Tests of the Metal

Laboratory tests usually are made with small metal specimens of specified shape and size. Some typical test specimens are shown in Figure 14.1. Such specimens have the advantages of being relatively inexpensive, small, and reproducible. They are intended to evaluate the corrosion response characteristics of the metal. For example, they should show how "Steel A" responds differently from "Steel B," or how some other metal would perform under these same test conditions.

Even for the relatively simple types shown (disk, panels, and tensile specimens), a variety of sizes and shapes are employed. Certain specimens contain the full thickness of the product, such as the flat tensile specimen and the cast and sheet panels. Others, such as the round tensile specimens and the stepped panel, are machined from a particular region of thicker products.

In protective coating evaluations, panels generally have well-rounded edges to facilitate uniform coating formation. Sometimes coated or clad specimens are intentionally damaged, one form of coating damage being various scribe marks through the film. Other forms of damage might be impact

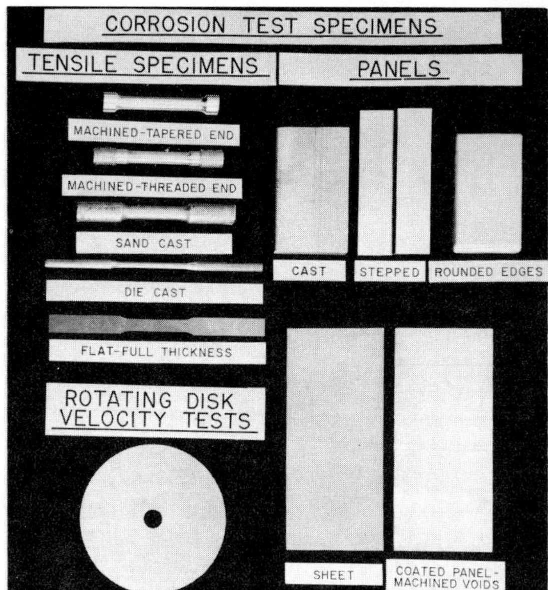

FIGURE 14.1 — Some of the many types of specimens used to evaluate the general corrosion characteristics of metals: left, tensile specimens; right, panels.*

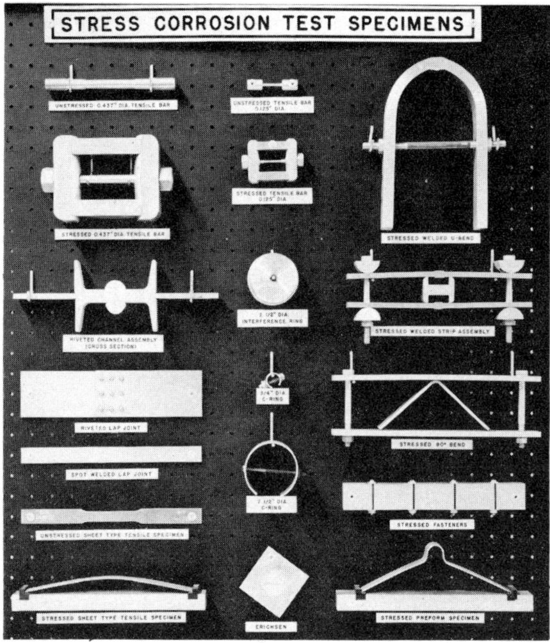

FIGURE 14.2 — Some of the types of specimens used to make stress corrosion tests.

blows to cause crazing, or countersunk holes.

Other types of specimens are used to evaluate the effect of applied stress on corrosion (Figure 14.2). These types of specimens, however, are designed for evaluation of the resistance of the alloy, without much regard to its end use. Certain types, such as the two tensile bars and the sheet tensile specimen, are exposed unstressed as well as stressed. This permits the determination of the effect of stress on reduction of strength by corrosion. Numerous other specimens are also used to evaluate specific products or methods of stressing.

Tests of Fabricated Items

Fabricated items also are evaluated in laboratory corrosion tests. Usually, this is done on a limited basis to establish the correlation between tests on cut or machined specimens and the fabricated item to see if results with small test specimens are giving valid answers. This may be necessary when the final product is so complex that the investigators cannot be sure that the test data obtained for each small component will give a true evaluation of the behavior of the final assembly.

Field Tests

Field tests, on the other hand, take us one step closer to end use, and normally range in duration from several months to many years. Some tests of materials have been conducted for more than thirty years (Figure 14.3). Field tests, for example, may consist of exposure to the corrosive conditions of:

1. Natural Environments
   a. Atmospheres (seacoast, industrial, rural)
   b. Waters (distilled, fresh, salt, flowing, stagnant)
   c. Soils (dry, moist, clay, sandy)

2. Particular Industrial Applications
   a. Architectural
   b. Processing plants
   c. Transportation vehicles
   d. Storage tanks

3. Particular Substances
   a. Concrete
   b. Fertilizers
   c. Petroleum fuels
   d. Flue gas
   e. Acids
   f. Bases
   g. Solvents

FIGURE 14.3 — Atmospheric corrosion test racks exposed at Kure Beach, NC. [SOURCE: International Nickel Company.]

*All figures from ALCOA unless otherwise stated.

FIGURE 14.4 — Heavily biofouled test racks removed from the seawater tidal zone at Wrightsville Beach, NC. [SOURCE: International Nickel Company.]

FIGURE 14.5 — Undervehicle corrosion testing with coupons. [SOURCE: Automotive Corrosion by Deicing Salts, R. Baboian, Ed., National Association of Corrosion Engineers, Houston, TX, p. 187, 1981.]

Examples of such tests are shown in Figures 14.4 through 14.6. For the most part, these tests use small cut or machined specimens, but assemblies of simulated components are also tested.

Field tests are often conducted for many of the same reasons that laboratory tests are made. New and promising materials must be evaluated, changes in the environment and the effect on materials of construction must be monitored, and a constant appraisal of the corrosion occurring in many environments is necessary. Such testing is obviously of great value when making plans for maintenance of equipment, and can improve the economics and safety of a system. Field tests conducted in operating equipment should provide a more accurate assessment of the existing corrosive conditions than any laboratory test that can be performed.

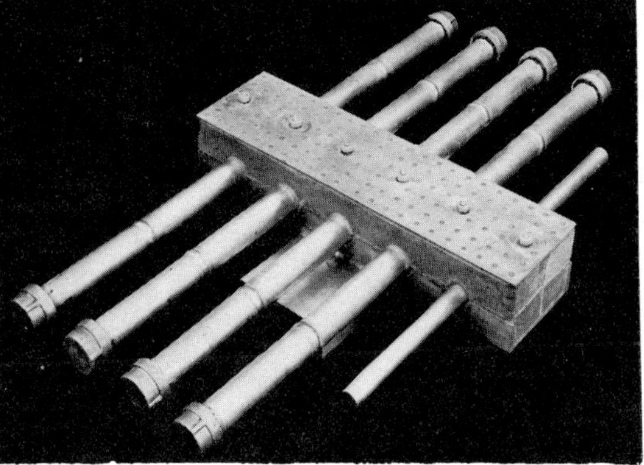

FIGURE 14.6 — How metal shapes are exposed to concrete. Performance of these shapes can be evaluated not only with respect to the enclosed environment where they are fully covered, but also at the margins where they protrude into the atmosphere.

Field tests can be conducted by three basic means: (1) exposure of specimen materials, (2) electrochemical measurements, or (3) examination of the equipment before and after use.

For the process industries, the first procedure has been standardized as well as possible.[2] Various means of introducing coupons of materials of construction into operating equipment have been devised.[2] For equipment operating under considerable pressure, special attachments are available from corrosion equipment suppliers.

The important aspect of the field testing of specimens is to ensure that the materials are fully and freely exposed in the environment. The exposure site must be analyzed to understand the factors contributing to the corrosion experiences (velocity, temperature, environment composition, etc.). A sufficient number of sites should be selected to ensure that variations in the exposures will be revealed. Sometimes the installation of a sidestream on operating equipment is an economical, convenient method for field testing. This may be particularly true when attempting to appraise the effect of inhibitor additions or other attempts to alter the environment.

The following paragraphs describe means of conducting the tests of the second and third category. Most of the tests utilize sophisticated electronic instruments and sensors. It is beyond the scope of this chapter to discuss the operating principles and procedures for these instruments.

Acoustic Emissions[3]

Active corrosion has been found to emit bursts of energy that can be detected by acoustic (sound) sensors.[3] This testing technique is used in the aircraft industry to detect active cracking (stress or fatigue) in complex structures such as wing assemblies. By using an array of sensors and multichannel monitoring equipment, the location of active cracks can be pinpointed.

Ultrasonics[4]

Ultrasonic techniques are used in quality control to measure and locate flaws in metal.[4] Ultrasonic techniques are also used to measure thickness of metals. Such thickness measurements can detect metal losses caused by corrosion and erosion. Ultrasonic thickness measurements can be made from the outside of vessels and pipelines during equipment operation.

Instantaneous Corrosion Rates

Linear polarization measurements can be used to determine instantaneous corrosion rates. This type of measurement utilizes test probes which can be inserted into electrically conductive process streams or can be used in containers of electrically conductive process fluids in the laboratory. The stream must be "clean" to prevent fouling of the electrode surfaces.

Linear polarization measurements are often sensitive enough to detect process changes and upsets. In cases where test probes are inserted in process equipment, the output from the electronic readout equipment can be used to trigger an alarm. This is one of the few positive uses for corrosion.

Resistance Probes

Corrosion can be detected by measuring the change in resistance of a wire as the wire corrodes and loses cross section. Systems of probes and electronic readout equipment are available which relate resistance change with time to indicate corrosion rate. The response of this equipment to changes in corrosion conditions is not as rapid as the response of the linear polarization-type equipment. Thus, the resistance method is used in process monitoring to activate alarms where less rapid changes require monitoring. This type of corrosion probe will operate in most any environment, even those environments that are not electrically conductive.

Electrochemical Potentials

Electrochemical potentials can be measured on any metal surface that is in contact with a conductive fluid (including soils and other moisture-containing materials). Potential measurements are used to locate areas that may be corroding. The extent of cathodic protection can also be evaluated by this method. In the field, potential measurements are used to locate metal surface areas that may require corrosion investigation actions.

Eddy Currents

A number of instruments that utilize magnetic fields are used to locate loss of metal or flaws in metal. These instruments are often called eddy current testers.[5]

Hydrogen Probes

In using hydrogen probes, a thin-wall steel cylinder is attached to a pressure gauge. When exposed to an environment where significant atomic hydrogen is generated on the exterior surface, but does not combine readily (*e.g.*, sulfur compounds present), the hydrogen penetrates the metal shell and increases the internal pressure in the cylinder. The pressure rise with time can be correlated with corrosion rate.

Others

Visual inspection of used equipment is always valuable as a source of information. Various aids for visual inspection are available, such as small, lighted magnifiers, pit gauges, portable hardness testers, penetrant dye processes to find cracks, portable metal analyzers, and other specialized equipment.

Final testing often takes the form of trial use of noncritical parts, pilot plant operation, or experimental models to determine performance under actual service conditions. Such testing involves actual components and is limited to alloys, finishes, and fabrication techniques which have been found promising in laboratory and field evaluations.

Test Objectives

Corrosion tests can also be subdivided according to objective. These objectives determine the types of environments an investigator will select. Some corrosion tests, as a function of their test objective, are given in Table 14.1.

Fundamental Investigation

In fundamental corrosion research, the investigator seeks to determine how or why a particular form of corrosion occurs. The investigator's aims are not necessarily tied to any particular product or use, and seldom is there a set time for comple-

TABLE 14.1 — Objectives of Tests

Type of Test	Objective, Timing
Fundamental Investigation	Long Range (many years)
Alloy or Product Development and Improvement	Short Range (up to a few years)
Material Selection	
Quality Control	Immediate (within a year)
Plant Maintenance	
Failure Analysis	

tion of the objective. Tests in this area are almost exclusively conducted in a laboratory using small specimens and highly specialized techniques suited to the investigation.

Alloy or Product Development and Improvement

The researcher in the field of alloy or product development or improvement seeks to make better for tomorrow the materials and products of today. He wants the bridge, automobile, or home of tomorrow to be more resistant to deterioration by corrosion than those now in use. His aims are tied directly to related applications with the intent of achieving improvement in the near future.

The reseacher will start with small specimens and laboratory tests, and, as the material shows promise, will proceed through all phases of testing.

Material Selection

The engineer who designs a bridge, automobile, or home must decide which of many available materials will be used. Often, data developed by the supplier's applications engineering staff must be relied upon to provide this information. If this is insufficient, comparative corrosion tests must be conducted among candidate materials for the specific application.

One area which must be considered is the compatibility of different alloys with contacting metallic or nonmetallic materials. Actual parts testing may be involved, but test durations usually are short-term and are geared toward answering questions about a particular application or environment. As such, the tests conducted must be directly correlated with the end use of the material.

Quality Control

After a choice has been made of the material for an application, the fabricator and purchaser then need to know if the quality of in-coming material is as specified. Quality control corrosion tests[(1)] ensure that the material meets specifications. Such tests must be sufficiently rapid to avoid undue delay of metal shipments. Corrosion tests of this type do not require any particular correlation to end use. Alone, they do not necessarily tell us that this alloy will make a better product. They simply ensure that the metal shipped has the same corrosion resistance as the metal previously tested.

Plant Maintenance

As indicated previously, corrosion tests are important in maintaining the reliability of buildings and equipment already in use. Periodic tests determine whether a material fulfills design requirements. Such tests can involve cut specimens, coated and uncoated, exposed under operating conditions. They might also involve evaluations of parts removed from the actual building or operating unit. Tests of this type also yield practical information for the selection of materials for future buildings.

Failure Analysis

Failure analysis represents the "post-mortem" of corrosion testing. Failures occur either because of some design or application error, or because even the best materials and methods available can be inadequate for the demands made of them. By examining failures, an investigator seeks to determine cause, and, hopefully, to develop a remedy. The types of tests that can be made on such parts are numerous and require experience in interpretation. Standard methods of corrosion evaluation, such as weight loss or loss in strength, may not be possible or may be only approximate.

Usual procedures involve visual and microscopic examination of the corrosion or fracture (in a failed part), plus chemical analyses of: (a) the metal, (b) corrosion products, and (c) any foreign matter. These trouble-shooting techniques are important because they yield firsthand information about the performance of a material under actual operating conditions. Excellent examples of such analyses are provided in a book available from the American Society for Metals.[6]

Procurement of Test Materials

The first step in corrosion testing is procurement of the material to be tested. In some cases, such as quality control tests or failure analyses, the type and lot of material to be tested are predetermined. In most other cases, there is considerable freedom of choice.

Commercial Alloys

If commercial alloys are to be tested, the best procedure is to obtain mill-fabricated material, representative of current production. If possible, a fabrication history should be obtained listing the major fabricating steps, together with a precise analysis of the metal composition. An example of a form to record fabrication history for an aluminum extrusion is shown in Table 14.2.

At the very least, the material should be certified as being within the composition limits specified for the alloy and meeting the strength or hardness specification. Microscopic examination also may be necessary to ensure that the material is in proper metallurgical condition. Such basic information may prevent misleading conclusions as a result of off-composition or improper fabrication.

---

(1)Such as the Huey Test (5 to 48 hrs. in boiling 65% nitric acid) and Strauss Test (72 hrs. in boiling 3% cupric sulfate, 10% sulfuric acid) for stainless steels, or the Intergranular Corrosion Test [6 hrs. in 1% hydrogen peroxide, 5% sodium chloride at 86 F (30 C)] for heat treatable aluminum alloys.

TABLE 14.2 — Fabrication History for Standard Evaluations of Aluminum Alloy Extrusions[(1)]

EXTRUDED SECTION:	**DATE:**
ALLOY:	**LOT NO:**

A. MELTING AND CASTING OPERATIONS

1. Charge Components:

2. Alloying Procedure:

3. Fluxing Procedure:

4. Casting Data:

Casting Method	Ingot Diameter	Pouring Temperature	Metal Head	Lowering Rate	Water Volume	Skimmer

5. Cast Analysis: (percent of alloying elements, remainder = aluminum)

Cast No.	Cu	Fe	Si	Mn	Mg	Zn	Ni	Cr	Ti	Be	Pb

6. Gas Determinations:

7. Etch Results:

B. PREHEATING AND EXTRUSION OPERATIONS

1. Preheat: Soaking Temperature Total Cycle Time	Time at Temperature
2. Reheat: Furnace Setting Exit Time	Cycle Time
3. Extrusion: Type of Die Maximum Effort Runout Length	Ingot Size Maximum Exit Speed End Scrap, Front Rear

C. FINISHING OPERATIONS

1. Solution Heat Treatment:	Furnace Type
Soak Temperature, Maximum	Minimum
Time at Temperature	Cycle Time
Type of Quench	Rate of Quench

2. Stretcher Straightening Operation:
No. of inches per foot: inches per feet. inches per feet.

3. Additional straightening operations (give sequence):

4. Precipitation Heat Treatment (aging) Furnace Type

Soak Temperature, Maximum	Minimum
Time at Temperature	Cycle Time

5. Etch Results:

D. MECHANICAL TENSILE PROPERTIES

Test Direction:

Test Location	Tensile Strength (psi)	Yield Strength (psi)	Elongation (% in 4D)
FRONT			
MIDDLE			
REAR			

[(1)]This table shows an example of one method used to report the fabrication history. A standard form is prepared and the pertinent data is then merely filled in. The data required will, of course, vary with the type of metal and product being made. The more important data, such as ingot size, composition, times and temperatures of heat treatments, and tensile properties are required for all metal systems and products.

Nonstandard Alloys

When nonstandard alloys or tempers are to be evaluated, it may be economically impractical to fabricate them using large production equipment. In such cases, small laboratory casts or heats are made, sometimes weighing only a few pounds. While this is a valid procedure for initial tests, before a promising alloy candidate may "go commercial," it should be evaluated several times using specimens from quantities of material large enough to be representative of production variations. Evaluation of several production lots is necessary because, all too often, promising results from a single lot of material are not reproduced on subsequent fabrication runs.

Metal Form

Another preliminary decision is the form of the metal to be tested. Metals are available in two basic forms: *cast* and *wrought* (worked). These two forms usually cannot be interchanged in testing. The various methods of casting (die, permanent mold, sand mold, etc.) and of working (drawing, extruding, forging, rolling, etc.) will affect grain structure and homogeneity which, in turn, affect corrosion resistance. The metal procured should be, as nearly as possible, the type that will be used in the final product.

If it is anticipated that the final use of a metal will be in a cast form, then cast specimens should be evaluated. The same applies for wrought metal specimens. Data obtained from cast and wrought metal forms cannot be interchanged reliably.

Specimen Preparation

Size and Shape

Having selected and procured the test material, the next step is to prepare test specimens. The size and shape of the specimen will be limited by the material being tested and the test environment. The type of specimen used must permit ready evaluation. If several characteristics are to be evaluated, more than one type of specimen may be required. In addition, the following considerations can be important.

Visual Examination

Visual examinations of the specimens are made in all corrosion tests. When the appearance of the end product is important, such as for decorative trim or architectural applications, a fairly large surface area should be used [about 232 cm^2 (36 in^2) or larger] to permit a reliable appraisal in the event that corrosion is not uniform. Specimens of only a few square inches can give misleading results.

Depth of Attack

Specimens used to evaluate corrosion by measurement of the depth of attack should be thick enough so that they are not perforated by the corrosion. Other than this thickness consideration, no particular size or shape is required, but the size or area of the specimen will determine the amount of corrodent needed. Also, the specimen should be large enough, or a sufficient number of small specimens should be exposed, to include all important metallurgical or manufacturing variables. Specimens for this purpose usually are kept as small as possible for convenience, and because there are no strict requirements, data often can be obtained from specimens designed for some other test.

Weight Loss or Gain

Weight change measurement also does not require any definite size or shape, but a large area-to-volume ratio (A/V) is used for better sensitivity. Usually, a flat square or rectangle is used to simplify measurement of surface area. The specimen is kept relatively small to permit simple and accurate weight measurements. It is accepted practice to use a standarized size and shape for all specimens in a given test series so that the same surface area is always exposed and there can be the same degree of accuracy in measurement and calculations. A standard specimen frequently used in American Society for Testing and Materials *weathering* evaluations is a 10 × 23-cm (4 × 9-in.) panel of 0.16-cm (1/16-in.) thick rolled sheet.

Loss in Tensile Properties

If loss in tensile properties is to be evaluated, the best procedure is to use one of the standard ASTM tensile specimens, such as those shown in Figure 14.1.

Round tensile specimens are machined prior to exposure. When sheet or thin plate [up to about 1.3 cm (1/2 in.) in thickness] is to be tested, a full thickness, flat tensile specimen of the type shown in Figure 14.1 can be used. In this case, the tensile specimens either can be fully machined before exposure or they can be machined from a corroded panel after completion of the corrosion test.

Full thickness specimens machined before exposure will give an earlier indication of the effect of corrosion; but the indicated degree of loss in tensile properties, particularly elongation, usually is unrealistically high because of notch effects produced by corrosion of the edges of the specimen (Figure 14.7). When edge corrosion is not critical in actual applications (*e.g.*, roofing sheet), a more realistic appraisal of the loss in strength and elongation can be obtained by machining full thickness specimens from a corroded test panel, and in this way avoid edge corrosion effects in the appraisal.

Stress Corrosion Tests

The selection of a specimen for stress corrosion tests is complex, but is primarily dependent upon the ability to apply stress uniformly in a specific metallurgical direction. (A more detailed discussion of these factors is given in Chapter 6.)

FIGURE 14.7 — How machining sequence affects corrosion data. Cross sections of sheet tensile specimens: (A) machined from a large panel after corrosion, % reduction in tensile properties, strength 10, elongation 30; (B) Machined from sheet before corrosion exposure, % reduction in tensile properties, strength 16, elongation 70.

Corrosivity of the Test Environment

A second factor to be considered in conducting a corrosion test of a specimen is the corrosivity of the test environment. Shorter exposure times and thicker specimens are required when the test conditions are highly corrosive. If too small a specimen is used and it is completely dissolved, perforated, or fragments from it are lost, only limited information can be obtained. Conversely, it is also a mistake to use a specimen that is so thick that the overall effect of corrosion is too slight to be accurately determined.

Suitability to Other Test Purposes

Finally, if factors other than the metal and the corrosive environment are to be evaluated, the investigator must be sure that the specimen is suited to the particular test purpose. For example, if protective paints or metallic coatings are to be evaluated, the corners of the specimen should be well rounded before coating. Films are always thin or otherwise disturbed at sharp corners, and this presents unrealistically weak points for corrosion to begin. If galvanic corrosion is to be evaluated, the cathode/anode size ratios and geometry must be known and controlled.

Specimen Identification

An investigator *never relies on memory* to ensure identification. This is especially important when many specimens are tested, when tests are conducted over several months, or when several persons process the same specimens.

A suitable identification, usually a specimen number, should be given to each specimen. The first part of the number should be the identification given to the particular lot of metal from which the specimen was obtained. The second part of the number, sometimes called a dash number, is the identification of the particular specimen. When several specimens are taken from the same lot of metal, the various dash numbers may be assigned in numerical order. Frequently, a code letter is used to indicate a particular type of specimen or environment. In such a case, the various specimens of the same type or for the same purpose are then numbered consecutively. The important feature is that the investigator must be sure each specimen can be fully identified.

Whenever possible, the identification number should be marked on the specimen. The mark on the specimen should be such that it can *withstand the entire corrosion test.* The most common and best method is an impression stencil, except for cases where it is known that the performance of the specimen would be affected by the stresses or surface damage introduced. (Stress corrosion cracking can often be detected around these numbers.)

When only minor corrosion is expected, scribing with an engraving tool, indelible ink, or paint markers may be adequate. If extensive corrosion is expected, the specimen number should be protected with a suitable inert masking material such as wax, tape, or paint. Another method is to attach a nonmetallic marker that is not affected by the corrosive medium. Strips of plastic sheet, scribed with an engraving tool and fastened with plastic cord are excellent for this purpose.

If test conditions are such that paint or other "foreign" substances might affect the results, they should not be used. In such cases, specimen identity is retained by marking the rack or container and recording the location of the specimen. The common "lead" pencil should never be used for marking specimens, as the clay-graphite mixture is electrochemically reactive and can initiate sites of corrosion on most metals.

Replicate Specimens

A certain amount of *data scatter* is inevitable in any testing procedure. The amount of unavoidable scatter depends upon: (1) the uniformity of the material tested, (2) the accuracy in preparing the test specimens, and (3) the stability of test conditions.

All these factors influence the accuracy of corrosion testing. Therefore, good testing procedure will provide some method of cross-comparison or double-checking to eliminate the possibility of an incorrect conclusion based on a single nontypical result.

The use of a single specimen may be a valid procedure, provided more than one measurement can be made on the specimen. For example, loss in strength can be determined at several places in a single, corroded, 10 × 23-cm (4 × 9 in.) panel. Using a single specimen may be a poor procedure, however, if it provides only one value, such as loss in strength on a premachined tensile specimen, or failure of a single stress corrosion specimen. This becomes especially important if it is known that the test procedure can vary. In such tests, a cross-check is provided by the use of multiple specimens.

Generally, three or more specimens are used to establish a trend. Duplicate specimens are not suitable because they can result in two widely different values which do not clarify the results.

The number of specimens used in a test must be a compromise among: (1) the accuracy needed (statistical evaluation), (2) the cost of a test, (3) the known reproducibility of the test, and (4) the number of variables investigated.

Machining of Specimens

Many products, such as sheet, certain structural shapes, tube, and pipe are used in service just as they are received from the supplier. Consequently, if their surface condition is fairly consistent, it may be possible to corrosion-test these products after simple degreasing. When the surface is inconsistent because of dirt, scale, or oxide, steps must be taken to reach a more uniform metal surface. Generally, an acid pickle or a caustic etch is used, depending on the metal.

Other products such as plate, rod, bar, or forgings are generally machined into the final configuration before they are used. Such products are best tested with a machined surface, employing good shop procedure to minimize roughness variations.

Close dimensional tolerances are required for specimens used in tests to evaluate characteristics such as loss in strength or resistance to stress corrosion cracking. Accordingly, such specimens should be made by experienced machinists. A few of the more important considerations during machining are:

1. Properly sharpened tools and appropriate feed rates to avoid heating or mechanically damaging the specimen.
2. Noncorrosive cutting lubricants (especially important if water-soluble oils are used—in certain cases, dry machining may be necessary).
3. Dissimilar metals such as steel, brass, and aluminum should be finished on separate pieces of equipment to prevent contamination of one metal with particles or chips of another metal. This is especially important when using grinding equipment.

A sawed edge may be used if the specimen is evaluated merely for depth attack on the surface. Edges should be filed, machined, or ground smooth if edge attack is to be determined or if weight loss will be used in the final determination. Sheared or blanked edges must be machined, ground, or filed to remove the layer of severely cold-worked metal resulting from this operation. The amount of metal removed should be at least equal to the thickness of the piece. Drilled or reamed holes can be left as is, but punched holes should be treated as sheared edges, unless the purpose of the investigation is to evaluate the effect of cold work.

The usual procedure in preparing specimens is to keep the *area of cut edges small* in comparison to the *surface area* because this is generally the case in actual parts. In any case, the edge-to-surface area ratio should be known and kept uniform from test to test.

Simulated Machining of Specimens

If machine shop facilities are not available, only the simpler type specimens such as coupons or panels can be prepared. These can be evaluated by depth of attack or weight loss. The usual method of simulating a machined surface is to polish the metal on a suitable abrasive cloth or paper, or to scrub it with a fine abrasive such as wet pumice. Such specimens are most often cut from sheet or thin plate.

Other methods of producing specimens of very hard metals are electrical discharge (EDM) or electrochemical machining (ECM) techniques. In these somewhat complicated methods, special precautions are required.

Degreasing and Final Measurements

After machining, if weight loss or loss in strength after corrosion is to be determined, the specimens should be degreased just prior to initial measurements of weight and dimensions.

The easiest method of removing oils and greases (except silicone oils) is to use a solvent vapor degreaser. If such equipment is not available, the specimens should be wiped with a fast-drying, low-residue solvent such as alcohol, acetone, or methyl ethyl ketone (MEK). Solvents which are highly flammable (gasoline), toxic (carbon tetrachloride), or which leave a film residue (kerosene) should not be used.

The specimens should be weighed and measured as carefully as possible because subsequent data can be no more accurate than the original data. When the accuracy needed is dependent upon the expected weight loss, measurements should be carried to enough significant figures to keep the possible error in calculations less than 1%. In some cases, particularly for large or heavy specimens, measurement to the first decimal place is sufficient. However, measurements generally are made to as many decimal places as is practicable, *e.g.*, to the nearest 0.01 mm (0.001 in.) and 0.0001 gram for small specimens measured with a micrometer and an analytical balance, respectively.

To remove contamination caused by handling, the specimen should be degreased again just prior to exposure to the corrosive medium. After this final degreasing, specimens should be given a "white glove" treatment and handled only at edges in order to prevent fingerprinting or other contamination which can cause accelerated corrosion. If, for some reason, the specimens cannot be exposed in a reasonably short time, they should either be stored in a desiccator or coated with a protective oil and stored in a dry, dust-free place. If the latter method is used, the specimen will, of course, have to be degreased again just prior to exposure.

Controls

A vital consideration in the corrosion testing of metals is the use of suitable *controls,* both for the metal and for the test environment. This aspect of corrosion testing is perhaps the greatest single source for error—even by experienced investigators.

Control of the Metal

Control data for the uncorroded metal consists of measurement of those properties which can be affected by corrosion so that they may be used to assess the effects of corrosion. Typical metal properties used as controls are strength, electrical conductivity or resistance, and reflectivity. Specimens used to obtain control data should be from the same vicinity in the metal stock and of the same type as those used for corrosion tests. Also, the same method of measurement and degree of accuracy should be used for both sets of specimens. This ensures that any differences in data before and after exposure represent only the effects of corrosion.

When the corrosion tests are of a short duration, or when the metal is such that the characteristic being evaluated is not changed during long-term storage, the original control data may be obtained just once prior to any corrosion tests and used for reference when needed. Sometimes, certain characteristics will change in time, even in the absence of corrosion. In such cases, a set of control specimens must be provided for each period of corrosion testing. This set is then tested at the same time as the corroded specimens. Some metals, for example, gradually change in strength over long periods of time at *room* temperature, and the strength of most metals is affected by time at *elevated* temperatures. A true measure of the effect of corrosion, in such cases, can be obtained only by tensile-testing corroded and uncorroded specimens at the same time (Figure 14.8).

Bar graphs in Figure 14.8 show the corroded strengths of the two alloys after one and four years of exposure to a corrosive atmosphere. After one year exposure, both alloys show the same strength and a similar reduction from the strength of uncorroded metal. After four years' exposure, the cor-

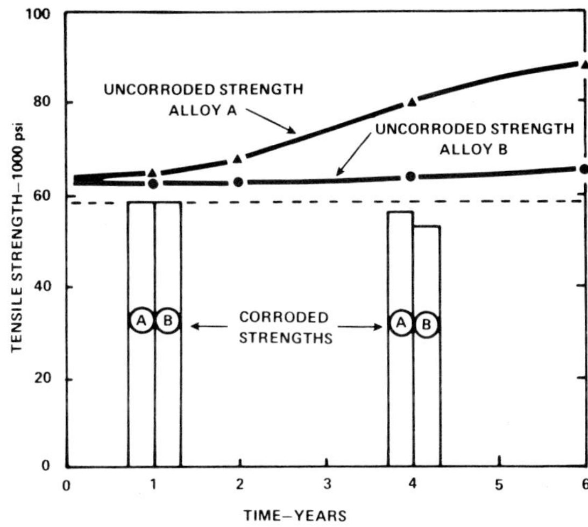

FIGURE 14.8 — Shows importance of exposing specimens of the same age when determining strength reduction in alloys whose strength changes with time. Alloy A has been appreciably strengthened by natural aging, while Alloy B, not naturally aged, is weaker.

roded specimens of Alloy A are somewhat stronger than those of Alloy B, but their reduction from the strength of uncorroded metal of the same age is considerably greater. If the corroded strength of Alloy A had been compared with the strength of new metal, an erroneous conclusion would be made as to the effect of corrosion on this alloy.

Specimens to be used for periodic controls should be protected from corrosion and stored at the same temperature at which the corrosion tests are conducted. In long-term field tests, where temperature will fluctuate and cannot be controlled (such as tests in natural atmospheres), it may be desirable to store the controls on-site in dry, air-tight containers. This ensures that both the corrosion specimens and the control specimens are exposed to the same time-temperature conditions.

In tests at elevated temperatures, special precautions may be necessary to ensure that the control specimens themselves do not undergo high-temperature corrosion. (Special aspects of high-temperature corrosion are given in Chapter 13.)

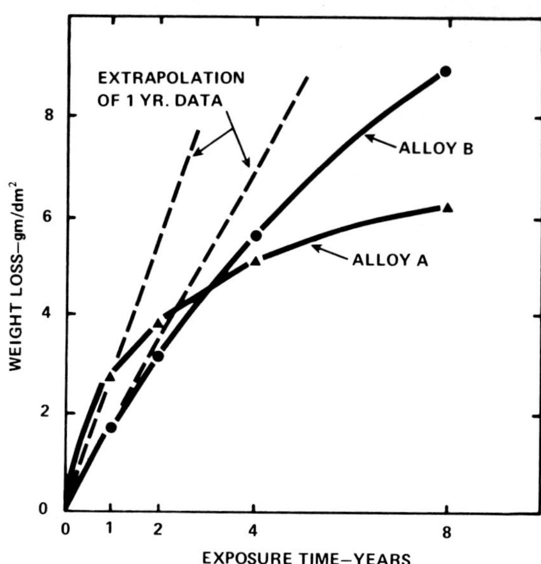

FIGURE 14.9 — Shows why data from several exposure periods are essential. Better resistance of Alloy A was not evident until fourth year of testing. Extrapolation rates based on one year data are in error for both alloys.

Control of the Test Environment

The other required control is a method of determining whether the corrosive medium maintains the intended degree of corrosivity. When an alloy is repeatedly tested by the same method, as in quality control tests, the investigator may be sufficiently experienced to know whether the visual appearance and resultant data are of the order expected. Often this is not so, especially with new alloys or products, when corrosion performance is unpredictable.

The usual procedure is to expose a well-established, previously tested alloy as a control to determine whether the results on it are consistent with past experience. If they are, the new alloy is considered to have received a valid exposure. However, if the effects of corrosion on the well-known alloy are not typical, the investigator is alerted to the possibility of an error in test environment and the need for a rerun.

Another control method is to expose specimens for several periods of increasing duration. At least three periods of exposure usually are used (Figure 14.9). Changes in the corrosion rates of metals can be caused either by a change in the condition of the metal itself or by a change in the corrosive environment. A properly designed corrosion test should enable the investigator to determine which of these two factors was operative, or whether both had an effect. Comparison of the data on the different groups of specimens then permits evaluation of whether the corrosion rate of the metal was constant (linear rate) or varied with time (either an accelerating or a self-limiting rate).

Chemical, electrical, or other physical checks on the test medium also help to establish its variablility.

Specimen Arrangement for Corrosion Tests

The final point to consider, before discussing the corrosion tests themselves, is the manner in which specimens are arranged or framed for exposure.

In laboratory beaker tests, usually one specimen per container is desirable. If more than one specimen of the same lot of metal is used, multiple specimens may be run in the same container (Figure 14.10), provided the volume of corrosive medium is sufficient to maintain its original properties throughout the exposure.

For plant and field tests, some sort of framing sequence is needed, depending upon the type of specimen and test method. An example of specimens racked for exposure is given in Figure 14.11. The important considerations are as follows.

1. All of the specimens must be isolated from one another and from direct contact with the racks. Inert, nonabsorptive racks or spacers of plastic or porcelain usually are used for this purpose.

2. The specimens should be arranged so that corrosion products from one specimen do not contaminate other specimens.

3. The corrosive environment should be equally accessible to all specimens.

4. In liquid immersion tests, specimens of different metals should not be exposed in the same container because one metal can affect the corrosion of another, even though they are not in contact. In fact, even in the same metal system, it may be desirable to separate one alloy type from another. For example, aluminum alloys containing copper are often separated from copper-free aluminum alloys.

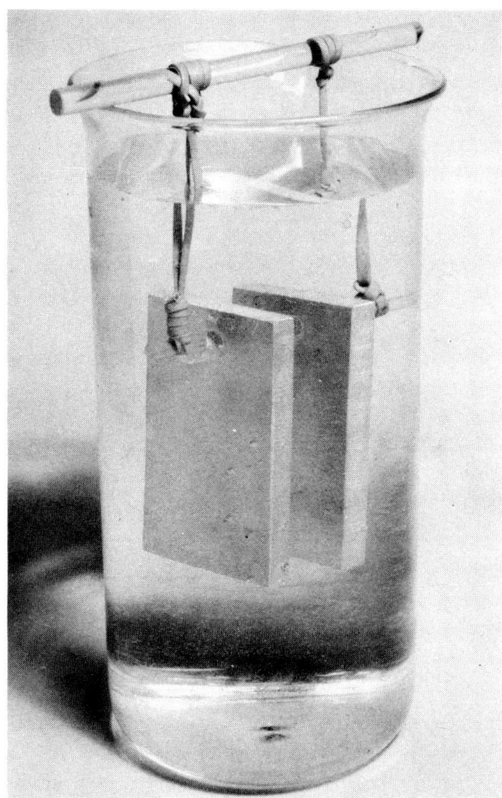

FIGURE 14.10 — Beaker test exposing two specimens. Note specimens are electrically insulated from each other and that supporting rod is suspended from a nonconductor.

FIGURE 14.11 — Typical specimen exposure rack. Note specimens are electrically separated from each other and from rack. Specimens with bolted-on straps may register galvanic, edge, or crevice effects and the influence of materials such as gaskets. Sequence of specimens should be carefully recorded.

In cases where metal containers are used to hold the test media, special precautions are required to prevent metal-to-metal contact between the container and the specimens.

Accelerated Test Environments

Four common types of accelerated atmospheric corrosion tests are exposure to: (1) dispersed fog spray, (2) high humidity, (3) simulated atmosphere, and (4) immersion in a liquid.

Dispersed Fog Spray Tests

Several types of dispersed fog spray tests are called for in material procurement specifications. All are operated on the same general principle. In some tests, the specimens are exposed to a continuous spray; in others, the spray is operated intermittently. Other principal variations are in the composition of the liquid used to make the spray and the test temperature. Common solutions are:

1. Pure water;
2. 3.5, 5, and 20% solutions of sodium chloride (NaCl); and
3. Sodium chloride solutions acidified to a specified pH, or with metal ions added to the solution which will plate out on certain areas of the metal and accelerate localized corrosion.

Some tests are run at room temperature, but most operate at 35 to 50 C (95 to 120 F).

A principal use of spray tests is to obtain a rapid evaluation of the protection afforded by coatings such as paint, anodic films, and surface conversion coatings. A common practice is to damage the coating intentionally by a controlled score mark (Figure 14.12) before exposure. This shortens the test by eliminating the time required to cause a natural breakdown. It also evaluates the resistance of the coating to failure by disbonding. If several coatings are compared, it ensures that all coatings have the same degree of damage, and that all damaged areas are exposed for the same length of time.

The two considerations to remember in fog spray testing are as follows.

1. The spray should never impinge directly upon the specimens. Instead, the spray is directed against a baffle prior to dispersion throughout the test chamber.

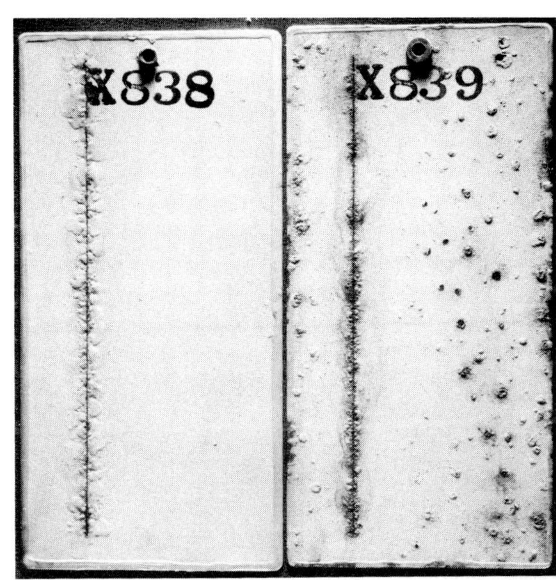

FIGURE 14.12 — Painted mild steel panels after exposure to salt spray. Scoring on both panels permitted discriminating more positively between the better coatiang on X 838, and the poorer coating on X 839.

2. Because the spray is a finely dispersed mist or fog, it will settle upon and wet only the upper surfaces of specimens. For this reason, specimens in spray tests should be exposed at an angle. Angles of inclination specified in certain ASTM test methods are 15, 30, and 45 degrees off vertical.

Humidity Tests

Another frequently used corrosive medium is high relative humidity at various temperatures. Tests of this sort are used to evaluate staining characteristics or temporary protective coatings. Another use is to test composite construction materials consisting of metals joined to or contacting absorbent nonmetallic substances such as paper, wood, or insulation, which can absorb moisture and hold it (poultice conditions) against the metal surface.

Commercial test equipment is available that can be set for any combination of percent relative humidity and temperature. When only a few small specimens are to be exposed, an inexpensive method is to store them above water in a sealed glass container (Figure 14.13). The vessel is then stored in a controlled temperature location. When lower humidities are desired, they can be obtained by using a mixture of water and sulfuric acid (Table 14.3) rather than pure water. Another method is to use salt solutions which produce specific relative humidities (Table 14.4).

Sometimes the best test procedure for high-humidity conditions is to cool metal specimens below the temperature of the test chamber before

FIGURE 14.13 — Specimens exposed to high humidity in sealed container.

TABLE 14.3 — Constant Humidity with Sulfuric Acid Solution[1]

Desired Percent Relative Humidity	Specific Gravity of Required Acid-Water Solution	cc of Acid to be Added[2]
100	1.00	none
95	1.09	69
90	1.14	135
85	1.17	190
80	1.20	229
70	1.25	309
60	1.29	379
50	1.34	469
40	1.37	586
30	1.44	766
20	1.49	945
10	1.59	1422

[1]Listed above are concentrations of sulfuric acid-water solutions that provide different percent relative humidity at 20 C (68 F).
[2]To a liter of water to give a solution of the required specific gravity.
[SOURCE: Handbook of Chemistry and Physics, 42nd Edition, The Chemical Rubber Publishing Co., Cleveland, Ohio.]

TABLE 14.4 — Constant Humidity with Salt Solutions[1]

Percent Relative Humidity	Solid Phase
100	None
95	$Na_2SO_3 \cdot 7H_2O$
90	$ZnSO_4 \cdot 7H_2O$
86	$KHSO_4$
80	NH_4Cl
75	$NaClO_3$
66	$NaNO_2$
58	$NaBr \cdot 2H_2O$
52	$NaHSO_4 \cdot H_2O$
42	$Zn(NO_3)_2 \cdot 6H_2O$
32	$CaCl_2 \cdot 6H_2O$
20	$KC_2H_3O_2$
10	$ZnCl_2 \cdot 1/2H_2O$

[1]The above table shows the percent humidity at 20 C (68 F) within a closed space above a super-saturated aqueous solution of the given solid phase. Solutions of these types can be used to provide a constant relative humidity in tests of the type illustrated in Figure 14.13.
[SOURCE: Handbook of Chemistry and Physics, 42nd Edition, The Chemical Rubber Publishing Co., Cleveland, Ohio.]

they are exposed. When this is done, the moisture in the air will condense on the specimen and wet the surface. In any case, a standard preconditioning procedure should always be used on the metal to be exposed so that reproducible condensation conditions are acheived. In cases where *no condensation* is desired, the specimens can be warmed above the temperature of the test chamber.

Simulated Atmosphere Chambers

Of considerable interest in recent years is the laboratory simulation and acceleration of atmospheric pollution effects used to evaluate degradation of materials. These tests involve injection of a variety of gases, vapors, and particulate matter into a controlled pressure chamber with controlled condensation on metal specimens. Sometimes, ultraviolet light is used to produce photochemical changes in the gases or vapors, such as are known to occur under present-day smog conditions.

The space program also requires development of equipment to duplicate the special conditions of gas composition, light, temperature, and pressure that are encountered in space flight. Other atmospheres, such as flue gases or chemical vapors, must be simulated—often at elevated temperatures.

Immersion Tests

Probably the most common corrosion test method is immersion in a liquid. Obvious differences in test procedures are the solutions used, agitation rates, and temperature.

Another variable is the method of immersion, as in the following examples.

1. Total immersion (specimen completely immersed in the solution).

2. Alternate immersion (specimen immersed in the solution for a period, then removed from the solution, allowed to dry, and then recycled (Figure 14.14). Other alternate immersion machines are in use which either raise and lower the specimens vertically in and out of the solution, or pump the solution in and out of a tank in which the specimens are placed. Equipment of this type can be regulated to provide any combination of immersion and drying times.

3. Partial immersion (lower portion of the specimen immersed in the solution and the upper half in air—or vapor, in the case of a closed system). This method actually evaluates three conditions of exposure: the immersion zone, the liquid-air interface line, and the vapor or air zone. Examples of

FIGURE 14.14 — This large alternate immersion testing machine is timed to rotate the racks 60 degrees every 10 minutes, thus immersing specimens for 10 minutes every hour and exposing them to air for 50 minutes.

FIGURE 14.15 — Cross-sectioned "Guinea Pig" one-quart shipping containers used to simulate such things as barrels and storage containers. When 3/4 full, 1/4 is exposed to vapor phase. Variables include material, weld composition, exposure time, and temperature. Container A shows no appreciable attack by liquid or vapor; B, no appreciable liquid attack but vapor staining; C, no appreciable vapor attack, random pitting by liquid; and D, no appreciable liquid attack, pitting attack by liquid at liquid-vapor interface.

specimens exposed by partial immersion are shown in Figure 14.15.

The type of immersion test used depends to some extent on whether the end product will be completely, partially, or intermittently immersed in a solution. Frequently, none of these methods of testing conform exactly to service conditions, and the tests are intended merely to give a quick indication of the degree of susceptibility that can be expected in use. In such cases, the method used is the one which will give the most rapid, *yet reliable,* indication.

Although rapid testing is desirable, reliability is even more important. It has been said that "accelerated tests are of no use if they only serve to obtain a wrong answer faster." Reliability of a test method is obtained by correlation of the relative performance with service or with field test data (Figure 14.16).

Evaluation of Plastics

Plastic materials of construction, because they differ in structure from metals, cannot be evaluated successfully by most criteria applied to metals.

The materials degrade from the effect of absorbed species from a solution. These are absorbed at rates defined by Fick's laws of diffusion. A

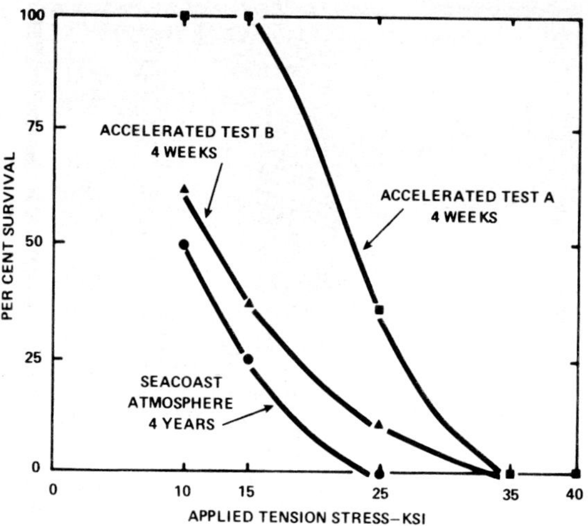

FIGURE 14.16 — A base of data from actual exposure is needed to validate accelerated test predictability. Test B results are in reasonably good agreement with the base, but Test A is misleading at low applied stress levels.

number of interactions between the resinous matrix and the absorbed compound are possible, such as breaking of the polymer chain, breaking of cross-links in the chains, oxidation of the polymer, reaction with terminal groups of the chains, or simply lubricating the area between chains to weaken the strength of the plastic.

One of the most important aspects of testing and using the plastic materials is to realize that the plastic can selectively remove one or more constituents from an environment. This means that a continuous supply of only parts per million of an aggressive chemical in a stream will produce the same result as exposing the plastic to 100% of that chemical.

Also understand that the plastic materials are susceptible to many of the types of failures described for the metals in Chapter 5, such as fatigue, stress cracking, notch sensitivity, etc.

Many plastics are composite materials (*i.e.*, combinations with fibers, such as glass), and the variables introduced, in addition to resistance of the resin matrix itself, may include the following.

1. Extent to which the fibers are "wet" by the plastic
2. Strength of the fibers
3. Mode of manufacture, *i.e.*, whether fibers are single filaments wound about a mandrel, or fabrics impregnated with the resins, etc.
4. Characteristics of the layers of the fabrication, *i.e.*, whether or not glass fibers are exposed to the corrosive
5. Internal or external strengtheners or stiffeners.
6. Percentage of resin to fiber, and others.

Types of Tests Applied to Plastics

Among the many types of tests applied to plastics are the following.

1. Visual (changes in appearance versus unexposed samples)
2. Tensile strength
3. Flexural strength
4. Flexural cycling
5. Impact resistance
6. Notch sensitivity
7. Heat stability
8. Abrasion resistance
9. Static buildup
10. Coefficient of expansion
11. Chemical stability
12. Weight changes
13. Hardness
14. Dimensional change
15. Volume change
16. Flame propagation rate
17. Biological effects
18. Creep rates
19. Stress cracking resistance

Obviously, not all of these tests are necessary in evaluating plastic materials for every use.

For many purposes, the change of hardness of a plastic during an exposure can be as good an indicator of the resistance of the plastic to the environment as any other mechanical test. However, working with the plastic materials at ambient temperatures is similar to working with metals at high temperatures. Thus, creep rates at given stress levels while exposed to the environment is the best of all corrosion tests to be used if the cost of the test can possibly be justified.

Whatever the test procedure employed, a test period of at least 100 days should be used. Changes in the properties of a plastic do not change linearly with time. Thus, appraisal of the properties at various times during the test period is advisable if possible.

Testing of Other Nonmetals

Other nonmetals, such as glass, ceramics, concrete, wood, stone, graphite, or mica can be tested for resistance to corrosives by more or less the same means as can metals and plastics, with similar reservations. Glass, for example, which is used for piping and as a lining to handle corrosive solutions, can be tested with comparative ease with results that have a high degree of precision.

Ceramics, because of the wide range of applications to which they are put, ranging from high-temperature, high-velocity gas environments, to ambient temperature aqueous environments, may require tests selected to evaluate them precisely for their intended use. Concrete, also a mixture, can be tested by many of the same methods used for other materials. For example, reinforced concrete exposed to salt water with severe cycling of temperatures, as in dock piling in arctic ports, must be tested for its ability to protect the structural steel it contains.

Tests are available for and are being used to evaluate nearly all inorganic materials, quite often with highly reproducible results. Usually, however, they involve long exposure times and do not always provide answers which permit extrapolation of performance data to periods greatly in excess of test periods. The tester must select the available parameters which will give results useful for the specific application.

Various tests are available for such materials as glass, ceramics, wood, and stone, but true evaluation of their properties often can be determined only from long-term exposure in the environments of interest (refer to ASTM Standards).

Corrosion Test Equipment

Commercial Test Equipment

When accelerated corrosion test procedures of a specific type are found to be highly reproducible and provide reliable information in less time than field testing, a demand for standardized equipment is developed. If enough metal producers, testing laboratories, and users of metal products are interested in the standardized equipment—especially if it is described in a performance specification—a manufacturer of corrosion test equipment sells a test unit. Some of the commonly used ones are as follows.

1. Salt spray cabinets (Figure 14.17)
2. Humidity cabinets (Figure 14.18)
3. Smog or gas chambers (Figure 14.19)
4. Weather-O-Meters (Figure 14.20)
5. Fade-O-Meters
6. Incubation ovens (for bacteriological studies).

In addition, some commercial organizations will build special equipment or modify a standard unit to accomodate a specific need. Some are programmed to produce an automatic sequence of varying conditions.

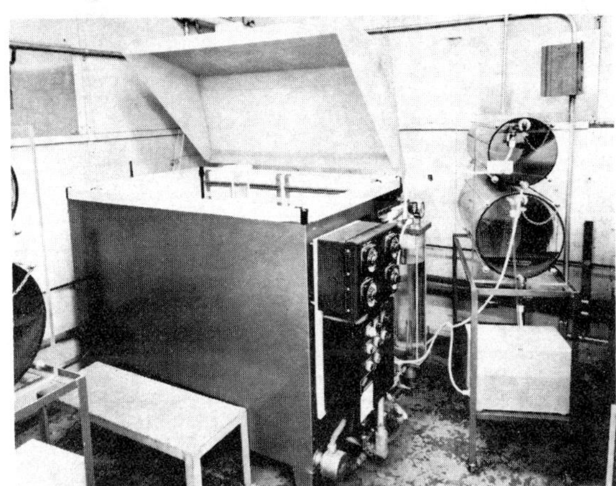

FIGURE 14.17 — Salt spray test cabinet with constant level solution reservoir. Tests using equipment of this type are used to evaluate alloys and coatings and are covered by military, federal, and ASTM specifications. [SOURCE: G. S. Equipment Company.]

FIGURE 14.18 — Specially designed humidity cabinet with dual chambers designed to operate at different temperatures and humidities. Removable dividing partition can be a test panel, simulating for example, an external wall in contact with indoor and outdoor environments. Test aim is to evaluate effect of metal-insulation composites on moisture transfer.

FIGURE 14.19 — "Smog" chambers used to study reactions involving products from internal combustion engine exhaust. Chamber at right has bank of ultraviolet lights to simulate sunlight and initiate photochemical reactions in exhaust gases.

Special Test Equipment

Often, commercial test equipment is not available to permit making the specific corrosion study needed. In such cases, the corrosion engineer usually designs or outlines the sort of special equipment which must be constructed to do the job. Such equipment or procedures may be as simple as laying weighted glass rods of various diameters on an immersed metal panel, or as complicated as an automatic cycle-programmed unit which alternately sprays a solution, turns on ultraviolet lights, bakes out by heating, freezes, and dries the surfaces of the specimens.

In any case, construction of equipment or use of special test procedures requires careful attention to details and a thoughtful approach to the work.

FIGURE 14.20 — A type of test cabinet with controls permitting simulating a wide range of climatic conditions for making accelerated tests.

Sloppy workmanship or carelessness can cause costly and time-consuming repetition.

Auxiliary Equipment

Conducting corrosion tests in the laboratory requires familiarity with a variety of auxiliary equipment. Because corrosion testing involves metallurgical, chemical, and electrical phenomena, some of the tools used in each of these fields are usually found in the well-equipped corrosion laboratory. Most of these tools are measuring devices of one sort or another, as indicated in Table 14.5.

In addition to this auxiliary equipment, a corrosion laboratory will be equipped with a variety of carpentry, metal-working, and plumbing tools. A corrosion technician must become familiar with these tools through experience.

Effects of Variables

As expected, results of a corrosion test on metals can be affected significantly by variations in environmental factors. The single or combined effects of these factors are not always known, but it is essential that the corrosion investigator do everything he can to make a test reproducible and specific with regard to the following.

1. Temperature
2. Solution concentration
3. Agitation
4. Aeration
5. Ultraviolet light
6. Bacteriological effects

TABLE 14.5 — Types of Equipment Used for Corrosion Tests

Measurement	Type
Gravimetric (weight)	Balance sensitive to one ten-thousandth of a gram.
Linear	Micrometer; metal rule.
Electrical	Sensitive DC milliammeter: electrometers; recorders (various); potentiometer; ohmmeter. More sophisticated equipment might include thè potentiostat (which maintains a steady potential while allowing current variations), a solution-potential half-cell (which serves as a reference for solution potential measurements on metal samples immersed in a solution), and a variety of specialized AC and/or DC equipment used for specific corrosion tests (Figures 14.21 and 14.22).
Fluid Flow	Flow meters; calibrated orifices (a variety of agitation and aeration devices are used).
Temperature	Thermometers; thermocouples; temperature-sensitive crayons and papers; controllers (water, oil, or air baths); hygrometers (temperature-humidity sensors).
Pressure	Manometers (in the test system); barometer (the environment); pressure gauges.
Chemical	pH meter (Figure 14.23), litmus papers, indicator solutions, hydrometers, specific ion electrodes. More complicated chemical appraisals usually are conducted in a chemical laboratory.
Visual	Visual binocular transmission microscope (Figure 14.24) (for bacteriological studies) and a variety of grids and special light boxes to facilitate reproducible examinations; metallograph.

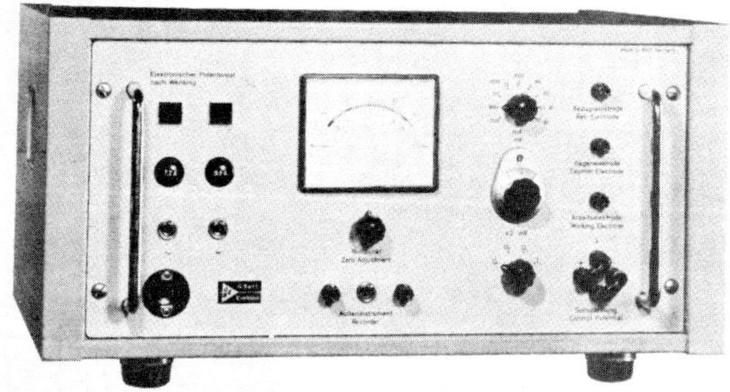

FIGURE 14.21 — Typical potentiostat.

Temperature

The temperature of the test solution should be controlled to within ±1.0 C (±2 F) in any laboratory corrosion test.

In certain cases, the temperature of the solution differs from the temperature of the metal. In such cases, it is important to know the actual metal temperature. This can be measured by use of a "dummy-control" specimen, equipped with a peened-in thermocouple. In such a case, the thermocouple contact point and leads should be masked by a sleeve or suitable coating. The difference be-

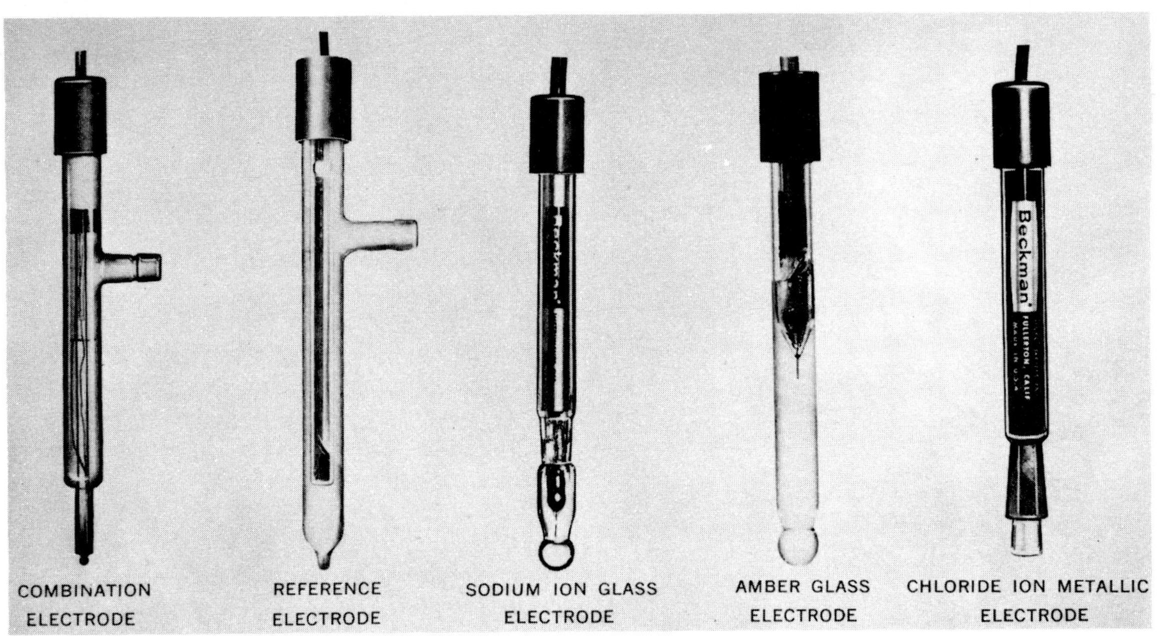

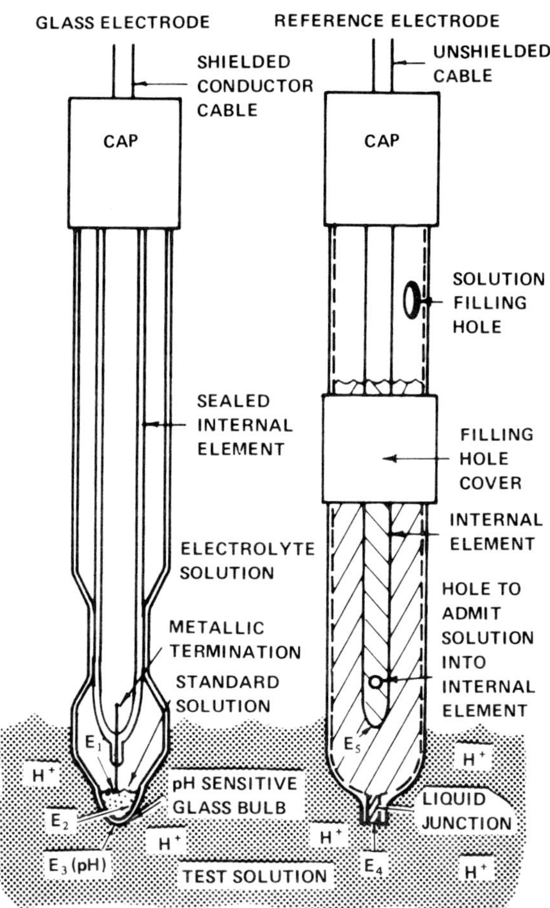

FIGURE 14.22 — (Top) Types of reference electrodes for making electrical measurements during corrosion tests. (Bottom) Schematics of glass and reference electrodes: (E_1) potential between internal element and standard solution; (E_2) potential between standard solution and inner surface of glass membrane; (E_3) potential between outer surface of pH-sensitive glass bulb and test solution—this is the only potential in electrode system that varies in accordance with change in hydrogen ion concentration, or pH, of test solution; (E_4) potential of liquid junction between electrolyte and test solutions; and (E_5) potential developed at internal element in electrolyte solution.

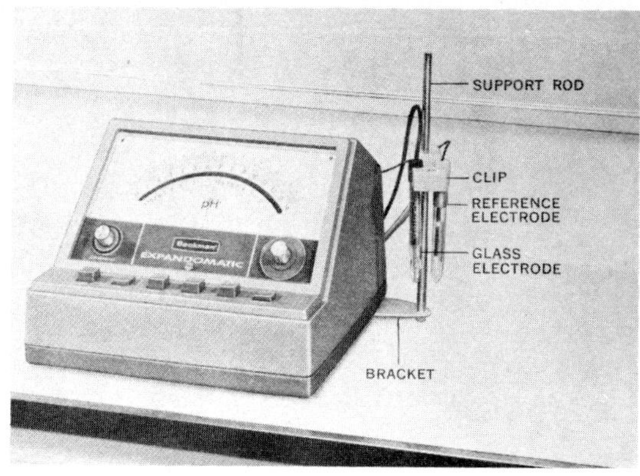

FIGURE 14.23 — Typical pH meter with electrodes.

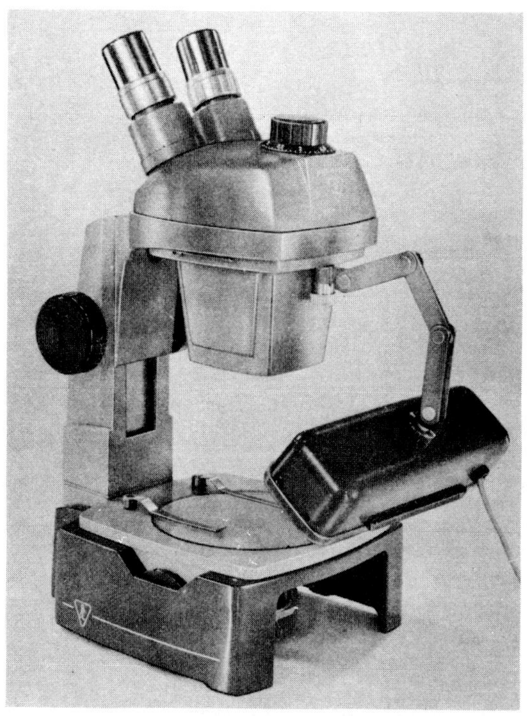

FIGURE 14.24 — High-resolution binocular microscope for inspecting and measuring corroded surfaces.

tween ambient and metal temperatures becomes especially important in a vapor or humidity test.

Vapors condense on surfaces which are colder than the environment. Consequently, near 100% humidity at 37.8 C (100 F), condensation may have little or no effect on metals warmer than 37.8 C (100 F). In the case of aluminum alloys, condensation produces *water staining,* a surface discoloration. If the aluminum alloy is above the humidity test temperature, condensation (and hence staining) does not occur. In the case of steel, rusting effects can similarly be encouraged or discouraged.

When appraising the effect of heat transfer through a metal, the simple exposure of a specimen in the appropriate solution at a given temperature may not provide the desired information. The metal surface should be heated on one side and cooled on the other to obtain proper results. This is particularly true when determining the corrosion or pitting rate of a metal surface to be used to boil a liquid.

Higher temperatures [above 212 F (100 C)] generally produce more intense corrosion. While the condensation in these cases is not a problem, the actual temperature can determine the thickness of the corrosion product oxide layer and extent of attack.

Solution Concentration

The effect of solution concentration may be an even more important variable than small variations in temperature. Often, metals are corroded by dilute solutions and not corroded by more concentrated ones.

Solutions react with metals in a variety of ways, including the following.

1. Some plate out metallic components (cations) on cathodic areas and concentrate anionic (negatively charges ions) components at anodic (corroding) areas.
2. Some solutions corrode metals vigorously for a time and then cease action abruptly.
3. Some organic solutions do not react for a time (induction time) and then react vigorously with the metal.

Hydrogen ion (H^+) concentrations and hydroxyl ion (OH^-) concentrations reflect the acidity or alkalinity of a solution (pH). These values can be quite variable on local areas of the metal surface and in the solution, depending upon initial solution concentration, agitation, and corrosion activity.

The important considerations are:

1. The gross solution concentrations must be known and controlled before, during, and after the test;
2. The volume of the solution should be sufficient to avoid expending it before the test is completed; and
3. The gross solution concentration can be quite different from the solution concentrations on the metal surface.

Parts per million of contaminants in a solution may be responsible for essentially all of the corrosion occurring, and their effect must be recognized. This is particularly true when evaluating the plastic materials.

Agitation

In order to assess the comparative corrosion effects on metals of quiescent and moving solutions, they are mildly stirred or otherwise circulated by pumps or jets. Agitation changes the concentrations of the solution on the local anodes and cathodes and provides a more homogeneous test environment.

Simple stirring of a solution is not adequate to determine the actual effect of velocity where liquid flow rate across the specimen face must be under control. Use an appropriate standard test when evaluating the effect of this parameter.[7]

In gaseous test environments, it usually is necessary to provide some agitation, to be certain that vapor concentrations are uniform in the test chamber. In humidity tests, where vapor pressures can vary markedly in a quiet test chamber, air flow ensures equilibrium.

Aeration

Aeration, or oxygenation, has a marked effect on the corrosion of metals. Consequently, oxygen is often added by bubbling it through the solution. Fine gas bubbles are better than large ones because there is more opportunity for the gas to dissolve in the solution during the time it rises to the solution surface.

The counterpart of aeration is deaeration, which can be controlled in a number of ways. Oxygen can be discouraged from entering the test solution by use of a closed system equipped with a trap (Figure 14.25). Another method is to deaerate by introducing an excess of an inert gas, such as nitrogen, into and above the test solution. Oxygen scavengers (Chapter 7), such as hydrazine, are also available for use in minimizing the free oxygen content of solutions.

Under special test conditions, other gases such as carbon monoxide, carbon dioxide, hydrogen sulfide, ozone, nitrous oxide, and sulfur dioxide are metered into liquid or gaseous test chambers to produce special effects on metal corrosion reactions.

Ultraviolet Light

Some chemical reactions are altered photochemically; that is, light of certain wavelengths can alter the chemistry of solutions, gases, or corrosion reactions. Of special interest is the photochemical oxidation of nitrous oxide to peroxyacetal nitrate in the atmosphere, which contributes to air pollution (*e.g.,* Los Angeles smog). In this case, the sunlight acts on the combined nitrous oxide and hydrocarbons from automobile exhausts (Figure 14.19).

Although ultraviolet light has been used for many years in destructive tests on fabrics and paint films, its effect on electrochemical corrosion has been neglected by most investigators. This influence, as well as its biocidal (killing) action on fungi and bacteria, promise to be the subjects of much more corrosion research in the future.

FIGURE 14.25 — Shows how liquid trap keeps oxygen out of test vessel vapor space.

Bacteriological Effects

More and more attention (especially in the Tropics) is being given to the influence of minute plant-animal species which abound in all but the most hygienic test facilities. General examples are corrosion by sulfate-reducing bacteria under anaerobic (oxygen-free) conditions, crude oil tank bottoms, and aircraft jet fuel tank corrosion.

In any one set of environmental conditions, several species of fungi or bacteria often may grow and reproduce in solutions, nonmetallics, and on metal surfaces. Their metabolic (life cycle) process can produce concentrations of organic or inorganic acids. Other organisms produce slime films which can screen oxygen or concentrate metal ions. Each of the innumerable species has its peculiar requirement for food, oxygen, and temperature. Variations in corrosion test results are caused by changing growth rates and balance of bacterial species.

Although detailed testing for biological corrosion is beyond the scope of this chapter, it is important to recognize that corrosion tests, particularly long-term ones where organic nutrients are present, may be affected by biochemical influences in moist air or aqueous environments.

Other Effects

Other accelerated test variables such as pressure, special cyclic effects, corrosion fatigue, stress relaxation, etc., are factors which may or may not have an important effect on corrosion test programs. Experience with these and other factors shows the need for care in repeatedly evaluating the test conditions in order to ensure test reproducibility.

Conducting the Test

Records

It is essential that a detailed record be kept for each specimen to help in the eventual analysis of the test. The laboratory notebook and/or other permanent record forms are used for this purpose. When possible, these records should be in a tabulated form which permits easy insertion of test results as they are obtained. Meaningful data should be reduced to the minimum and recorded promptly to avoid accumulation of unnecessary or confusing paperwork.

Interim Inspections

Usually, an exposure time is used which experience has shown will yield meaningful results. Because each test probes the unknown, however, the experimenter must be alert for unexpected trends. During the test, frequent examinations are made and notes recorded of any unusual effects on the specimens or in the test environment. Interim inspections often will permit corrections of problems with the specimens or environment, in the case of laboratory tests.

A regular schedule of examinations usually is established at the beginning of the test, with more frequent inspections in the beginning. Appropriate records of specimen condition and environment stability are often necessary to explain unexpected results. Personal judgment must be used to decide whether the original plans for interim inspections were correct. Often, progress reports are advisable to summarize segments of the work separately.

The Unexpected

When a test contains appropriate comparison specimens, it is unlikely that small environmental variations will be cause for alarm. Sometimes, when test "accidents" occur, the shrewd investigator can turn the mishap into a "lucky" accident. Many valuable inventions have been the result of such accidents which were observed, evaluated, and reproduced by intelligent investigators who did not hastily discard unexpected results. After all, if the answers are already known, why conduct the test?

Concluding the Test

Obviously, a decision to terminate an investigation is easiest when there is a fail, no-fail criterion. If an earlier investigator has established the proper test durations, your work is simplified.

Innovative tests, on the other hand, rarely signal that the program is completed. The investigator must make such a judgment after weighing results and interpreting trends. The danger of ending the test too soon is obvious when one realizes how many corrosion tests produce nonlinear results. The use of duplicate or multiple specimens helps to predict the long-range results by giving periodic indications of progress. When the graph of results is anything but a straight line with respect to time, periodic reviews are necessary to develop an understanding of what is happening.

When satisfied that there is little to gain by continuation of the test, the investigator terminates it. Often, the decision is hastened because of: (1) the shortage of test space, (2) backlog of other programs, (3) requests by those who need information, and (4) curiosity.

The investigator must weigh these pressures and arrive at a decision. Once it has been decided to end the test, post-test appraisals of the specimen often will prevent its re-exposure. A new test would then be required if the data obtained were insufficient.

Post-Test Appraisals

Before Cleaning

When laboratory tests are concluded, it is advisable to perform certain tasks rather quickly in order to avoid changes in specimen appearance or additional corrosion. Depending upon the goals, initial steps may include: (1) visual examination before cleaning, (2) photographs before cleaning, (3) careful collection of corrosion products for chemical or other analyses, and (4) saving test solutions for chemical study.

Visual examination before cleaning should included in test procedures for such a contingency. distribution of corrosion products (the relative severity and extent of corrosion will be recorded after cleaning). Also, it should be noted whether the identification number of each specimen has survived the exposure. If not, one should be able to determine the number from the location of the specimen in the frame or other cross-check methods included in test procedure for such a contingency.

Before discarding any test solutions, the investigator must decide on the possible need for a chemical analysis to evaluate how the solutions may have changed during the test.

Analyses of Products and Solutions

Selection of techniques for analysis of corrosion products and solutions is a rather complex procedure requiring background experience. Some typical methods include the following.

1. Analyses for pH and certain ion concentrations by wet chemical methods.

2. Spectrographic analysis of the ash after burning at 500 C (932 F) in a platinum crucible (this determines inorganic element concentrations).

3. X-ray diffraction analysis of dried powders (by placing them in an X-ray beam and observing the characteristic beam reflections on a photographic plate or a Geiger counter).

4. Measurement of solution conductivities (the ease of passing an electrical current through a known volume).

Other methods of evaluation are sometimes used, depending upon the goals of the program.

Cleaning the Specimens

Before the extent of corrosion on metal specimens can be completely evaluated, it is usually necessary to clean them according to approved methods. Some typical procedures are shown in Table 14.6. It is essential that special care be exercised in this procedure in order that the cleaning will: (1) not remove any metal, (2) remove *all* adherent corrosion products, and (3) facilitate the program of evaluation which will follow.

In some instances, it may be desirable to weigh the specimen before removing the corrosion product

TABLE 14.6 — Some Approved Methods for Nonelectrical Removal of Corrosion Products for Loss of Weight Tests

Magnesium Alloys

Chromic acid (20 Wt%)
Silver nitrate (in distilled water) (1 Wt%)
Time—1 min immersion. Rinse in water and dry.[(1)]

Aluminum Alloys

1. Chromic acid—20 grams/liter
   Phosphoric acid (85%)—40 grams/liter
   Temperature—93 C (200 F)[(2)]

or

2. Immerse in concentrated (S.G. 1.42) nitric acid at room temperature. Scrub with bristle brush under running water and dry.[(1)]

Steel

Immerse in 20% sodium hydroxide containing 200 gr/liter of zinc dust until clean. Scrub and rinse.

Zinc

Immerse in warm (60 to 80 C) ammonium chloride solution (10%) for several minutes. Rinse in water and scrub with a soft brush; then immerse 15 to 20 seconds in a boiling solution of 5% chromic acid + 1% silver nitrate. Rinse in hot water and dry.[(1)]

Copper

1. Dip two to three minutes in hydrochloric acid (1:1) or sulfuric acid (1:10) at room temperature. Scrub with bristle brush under running water and dry.[(1)]

or

2. 10% HCl containing 1% Rodine No. 50[(3)] for three minutes room temperature.

[(1)]ASTM Std. G1-79 Preparing, Cleaning, and Evaluating Corrosion Test Specimens.
[(2)]National Defense Res. Com., Combs & Lucks.
[(3)]Amchem Products Co., Inc., Ambler, PA.

to obtain a *weight gain* measurement. This is done when all the corrosion product is intact—as in high-temperature corrosion experiments on some metals.

Weighing

After rinsing and thoroughly drying the specimens, they should be weighed carefully, using the same equipment and care as used before the corrosion test.

A laboratory notebook should be used to tabulate final weights, as well as any changes in weight produced by the test. Usually, the level of *significant numbers* has been predetermined. For example, if a 30-gram sample of metal is tested, the weights may be recorded to the sixth significant number. That is, the record may show:

	Grams	
Original Weight:	30.0186	
Final Weight:	27.3243	
Weight Loss (△W)	2.6943	(14.1)

In the case of a 500-gram sample, it may be decided that accuracy to 0.1 gram is sufficient. Hence:

	Grams	
Original Weight:	**531.6**	
Final Weight:	**486.2**	
Weight Loss (△W)	**45.4**	**(14.2)**

These and other data obtained will be discussed in more detail later.

Visual Examination

Visual examination of cleaned specimens is not made until weighing has been completed. This avoids unnecessary handling which could affect weight.

Generally, the appearance of the test specimens should be recorded in the laboratory notebook in a form which will facilitate a stepwise appraisal of each specimen. Typical examples of frequent observations on cleaned specimens are listed below.

1. A stress corrosion specimen represents a fail, no-fail investigation, ordinarily requiring only a record of the time to failure. However, any unusual failures, such as failure outside of the most highly stressed region, should be noted. A general examination may be desirable on specimens which complete a predetermined exposure period without failing.

2. A panel, bar, or coupon specimen should be examined with some of the following questions in mind.

(a) Was corrosion uniform, or was it located at pits, edges, identification marks, or mounting areas?

(b) Is there any pattern to the corrosion? Does it follow work lines, edges, etc.?

(c) What is the shape and texture of the corrosion sites (blisters, exfoliation, minute pits, broad smooth pits, brightness)?

When the sites of attack are numerous, the tedious job of counting can be eliminated by using rating charts, as shown in Figure 14.26. A weighting factor, to account for the form of the corrosion, might also be used. Caution should be exercised in conversion to numerical values, so as not to imply greater accuracy than warranted.

3. An assembly of components should be examined in a few additional ways.

(a) What is the galvanic effect between parts (accelerated or decelerated corrosion)?

(b) How far does the effect extend?

(c) Are crevices (faying surfaces) attacked?

If the investigator is sufficiently experienced, a more detailed tabulation of visual appearance may be desirable before selecting the column headings which will be permanently recorded in the notebook and ultimately appear in the test report.

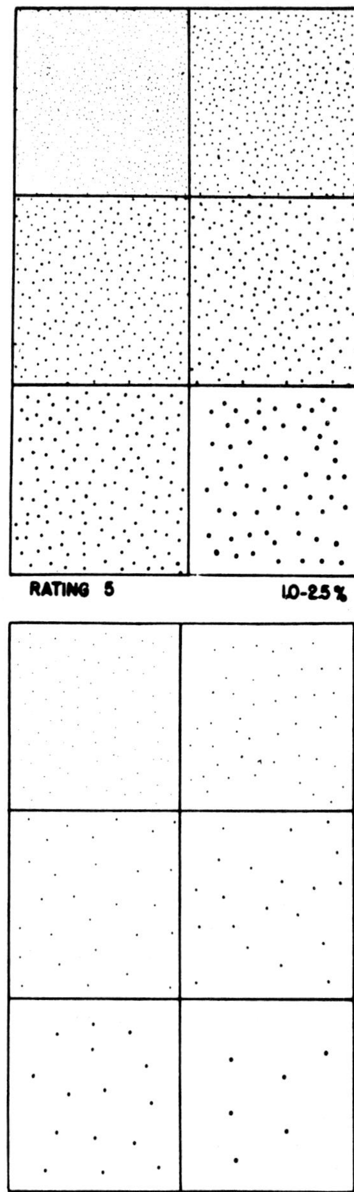

FIGURE 14.26 — Rating charts that eliminate tedious counting of pitting sites.

Photographs

Photographs or X-ray radiographs can be used to supplement the visual examination report. Entire specimen groups, representative ones, or even a magnified view of a typical area of one panel can be quite valuable in reducing the extent to which the visual examination must be recorded. Radiographs, particularly useful for flat sheet specimens, help to show severity (depth) of the corrosion sites, as well as their locations.

All photographs should be accompanied by descriptive identifications of the specimen(s) and magnifications used.

Mechanical and Physical Tests

The next step in the appraisal may be to obtain data required to establish effects of the test on mechanical properties. Usually, ultimate strength

TABLE 14.7 — Mechanical Test Data for Corrosion Test Panels Exposed 12 Weeks to 3.5% Sea Salt by Alternate Immersion[(1)]

		Unexposed Specimens					
Specimen Number	Steel	Cross Section Dimensions, in	Tensile Strength lb	Tensile Strength lb/in^2	Yield Strength lb	Yield Strength lb/in^2	Elongation in 2" (%)
17965-B-1	Mild	0.0520 × 0.493	1055	41,210	475	18,600	33.0
-2	Mild	0.0515 × 0.495	1040	40,790	470	18,400	35.0
	Average	—	—	41,000	—	18,500	34.0
16566-B-1	Alloy A	—	—	—	—	—	—
-2	Alloy A	—	—	—	—	—	—
	Average	—	—	76,815	—	50,350	22.5
16567-B-1	Alloy B	—	—	—	—	—	—
-2	Alloy B	—	—	—	—	—	—
	Average	—	—	79,035	—	49,850	19.3

		Exposed Specimens[(2)]			Calculated % Loss by Corrosion		
Specimen Number	Steel	Tensile Strength lb/in^2	Yield Strength lb/in^2	Elongation in 2" (%)	%-TS	%-YS	%-El
17965-B-1	Mild	—	—	—	—	—	—
-2	Mild	—	—	—	—	—	—
	Average	30,605	13,250	23.0	25.3	28.4	32.4
16566-B-1	Alloy A	—	—	—	—	—	—
-2	Alloy A	—	—	—	—	—	—
	Average	51,660	35,150	11.3	32.7	30.2	50.0
16567-B-1	Alloy B	—	—	—	—	—	—
-2	Alloy B	—	—	—	—	—	—
	Average	58,735	37,950	11.8	25.7	23.9	38.8

(1) Detailed values are shown only for the first two specimens. The other data are shown only as average results, for simplicity.

(2) The exposed specimens are those which were corrosion tested. They were cut from the same material for which the "unexposed" data are cited. The calculations, therefore, represent the losses as a percentage of original values.

(tensile strength) and elongation (stretching caused by tensile testing) tests are compared on specimens which have and have not been corrosion tested. The differences are attributed to corrosion. Generally, the final data are recorded as percent losses. An example is given in Table 14.7.

Special note is taken of any unusual failure observed during testing (failure outside gauge length, at a scratch, etc.) in order to decide whether these data should be ignored in the final appraisal.

In addition to or in lieu of the ultimate strength and elongation changes, tests are sometimes made to evaluate changes in electrical conductivity. As with any evaluation tool, it is necessary to know how corrosion can influence the conductivity values obtained (such as by intergranular corrosion or dezincification).

Microscopic Studies

One of the best corrosion test evaluation tools is the metallurgical microscope or metallograph. A careful selection must be made of metal areas to be studied. Special samples may be corroded specifically for this purpose, or remnants of test specimens, previously evaluated by other methods, may be used. A skilled metallographer then sections through the corrosion sites, polishes the cut surfaces, and examines them for:

1. Type of corrosion (relationship of corrosion with metallurgical structure),
2. Corrosion depth,
3. Path of corrosion, and
4. Other changes caused by corrosion.

The metallographer uses a variety of pro-

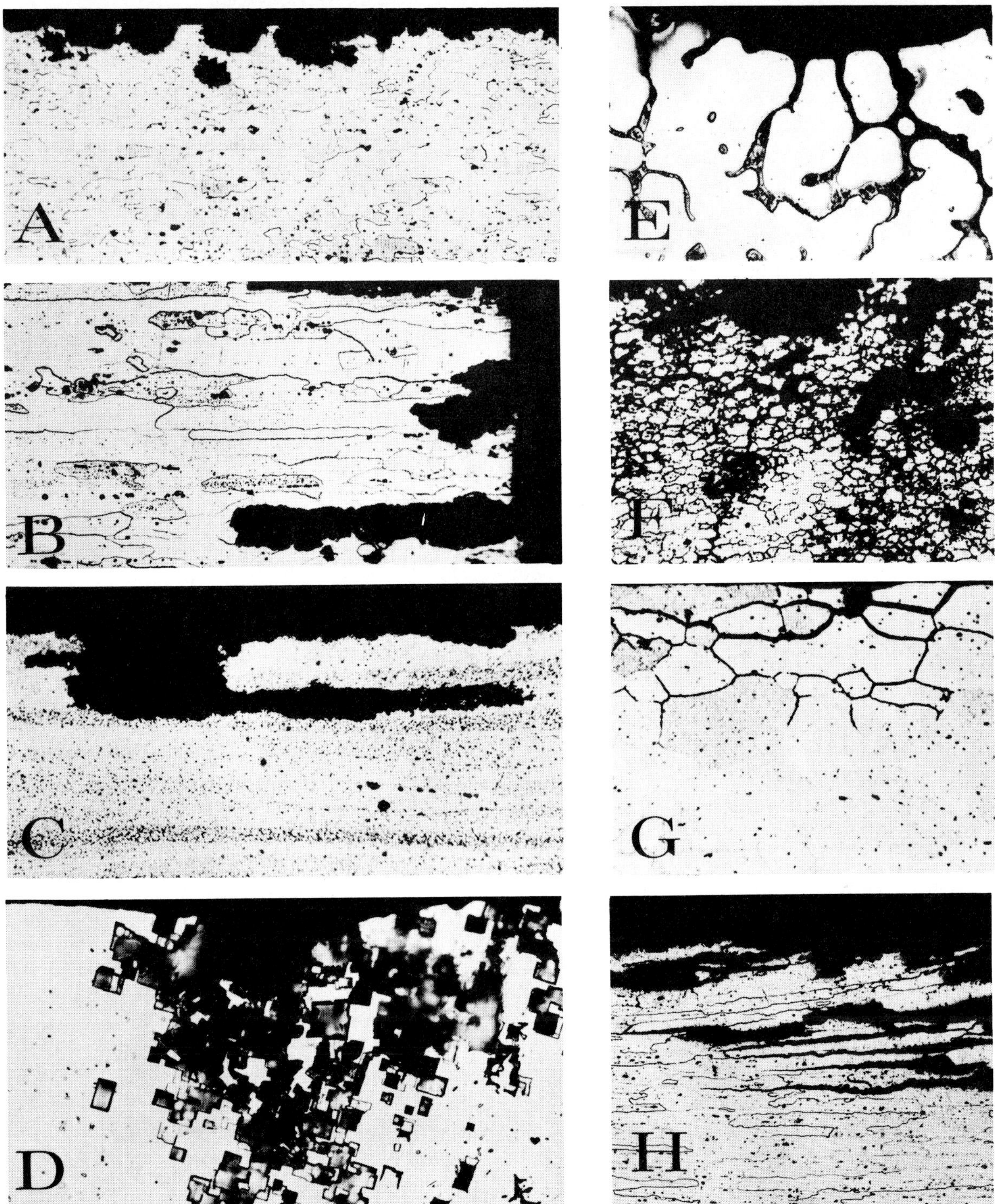

FIGURE 14.27 — Examples of several types of corrosive attack on aluminum alloys. Pitting corrosion: (A) surface pitting; (B) edge pitting; (C) undermining pitting of surface; (D) "cubic" pitting attack. Intercrystalline corrosion: (E) interdendritic attack in cast structure; (F) interfragmentary attack in wrought, nonrecrystallized structure; (G) intergranular attack in wrought, recrystallized structure; and (H) exfoliation attack. A, B, and H, X100; others X500.

cedures, including cutting, polishing, chemical or electrochemical etching, and illumination techniques to permit high-magnification views of the corrosion damage. Generally, the metallographer works closely with the investigator to prepare appropriate photomicrographs for use in the investigator's report.

Experience is necessary to select and interpret photomicrographs. From proper high-magnification photographs of surfaces and cross sections, the trained investigator often is able to identify factors relating to the mechanism of corrosion, as well as the type, depth, and metallurgical condition of the metal, and the presence of any unusual conditions.

Figure 14.27 (composite) shows a few of the types of corrosion which may be encountered.

Calculations

After the weight change data have been obtained, it is necessary to convert the raw information into suitable units. Obviously, units may be different in foreign countries or for special criteria. Where possible, however, unique methods of reporting should be avoided. Unusual data forms distract the reader and thereby diminish the effectiveness of communication.

The following are some commonly used ways to report corrosion data.

Percent Change. Applicable to loss in weight, strength, elongation, conductivity, etc., on a corroded specimen after cleaning.

$$\frac{\textbf{Original Value} - \textbf{Final Value}}{\textbf{Original Value}} \times 100 = \textbf{Percent Change} \qquad (14.3)$$

Corrosion Rate. Involves two types of expressions.

Engineering Expressions: microns per year (μm/y), mils per year (mpy), or inches per year (ipy).[(2)]

Laboratory Expressions: milligrams/square decimeter/day (mdd) or mm per year.

Sample Computation: If a specimen having a surface area of 6.4516 dm^2 (100 in^2) lost 7.3216 grams (7321.6 milligrams) in 50 days:

(a) The corrosion rate in mdd would be

$$\textbf{C.R.} = \frac{\textbf{mg}}{(\textbf{dm}^2) \times (\textbf{days})}$$

$$= \frac{\textbf{7321.6}}{(\textbf{6.4516}) \times (\textbf{50})}$$

$$= \textbf{22.697 mdd} \qquad (14.4)$$

(b) For steel (density 7.87) the corrosion rate also may be expressed in mpy (mils per year) as

$$\textbf{C.R.} = \textbf{mdd} \times \frac{\textbf{1.437}}{\textbf{D}}$$

$$= \textbf{22.697}\,\frac{\textbf{1.437}}{\textbf{7.87}}$$

$$= \textbf{4.14 mils per year.} \qquad (14.5)$$

Direct calculation of corrosion rates from weight loss is shown in Table 14.8.

Depths of Corrosion. Calculated—as above; Measured—Can be useful in certain corrosion tests which produce *uniform* corrosion. However, by itself, this is rarely an accurate method of reporting mpy or ipy. The exception is maximum pit depth where pitting penetration is a factor.

(2)ipy really means cubic inches of metal lost per square inch of metal surface exposed for one year, or average depth of metal loss from initial surface.

TABLE 14.8 — Calculating Corrosion Rate from Weight Loss Data

$$\frac{\text{Wt. Loss (gms)} \times 534}{\text{Area (in}^2) \times \text{Time (hrs.)} \times \text{d(gms/cc)}} = \text{IPY}$$

or

$$\frac{\text{Wt. Loss (gms)} \times 8760}{\text{Area (m}^2) \times \text{Hrs.} \times \text{d(gms/cc)}} = \mu\text{m/y}$$

Metal	Density Grams/cc	Metal	Density Grams/cc
Aluminum 3003	2.73	Lead	11.34
Brass, red	8.75	Monel	8.84
Brass, yellow	8.47	Nickel	8.89
Bronze, aluminum	7.58	Silicon Cast Iron	7.00
Bronze, tin	8.78	Steel	7.85
Bronze, silicon	8.53	304 Stainless Steel	7.93
Cast Iron	7.00	Tantalum	16.60
Copper	8.91	Tin	7.30
Cupronickel	8.93	Titanium	4.54
Hastelloy C	8.94	Zinc	7.14
Inconel	8.51	Zirconium	6.53

To convert IPY to mpy, multiply IPY by 1000.
To convert mpy to μm/y, multiply mpy by 25.4.

Theoretical Weight Loss. This method can be used to calculate loss of metal, where current flow is measured from the metal, by Faraday's Law of Electrolysis:

$$\textbf{W} = \textbf{Ite} \qquad (14.6)$$

where: W = weight loss in grams
I = current in amperes
t = time in seconds
e = electrochemical equivalent (Table 14.9).

A rough rule of thumb is that a constant current density of 1 mA/dm^2 will produce 105 μm/y corrosion on an anode of aluminum, or about 115 μm/y on iron (1 mA/in.^{2} = 65 mpy on aluminum or 71 mpy on steel).

The theoretical weight loss calculation is a convenient method of estimation, but requires continuous monitoring of current flow. Also, some sophisticated metal-solution interface reactions tend to make this evaluation method unsatisfactory. It is often used, however, to predict the corrosion rate during a test which ultimately will be evaluated by other methods.

TABLE 14.9 — Electrochemical Equivalents (For Use in Determining Theoretical Weight Loss)

Metal	Valence	Electrochemical Equivalent, e grams/ampere/second
Aluminum	3	0.000093
Copper	2	0.000329
Copper	1	0.000659
Iron	3	0.000193
Iron	2	0.000289
Tin	4	0.000615
Zinc	2	0.000339

[SOURCE: Handbook of Chemistry and Physics, Chemical Rubber Publishing Company, Cleveland, OH, 1960.]

In general, the sort of calculation used for each corrosion test requires that:

1. You know the metal characteristics,

2. You know the limits of your calculation methods, and

3. You keep your goals in mind.

The Notebook

All entries and calculations should be shown carefully in a permanent notebook. Although each company has its own requirements, everyone agrees that a notebook is a very important document and often serves to establish initial records for patent protection. Most patent attorneys will advise an investigator to write clearly, make a carbon copy of each entry, and have each page of the notebook witnessed and dated by someone who understands the entry. The carbon copy pages usually are stored at another location in order to minimize the possibility of loss of records.

Tabulations

In preparation for the report, it is vital that the data be tabulated in a concise form or drawn as suitable graphs. From these data, results will be itemized and conclusions drawn in the report.

A typical tabulation of corrosion test data is shown in Table 8.3 (Chapter 8) of this book.

Communication—The Report

The written report serves to communicate results of any corrosion test. It should be as brief as possible, but must include the "why, how, what, and where" of the test program. It should clearly state conclusions and make appropriate recommendations.

Because specific rules or procedures for report writing vary from company to company, this section will present only basic considerations for reporting.

If the report is written according to a plan, it will help the investigator as well, leading him through a process of self-interrogation to minimize omissions.

These reports are usually broken into parts or sections. Reports typically contain one or more of the following sections: (1) Synopsis or Abstract; (2) Introduction; (3) Object; (4) Materials; (5) Procedure; (6) Results; (7) Summary; (8) Conclusions, Status, or Recommendations; and (10) Appendix.

Summary

In this chapter, an effort has been made to point out the major objectives of corrosion testing programs, describe some of the variables that must be taken into consideration, and the means whereby these variables are taken into account in assessing results of tests. In addition, attention has been given to the kinds of equipment that may be used and special precautions that must be taken in evaluating the results they give. The importance of record-keeping and reporting have been reviewed. Because of the complexities of the corrosion process and the wide variety of materials and environments that must be taken into account, it is obvious that the topic of corrosion testing covers far more than has been given herein. Those who may be concerned with other types of testing, *i.e.,* electrical tests in soils, testing of inorganic materials (plastics, glass, wood); testing of composites, *i.e.,* where metals and inorganics may be in combination, are referred to specialized studies on these topics. Many of these specialized types of tests are described in other chapters in this book.

References

1. ASTM, ASTM Laboratory Standard G31.
2. ASTM, Conducting Plant Corrosion Tests, ASTM Standard G4-68.
3. Kelly, Michael P. and Schlamp, Robert J., Detecting Structural Degradation by Acoustic Emission, NACE CORROSION/77 (Paper No. 141), March 14-18, Houston, TX, 1977.
4. Toth, J. M. and Ross, B. J., The Involute Search Unit—A New Concept in the Ultrasonic Inspection of Pipe, Materials Evaluation, August, Columbus, OH, 1981.
5. Bradshaw, James M., Vertilog: A Downhole Casing Inspection Service, NACE, CORROSION/76 (Paper No. 44), March 22-26, Houston, TX, 1976.
6. ASM, Metals Handbook, Failure Analysis, Vol. 12, 1976.
7. NACE, Method of Conducting Controlled Velocity Laboratory Corrosion Tests, NACE Standard TM-02-70, 1970.

Bibliography

Ailor, W. H., Ed. Handbook of Corrosion Testing and Evaluation. John Wiley & Sons, New York, N.Y., 1971.

ASTM. ASTM Standards G4, G31, G46, G50, A262, etc. (those relating to corrosion testing and evaluation).

Bregman, J. I., Corrosion Inhibitors. The MacMillan Co., New York, N.Y., 1963.

Champion, F. A. Corrosion Testing Procedures. Chapman and Hall, London, England, 1963.

Fisher, A. O. Tools and Techniques for Laboratory Corrosion Testing. NACE, Houston, TX, 1964.

Holister, G. S. and Thomas, C. Fiber Reinforced Materials. Elsevier Pub. Co., Ltd., New York, N.Y., 1966.

Holmes, Raymond G., Ed. Source Testing Manual. Air Pollution Control District of Los Angeles, Los Angeles, CA, 1966.

Mallinson, John H. Chemical Plant Design with Reinforced Plastics. McGraw-Hill Book Co., New York, N.Y., 1969.

NACE. NACE Recommended Practices—Test Methods. Houston, TX.

Schweitzer, Philip A. Handbook of Corrosion Resistant Piping. Industrial Press, Inc., New York, N.Y., 1969.

VanDelinder, L. S. Extended Screening Tests of Plastic Materials of Construction in Organic Chemicals. Materials Protection, Vol. 2, No. 5, pp. 30-46 (1963).

Chapter 15

Design and Failure Analysis

DESIGN AND FAILURE ANALYSIS

This final chapter reviews information presented in the previous chapters and emphasizes in particular the requirements of good design and failure analysis. The practical data cited in this chapter can be used to analyze and improve conditions which will minimize corrosion failures and extend the useful life of many materials that must be exposed to a "hostile" environment.

What Constitutes Failure?

The term *failure* is literally defined as "a falling short, a deficiency or lack, an inability to perform...." For the corrosion engineer, the term *failure* is defined in terms of how well a material fulfills all aspects of the functional requirements of the application for which it was selected.

It is not enough, however, to merely provide functional capability. The best choice is the material which fulfills the required function most economically, taking into account safety, initial cost, maintenance costs, reliability, return on invested capital, product quality, need for inhibitors, product degradation, product loss, unscheduled shutdowns, etc.

Consequences of Failure

Excessive Maintenance Costs

The cost of maintaining both the plant and equipment is an operating expense that directly reduces profit. Therefore, any reduction in that expense would appear desirable from a profit standpoint. When taxes are taken into account, however, an economic decision must be made as to whether it is better from a standpoint of overall profitability to maintain and protect a lower cost material (as an expense), or to invest in higher cost capital equipment.

Because company managements will appraise the recommendations of corrosion engineers on the basis of overall profitability, corrosion engineers are strongly urged to gain a working knowledge of engineering economics. Such knowledge will help define "excessive maintenance cost."

A typical example of a decision based on maintenance costs involves soda ash plants. In these plants, exposure to sodium chloride, calcium chloride, high humidity, etc., produces an environment extremely corrosive to unprotected steel. It is practically useless to attempt maintenance of protective coatings on steel grating-type stair treads because pedestrian traffic damages the coatings, permitting corrosion and therefore the exposure of additional metal to attack. Thus, aluminum stair treads, due to their higher degree of corrosion resistance, are commonly employed in soda ash plants. Care must be exercised, however, to avoid galvanic action between the aluminum tread and steel structural members.

Unscheduled Shutdowns

It is good engineering practice to plan periodic shutdowns of process equipment for inspection and maintenance purposes. This permits orderly repair and reconditioning without disruption of operations or inconvenience to customers. Unfortunately, emergencies sometimes arise which require unscheduled shutdowns.

Consider the example of a large central station steam power plant which was forced to shut down a unit because of excessive leakage from condenser tubes. Power plants are committed to supply power on a continuous basis at a fixed price to a large cross section of domestic and industrial customers. Loss of the use of a major power unit forces the power company to purchase power from other power companies in order to supply their customers. They usually must pay a premium price of approximately $20,000 per day or more for such back-up power.

In addition, replacing condenser tubes in a large unit on an emergency basis can, in itself, become alarmingly expensive. A modern surface condenser may require upwards of 150 miles of tubing. Rarely will such a large quantity be held in inventory for immediate shipment. Therefore, additional delays may be involved for the manufacture of tubes. After tubing has been received, another 2 to 3 weeks could be required for installation and testing.

The unscheduled shutdown of a chemical plant or refinery may not only involve disruption of activities at the plant where failure occurs, but it also may interrupt operations of several other plants which depend on the first plant for their supply of raw materials. Consequently, it sometimes is considered necessary to provide large storage capacity for certain products as a hedge against disruption of operations during an unscheduled shutdown.

It is clear that unscheduled shutdowns, as well as a fear of them (which results in investment in duplicate facilities for standby), represent an immense expense to industry, which justifies a considerable effort to avoid them.

Economic Estimates

A primary goal of the corrosion engineer is to ensure maximum return for the money invested. To do this, it is necessary to have some means of appraising the worth of one expenditure as compared with another, *e.g.,* Should we change the material of construction to Monel, or continue to replace the item with steel every nine months? The answer cannot be based on the initial investment in each material, but obviously must relate to how long each one will last. Yet cost per year is only a part of the answer. Is the item to be charged as a maintenance expense, or as a capitalized investment? What is the worth of the money tied up during this operating period? An accurate answer to these questions requires economic calculations.

In the past, economic calculations were laborious procedures which were properly done only by those who understood the interrelationship of all factors involved. Thanks to Jelen, Dillon, and others, however, such estimates can now be made quickly through the use of formulas. (These formulas are given in a NACE Standard.[1]) The effects of the initial cost, life, rate of depreciation, taxes, and value of money to the employer have been combined into easily used equations which take only a couple of minutes to understand, and do not require a knowledge of economics to use.

When proposals are made with recommendations that are based on economic calculations, those controlling the purse strings will listen. This provides them with an objective basis on which to make a decision regarding where to spend the limited money available, and adds more credibility to your position. It is important to use all the tools available in your effort to do a better job.

Why Materials Fail

Failures of materials in service usually are traceable to misapplications resulting from:

1. Choice of the wrong material,
2. Improper treatment or fabrication of the material,
3. An inadequately controlled (or defined) environment, or
4. Improper design.

Materials Parameters

Corrosion literature is filled with data on the performance of various materials in myriads of chemical environments. Engineers must constantly be on guard when considering such information to be certain not only that the chemical environment is adequately defined, but also that both the particular alloy (and heat treatment) and the character of attack are fully described.

For example, corrosion rates for "aluminum" often are given without further alloy designation. There are approximately 200 commercial compositions of aluminum alloys available. Their level of corrosion resistance varies widely with composition. The wrought alloys of the 1000, 3000, 5000, and 6000 series (Chapter 4) are roughly similar in their corrosion behavior and have far better corrosion resistance in most chemical environments than have alloys of the heat-treatable 2000 and 7000 series. The corrosion behavior of the 2000 and 7000 series alloys is strongly influenced by heat treatment practices. Merely by varying heat treatment, the character of attack on the 7075 (A97075) alloy in 3.5% NaCl solution can be changed from intergranular to pitting.

The stainless steels likewise tend to be treated in the literature as a "class" and not as individual alloys. Naturally, this leads to much confusion and some misapplications. It is therefore necessary to differentiate between the austenitic and ferritic steels, as well as between stabilized versus non-stabilized grades of austenitic stainless steels (Chapters 3 and 4).

The technical literature is far more reliable in deciding which materials cannot be used than in deciding which can be used. On the other hand, the literature can narrow the choice of materials to a manageable number of alternatives from which a final choice can be made more quickly.

Materials suppliers are excellent sources of guidance in the selection of proper alloys. Major materials suppliers often maintain staffs of consultants to advise customers on the proper use of their materials. These suppliers recognize that a "bad" application can result in damaging publicity.

Causes of Corrosion Currents

The sources of the electrochemical driving force for corrosion reactions are of considerable interest and merit special attention. Mears and Brown[2] have summarized 18 mechanisms by which differences in potential may develop on metal surfaces. This article is strongly recommended for further study. Chapter 2 of this text also discusses many of these causes. Causes of corrosion currents listed by Mears and Brown are given in Table 15.1.

Forms of Corrosion Attack

Much can be deduced from examination of materials which have failed in service. It is often possible by visual examination to decide which corrosion mechanisms have been at work and what corrective measures are required to solve the problem. However, a more complete analysis of a problem should be made. Too many offhanded analyses have been made with costly results. Even metallurgical examinations may require sophisticated tools to differentiate between a corrosion fatigue crack and a stress corrosion crack. However, this is not meant to denigrate the value of visual inspection.

The important point is to understand that a variety of corrosion mechanisms can occur, that these are caused by different combinations of fac-

TABLE 15.1 — Causes of Corrosion Currents

Impurities in the metal
Orientation of grains
Grain boundaries
Differential grain size
Differential thermal treatment[(1)]
Surface roughness
Local scratches or abrasions[(1)]
Difference in shape
Differential strain
Differential pre-exposure to air or oxygen
Differential concentration or composition of solution[(1)]
Differential aeration[(1)]
Differential heating
Differential illumination
Differential agitation
Contact with dissimilar metals[(1)]
Externally applied potentials
Complex cells

[(1)]Particularly likely to determine site of attack.

TABLE 15.2 — Types of Corrosion

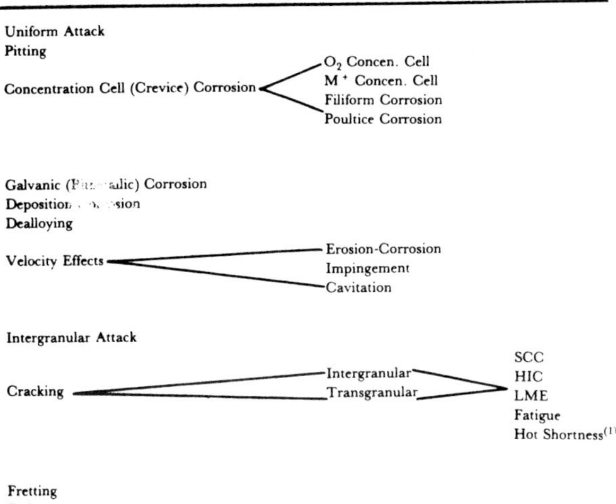
Uniform Attack
Pitting
Concentration Cell (Crevice) Corrosion — O_2 Concen. Cell; M^+ Concen. Cell; Filiform Corrosion; Poultice Corrosion
Galvanic (Bimetallic) Corrosion
Deposition Corrosion
Dealloying
Velocity Effects — Erosion-Corrosion; Impingement; Cavitation
Intergranular Attack
Cracking — Intergranular; Transgranular — SCC; HIC; LME; Fatigue; Hot Shortness[(1)]
Fretting
Fatigue

[(1)]Metallurgical problem not associated with corrosion.

tors, and that corrective procedures are widely varied. A summary listing of the types of corrosion discussed in this text is given in Table 15.2.

The ability to recognize these forms of corrosion, along with their implications, is of great value to anyone charged with the responsibility of operating processing plants and equipment. Consequently, examination of samples prior to cleaning is highly desirable.

Uniform Attack

Although representing potentially the greatest metal weight loss, uniform attack is predictable as a function of time; hence, it may be less troublesome than other forms of corrosion. Use of a *corrosion allowance* is one of the simplest methods of dealing with uniform attack. Unfortunately, the literature does not always describe the character of corrosive attack when discussing corrosion behavior of materials. It would be dangerous to assume that attack will be uniform, unless direct evidence is available to verify this assumption.

Galvanic Corrosion

If two dissimilar metals are electrically connected and exposed to a common electrolyte, a current will flow between the electrodes in accordance with Ohm's Law:

$$I = \frac{E}{R_a + R_c + R_e + R_w} \tag{15.1}$$

I = Corrosion current, amps.
E = Potential difference between anode and cathode, volts.
R_a, R_c, R_e, R_w = Resistances at anode, cathode, in electrolyte, and in return circuit, respectively, ohms.

The weight of metal dissolved at the anode is related directly to the current flowing, according to Faraday's Law (Chapter 14). Any procedure that reduces the potential difference, E (*e.g.*, cathodic protection), or increases the sum of the resistance, R (*e.g.*, protective coatings), or both, reduces the corrosion current. It is possible to eliminate galvanic corrosion by cathodic protection, protective coatings, inhibitors, electrical isolation of the electrodes (breaking the circuit), or using an environment free of electrolytes.

Galvanic corrosion usually is characterized by special attack on one metal, while the other metal is relatively free of attack (Figure 15.1). Galvanic cells also can result from contact of two identical metals if one is highly stressed.

Area Effects

Most metals corrode under *cathodic control, i.e.*, the polarization curve (potential vs current) of local cathodes is considerably more steeply inclined than that of the anodes. Under these circumstances, the anode current density (which is a measure of corrosion intensity) is very sensitive to the relative sizes (areas) of anode and cathode.

FIGURE 15.1 — Effect of galvanic corrosion caused by contact between dissimilar metals.

TABLE 15.3 — Representation of Area Size Effects on Corrosion Current Density
(Assumes No Polarization)

Areas		Relative Anode Current Density
Anode	Cathode	
□	□	1
□	□ □	2
□ □	□	1/2
□	▽	1/2

The effect of varying the relative areas on the anode current density can be represented as shown in Table 15.3 (assuming constant cathode current density). Thus, it is seen that reducing the area of the cathode relative to the size of the anode reduces the anode current density, but reduction of anode size without reducing the cathode area will intensify the attack. (Methods of combatting galvanic corrosion are given in a later section of this chapter.)

Misapplication of EMF Series

There is a prevailing fallacy that by consulting a standard table of electrode potentials it is possible to predict accurately the likelihood of galvanic action. In fact, such a list as the Standard EMF Series can be of only qualitative value because the reported electrode potential is sensitive to compositions of the electrode and the electrolyte and of temperature, as well as other factors such as degree of agitation, presence of depolarizers, etc.

An indication of the influence of the electrolyte composition on anode solution potential is illustrated by the fact that in laboratory tests, the potential of commercially pure aluminum varied over a range of approximately 1.6 volts, depending on the choice of electrolyte. Corrosion engineers concerned particularly with marine exposures are inclined to prefer a more practical EMF table based on solution potentials of alloys measured in real or simulated seawater. The relative order of activity in seawater is different, in many cases, from the order shown in the Standard EMF Series.

Intergranular Corrosion

Intergranular corrosion is preferential corrosion at grain boundaries of a metal. Although the detailed mechanism of intergranular corrosion varies with the metal system, its physical appearance in most systems (microscopic) is quite similar. The effects of this form of attack on mechanical properties may be extremely harmful.

Aluminum Alloys

The aluminum-copper alloys (*e.g.,* 2024, sometimes referred to as duralumin alloys), containing about 4% copper, exhibit intergranular attack under certain circumstances. The occurrence of intergranular corrosion may be sensitive to precipitation and the prior metallurgical history of the alloys, as was described in Chapter 3.

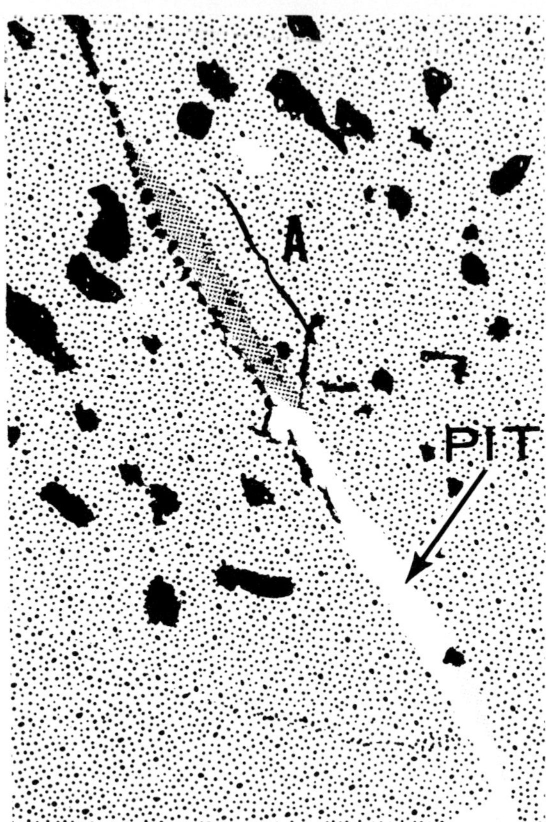

FIGURE 15.2 — Showing consequences of precipitation at grain boundaries in CuAl alloys. In Zone A, one alloy constituent is reduced in local concentration with respect to bulk alloy. In the case illustrated, the copper-rich phase (dark) is present at grain boundaries and in the grains. Specimen was aged four hours at 375 F (190 C). X50,000.

After fabrication of Al-Cu alloys, it is customary to solution heat treat at approximately 490 C (920 F) and water quench. This class of alloys will age harden at room temperature. If, however, the quench is not severe (*e.g.,* quenching in hot or boiling water), $CuAl_2$ precipitates as a "string of pearls" along grain boundaries, leaving a zone adjacent to the grain boundaries that is depleted of copper (Figure 15.2).

This depleted zone is anodic both to the main body of the grain and to the precipitated $CuAl_2$. Therefore, the depleted zone is attacked selectively, giving rise to intergranular attack. Such compositional variations usually result from a reaction between one alloy constituent and other elements in the alloy (either as impurities or alloy additions) to form precipitates. While such precipitates may occur either at grain boundaries or within grains, or both, the grain boundary case is particularly important because intergranular corrosion may result.

Similar behavior will be observed even in properly quenched alloys of the 2000 type if they have received a subsequent (post-quench) heat treatment, as had the specimen shown in Figure 15.2. For example, a zone susceptible to intergranular at-

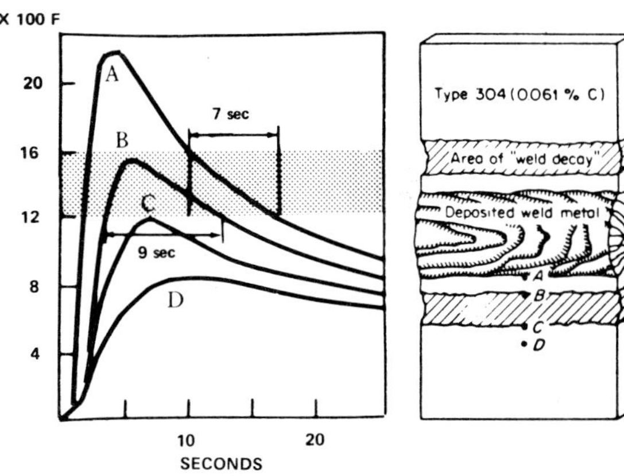

FIGURE 15.3 — Shows temperatures reached during electric arc welding of Type 304 stainless steel at which metal becomes sensitive to "weld decay" intergranular attack.

tack often will develop in the heat-affected zone adjacent to fusion welds. For this reason, care must be exercised in the choice of assembly methods for this class of aluminum alloys.

Stainless Steels

The austenitic stainless steels may also be "sensitized" to intergranular attack by heating in the temperature range 400 to 900 C (750 to 1650 F). A short time (minutes) above 750 C (1380 F) will produce sensitization; longer times (hours) are required to produce sensitization at lower temperatures. The commonly accepted mechanism attributes intergranular corrosion to a difference in potential between a chromium-depleted zone (at the grain boundaries) and the chromium-rich central zones of the grains, as was previously described in Chapter 3. *Weld decay* is an example of intergranular attack adjacent to a stainless steel weld (Figure 15.3).

High-nickel alloys are attacked by sulfur and sulfur-bearing gases (H_2S or SO_2) above about 315 C (600 F). Sulfides form along grain boundaries and cause brittleness, as shown in Figure 15.4. Sulfur forms low melting Ni_3S_2 eutectics. As was previously discussed in Chapter 13, reducing atmospheres containing sulfur are more aggressive to high-nickel alloys than are oxidizing atmospheres.

Pitting

Localized attack, where the rate of attack is much greater at certain areas on a surface than at others, is called *pitting*. This is one of the more dangerous forms of corrosive attack because it may cause failure by perforation without great loss in overall weight. Corrosion rate data appearing in the literature must be viewed with considerable suspicion, unless complete information regarding the geometrical character of attack is fully known. Some authors make use of a *pitting factor* to provide this information. The pitting factor is the ratio of the depth of the *deepest* pit to the *average* depth of pitting. Ideally, such measurements would be performed

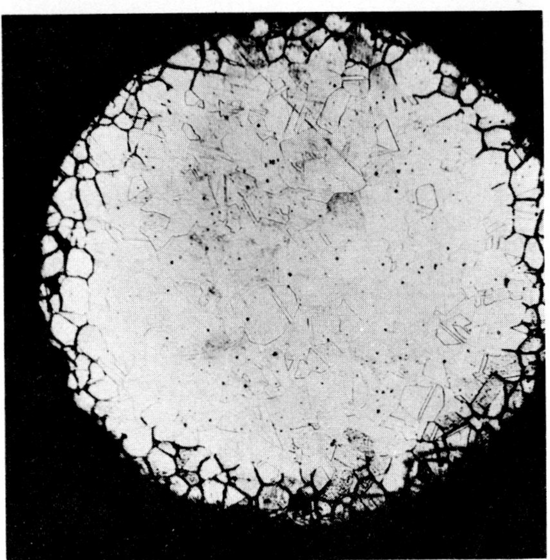

FIGURE 15.4 — An example of intergranular attack on a high-nickel alloy by a molten nickel-sulfur eutectic.

TABLE 15.4 — Behavior of Type 316 Steel 13 Months in Seawater[1]

(Specimen Size 8 × 12 × 1/4 in.)

Exposure	Wt Loss, %			Max Pitting Depth, in		
	1	2	Mean	1	2	Mean
Quiet water, heavy fouling	9	17	13	0.105	0.040	0.073
Flowing at 4-5 ft/sec, little fouling	0	0	0	0.004[2]	0.000	0.002

[1]Average penetration in inches, based on weight loss; therefore, pitting factor in quiescent water is about 70.
[2]Single shallow pit.

metallographically. Often, however, the *average* depth of pitting may be estimated from weight-loss data (Table 15.4).

Mitigation of Pitting

Pitting attack may be eliminated or reduced by any of the following, singly or in combination.

1. Cathodic protection
2. Alteration of environment
3. Coatings
4. Periodic cleaning
5. Alloy selection

Concentration Cells

A difference in potential will be observed if electrically connected specimens of the same metal are immersed in solutions having different concentrations of ions of the metals. This is one common type of concentration cell.

Another type is the oxygen concentration cell. This type and the metal ion cell can oppose one another. For example, variations in oxygen concentration can establish oxygen concentration cells. In solutions containing metal ions, the solubility of oxygen may vary, depending on the concentration of metal ions.

For this reason, although it would be expected that the electrode immersed in the solution with the lower concentration of metal ions would be anodic to an electrode immersed in a more concentrated solution, the greater solubility of oxygen in the dilute solution may mask the electrochemical effect of this difference in metal ion concentration. It may even cause the electrode in the solution dilute in metal ion to become the cathode. Therefore, in real solutions, when differences in chemical or oxygen concentration exist, one cannot predict with certainty which electrode will be the anode, or the magnitude of the current flow.

If electrically connected electrodes of the same metal (and having the same metallurgical history) are immersed in electrolytes of the same initial composition and concentration of ions (not necessarily ions of the metal), and the solutions are connected by a salt bridge, initially, no current flow would be expected. If air or oxygen is bubbled into the electrolyte around one of the electrodes, this electrode will shift in potential to the degree that it becomes cathodic to the other electrode. This is known as a *differential aeration* cell or an oxygen concentration cell.

There are many service examples of this type of cell. Differential aeration cells can be caused by crevices, lap joints, dirt and debris, moist insulation, etc. Under these conditions, the oxygen-starved areas are anodic, while the areas with free access to oxygen are cathodic. Other common terms for this type of corrosion include crevice corrosion, oxygen screening, and poultice action.

Selective Depletion of Alloys

The term *dezincification* applies to a phenomenon, usually associated with brasses containing more than 15% Zn, in which a porous copper surface zone develops as a result of the depletion of zinc. Often the gross appearance and size of the part which has suffered dezincification is unchanged except for the development of a copper hue. The part, however, will have become weak and embrittled, and therefore subject to failure without warning. To the trained observer, dezincification is readily recognized under the microscope, and even with the unaided eye, because the red copper color is easily distinguished from the yellow of brass.

There are two general types of dezincification. The more common is the *layer* type in which dezincification proceeds uniformly, as shown in Figure 15.5. The second type is referred to as the *plug* type and occurs at localized areas, as was previously illustrated in Chapter 5. The layer type is frequently observed in high brasses, whereas the plug type usually occurs in low brasses. (The terms *high* and *low* brasses refer to zinc content.)

The conditions of exposure also tend to influence the type of dezincification. High temperatures and aggressive corrosion conditions seem to favor plug-type attack in either class of alloy. Conditions generally conducive to dezincification include a good electrolyte (*e.g.,* seawater), slightly acid conditions, presence of carbon dioxide, and appreciable oxygen.

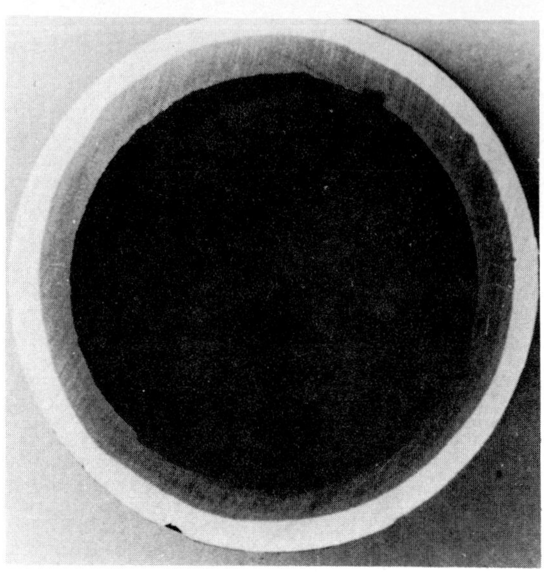

FIGURE 15.5 — Layer-type dezincification in a tube.

Dezincification does not occur in red brass (below 15% Zn). It may be controlled in the yellow brasses by additions of tin, arsenic, antimony, or phosphorus. Admiralty metal heat-exchanger tubes are "inhibited" by use of such alloy additions.

Selective leaching processes also are observed in other metallic systems. Cobalt can be removed selectively from Stellite No. 1[(1)] by contact with sulfuric acid slurry containing ferrous sulfate. Raney nickel catalyst is prepared by selectively leaching the aluminum from a 60% Al-40% Ni binary alloy using sodium hydroxide. The resultant product is a nickel sponge of high chemical activity.

Cast iron pipe may suffer graphitic corrosion (known as *graphitization*) as a result of selective dissolution of ferrite by corrosive action of certain soils, leaving a graphite structure. When this has occurred, it is possible to literally carve your initials in the pipe.

These selective corrosion phenomena have also been referred to as *dealloying* or *parting*.

Stress Corrosion Cracking

Stress corrosion cracking is a mechanical-chemical process leading to the cracking of certain alloys at stresses considerably below their tensile strengths. A susceptible alloy, the proper chemical environment, plus an enduring tensile stress are required. Cracking may proceed either intergranularly or transgranularly. Pure metals are considered immune from stress cracking.

It is likely that there are no alloy systems which are completely immune to stress cracking if the

[(1)]Cabot Corp., Kokomo, IN.

TABLE 15.5 — Some Stress Corrosion Environments for Metals[(1)]

Aluminum Alloys	NaCl-H_2O, NaCl solutions, seawater, mercury
Copper Alloys	Ammonia vapors and solutions, mercury
Gold Alloys	$FeCl_3$ solutions, acetic acid-salt solutions
Inconel	Caustic soda solutions
Lead	Lead acetate solutions
Magnesium Alloys	NaCl-$K_2Cr_2O_7$ solutions, rural and coastal atmospheres, distilled water
Nickel	Fused caustic soda
Carbon Steels	NaOH solutions; NaOH-$NaSiO_3$ solutions; $CaNO_3$, NH_4NO_3, and $NaNO_3$ solutions; mixed acids (H_2SO_4-HNO_3); HCN; H_2S; seawater; NaPb alloy
Stainless Steels	$BaCl_2$, $MgCl_2$ solutions; NaCl-H_2O_2 solutions; seawater, H_2S, NaOH-H_2S solutions
Titanium	Red fuming nitric acid

(1)For a list of more than 100 environments that may cause stress corrosion cracking of metals and alloys, refer to Corrosion Data Survey, 1974 Edition, NACE, Houston, TX, p. 268.

proper environment can be found. Table 15.5 lists environments in which stress corrosion has been observed for at least some alloys of the systems listed. It is emphasized that the use of general classes of alloys in this table does not imply that all alloys of a given material will be equally susceptible, or that there are none in the class which may be immune to the environments listed. (Chapter 6 gives more information on stress corrosion.)

Analyzing Corrosion Problems

The corrosion engineer is often confronted with a corroded sample taken from an apparatus which has failed. Usually, some *stopgap* action must be immediately taken so that the apparatus can again be placed in service. It must then be decided what further action should be taken to avoid a recurrence of the failure. Often, this involves planning and conducting a test program under simulated conditions to determine which factors were most responsible for the failure.

The first step is to learn as much as possible from the existing sample. Care must be taken in handling the sample to avoid destroying valuable evidence. Much can be deduced from the physical appearance of the corrosion as viewed with the unaided eye, or with a low power magnifying glass or microscope. The geometry of attack, color, and form of corrosion products, etc., all provide valuable clues.

Other important information can be obtained only by altering the shape of or partially destroying the sample. For example, metallographic observation of type and depth of attack usually will require cutting and polishing the sample. In planning the sequence of investigation, all visual examinations, collection of corrosion products for chemical and X-ray analysis, etc., should be completed before cutting the sample.

The subject of corrosion testing and specimen evaluation was described in some length in Chapter 14. For further suggestions on the conduction of corrosion tests and failure analysis, the reader is referred to other textbooks.[2,3,4,5,6]

The preceding section has made comment on certain types of corrosion mechanisms as examples of problems that must be analyzed. The following sections will look at these and other types of attack as they relate to design problems.

Designing to Prevent Corrosion

Engineered structures should be designed so as to provide the desired functional qualities for the required period of service. This implies that a structure should neither be underdesigned nor overdesigned. Thoughtful design requires more than the provision of adequate strength. The part must also last for a given period of time.

It should be remembered, however, that designing for too long a life also may constitute overdesign if it involves the use of higher cost materials. Therefore, the corrosion behavior of the materials of construction must be considered carefully under the conditions of service. It is relatively easy to obtain quantitative information on physical and mechanical properties such as tensile strength, yield strength, impact values, fatigue limit, effect of temperature on properties, etc. Truly representative corrosion data, however, are much more difficult to obtain.

Utility of Test Data

Most published data have limited usefulness because the character of attack often is omitted. Another factor to keep in mind is that published data usually are obtained under carefully controlled laboratory conditions using high-purity, reagent-grade materials. This is in contrast to plant conditions using commercial-grade materials which are less pure.

If laboratory tests indicate that a given chemical is seriously corrosive to a particular material of construction, a decision can usually be made to eliminate this material from further consideration. Favorable laboratory or field test data, on the other hand, are not positive assurances of good performance of a material in service, unless the data were developed under precisely the same circumstances it will encounter in use. Even two different petroleum refineries belonging to the same oil company and using the same basic processes find differences in performance of materials traceable to differences in local conditions, *e.g.*, different source of crude, different mixture of crudes, and different pretreatment because of different end products (lube oil, asphalt, fuel oil, etc.).

In a soda ash plant, consideration was being given to the use of aluminum heat-exchanger tubes and shells for the ammonia recovery condensers. This is a standard aluminum application in many plants of this type. Sheet specimens of aluminum alloys were carefully mounted in the existing cast iron shells using brackets coated with electrical insulating material to avoid dissimilar metal attack.

After six months of exposure, examination of the specimens showed no discernible attack. Therefore, because this evidence seemed to confirm experience in other soda ash plants, a decision was made to build these larger, water-cooled condensers of aluminum alloys.

One week after the plant went in service with the new condensers, it was necessary to shut the plant down because the new condensers had failed. While the walls of the condensers were unaffected by corrosion, the tubes were completely perforated from the condensate side (exterior).

The evaluation tests using aluminum coupons had not taken into account the influence of heat transfer on corrosion behavior. Actually, the specimens accurately simulated only the sidewall conditions. In this particular plant, the ratio of CO_2 to NH_3 in the moist gaseous stream was lower than was customary in other soda ash plants; consequently, the inhibiting effect of CO_2 was incomplete. This illustrates the type of local situation which complicates the task of material selection.

Distribution of Attack

The character and distribution of attack is important in evaluating the potential danger of corrosion damage. Therefore, in contemplating a design, it is useful to consider the likely character of attack under the prospective operating conditions. For example, pitting attack could result in premature failure of heat-exchanger tubing or chemical storage tanks or drums. For these services, materials which suffer uniform attack are much more desirable. By contrast, for many structural applications, a material subject to localized pitting may be at least as suitable as one which suffers general attack (assuming stress raisers are not important in the particular use).

Effect of Metals on Environment

When designing equipment for handling chemicals, synthetic fibers, or paper, it may be desirable to select a material whose corrosion products are not highly colored. Food handling equipment or apparatus for microbiological processes will require materials whose corrosion products are nontoxic. In flowing systems, use of metals producing voluminous corrosion products can result in the plugging of tubes, screens, and valves. Certain sensitive substances, such as highly refined vegetable oils and the like, require materials which will not adversely affect product stability.

Product Degradation

The naval stores industry, located mainly in the southeastern and south-central parts of the United States, has an interesting materials selection problem. The raw material for this industry is pine gum collected from long-leaf yellow or slash pine trees, or extracted from pine stumps. Typical products of this industry include rosin, turpentine, dipentene, etc. These products are not especially corrosive to mild steel, although the cleansing action of turpentine and some of the processing solvents can make steel more vulnerable to atmospheric attack.

The real problem is catalytic degradation of the product from contact with metals. Rosin is priced on the basis of its paleness in color. The paler it is, the higher the price it commands. If small metal cups of copper, steel, aluminum, and stainless steel were placed on a table and each was filled with molten rosin, in a matter of minutes the rosin in the copper and the mild steel cups would turn black; whereas the rosin in the aluminum and the stainless steel cups would remain pale amber in color. As a consequence, the materials used for construction of process equipment, storage, and shipment of naval stores are usually either aluminum or stainless steel alloys, despite their initial cost premium over steel.

The textile and paper industries also are concerned with the color of their finished products. Since the corrosion products of many metals are highly colored, care must be exercised in the selection of materials to be used in contact with the finished products. Mills of this type generally are highly automated, so accidental product contamination may go unnoticed until the finished product reaches a customer's plant.

For example, undesirable rust spots were discovered on paper stock being processed at a paper specialties plant. Careful investigation revealed that a steel electrical conduit passed above the paper machine in the paper mill. Corrosion of the steel in the humid paper mill environment permitted iron corrosion products to fall and contaminate the paper.

Edible products also must be protected from degradation. Vegetable oils tend to become rancid in contact with some materials. Copper-base alloys have significantly undesirable effects on edible oils. Accordingly, such products normally are handled in organic coatings, nickel, stainless steels, plastics, or aluminum vessels.

Corrosion of certain metals can introduce toxic reaction products into process streams. For this reason, in the concentration of sap in the maple sugar making process, use of lead equipment or lead-base solder for fabrication of steel equipment is forbidden. Similarly, lead alloys are forbidden in the processing of edible gelatin.

Crevice Corrosion

The form of corrosion which produces the greatest number of failures of equipment due to poor design is that of concentration cell action. The two major design deficiencies causing such attack are: (1) the presence of crevices, and (2) pockets to retain deposits on a metal surface. Much has been written on the subject, and sketches of proper design configurations have long been available.[7,8]

The use of angle iron construction, skip welding, high outlets in tank bottoms, "dead" flow areas in piping stubs, heat exchangers, and vessels must be avoided.

Crevices are present in most types of equipment. Examples are gasketed flanges, rolled joints between tubes and tube sheets in heat exchangers, faying surfaces between tanks and supporting structures, etc. Moisture and chemical solutions may be trapped within crevices and held stagnant. These chemical differences in environment cause potential differences essentially similar to those encountered in galvanic corrosion.

Under these circumstances, two types of corrosion reactions proceed. The difference in oxygen concentration between the bulk solution and the trapped solution in the crevice produces oxygen concentration cells. After oxygen is depleted, a second process dominates which causes the anodic region to become more acidic. This results in great intensification of attack.

A number of procedures may be used to avoid crevice corrosion.

1. Use butt joints in preference to lap joints.
2. Eliminate or carefully seal lap joints so that they will not "open up."
3. Provide complete drainage.
4. Inspect regularly; clean thoroughly.
5. Avoid the use of packing materials or thermal insulation which can hold moisture in contact with metals.

Contact with Nonmetals

Severe attack of metal can occur if it is in contact with a nonmetal such as cork, hairfelt, wood, cloth, or paper (especially severe when impregnated with certain flame proofing or bacteriocidal chemicals). The cause of such attack is usually the formation of differential aeration cells; the term *poultice corrosion* was used to describe this in Chapter 5. Such attack can be prevented by one or more of the following.

1. Using nonmetallic materials known to be without corrosive action.
2. Preventing nonmetallic materials from becoming wet.
3. Providing adequate drainage and air circulation.
4. Coating the metal surfaces.
5. Using water-impervious mastics for attachments of nonmetallics.
6. Impregnating nonmetals with inhibitors.
7. "Building in" cathodic protection by use of metal coatings (or clad aluminum products).

Practical Problems

It is good engineering practice to provide adequate drainage. In the design of tanks, the drain should be at the lowest point (the bottom should slope toward the drain). The design of tank supports also is of vital importance.

Crowned *pads* for flat bottomed tanks will encourage free drainage of contents, particularly where a sump is provided. Crowning also encourages free drainage of liquids from under the tank floor under variations of hydrostatic head, which can pull corrosive liquids under the tank. Use of pads slightly smaller in diameter than the tank base will reduce the danger of moisture and chemicals being drawn under tanks by surface tension effects. *Drip skirts,* as shown in Figure 15.6, also are helpful in preventing corrosion under these circumstances.

Tubular heat exchangers should be constructed of floating head or hairpin construction when possible so that replacement of entire tube bundles may be made quickly. Many companies keep spare bundles in stock (sometimes installed on a by-pass around critical operations) so that process continuity may be maintained while repairs are being made.

Structural shapes and stiffeners should be incorporated in a design in such a way as to avoid en-

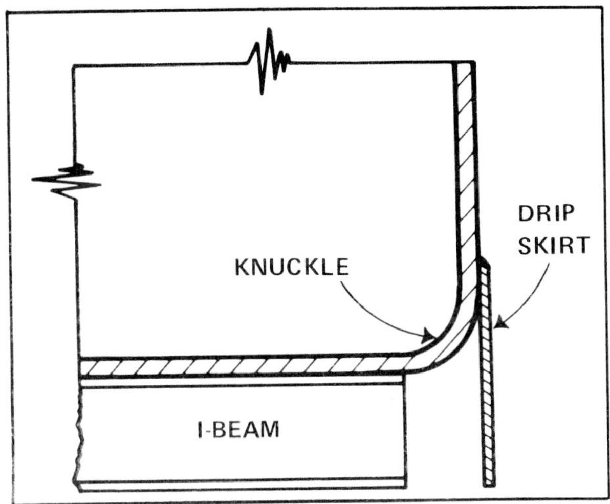

FIGURE 15.6 — Drip skirt can be used effectively to protect the external bottom-to-I-beam grillage crevice of storage tanks from drips and weather.

trapment of moisture and to provide free drainage. One major oil refinery installed approximately 65 km (40 miles) of aluminum instrument air line tubing serving a catalytic cracking unit. The tubing was clustered where possible and was supported in steel channels.

Unfortunately, these supporting channels were, in many cases, installed with the flanges pointing upward. This allowed rain water, insulating material, test solutions, cleaning solutions, etc., to collect in the channels containing the aluminum tubes during the construction process. When final tests were made, many of the aluminum tubing runs would not hold pressure. Examination of the tubing revealed extensive pitting and perforation of a number of the tubes. If adequate drainage had been provided in the design, it is likely that no such difficulty would have been encountered.

Corrosion of a large brewery fermenter had resulted in considerable loss of brew. Close examination revealed that corrosion started from the *outside* of the aluminum fermenter under vertical, hat-shaped stiffeners used to stay the sidewalls of this rectangular, open-top tank against hydrostatic loadings. Apparently, for aesthetic reasons, the tank fabricator had plugged the bottom of each of the sidewall stiffeners with a welded plate. The tops of the stiffeners were above eye level and so were left open. Moisture and dirt accumulated and corrosion resulted, causing failure. The cost of repairs was almost as much as the original cost of the tank.

It is apparent that good corrosion performance of materials starts with good initial material selection and good design, and is ensured by proper installation and operation, effective maintenance procedure, and frequent inspection.

Galvanic Corrosion

The problem of galvanic corrosion has been discussed previously. It is important to remember not to use dissimilar metals exposed to an electrolyte unless necessary.

It is good engineering practice, where dissimilar metals must be used, to use metals of as nearly similar solution potential as feasible under the exposure conditions. Even so, it may be necessary to take special precautions to prevent or reduce galvanic attack. Such precautions include the following.

1. Electrical isolation
2. Use of organic protective coatings
3. Cathodic protection
4. Use of inhibitors
5. Use of easily replaceable parts
6. Use of special design (*e.g.*, remove the dissimilar metal joint from contact with the electrolyte)
7. Use of transition joints
8. Design of joints so that anode area:cathode area is large

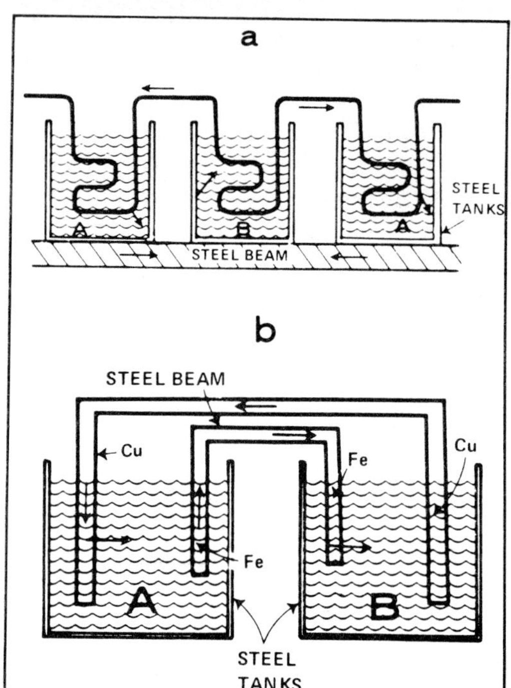

FIGURE 15.7 — (a) Real situation: Continuous copper coils are physically separated from steel tanks which are electrically connected by the steel beam. Current from tank B was collected by coil and discharged in tank A. (b) Schematic analog: Shows equivalent circuit of the situation in (a).

9. Seal out moisture

Galvanic cells often may assume subtle forms. For example, complex cells may be encountered. Figure 15.7a illustrates one type of complex cell. Figure 15.7b is its schematic analog.

In this example, the copper coils were interconnected and electrically insulated from the steel tanks. The steel tanks, in turn, were interconnected through the steel beam. Despite the lack of a direct dissimilar joint between steel and copper, there was sufficient leakage of current to cause a corrosion problem. Current flowed from the steel tank containing solution B to the copper coil. The copper coil was polarized to a sufficiently anodic potential that those segments immersed in the steel tanks containing solution A corroded where current leaked to these steel tanks. Fortunately, experience seems to indicate that corrosion caused by complex cells is generally less severe than that caused by galvanic cells.

Deposition Corrosion

Another subtle form of galvanic cell results from contamination of solutions by corrosion of upstream equipment. For example, corrosion of a brass pump can result in corrosion of aluminum or steel piping downstream from it by this mechanism. One practical solution to this problem is the use of heavy metal traps or "waster sections" of clad aluminum pipe which scavenge the troublesome ions from the flowing solution (Figure 15.8). Another approach to this problem is the use of glass-reinforced plastic pipe.

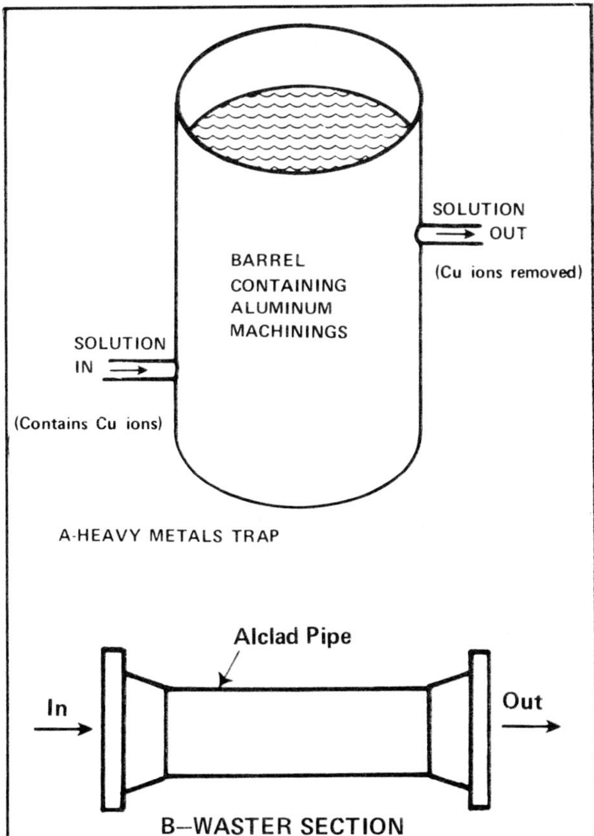

FIGURE 15.8 — Methods of removing troublesome ions from solution. (A) Heavy metals trap: Solutions containing copper ions enter barrel filled with aluminum shavings, which trap them. (B) Waster section: Aluminum clad pipe inserted in a system removes heavy metal ions. When substantially corroded, it is replaced with another section of aluminum pipe.

FIGURE 15.9 — Shows the wedging effect of corrosion products in breaking lap joints.

Fabrication Parameters

Residual and Fabrication Stresses

Often, even when the proper materials are selected for the anticipated service conditions, some will fail because of improper fabrication. The assembly of parts for transportation and airframe structures, for example, must be done under closely controlled conditions to avoid introduction of high assembly stresses which might lead to stress corrosion cracking.

For this reason, special care is exercised in *fit-up*. Fasteners are selected which do not introduce localized regions of residual tensile stress (*e.g.*, avoid tapered drive pins). Inasmuch as stress corrosion requires an enduring tensile component of stress, sometimes it is desirable to shot peen critical aircraft structures to ensure the presence of residual compressive stresses at the surface.

Residual stresses can be introduced by service conditions. Differences in thermal expansion between dissimilar metals forming parts of the same structure may result in either tensile or compressive failure, or in corrosion fatigue.

Corrosion products usually occupy more volume than the metal from which they were formed. Corrosion product buildup between the shanks of fasteners and fastener holes or between the faying surfaces of lap joints can cause pressure and eventual failure of the joint (Figure 15.9).

Surface Contamination

Roll forming of plates is a common source of surface contamination caused by chips, shavings, and other foreign particles which are embedded in the surface of the metal. Special procedures are used to avoid surface contamination. These include cleaning the rolls, filtering rolling lubricants, coating rolls with heavy paper, and coating plates with protective tape. High-purity aluminum plates used for fabrication of hydrogen peroxide storage tanks usually are coated with protective tape at producers' mills to avoid surface contamination during shipment and fabrication. Such surface contamination, in addition to being a potential corrosion hazard, might contribute to catalytic breakdown of the peroxide.

Protection of Joints

Welded Joints

Nonstabilized grades (Chapter 3) of austenitic stainless steel (Types 304, 316, etc.) sometimes must be quench-annealed following fusion welding to avoid sensitization to intergranular attack caused by carbide precipitation. Therefore, fabricated parts which cannot be heat treated readily after welding should be specified in low-carbon or stabilized grades such as Types 321 and 347 (S32100 and S34700). Even the stabilized grades can suffer a form of intergranular attack (knife-line attack) under specific circumstances.

In handling chemical products such as fatty acids containing minute traces of sulfuric acid in aluminum piping systems, alloy selections for pipe and weld wire should be made with care. The 6000 series aluminum pipe alloys often are welded with 4043 (Al-5% Si) weld wire. This combination can suffer knife-line attack adjacent to the welds in the presence of dilute mineral acids. A better combination under these conditions is 3003 (A93003) alloy pipe with 1100 (A91100) alloy weld wire.

Solders

Properly designed soldered joints, if kept dry, generally are successful. The solders used for aluminum alloys, however, usually are alloyed with heavy metals in order to achieve the desired melting point. As a consequence, a dissimilar metal couple results when aluminum soldered joints are immersed in an electrolyte, and corrosion can be accelerated.

Brazing

Care should be taken to make certain that all brazing flux residues are removed from brazed assemblies. These residues often are hygroscopic and draw moisture from the air, thereby establishing their own corrosive solutions.

Coatings

Protective coatings are among the most widely used corrosion preventives. A good quality coating, correctly applied over a properly prepared surface, will give excellent performance. However, application of a coating system without proper surface preparation is a waste of time and money. Major suppliers of protective coatings provide excellent consultation services on their products. Discussion of coating problems with reputable suppliers is recommended. (Chapter 12 provides a great deal of information concerning protective coatings.)

Influence of Environmental Factors

It is not always possible to anticipate the actual environment in which a metal will operate. Even if initial conditions were known completely, there often is no assurance that operating temperatures, pressures, or even chemical compositions will remain constant over the expected life span of equipment. Under these circumstances, one is tempted to ask, "Why bother?"

Experience has shown, however, that even when impossible circumstances apparently are present, substantial cost savings are possible by intelligent use of corrosion data and good design practices. Table 15.6 lists some of the more important environmental and mechanical factors which should be considered in designing to prevent corrosion and in analyzing failures of existing apparatus.

TABLE 15.6 — Principal Environmental Factors Influencing Corrosion

1. Temperature
2. Pressure
3. Velocity
4. Impurities
5. Mechanical Effects, Corrosion Fatigue, Fretting
6. Biological Factors
7. Stray Currents

Temperature and Pressure

In most chemical reactions, an increase in temperature is accompanied by an increase in reaction rate. A rough rule-of-thumb suggests that the reaction rate doubles for each ten degree (C) rise in temperature. Although there are numerous exceptions to this "rule," it is important to take into consideration the influence of temperature when analyzing why materials fail, and in designing to prevent corrosion.

Figure 15.10 illustrates the influence of temperature and composition on the corrosion behavior of materials exposed to sulfuric acid solutions. This isocorrosion chart is drawn for a corrosion rate of 500 μm/y (20 mpy) or less. The upper (high-temperature) boundary of each area represents a corrosion

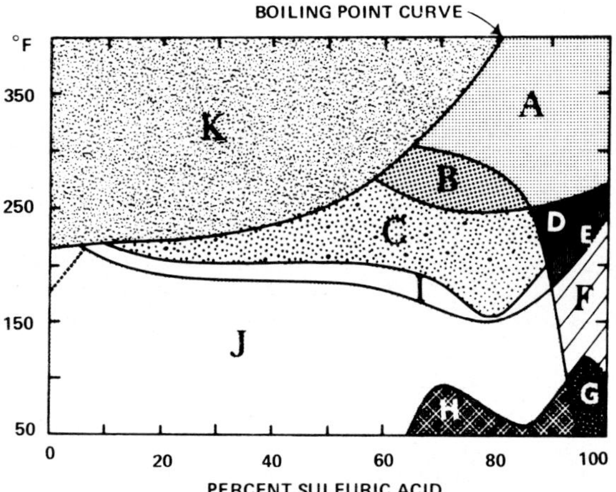

Zone	Metals
A	Duriron[1]
B	Lead, Duriron
C	Lead, Ni-Mo, Duriron
D	Duriron, high Ni-Mo alloy
E	Duriron, high Ni-Mo alloy, high Ni-Mo-Cr alloy
F	Duriron, high Ni-Mo alloy, high Ni-Mo-Cr alloy, Durimet 20
G	All except lead
H	Iron, lead, high Ni-Mo alloy, high Ni-Mo-Cr alloy, Durimet 20, Duriron
I	Lead, Ni-Mo-Cr, Ni-Mo, Duriron
J	Lead, Ni-Mo, Ni-Mo-Cr, Durimet-20,[1] Duriron
K	Vapor space at 1 atm pressure

(1) Tradenames of The Duriron Company, Dayton, Ohio.

FIGURE 15.10 — Shows zones in which metals listed below have corrosion rates of 20 mils per year or less at the indicated temperatures and concentrations.

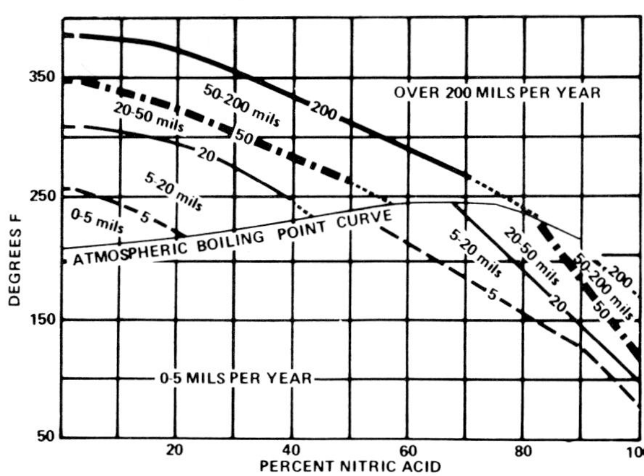

FIGURE 15.11 — Corrosion rates of quench-annealed 18-8 stainless steel versus temperature, pressure, and percent nitric acid in aqueous solutions.

rate of 500 μm/y (20 mpy). The materials listed in each area are metals having a corrosion rate of 20 mpy or less under the indicated conditions of temperature and acid concentration. The boiling point curve for atmospheric pressure is shown because at temperatures above this curve, the metal is no longer in a liquid environment at 1 atmosphere pressure.

It is possible to examine the combined influence of temperature, pressure, and chemical concentration. Figure 15.11 shows the behavior of Type 304 stainless steel in nitric acid aqueous solutions at elevated temperatures and pressures.

Changing the temperature of a solution can influence the corrosion tendency. For example, historically, household hot water heater tanks were made of galvanized steel. The zinc coating on the mild steel base offered a certain amount of cathodic protection to the underlying steel, and the service life (usually judged by how long it took to produce *red water, i.e.,* rusty water) was considered adequate. Water tanks seldom were operated above 60 C (140 F).

With the development of automatic dishwashers and automatic laundry equipment, the average water temperature was increased so that temperatures of about 80 C (175 F) are not unusual in household hot water tanks. Coinciding with the widespread use of automatic dishwashers and laundry equipment was a sudden upsurge of complaints of short-life of galvanized steel water heater tanks. Electrochemical measurements showed that in many cases, iron was anodic to zinc at 77 C (170 F), whereas zinc was anodic to iron at temperatures below 60 C (140 F). This explained why zinc offered no cathodic protection at 77 C (170 F), and why red water and premature perforation of galvanized water tanks had occurred. The problem was reduced by use of magnesium anodes, protective coatings, and development of new alloys.

Influence of Velocity

Relative movement of metal parts with respect to their liquid environments can stimulate increased attack. *Erosion-corrosion* is mechanically accelerated corrosion. The mechanical factor may result from abrasion, impingement, turbulence, cavitation, etc. Metals generally owe their corrosion resistance to a tightly adherent, protective film on the metal surface. This film may consist of reaction products, adsorbed gases, or both. Any mechanical disturbance of this protective film could stimulate attack of the underlying metal, until either the protective film is reestablished, or the metal has been corroded away.

Erosion-corrosion is encountered most frequently in pumps, valves, centrifuges, elbows, impellers, inlet ends of heat-exchanger tubes, agitated tanks, etc. Experience with lead valves is an illustration. Hard lead specimens exposed in hot, stagnant, dilute sulfuric acid suffered negligible attack. However, throttled valves of the same material exposed to high-velocity effects failed in less than a week.

Locations in flowing systems where there are sudden changes in direction or flow cross section, as in heat exchangers where water flows from the water boxes into the tubes, are likely places for erosion-corrosion. Under these conditions, which stimulate some corrosion of the metal surface, the effects of flow velocity may be to displace the corrosion product, thereby exposing fresh metal to the corrosion action of the solution. This action leads to a much increased corrosion rate (Figure 15.12).

Cavitation and Impingement

Cavitation is the formation and collapse of voids in fluids (*e.g.,* water) because of the momentary occurrence of low pressure. It occurs whenever the absolute pressure at a point in the liquid stream is reduced to the vapor pressure of the fluid. Cavitation damage is the wearing away of metal by mechanical damage resulting from repeated impact blows produced by collapse of voids within a fluid.

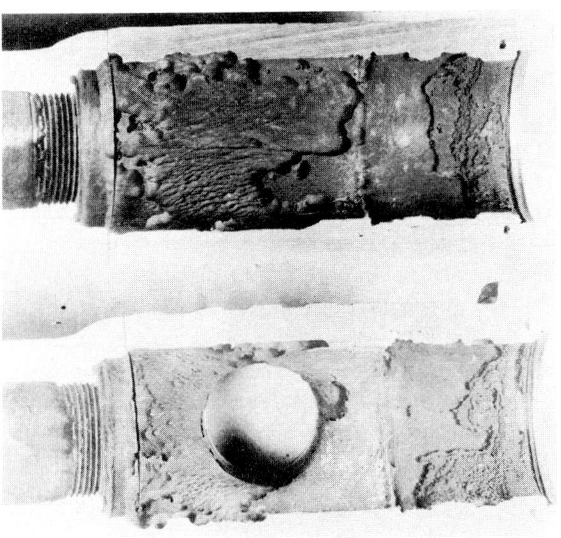

FIGURE 15.12 — Inside surfaces of a pipe joint showing effects of erosion of corrosion products by rapidly moving or turbulent liquids. A good example of erosion-corrosion.

FIGURE 15.13 — A pump impeller subjected to cavitation enhanced corrosion caused by the collapse of cavities in the liquid at the metal-liquid interface.

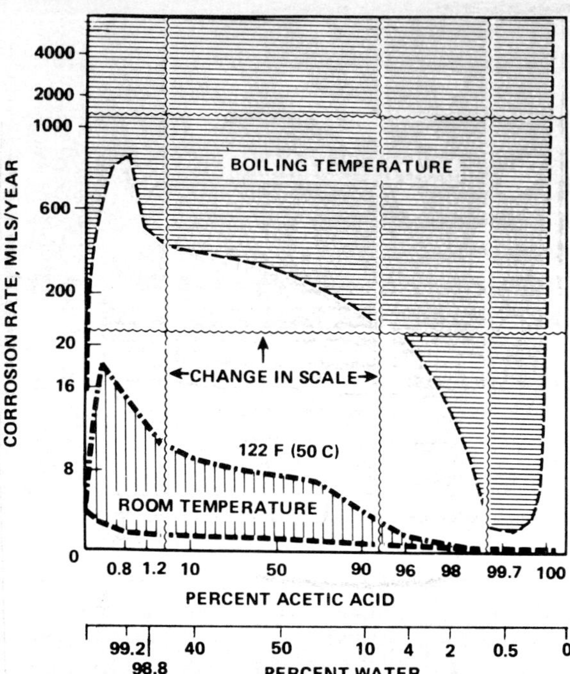

FIGURE 15.14 — Behavior of 1100-H-14 aluminum in acetic acid solutions. The corrosion rate in the 90 to 100% range is controlled by water content. The curve can be applied generally to aluminum alloys. [SOURCE: Alcoa.]

Pump impellers (Figure 15.13) and ship propellers often show cavitation damage effects.

Wet steam traveling at high velocities (3000 to 4000 feet per sec) can cause considerable damage to power plant condenser tubes, turbine blades, valve seats and disks, piston rings in engines, etc. Removal of moisture droplets from the steam is helpful, but not always possible. Consequently, the practical solution of steam impingement problems usually involves a combination of several procedures: improved flow design, reduction of water droplet content of the steam, and the use of materials having higher resistance to erosion-corrosion damage (*e.g.*, stainless steels) in areas where impingement is anticipated.

Impurities

Often, substances present only in minute amounts have more influence on the corrosion behavior of materials than do substances present in much greater quantities. Sometimes such trace quantities of impurities serve to accelerate attack; at other times they may act as inhibitors. Figure 15.14 shows the behavior of commercially pure aluminum in acetic acid at various temperatures and concentrations. In the concentration range between 90 to 100% acetic acid, the water content of the acid controls the rate of attack. Violent attack occurs at elevated temperatures if the water content is reduced below about 0.2%. Apparently, so long as at least 0.2% moisture is present, the protective film (which is what gives aluminum its corrosion resistance) is maintained. Therefore, a trace of moisture (in this case) acts as an inhibitor.

The introduction of small amounts of ions of metals such as copper, lead, or mercury can cause severe corrosion of aluminum equipment. For example, corrosion of upstream copper alloy equipment can result in contamination of cooling water. Under these circumstances, copper can plate out (cathodic reaction) on downstream equipment and pipe, setting up local galvanic cells which can result in severe pitting and perhaps perforation.

Mercury is a corrosion hazard for most metallic materials of construction. Mercury causes severe pitting of aluminum and lead and stress cracking of copper-base alloys (Figure 15.15). Common sources of mercury contamination include blown manometers, broken thermometers and seals, or the use of mercury-containing biocides or wood preservatives.

Corrosion Fatigue

The reduction by corrosion of the ability of a metal to withstand cyclic or repeated stresses is known as *corrosion fatigue.* The first well-known example of corrosion fatigue occurred during World War I when considerable difficulty was experienced as the result of failure of steel tow-lines for mine-sweeping paravanes. The seawater provided a cor-

FIGURE 15.15 — Effect of mercury contamination in cracking a brass tube. Very small concentrations of mercury may produce this effect.

rosive environment and the vibration of the steel lines under tension as they pulled the paravanes through the water provided the cyclic stress component. The substitution of higher tensile strength wire rope did not materially lengthen service life. However, the use of cathodic protection by the substitution of galvanized steel wire rope for bare wire corrected the problem. This response to cathodic protection is evidence that the mechanism is at least partially electrochemical.

Corrosion fatigue continues to be a serious cause of failure, and results in major expenditures by industry to correct the damage. For example, the petroleum industry encounters major trouble with corrosion fatigue in the production of oil. The exposure of drill pipe and of sucker rods to brines and sour crudes encountered in many producing areas results in failures which are expensive both from the standpoint of replacing equipment and from loss of production during the time required for "fishing" and rerigging.

Figure 15.16 compares the fatigue performance of steel tested in a corrosive environment with steel tested in air. The number of cycles to cause failure at a given stress is shown to be substantially reduced by superimposing the corrosive environment.

A part which has failed by corrosion fatigue may not exhibit general attack. A distinguishing feature of some corrosion fatigue is the presence of numerous cracks in addition to the crack which caused failure.

For uniaxial stress systems, there will be an array of parallel cracks which are perpendicular to the direction of principal stress. Torsion loadings tend to produce a system of crisscross cracks at roughly 45° from the torsion axis. Corrosion fatigue cracks found in pipes subjected to thermal cycling usually will show a pattern made up of both circumferential and longitudinal cracks (Figure 15.17).

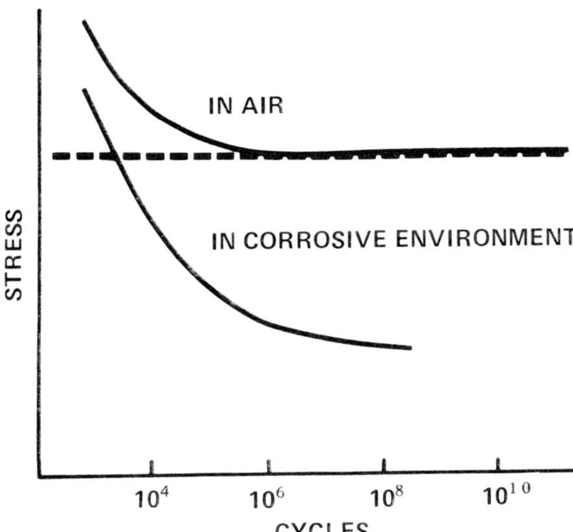

FIGURE 15.16 — Typical stress versus number of cycles curve for steel subjected to cyclic stress. Note how corrosive environment reduces the number of cycles required to produce fatigue failure.

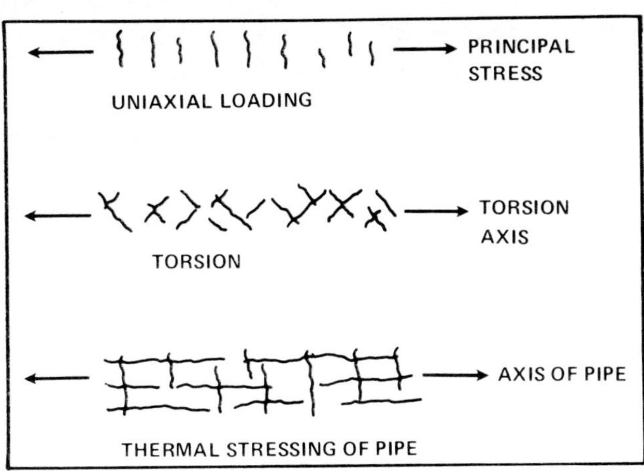

FIGURE 15.17 — Typical fatigue crack arrays for various stress systems.

There are several effective remedial procedures for reducing or preventing corrosion fatigue. Cathodic protection is useful in aqueous solutions, and under ideal conditions can increase the fatigue limit to that predicted from fatigue tests *in vacuo*.

Inhibitors also are effective if properly applied. The supply of inhibitor must be uninterrupted, otherwise it is possible to stimulate localized attack which can form focal points for corrosion fatigue. Putting the outer surface in compression, as by shot peening, has been shown to be effective in some cases.

Fretting Corrosion

Fretting corrosion is defined as metal deterioration caused by repetitive slip at the interface between two surfaces in contact. Conditions necessary for the occurrence of fretting are: (1) the interface must be under load, and (2) vibratory or oscillitory motion of small amplitude must result in the surfaces striking or rubbing together.

The results of fretting are as follows.

1. Metal loss in the area of contact
2. Production of oxide debris
3. Galling, seizing, fatiguing, or cracking of the metal
4. Loss of dimensional tolerances
5. Loosening of bolted or riveted parts
6. Destruction of bearing surfaces

One outstanding example of fretting corrosion involved rail or water shipment of automobiles. Serious fretting corrosion occurred in the wheel bearings when cars were held in position by chocking the wheels. Supporting the cars on their axles greatly reduced (but did not completely eliminate) the problem. In normal operation, of course, the slip between mating parts in bearings is too great (complete revolutions) to cause fretting. Another commonly experienced example of fretting occurs on bolted tie-plates on railroad tracks.

The mechanism of fretting corrosion involves both chemical and mechanical factors. The relative motion scrubs off tiny high points or surface irregularities on the metal surface, and these, in turn, are chemically converted to oxide. They also may be sites for galling, in which case the mechanical action results in tearing out small bits of metal from one or the other of the interfaces and these oxidize to form debris, etc. Listed below are four means of mitigating fretting damage.

1. Use mating surfaces of differing character, *e.g.,* soft metal against harder metal.

2. Avoid small relative motion between surfaces, *e.g.,* roughen interface and/or increase load. Alternatively, greatly increase relative motion.

3. Exclude air, *i.e.,* use cements or sealants.

4. Lubricate the faying surfaces either by use of lubricants such as molybdenum sulfide, or by making one of the contacting surfaces of a low-friction material such as Teflon.

Biological Effects

Buried iron pipes are known to have suffered severe corrosion as the result of bacterial action. In unaerated soils, this attack is attributed to the influence of the sulfate-reducing bacteria (*Sporovibrio desulfurican*). The mechanism is believed to involve both direct attack of iron by hydrogen sulfide and cathodic depolarization aided by the presence of the enzyme, hydrogenase. Even in aerated soils, there are sufficiently large variations in aeration that the action of sulfate-reducing bacteria cannot be neglected. For example, within active corrosion pits, the oxygen content becomes exceedingly low.

In aerobic environments (presence of oxygen), the species *thiobacillus* accounts for most of the microbiological corrosion. This species converts sulfur to sulfuric acid, which stimulates attack.

Clearly, the possibility of bacterial action is an important consideration in assessing the likelihood of corrosion, particularly of buried or submerged steel pipes and structures.

Stray Currents

"Stray" direct current can have a strong influence on the life of buried or submerged metal structures. Cathodic protection occurs at locations where current enters the structure. Severe, localized corrosion occurs where the current leaves the structure and metal goes into solution as ions.

Where several buried or submerged structures are in close proximity, *e.g.,* a network of buried pipelines, special attention should be given to the possibility that leakage of current between them could cause localized corrosion. If nonferrous structures of aluminum, zinc, or lead are involved, there is danger of overprotection (*i.e.,* cathodic corrosion), even at locations where the structure receives current. Similarly, coated steel structures can have their protective coatings "popped off" by hydrogen evolution if the protective currents are too great.

Each situation involving stray currents requires individual study. If an entire buried structure can be brought to the same potential, or if provision can be made to discharge leakage currents through either inert ground rods or replaceable structures such as galvanic anodes, damage to the structure will be minimized.

A summary checklist of some common potential problems facing the corrosion engineer when considering the design of a new installation is provided in Table 15.7. This should be helpful once it is understood why the questions are asked.

TABLE 15.7 — Some Design Considerations to Control Corrosion

Plant Construction:
Prevailing wind direction OK?
Method of using cooling water defined (open recirculated, salt water, double exchange, ponds, air cooling)?
Cooling tower downwind?
Closed construction used (pipe or box stanchions, etc.)?
Exterior designs free of pockets?
Tank bottoms protected (cathodically or sealed, depending on type)?
Metal-to-metal contacts protected?
Equipment designed properly for coating?
Will hydrotesting corrode the steel or leave chloride on stainless?
Is a system of material verification needed?
Will all steel be cleaned and primed before erection?
Are temporary preservatives needed for some items?
Should cathodic protection be installed now?
Is electrical equipment properly grounded to prevent stray currents?

Equipment Design:
No "dead" spaces in lines, tanks, vessels?
All possible crevices sealed or flushed?
Inlets-outlets properly contoured? (If not, is protection provided?)
Offsets on mechanical equipment properly filleted?
Other potential fatigue sites?
Can equipment be cleaned adequately if needed?
Will end grain attack be a problem?
Homogenizing anneal needed?
Stress relief required? (Forestall SCC?)
Are weld deposits compatible with parent metal in chemistry and hardness?
Corrosion allowances added to metal thickness?
Possibility of liquid metal embrittlement (*e.g.,* zinc on stainless)?
Desiccating vents needed?

Temperature:
Evaporation area—vaporization or ebullition cause problem?
Hot spots—from stream injection point?
Hot spots—from impinging flame?
Cold spots—attachment of external metal?
Cold spots—condensation zone a problem? (If so, keep +10 C above dew point.)

(continued)

TABLE 15.7 (continued)

Breakaway point exceeded (*e.g.*, organic acids on stainless steels)?
Freeze-thaw zones properly protected?

Velocity:
Excessive linear flow?
Turbulent flow (from excess velocity; from design, such as raised tube ends)?
Insufficient flow? (Passive metals need O_2. Vapor voids created?)
Adequate radii in bends?
Stream mixing tangential or in a tank?

Impingement and Cavitation:
Restrictions out of lines?
Adequate radii in bends?
Varying diameters in a line?
Pump problems analyzed?
Stream inlets baffled?

Bimetal Problems:
Galvanic couples in an aerated electrolyte?
Heavy metal (Fe, Cu) upstream of aluminum or steel?
Differential expansion problems?

Environmental Control:
Neutralization, dehydration, or deaeration of streams considered?
Parameters of operation to reduce corrosion registered with operators?
Two-phase liquid systems adequately evaluated?
Gas-liquid interfaces evaluated?
Solids in any gas or liquid stream?
Adequate alarms for breakthrough of corrosive agent (ppm's can be disastrous)?
Will atomic hydrogen be present? What effect?
Could bacteria create a corrosive situation?

Testing:
Initial testing defined required materials?
Extra nozzles available on equipment for later test rack insertion?
Resistance wire probes of help?
Linear polarization instrument of help?
Can side stream be built for critical test area?
Corrosion test program planned firmly?
Ultrasonic bench-mark readings made on critical equipment?

Summary

No single text or academic course can give an engineer or technologist the answers to all corrosion problems. The avoidance of corrosion should be the long-range goal of all engineers and designers. The best way to avoid corrosion is the imaginative use of knowledge. Proper analysis of failures, intelligent use of technical information, and creative use of engineering principles, combined with a deep appreciation of the economic aspects involved will result in improved techniques of materials selection and will eliminate many needless and expensive corrosion failures.

References

1. NACE, Direct Calculation of Economic Appraisals of Corrosion Control Measures, Recommended Practice RP-02-72, 1972.
2. Uhlig, H. H., ed., Corrosion Handbook, John Wiley & Sons, Inc., New York, N.Y., 1963.
3. Fontana, M. G. and Greene, N. D., Corrosion Engineering, 2nd Ed., McGraw-Hill, New York, N.Y., 1978.
4. Champion, F. A., Corrosion Testing Procedures, John Wiley & Sons, Inc., New York, N.Y., 1965.
5. ASM, Failure Analysis, Metals Handbook, Vol. 12, 1976.
6. Dillon, C. P., Forms of Corrosion—Recognition and Prevention, NACE, Houston, TX, 1982.
7. Burton, W. H., Designing Process Equipment, Materials Protection, Vol. 6, No. 2, p. 22 (1967). (This entire issue relates to design and includes a Bibliography of Design in NACE Corrosion Literature on p. 60.)
8. Landrum, R. J., Designing for Corrosion Resistance, Process Industries Corrosion, NACE, Houston, TX, p. 40, 1975.

Bibliography

Evans, U. R. Metallic Corrosion, Passivity and Protection. Longmans, Green & Co., 1948.

Ind. Eng. Chem., Vol. 33, No. 8, p. 1001 (1941).

Shreir, L. L., Ed. Corrosion. Volumes I and II. John Wiley & Sons, Inc., New York, N.Y., 1963.

Speller, F. N. Corrosion Causes and Prevention. McGraw-Hill, New York, N.Y., 1951.

SUBJECT INDEX

T

U

V-W-Z

Other NACE Publications

The NACE Corrosion Engineer's Reference Book

Edited by R. S. Treseder

A valuable single-source reference book for corrosion engineers, scientists, and technicians. Includes detailed tables, graphs, and charts covering physical, chemical, mechanical, and performance properties of materials.

Corrosion and Its Control: An Introduction to the Subject

By J. T. N. Atkinson and H. VanDroffelaar

This publication introduces both the theoretical and practical aspects of corrosion. Contains seventeen chapters emphasizing the engineering aspects of corrosion and its control.

The NACE Book of Standards

A complete collection of NACE's 46 industry Standards recognized and used worldwide.

Forms of Corrosion—Recognition and Prevention

Edited by C. P. Dillon

Detailed case histories and illustrations of specific forms of corrosion in this handbook give the corrosion engineer the closest thing to first-hand knowledge possible.